EXTREME ULTRAVIOLET ASTRONOMY

The study of the universe in Extreme Ultraviolet (EUV) wavelengths is a relatively new branch of astronomy. Lying between the X-ray and UV bands, Extreme Ultraviolet has proved to be a valuable wavelength for the study of specific groups of astronomical objects, including white dwarf stars and stellar coronae, as well as the interstellar medium.

This text describes the development of astronomy in the EUV wavelength range, from the first rocket-based experiments in the late 1960s through to the latest satellite missions. Discussions of the results from the most important space projects are followed by an analysis of the contributions made by EUV astronomy to the study of specific groups of astronomical objects. Within this framework, the book provides detailed material on the tools of EUV astronomy, dealing with the instrumentation, observational techniques and modelling tools for the interpretation of data. Prospects for future EUV missions are discussed and a catalogue of known EUV sources is included.

This timely text will be of great value to graduate students and researchers. It is the first to give a complete overview of EUV astronomy, and comes at the end of a major phase of discovery in the field.

MARTIN BARSTOW is a Reader in Astrophysics and Space Science at the University of Leicester. His research focuses on the study of hot white dwarfs and the interstellar medium, and he has a strong background in the analysis of astronomical X-ray, EUV, UV and optical data. He served as Detector Scientist for the ROSAT Wide Field Camera, for which he received a NASA Group Achievement Award. He has been involved in the development and operation of EUV and X-ray instruments, including a novel high spectral resolution EUV spectrometer for flight on a NASA sounding rocket.

JAY HOLBERG is a Senior Research Scientist at the Lunar and Planetary Laboratory of the University of Arizona. He has worked extensively in UV and EUV astrophysics in the areas of white dwarfs, the interstellar medium, planetary atmospheres and planetary ring systems. He has conducted pioneering observations in the EUV and far-UV using a variety of spacecraft, including the Hubble Space Telescope, the Extreme Ultraviolet Explorer, and Voyagers 1 and 2 for which he received a NASA Group Achievement Award.

Cambridge Astrophysics Series

Series editors

Andrew King, Douglas Lin, Stephen Maran, Jim Pringle and Martin Ward

Titles available in this series

EXTREME ULTRAVIOLET ASTRONOMY

MARTIN A. BARSTOW

University of Leicester, UK

JAY B. HOLBERG

University of Arizona, Tucson, USA

CAMBRIDGE
UNIVERSITY PRESS

CAMBRIDGE UNIVERSITY PRESS
Cambridge, New York, Melbourne, Madrid, Cape Town, Singapore, São Paulo

Cambridge University Press
The Edinburgh Building, Cambridge CB2 8RU, UK

Published in the United States of America by Cambridge University Press, New York

www.cambridge.org
Information on this title: www.cambridge.org/9780521580588

First published 2003
This digitally printed version 2007

A catalogue record for this publication is available from the British Library

Library of Congress Cataloguing in Publication data

Barstow, Martin A. (Martin Adrian), 1958–
Extreme ultraviolet astronomy : EUV astronomy / Martin A. Barstow, Jay B. Holberg.
   p. cm.
Includes bibliographical references and index.
ISBN 0 521 58058 7
1. Ultraviolet astronomy.   I. Holberg, J. B.   II. Title.
QB474 .B37   2002   522'.68–dc21   2002073705

ISBN 978-0-521-58058-8 hardback
ISBN 978-0-521-03906-2 paperback

The authors dedicate this book to Dr C. Stewart Bowyer who, through a unique personal combination of foresight, tenacity and self-belief, has done more than anyone else to realise the field of Extreme Ultraviolet Astronomy.

We would also like to dedicate this work to our families, who have had to endure our absences for all the travel associated with our research together with late nights and lost weekends preparing scientific papers, proposals and reports.

Contents

Preface

This book is the first comprehensive description of the development of the discipline of astronomy in the Extreme Ultraviolet (EUV) wavelength range ($\approx$100–1000 Å), from its beginnings in the late 1960s through to the results of the latest satellite missions flown during the 1990s. It is particularly timely to publish this work now as the Extreme Ultraviolet Explorer, the last operational cosmic EUV observatory, was shut down in 2001 and re-entered the Earth's atmosphere in early 2002. Although new EUV telescopes are being designed, it will be several years before a new orbital observatory can come to fruition. Hence, for a while, progress beyond that reported in this book will be slow.

We intended this book to be for astrophysicists and space scientists wanting a general introduction to both the observational techniques and the scientific results from EUV astronomy. Consequently, our goal has been to collect together in a single volume material on the early history, the instrumentation and the detailed study of particular groups of astronomical objects. EUV observations of the Sun are not within the scope of this current work, since the Sun can be observed in far more detail than most sources of EUV emission, providing material for a book on its own. We have found it useful to deal with the subject in its historical context. Therefore, we do not have specific chapters on instrumentation but integrate such material into the development of the scientific results on a mission-by-mission basis. The overall framework can be divided into three main sections:

Early history of the subject leading up to the first orbital missions, which had an EUV capability but were not dedicated to EUV astronomy (chapters 1 and 2).

The first dedicated EUV astronomy orbital telescopes and the sky surveys carried out by them to produce reference catalogues of EUV sources (chapters 3 and 4). We include an integrated catalogue of all EUV sources known at the time of publication in appendix A.

EUV spectroscopy techniques and study of specific groups of astronomical objects: stars, the interstellar medium, white dwarfs, cataclysmic variables and extragalactic sources (chapters 5 to 10).

Since we are very active in the field of EUV astronomy, it is inevitable that much of the material included here has been drawn from our own work. However, we have made a concerted attempt to represent all of the many astronomers who have made significant contributions. It has not been possible to include all the EUV astronomy results published, as this would constitute several volumes. Therefore, we have had to carefully select representative material, which we hope gives the overall flavour of work in each subtopic. We have tried to make the bibliography extensive to compensate for these necessary omissions. As will be seen from the book, we have obtained a large number of figures and tables from the many original authors of scientific papers. It would be invidious to single out particular individuals from this list: rather, we would like to collectively thank all who have contributed to the content of this book. Specific acknowledgement of individual figures or tables can be found in the captions. We have been extremely pleased by the positive support received from everyone that we have asked for help. It is clear to us that this reflects the general friendly nature of the EUV astronomy community, with shared interests and common goals. It has been a great pleasure to work with you all over many years. Thanks to everyone for that.

There are a few individuals that deserve specific thanks. First, to Bob Stern who was originally a co-author but was forced to drop out owing to other commitments. Nevertheless, he played an important role in developing the proposal for the book. We hope you like the result, Bob. Also a number of people have done some sterling work in helping to generate figures or tables that we could not obtain directly from the original authors. These are: Reni Christmas, Graham Wynn, Elizabeth Seward, Nigel Bannister, Jim Collins and David Sing.

Finally, we have made an attempt at prescience by looking at the possible future of EUV astronomy in the final chapter (11) of our EUV Astronomy book. We hope it is a bright one and that this book can be part of its foundation.

Martin Barstow and Jay Holberg

Abbreviations

ACS	attitude control system
AGB	asymptotic giant branch
AGN	active galactic nuclei
ALEXIS	Array of Low Energy X-ray Imaging Sensors
ALI	Accelerated Lambda Iteration
ASCA	a Japanese X-ray astronomy satellite
ASTP	Apollo–Soyuz Test Project
BSC	Bright Source Catalogue
CEM	channel electron multiplier
CHIPS	Cosmic Hot Interstellar Plasma Spectrometer
CMA	channel multiplier array
CSM	Command and Service Module
CSPN	central stars of planetary nebulae
CV	cataclysmic variable
DEC	astronomical position coordinate: declination
DEM	differential emission measure
DM	dispersion measure
DS	deep survey
DSS	deep survey/spectrometer
ESA	European Space Agency
EUV	Extreme Ultraviolet
EUVE	Extreme Ultraviolet Explorer
EUVI	Extreme Ultraviolet Imager
EUVS	Extreme Ultraviolet Spectrograph
EUVT	Extreme Ultraviolet Telescope
EXOSAT	European X-ray Astronomy Satellite
FIP	first ionisation potential
FOS	Faint Object Spectrometer
FOV	field of view
FUSE	Far Ultraviolet Spectroscopic Explorer
fwhm	full width half maximum
GHRS	Goddard High Resolution Spectrometer
HEAO	High Energy Astronomical Observatory
HEW	half energy width

HR	Hertzsprung–Russell
HRI	High Resolution Imager
HST	Hubble Space Telescope
HUT	Hopkins Ultraviolet Telescope
IAU	International Astronomical Union
IEH	International Extreme-ultraviolet Hitchhiker
IPC	imaging proportional counter
IRAS	Infrared Astronomical Satellite
ISM	interstellar medium
IUE	International Ultraviolet Explorer
J-PEX	Joint Plasmadynamic Experiment
LANL	Los Alamos National Laboratory
LE	low energy
LEIT	low energy imaging telescope
LETG	low energy transmission grating
LIC	local interstellar cloud
LISM	local interstellar medium
LLNL	Lawrence Livermore National Laboratory
LTE	local thermodynamic equilibrium
LW	long wavelength
LWR	Long Wavelength Redundant camera on IUE
MAD	metal abundance deficiency
MAMA	multi-anode microchannel array
MCP	microchannel plate
MIT	Massachusetts Institute of Technology
MJD	Modified Julian Date
MPE	Max-Planck Institüt für Extraterrestriche Physik
MSSL	Mullard Space Science Laboratory
MW	medium wavelength
NEWSIPS	The new version of IUESIPS, the original processing of IUE data
NRL	Naval Research Laboratory
ODF	Opacity Distribution Function
OGS	objective grating spectrometer
ORFEUS	Orbiting Retrievable Far and Extreme Ultraviolet Spectrometers
OSO	Orbital Solar Observatory
PG	Palomar Green
PMS	pre-main sequence
PSPC	position sensitive proportional counter
RA	astronomical position coordinate: right ascension
RAP	Right Angle Program
RE	ROSAT EUVE
ROSAT	Roentgen Satellit
RXTE	Rossi X-ray Timing Explorer
SAA	South Atlantic Anomaly
SIC	surrounding interstellar cloud
SIMBAD	a database operated by the Centre de Donnes astronomiques de Strasbourg

SM	Service Module
SPIE	Society for Photoinstrumentation Engineers
SSS	Solid State Spectrometer
STIS	Space Telescope Imaging Spectrograph
SW	short wavelength
SWP	Short Wavelength Prime camera on IUE
TGS	transmission grating spectrometer
UNEX	University-Class Explorer
UVS	ultraviolet spectrometer
UVSTAR	Ultra Violet Spectrograph Telescope for Astronomical Research
VHF	very high frequency
WFC	Wide Field Camera
WIRR	Wind-accreditation Induced Rapid Rotators
WS anode	wedge-and-strip anode
WSMR	White Sands Missile Range
XMA	X-ray mirror assembly
XRT	X-ray telescope

Introduction to the Extreme Ultraviolet: first source discoveries

1.1 Astrophysical significance of the EUV

The Extreme Ultraviolet (EUV) nominally spans the wavelength range from 100 to 1000 Å, although for practical purposes the edges are often somewhat indistinct as instrument bandpasses extend shortward into the soft X-ray or longward into the far ultraviolet (far UV). Like X-ray emission, the production of EUV photons is primarily associated with the existence of hot gas in the Universe. Indeed, X-ray astronomy has long been established as a primary tool for studying a diverse range of astronomical objects from stars through to clusters of galaxies. An important question is what information can EUV observations provide that cannot be obtained from other wavebands? In broad terms, studying photons with energies between ultraviolet (UV) and X-ray ranges means examining gas with intermediate temperature. However, the situation is really more complex. For example, EUV studies of hot thin plasma in stars deal mainly with temperatures between a few times 10^5 and a few times 10^6 K, while hot blackbody-like objects such as white dwarfs are bright EUV sources at temperatures a factor of 10 below these. Perhaps the most significant contribution EUV observations can make to astrophysics in general is by providing access to the most important spectroscopic features of helium – the He I and He II ground state continua together with the He I and He II resonance lines. These are the best diagnostics of helium, the second most cosmically abundant element, with the line series limits at 504 Å and 228 Å for He I and He II respectively.

Sources of EUV radiation can be divided into two main categories: those where the emission arises from recombination of ions and electrons in a hot, optically thin plasma, giving rise to emission line spectra (figure 1.1); and objects that are seen by thermal emission from an optically thick medium, resulting in a strong continuum spectrum but which may contain features arising from transitions between different energy levels or ionisation stages of several elements (figure 1.2).

Examples of the former category are single stars and binary systems containing active coronae, hot O and B stars with winds, supernova remnants and galaxy clusters. Hot white dwarfs, central stars of planetary nebulae (CSPN) and neutron stars are all possible continuum sources. Cataclysmic binaries, where material is being transferred from a normal main sequence star (usually a red dwarf) onto a white dwarf, may well contain regions of both optically thin and optically thick plasma. In O and B stars the EUV emission will be dominated by emission from the shocked wind plasma only at short wavelengths while at longer wavelengths, below the Lyman limit, the continuum flux will be the most important. Apart from studying directly the EUV emission from astronomical objects, these same sources can potentially be used as probes of the interstellar medium. The absorbing

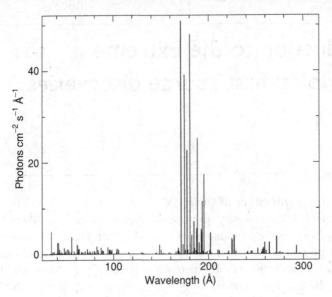

Fig. 1.1. An example of an emission line spectrum in the EUV, arising from a 10^6 K optically thin plasma. The strongest group of lines in the region 170 to 200 Å are mainly Fe transitions (Fe IX to Fe XII).

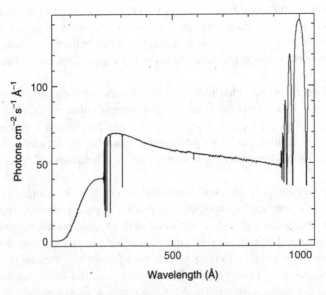

Fig. 1.2. Continuum spectrum in the EUV, formed in this example in the hot photosphere of a 50 000 K white dwarf. The absorption features near 1000 Å are the H I Lyman series and those near 228 Å correspond to the Lyman series of He II, with an assumed abundance (by number) He/H $= 1 \times 10^{-5}$.

effect of the interstellar medium can modify the flux received at the Earth allowing, if the properties of the radiating source are well understood, its structure and density to be studied.

EUV observations can be used to study a wide variety of astronomical environments and physical processes. The aim of this book is to present the scientific discoveries arising from cosmic EUV astronomy, discussing some of the underlying physics to illustrate the insights that can be gained from this work. Since this field has only recently become a fully-fledged branch of astronomy, we present the information in a historical context dealing with the development of both observational and instrumental areas of the subject.

1.2   The 'unobservable ultraviolet'

Until the early 1970s, the conventional view was held that EUV astronomy was not a practical proposition. Since most elements have outer electron binding energies in the range 10–100 eV, photons in the corresponding energy range will be strongly absorbed in any photon–atom interaction when

$$\chi_i < hc/\lambda \qquad (1.1)$$

where λ is the photon wavelength, χ_i is the ionization potential and c is the speed of light.

When $10 < \chi_i < 100$ eV, λ lies in the range 100–1000 Å, i.e. within the EUV band. As a result, the Earth's atmosphere is opaque to EUV radiation due to photoabsorption by N_2, O_2 and O. The $1/e$ absorption depth of the atmosphere is ≈ 130 km at 100 Å. Hence, ground-based EUV astronomy is inconceivable. As in X-ray and far-UV astronomy, this problem can be overcome by placing instrumentation above the atmosphere. However, a more serious problem is posed by absorption of interstellar gas which can potentially make the interstellar medium (ISM) opaque to EUV radiation over interstellar distances (Aller 1959). For example, the mean free path (τ, cm) of a photon in the ISM, where hydrogen is the most abundant element, can be estimated from the known neutral hydrogen density (n_H, cm^{-3}) and hydrogen absorption cross-section (σ, cm^2), where

$$\tau = 1/(n_H \, \sigma) \qquad (1.2)$$

In an influential paper, Aller (1959) argued that, based on the then knowledge of the distribution and density of hydrogen in the galaxy, the ISM would be completely opaque at wavelengths between the X-ray band and Lyman limit at 912 Å and EUV observations impossible. This calculation assumes an interstellar value of 1 atom cm^{-3} for n_H, inferred from 21 cm radiowave surveys. For a neutral hydrogen cross-section of $\approx 10^{-18}$ cm^2, averaged over the entire EUV range, the estimated mean free path (eqn (1.2)) is a mere 10^{18} cm (0.4 parsec) – well below the distance to any astronomical objects other than those in the solar system. Even taking the most optimistic value of the absorbing cross-section (3×10^{-20} cm^2 at 100 Å) gives a viewing distance of only 10 pc. Aller's basic conclusions, self-evident from the knowledge available in 1959, remained unchallenged for more than a decade consigning this region of the electromagnetic spectrum to the 'unobservable ultraviolet' (Harwit 1981).

In contrast to Aller's first, relatively crude estimate of the absorbing effect of the ISM on EUV radiation, Cruddace *et al.* (1974) calculated the interstellar absorption cross-section over

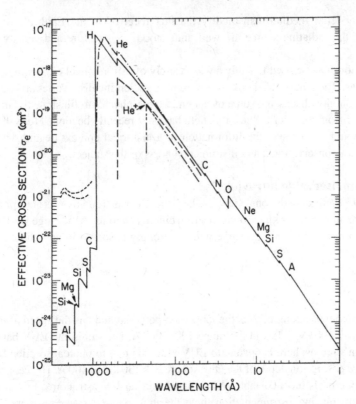

Fig. 1.3. Effective cross-section of the interstellar medium. —— gaseous component with normal composition and temperature; —·— hydrogen, molecular form; — — H II region about a B star; — — — — H II region about an O star; - - - - dust (from Cruddace *et al.* 1974).

the wavelength range 1 to 2000 Å (figure 1.3) and then determined the percentage absorption as a function of wavelength and of n_H. These calculations, expressed in figure 1.4 as the distance for 90% attenutation of the radiation, coupled with increasing indirect evidence for lower values of the value of n_H along at least some lines-of-sight indicated significant transparency in the EUV window and demonstrated the possibility of being able to 'see' EUV sources out to distances of a few hundred parsecs, at the shorter (100–500 Å) wavelengths. Figure 1.4 conveniently expresses a fundamental observational fact of EUV astronomy – for a given column of interstellar gas the distances at which sources can be detected rapidly shrinks as wavelengths increase. A direct consequence of this is that, in general, the number of sources that can be detected diminishes rapidly at longer wavelengths.

Radio measurements of interstellar hydrogen densities are really measuring the total amount of material along the line-of-sight, over distances of several kiloparsecs. Hence, the estimated volume densities are just averages within these long columns and may not be representative of values for the local ISM (LISM), within the distances of a few hundred parsecs critical for EUV astronomy. Furthermore, at least in the 1950s and 1960s, such measurements were carried out with beams several degrees in angular extent and were unable to detect any possible variations in line-of-sight absorption on scales smaller than this. In the early 1970s far-UV satellite experiments provided the first sensitive localised measurements

Table 1.1. *Neutral hydrogen densities* N_H *along the lines-of-sight to bright stars observed in the far-UV.*

Object	Distance (pc)	n_H (cm^{-3})	N_H (cm^{-2})	Reference
α Leo	22	0.02	1.4×10^{18}	Rogerson *et al.* 1973
α Eri	24	0.07	5.1×10^{18}	Rogerson *et al.* 1973
α Gru	28	1.16	1.0×10^{20}	Spitzer *et al.* 1973
δ Per	83	0.59	1.5×10^{20}	Spitzer *et al.* 1973
β Cen	90	0.11	3.0×10^{19}	Savage and Jenkins 1972

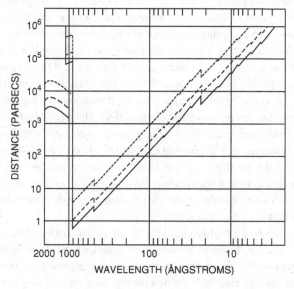

Fig. 1.4. Distance at which attenuation of the incident radiation reaches 90%, as a function of wavelength. An unionised interstellar medium of normal composition is assumed. —— $n_H = 0.2$ cm^{-3}; — — $n_H = 0.1$ cm^{-3}; – – – $n_H = 0.03$ cm^{-3} (from Cruddace *et al.* 1974).

of n_H over much shorter path lengths and narrower angular scales by using the absorption profiles of Lyman α (1216 Å) and Lyman β (1026 Å) in the spectra of bright, unreddened stars (i.e. without the presence of dust along the line-of-sight). Results from studies with Mariner 9 (e.g. Bohlin 1973), OAO–2 (e.g. Savage and Jenkins 1972) and Copernicus (e.g. Rogerson *et al.* 1973; Spitzer *et al.* 1973) showed that the ISM is far from uniformly distributed as had been supposed previously and that 'local' values of n_H could be very much lower than 1 atom cm^{-3}. Table 1.1 summarises several of these measurements, showing volume densities ranging from 0.02 to 1.16 atom cm^{-3}.

1.3 Early detectors for the EUV

To observe at wavelengths in the EUV, astronomers in the 1960s and 1970s needed new detector technology. Solar EUV astronomers used specially developed EUV/X-ray sensitive films for the Skylab telescopes and spectrographs. Although such films provided high spatial resolution, they had a number of insurmountable problems for cosmic EUV observations.

First and foremost, the film required returning to Earth for developing, thereby excluding use on unmanned, long-lifetime observatories. Also, such films were not particularly sensitive, and required exceptionally stable pointing (or very short exposures) to be of practical use on spacecraft.

Thus, to make progress in space-based EUV astronomy, sensitive, photon-counting detectors were needed. Two approaches were natural developments from shorter (X-ray) and longer (UV) wavelengths: proportional counters and photomultipliers.

1.3.1   Proportional counters

The proportional counter is based on the familiar *Geiger counter*, a device that gives an audible 'click' and/or deflection of a meter on detection of a radioactive decay product. The rate of clicking depends on the level of radioactivity, i.e. the number of decay products passing through the counter in each second. Such counters consist of a sealed metallic tube containing a gas such as xenon. The tube usually has a thinner area, or window, through which some less-penetrating radiation can pass. Along the central axis of the tube, a thin, very uniform wire is kept at a high voltage (typically $1\,\mathrm{kV}$ or more). When a gamma ray ($E > 50$–$100\,\mathrm{keV}$ or $\lambda < 0.1\,\text{Å}$) penetrates the gas volume in the Geiger counter, it may ionise a gas atom. The likelihood of this depends upon the gas pressure and composition. The freed electron is then accelerated towards the positively charged central wire, gaining energy and colliding with another gas atom. This atom is ionised as well, freeing a second electron. If the voltage on the central wire is high enough and the gas pressure is in a particular range, this process rapidly recurs until a charge 'avalanche', akin to a microscopic lightning strike, occurs between the centre wire and outer tube. Such an avalanche produces a discharge, which is then amplified electronically and converted into a measurable signal.

This type of counter signals when a highly energetic X- or gamma ray is detected, but gives no information about its energy. To measure energy, the counter needs to be operated at a somewhat lower voltage and possibly a different gas pressure. When these two parameters are adjusted correctly, the number of electrons created during subsequent ionisations after the initial collision of the gamma ray is *proportional* to the energy of the incoming gamma or X-ray photon. Hence the name *proportional counter*. A schematic diagram of such a proportional counter is shown in figure 1.5.

Because of their ability to count individual photons, and also retain information on the energy of the detected photon, proportional counters were rapidly adopted as the workhorse detector of X-ray astronomy in the 1960s and 1970s. However, their use for low-energy

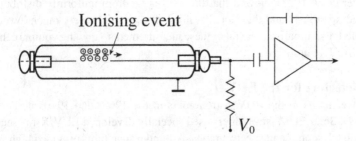

Fig. 1.5. Schematic diagram of a proportional counter (from Zombeck 1990).

X-rays ($\approx$10–100 Å) or EUV ($>$100 Å) required some improvements. First, the window used in high-energy X-ray proportional counters would absorb all the lower energy X-rays or EUV radiation. So thinner windows were required that would allow these low-energy photons through and yet be strong enough to contain the counter gas. Certain plastics, such as polypropylene, were found to have these properties. Unfortunately, when made thin enough to be transparent to soft X-rays, most were found to be permeable to the gas molecules, causing a slow leak in the counter. Thus a gas flow system was needed to maintain the fill pressure and gas purity, resulting in the so-called 'thin-window, gas-flow' proportional counter.

For the very softest X-rays and EUV photons, extremely thin windows and low gas pressures are required. In an early search for EUV sources near the north galactic pole, astronomers at the University of California at Berkeley developed counters with polypropylene windows as thin as 0.3 μm. In addition, the gas itself had to be chosen to be less absorbing in the EUV region, so organic molecules such as methane (CH_3) or propane (C_3H_8) were used. These counters worked reasonably well, but operating them at wavelengths much longer than 200 Å was not possible due to limitations in window materials and gas properties. In addition, the unique feature of the proportional counter – its ability to measure photon energy – diminished, as the uncertainty in the measured photon energy was about the same as the photon energy itself in the EUV. Other approaches have proved to have a wider application for EUV astronomy: these were the channel electron multiplier (CEM) and later, the micro channelplate (MCP).

1.3.2 *Photomultipliers and channel electron multipliers*

Photomultiplier tubes have been used for decades to detect visible and ultraviolet photons at extremely low intensity levels. A photomultiplier tube is an evacuated glass tube with a light-sensitive surface (a photocathode): a photon striking the photocathode will eject an electron (with some probability – typically a few to 20%). After the photocathode is a series of plates which, when struck by the photoejected electron, emit additional (i.e. secondary) electrons. The plates are connected to a high-voltage supply through a series of resistors. Thus the electrons emitted from successive plates 'see' an electric field which accelerates them towards the next plate in the series. At the end of the series of plates, or stages, a single electron will have been amplified into a pulse of 10^7–10^{10} electrons, which in turn is collected by a metal plate and sensed through external electronics. Thus, just as in the Geiger counter, a single photon of visible or UV light may be detected using the photomultiplier.

The CEM is a compact, windowless version of the photomultiplier. Instead of a sealed tube, the CEM itself must be used in a high vacuum (about a billionth of atmospheric pressure). The photocathode of the CEM may be a cone covered with a photosensitive material (in the EUV this may be MgF_2, CsI or KBr) with a hole at its apex. Attached to the hole is a small-diameter tube (frequently a few millimetres or less) of lead glass coated with a material that acts much like the electron-emitting plates in a photomultiplier (see figure 1.6). The leading part of the inner wall of the tube may also act as the photocathode. The tube is often curved, and a high voltage is applied to one end. Just as in the photomultiplier, an electron ejected when a photon strikes the conical photcathode is accelerated towards the tube at the apex of the cone. When it strikes the tube's surface, secondary electrons are emitted, which in turn are accelerated further along the length of the tube, striking the tube wall within a short distance. The process will be repeated until, as in the case of the photomultiplier, a large pulse

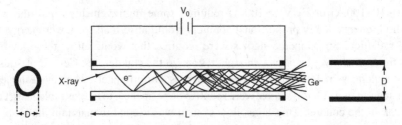

Fig. 1.6. Diagram illustrating the operation of a continuous-dynode electron multiplier (from Fraser 1989).

of about 10^6–10^7 electrons can be collected at the end of the tube and sensed electronically. The windowless nature of the CEM means that it may be used to detect photons at all EUV wavelengths.

1.4   Early experiments with sounding rockets

With the realisation that many galactic EUV sources could conceivably be detected from above the Earth's atmosphere, the Space Sciences Laboratory of the University of California (UC) at Berkeley embarked on a sounding rocket programme. Several experiments were flown between 1972 and 1974 to search for sources of EUV emission. These early experiments did not use imaging instruments, although nested gold-coated plane mirrors were employed as a flux concentrator. A mechanical collimator was used to constrain the field of view to $1° \times 50°$ full width half maximum (fwhm). The collecting area was divided into five overlapping segments each covered with a channel electron multiplier (CEM) photon detector.

A bandpass of 135–475 Å was defined using thin-film filters composed of aluminium and carbon. During the first flight of this instrument an area of sky approximately 1350 square degrees around the north galactic pole was surveyed, with a limiting sensitivity of 4.3×10^{-8} erg cm^{-2} s^{-1} (Henry *et al.* 1975). An improved second flight covered a similar area, achieving a flux limit of 2.9×10^{-8} ergs cm^{-2} s^{-1} (Henry *et al.* 1976a). No sources were detected on either flight. However, 3σ upper limits were obtained for the cataclysmic variables RX Andromedae, during flare, and U Geminorum, during quiescence.

Experiments were also carried out using five propane-filled proportional counters as detectors (Henry *et al.* 1976b). One disadvantage of these devices is the need for a gas-tight entrance window, restricting the useful wavelength range to the lower range of the EUV, where the absorption of the window is lowest. Although an additional wavelength-defining filter may then not be needed, counter windows usually need to be thicker than the thin-film filters employed with CEM experiments. Spanning a wavelength range 44–165 Å and with a field of view approximately $2° \times 15°$, some 3700 square degrees around the south celestial pole were surveyed, resulting in the detection of a single source with a flux of 1.3×10^{-9} ergs cm^{-2} s^{-1}. The most likely counterpart in the source error box was the cataclysmic variable VW Hydri.

Other investigators also carried out experiments with sounding rockets, including a group from California Institute of Technology. Using a spiraltron detector (a CEM coiled in a spiral) upper limits were obtained in the wavelength range 140–430 Å but no further sources were detected. The main limitations of all these sounding rocket payloads were the effective areas available with the existing technology, restrictions on payload dimensions and the short exposures (300–500 s) allowed by sub-orbital flights. A chance to increase the exposure times with a longer duration space mission appeared with the Apollo–Soyuz Test Project.

1.5   EUV astronomy on the Apollo–Soyuz mission

In 1975, an opportunity arose to perform an extended search for EUV sources of cosmic radiation on the Soviet–American Apollo–Soyuz mission (officially termed the Apollo–Soyuz Test Project, or ASTP). In July of that year, US and Soviet astronauts were to link up in space in a demonstration of international good will. NASA realised that, in addition to the political benefits of the flight, some science instruments could 'piggyback' on the Apollo capsule, and solicited proposals for experiments which could be flown over the 9-day duration of the mission. The UC Berkeley group received approval for two such experiments: the Extreme Ultraviolet Telescope (EUVT) and the Interstellar Helium Glow Experiment. The results from the latter experiment, which was intended to map out the distribution and motion of He and He^+ in the heliosphere, were described in detail by Bowyer *et al.* (1977a). In the remainder of this section, we will concentrate on results from the EUVT, including the first unambiguous detection of a cosmic EUV source, the hot white dwarf HZ 43.

The EUVT (figure 1.7) was, in reality, not an imaging telescope at all, but a 'light bucket', consisting of four concentric Au-coated paraboloidal mirrors which focused EUV radiation onto one of two channel electron multiplier (CEM) detectors, one with a 2.5° diameter field of view (FOV) and the other with a 4.3° FOV (see Bowyer *et al.* 1977b). Just in front of the detector was a six-position filter wheel, with four EUV filters (parylene N, Be/Par N, Al + C, Sn), one far-UV filter (BaF_2) and a 'blank' position consisting of a thin Al disc for use in determining instrumental background. The sensitivity of the EUVT for the four EUV bands is given in figure 1.8.

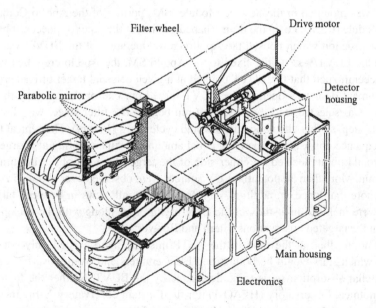

Fig. 1.7. A schematic diagram of the EUV telescope (EUVT) flown on board the Apollo–Soyuz mission in 1975 (courtesy S. Bowyer).

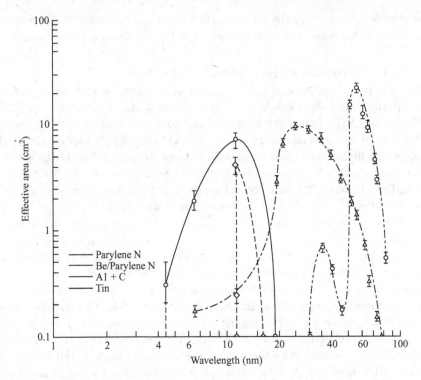

Fig. 1.8. Effective area of the EUV telescope (EUVT) on the Apollo–Soyuz Test Project (courtesy S. Bowyer).

The EUVT was mounted in the Service Module (SM) portion of the Apollo Command and Service Module (CSM), i.e. in the drum-shaped portion of the Apollo spacecraft behind the cone-shaped section which housed the Apollo crew. Operation of the EUVT was quite simple. Since the EUVT was rigidly fixed to the Apollo SM, the Apollo crew first had to orient the spacecraft such that the EUVT pointed at a given celestial target or background. Then an instrument door was opened, the EUVT turned on, one or the other of the two CEM detectors was selected, and the observing run begun. The filter wheel was designed to continuously step, so that all six positions would cycle in 6 s (10 rpm). The typical target observation sequence consisted of spending equal amounts of time pointing at a target and at two background points located on either side of the target, for total observing times of several to 20 min. More than 30 potential EUV sources were observed over about 15 Apollo orbits, some more than once. In addition, data from spacecraft slews in between targeted observations were included in a study of the EUV background, along with the background pointings from the targeted observations (Stern and Bowyer 1979).

It is worth noting that, at the time of the ASTP mission, there existed only educated guesses as to which stars would be both strong EUV emitters and would also have a low enough interstellar absorption so as to be detectable by the EUVT. Neither the first High Energy Astronomical Observatory (HEAO-1) nor the Einstein observatory X-ray missions had yet been launched, and, in fact, the first detection of a stellar corona (other than that of the Sun) associated with the binary system Capella (α Aurigae; Catura *et al.* 1975) had only just occurred. In retrospect, the target list provided a good indication of which stars would also be later detected during the Roentgen Satellit (ROSAT) and Extreme Ultraviolet Explorer (EUVE) missions (see chapters 3 and 4). The target list consisted of:

- nearby stars with potential steady or flaring coronal emission: α Cen, ϵ Eri, 70 Oph, 61 Cyg, α Aql, i Boo, Proxima Centauri, Wolf 424, UV Cet, and β Hydri, EV Lac, θ Oph, α PsA,
- hot white dwarfs or central stars of planetary nebulae: HZ 43, HZ 29, Feige 24, BD + 28°4211, NGC 7293, NGC 6853,
- cataclysmic variable stars or dwarf novae: SS Cyg, AE Aqr, VW Cep, DQ Her, Z Cha, VW Hyi,
- nearby pulsars: PSR 1929 + 10, PSR 1133 + 16, PSR 1451 − 68, and
- Jupiter(!)

The principal contribution of the EUVT on ASTP was the detection of four prototypical sources of EUV radiation: two hot white dwarfs, HZ 43 (Lampton *et al.* 1976a) and Feige 24 (Margon *et al.* 1976), a cataclysmic variable star (dwarf nova) SS Cyg (Margon *et al.* 1978), and the nearby flare star Proxima Centauri (Haisch *et al.* 1977). Only the two white dwarfs were detected as steady sources: in the original Lampton *et al.* (1976a) paper, the temperature of HZ 43's surface was estimated using a blackbody model to be $\sim 10^5$ K, although we now know, using a combination of optical and EUV spectroscopy, that the photospheric temperature of HZ 43 is closer to 50 000 K (see chapter 8). It is, however, still one of the hottest white dwarfs known, and remains to this day a 'standard candle' for all other EUV sources. Remarkably, HZ 43 is ≈ 65 pc distant, clearly demonstrating that the ISM was transparent enough, at least in some directions, to allow the observation of cosmic EUV sources. Subseqently, over 100 EUV emitting white dwarfs have been catalogued (Pounds *et al.* 1993; Bowyer *et al.* 1996), comprising the second largest population of EUV emitting objects. SS Cyg was caught in outburst in the shortest wavelength band of the EUVT: nearly 20 years later, the inexplicable spectrum of another dwarf nova caught in outburst, VW Hyi proved a challenge to theorists trying to understand the nature of the EUV emitting region (Mauche *et al.* 1996). Finally, the detection of Prox Cen during a flare was a harbinger of the hundreds of coronal sources seen by the ROSAT Wide Field Camera (WFC) and EUVE, not to mention the tens of thousands of stellar coronae detected in the ROSAT all-sky X-ray survey. Thus the ASTP mission demonstrated that EUV astronomy was feasible, that the 'unobservable ultraviolet' was a myth, and that future missions with greater sensitvity could begin to explore this wavelength region with some expectation of success.

1.6   After Apollo–Soyuz

Following the outstanding success of the EUVT on ASTP but without immediate prospect of a longer term space mission, further attempts at observations in the EUV were made with sounding rockets. Technological improvements led to the first true imaging EUV telescope flown by Berkeley. This payload made the first use of a focusing grazing incidence telescope. The use of grazing incidence telescopes in both X-ray and EUV astronomy gives a major improvement in sensitivity for two reasons. First, they can be used to increase the geometric area over which flux is collected without the need for very large detector areas. Second, being able to concentrate the image of a source in a small region of the detector yields a dramatic improvement in signal to background ratio, as most of the unwanted counts lie outside the source image.

Several designs for focusing optics have been used. The most useful are the three designs originated by Wolter (1952a,b). Of these the Wolter Type III (figure 1.9) is too complex to

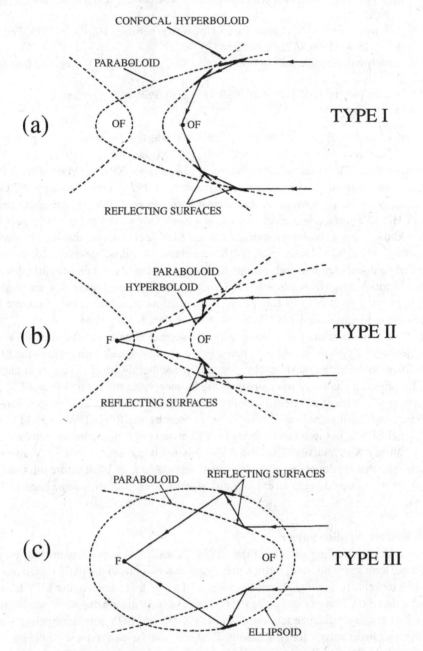

Fig. 1.9. The three configurations of X-ray (and EUV) imaging telescopes suggested by Wolter (1952a,b).

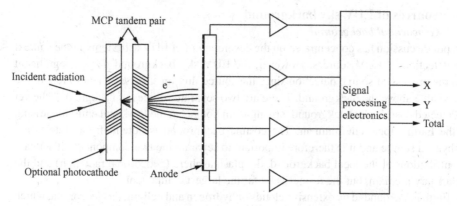

Fig. 1.10. Schematic diagram of an imaging microchannel plate detector.

be practical, besides having a small collecting area for a given telescope diameter, but both Type I and Type II optics have found application in EUV astronomy. The Type I is well-suited to wide-field survey applications since it suffers less from off-axis aberrations than the Type II. However, in the Type I the wide angle of the focused cone of light is difficult to use in a spectroscopic experiment and the Type II design has been preferred for such applications. Working at longer wavelengths than X-ray mirrors the tolerances on surface roughness are not so severe and, in the EUV, it has been possible to use lightweight mirrors, made from an aluminium substrate plated with nickel and overcoated with gold, to enhance the reflectivity.

To take advantage of the benefits of an imaging telescope requires a two-dimensional EUV detector. The main techniques used so far have been the microchannel-plate (MCP) detector and the imaging proportional counter. MCPs are arrays of single channel electron multipliers that have been miniaturised using technology similar to that used to manufacture optical fibres. Each plate comprises 10 million or more individual multipliers whose positions can be encoded by a range of output anode designs (e.g. figure 1.10). Alternatively, imaging proportional counters (IPCs or position sensitive proportional counters, PSPCs) can be used in the focal plane. Usually, arrays of wire grids are used to encode the positions in these devices (e.g. figure 2.11). However, as in the non-imaging versions, absorption of the incident EUV radiation by the counter window is a problem.

Berkeley's rocket-borne imaging telescope made use of a Wolter–Schwarzschild Type II mirror design (see section 2.3) and employed a MCP detector in the focal plane. EUV pass bands were determined by two filters with responses peaking at 300 Å and 500 Å. Despite the improved technical performance of the instrument only upper limits were obtained for the three white dwarfs observed – Sirius B (Cash *et al.* 1978), Feige 24 and G191–B2B (Cash *et al.* 1979). However, the factor of ten improvement over the limit obtained for Sirius B by ASTP is a good indication of the advantage of an imaging system.

A Dutch/Japanese collaboration developed a sounding rocket payload carrying two detection systems covering the band 50–270 Å. One of these was a one-dimensional X-ray focusing collector with PSPC detector while the second comprised eight large-area collimated proportional counters having a range of entrance windows (Bleeker *et al.* 1978). This payload obtained the first spectrum of an EUV source, the hot white dwarf HZ 43.

1.7   Sources of EUV sky background

1.7.1   *Geocoronal background*

So far, our discussion has concentrated on the development of EUV astronomy in the context of the detection of EUV sources. However, the EUV sky background plays a significant role in the process of source detection since the source flux must be sufficient to render the source visible above the background. There are two separate sources of background; the yet to be detected cosmic sky background, arising from hot gas in the ISM, and emission arising from the Earth's local environment. The cosmic background is potentially an interesting astrophysical source and it is therefore important to separate the two components. There are two contributions to the local background, the plasmasphere (geocoronal radiation) and the interplanetary medium, but the former is by far the largest component.

The Earth is surrounded by extensive clouds of hydrogen and helium, the geocorona, which result from the gradual loss of these gases from the Earth's atmosphere. The geocoronal background arises from resonant scattering of solar EUV radiation incident on these gases, which has an intensity of ≈ 1 erg cm^{-2} s^{-1}, mostly in the form of line radiation. The intensity of the scattered radiation will depend upon the number of atomic and ionic species present, the level of solar activity and the viewing geometry of the instrument. Below 1000 Å the lines of He I 584 Å and He II 304 Å predominate. Much weaker flux from He II 256 Å and He II 237 Å is also present. The general variation of these emissions with the solar zenith angle (the angle between Sun, Earth and satellite, see figure 1.11) and magnetic latitude was established in the mid 1970s by Meier and Weller (1974, He I) and Weller and Meier (1974, He II). In both cases, minimum intensity occurs when the satellite is at its nadir with respect to the Sun and when the view direction is along the Earth shadow. Maximum He I flux (≈ 50 Rayleighs, $1\,R = 7.96 \times 10^4$ photons cm^{-2} s^{-1} sr^{-1}) occurs during the daytime (solar zenith angle $<90°$) and appears to be independent of the view direction. At night the flux

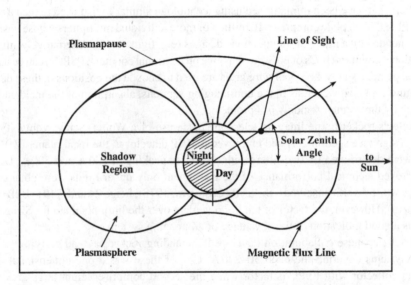

Fig. 1.11. Observing geometry for a spacecraft in low Earth orbit, defining the solar zenith angle and illustrating the region of the Earth shadow.

drops by two orders of magnitude or more to a minimum, looking along the Earth shadow. This night-time flux is the contribution from interplanetary helium (1–9 R, Weller 1981). Emission of He II (5 R maximum) is an order of magnitude below the He I flux, although it shows similar variation, but with some differences arising from the interaction between the ions and the Earth's magnetic field. The minimum contribution of He I in the Earth shadow is 0.03–1 R.

Above 1000 Å, the geocoronal spectrum is dominated by the emission lines of neutral hydrogen (1216 Å and 1025 Å) and those of neutral oxygen (1304 Å and 1356 Å). Although these lines lie outside the EUV, they are very bright and can still potentially contaminate EUV data if they 'leak' through any blocking filters. The lines of neutral hydrogen have been observed by satellite and sounding rocket experiments (Meier and Mange 1970; Young *et al.* 1971; Paresce *et al.* 1972; Thomas and Anderson 1976). At high altitude ($\approx$650 km), the intensity of Lyman α (1216 Å) radiation varies between about 10^3 and 2×10^4 R according to the solar zenith angle (Meier and Mange 1970; Thomas and Anderson 1976). Lines of OI may have similar intensity (Meier and Prinz 1971). These fluxes are considerably more intense than either of the strong He lines and must be efficiently rejected from any EUV instrumentation, by use of detectors that are blind to these far-UV wavelengths or suitable blocking filters.

1.7.2   The cosmic EUV background

EUV emission from hot gas is not necessarily confined to that from discrete stellar coronal sources or hot white dwarfs. Supernovae are expected to provide a source of hot gas in the ISM. Hence, apart from its absorbing effects on the emission from distant objects, it seems likely that the ISM may well be an EUV source in its own right, but a diffuse one, covering large areas of the sky. Real evidence for the existence of hot material, with temperature $\approx$$10^6$ K, is provided by the diffuse soft X-ray background (deKorte *et al.* 1976; Sanders *et al.* 1977; Marshall 1982; Snowden *et al.* 1995). These results also indicate that the emission is local. The brightness distributions are patchy, unlike the isotropic and predominantly extragalactic higher energy background (>1 keV), and the correlation with the neutral hydrogen column density is not consistent with that expected from simple photoabsorption of the galactic flux. In addition, a number of energy-dependent large-scale features, such as the North Polar Spur, are discernible. The presence of OVI absorption lines in the spectra of hot stars also provides evidence for material with temperatures in excess of 10^5 K (e.g. Jenkins 1978).

Few direct measurements of any diffuse cosmic emission component were made in the early EUV astronomy experiments, mainly due to the difficulty of separating out the often overwhelming geocoronal component. Hence, predictions of the expected flux relied heavily on theoretical models. For example, Grewing (1975) derived a blackbody model ($T \approx 1.1 \times 10^4$ K) of the background between 1750 Å and 504 Å, showing that the flux is the result of contributions from a large number of O and B stars. The Apollo–Soyuz experiment was able to carry out an extensive survey of the EUV background in the 50–775 Å range (Stern and Bowyer 1979). At the longer wavelengths, the cosmic background was indeed masked by the geocoronal flux but an intense background was apparently observed in the 110–160 Å band, at a level of $\approx$4 photons cm^{-2} s^{-1} sr^{-1} Å^{-1}. A subsequent analysis (Stern and Bowyer 1980) concluded that, barring the existence of an unreasonable number density of EUV sources, the uniformity of the observed flux indicated that the emission was produced by a diffuse source.

2

The first space observatories

2.1   Introduction

The first true space observatories incorporating imaging telescopes and providing access to the soft X-ray band, but providing some overlapping response into the EUV were flown in the late 1970s and early 1980s. With the Einstein and European X-ray Astronomy Satellite (EXOSAT) satellites, launched in 1978 and 1983 respectively, long exposure times coupled with high point source sensitivity became available to high-energy astronomers for the first time. This progress depended mainly on developments in optics and detector technology but also coincided with a more sophisticated understanding of the physical processes involved in soft X-ray and EUV emission and a better appreciation of the potential significance of observations in these wavebands. This chapter describes the Einstein and EXOSAT missions detailing both the telescope and detector technology. Developments in the understanding of the physical emission processes are outlined to set the context for discussion of the most significant results from these observatories, relevant to the broad field of EUV astronomy.

Parallel developments in far-UV astronomy, arising mainly from the International Ultraviolet Explorer (IUE) observatory, provide an important complement to the work of Einstein and EXOSAT. However, since the technology of IUE is quite different to that used at shorter wavelengths we concentrate solely on the appropriate scientific results. Finally, the Voyager ultraviolet spectrometer (UVS), while mainly designed for planetary studies, was used during cruise phases of the mission, providing the first stellar EUV spectra in the region just below the Lyman limit at 912 Å.

It is not the purpose of this chapter to give a comprehensive review of instrumental techniques for soft X-ray, EUV and UV astronomy, since these have been adequately covered in other publications. Rather we concentrate on providing a summary of the basic ideas, together with more detail on certain aspects that are important for an understanding of the technological development of EUV astronomy and that are relevant for interpreting the data from the variety of instruments employed. Those seeking more technical detail than is included here should refer to important texts on vacuum ultraviolet spectroscopy (Samson 1967), X-ray detectors (Fraser 1989) and numerous proceedings published by the Society for Photoinstrumentation Engineers (SPIE).

2.2   EUV emission processes

2.2.1   A physical understanding of emission processes

As discussed in section 1.1, there are two emission processes of fundamental importance to EUV astronomy: continuum emission, principally from stellar photospheres; and line emission from hot, optically thin plasmas such as stellar coronae. A third, less common, mechanism in objects with large magnetic fields is synchrotron emission. Here we consider those aspects of the emission processes which play important roles in EUV observations.

2.2.2   Stellar photospheres

Approximately 28% of the sources detected in the ROSAT and EUVE all-sky EUV surveys are due to photospheric emission from stars, nearly all from hot hydrogen-rich white dwarfs. The analysis of these EUV fluxes depends almost entirely on the ability of model atmospheres to predict the emergent fluxes at short wavelengths. Such models have steadily increased in sophistication since the first white dwarfs were detected in the EUV. The simplest, and first models used were blackbody energy distributions (Lampton *et al.* 1976a). Although the blackbody approximation is, in general, not appropriate for real stellar atmospheres, it is still used to estimate the EUV flux from the hot surfaces of neutron stars (Edelstein *et al.* 1995).

In wavelength units, the flux emitted by a blackbody of temperature T is given by the Planck function:

$$B_\lambda(T) = 2hc^2\lambda^{-5}\left(e^{\frac{hc}{\lambda kT}} - 1\right)^{-1} \tag{2.1}$$

If the blackbody is a stellar surface then the Eddington flux or spectral flux emitted per unit area per unit time per unit wavelength per unit solid angle is

$$H(\lambda)_{\mathrm{BB}} = \frac{1}{4}B_\lambda(T) \tag{2.2}$$

In units of $\lambda_{100} = 100\,\text{Å}$ and $T_{100} = 100\,000\,\text{K}$ the above relation becomes

$$H(\lambda)_{\mathrm{BB}} = 2.98 \times 10^{26}\lambda_{100}^{-5}\left(e^{\frac{14.4}{\lambda_{100}T_{100}}} - 1\right)^{-1} \tag{2.3}$$

where the Eddington flux is given in units of ergs $\mathrm{cm^2\,s^{-1}\,Å^{-1}\,sr^{-1}}$. The Eddington flux is introduced because it is a frequently used quantity which is the product of model stellar atmosphere calculations. It is related to the observed flux at the Earth by

$$f(\lambda) = 4\pi\left(\frac{R^2}{D^2}\right)H(\lambda) \tag{2.4}$$

where R is the stellar radius and D the distance to the star. It is also related to the effective temperature (T_{eff}) through Stefan's law,

$$4\pi\int_0^\infty H(\lambda)\,d\lambda = \sigma T_{\mathrm{eff}}^4 \tag{2.5}$$

where σ is the Stefan–Boltzmann constant.

The simplest realistic stellar atmosphere is that of a pure-hydrogen photosphere where, in principle, it is possible to compute the atmospheric structure and emergent flux, with a high degree of confidence. This is particularly true of hot ($T_{\mathrm{eff}} > 15\,000\,\text{K}$) H-rich (DA) white dwarfs, where the atmospheres are fully radiative and convection is not an issue. A good

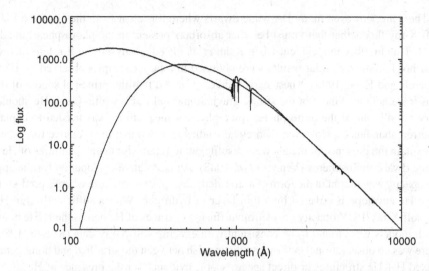

Fig. 2.1. Comparison of a blackbody spectrum with that computed from a stellar atmosphere model.

example of pure-hydrogen model atmosphere calculations for hot white dwarfs are those of Wesemael *et al.* (1980). In figure 2.1 we compare the flux from a pure-hydrogen model atmosphere with $T_{eff} = 60\,000$ K having a typical surface gravity of $\log g = 8$ with that of a 60 000 K blackbody. The structure evident in the pure-H model atmosphere, in contrast to the blackbody in figure 2.1, is due to the opacity of the hydrogen atom. Opacity is given by

$$\kappa_\lambda = \rho^{-1}\left(\frac{d\tau_\lambda}{dz}\right) \tag{2.6}$$

where ρ in the local mass density and $d\tau_\lambda/dz$ is the rate of change of optical depth with height in the atmosphere. Opacity has units of reciprocal surface mass density and depends on the local mean particle densities and atomic cross-sections. It includes scattering and absorption of photons and is often split into line and continuum contributions. It is the relatively low short-wavelength opacity of a pure-H atmosphere which explains why relatively cool DA stars such as Sirius B at $T_{eff} \sim 25\,000$ K can be observed at EUV and soft X-ray wavelengths. This was first discussed by Shipman (1976), who showed that while the optical emission from the photosphere of Sirius B was consistent with a gas temperature $\approx 25\,000$ K, the significant EUV and soft X-ray fluxes were emitted from deeper levels in the photosphere, at temperatures of $\approx 100\,000$ K.

In EUV astronomy, such pure-hydrogen photospheres are realized in a number of hot white dwarfs where it is possible to match the observed flux with that of a model atmosphere from the soft X-ray to the infrared. A good example of this is the analysis of the 50 000 K DA white dwarf HZ 43 by Barstow *et al.* (1995b). It is evident from the detailed studies of DA stars detected with ROSAT (Marsh *et al.* 1997b), that most white dwarfs with T_{eff} below 40 000 K must possess pure-H or nearly pure-H photospheres. These stars can thus all be successfully analysed in terms of relatively simple pure-H model atmospheres.

There are, however, many DA white dwarfs where it is clear from the observed EUV and soft X-ray fluxes that there must be other absorbers present in the photosphere, in addition to H. This first became evident when Kahn *et al.* (1984) analysed the Einstein fluxes from four hot DA stars. Similar results were obtained in a larger sample, observed by EXOSAT (Paerels and Heise 1989). These early studies assumed that the principal source of opacity was trace helium. Since, in the strong gravitational field of a white dwarf, He should sink very rapidly out of the atmosphere, some physical mechanism was needed to explain the inferred abundances. However, theoretical studies revealed that the radiative forces invoked to explain the presence of metals were insufficient to retain signficant quantities of He in hot white dwarf photospheres (Vennes *et al.* 1988). An alternative explanation for the apparent He opacity was sought in the form of a stratified atmospheric structure where a predominantly pure He envelope is covered by a thin layer of hydrogen. With a sufficiently thin H layer the soft X-ray/EUV opacity arises from diffusion of traces of He across the H/He boundary. Such models were found to be consistent with existing soft X-ray data (Koester 1989), but there was no observational evidence to distinguish between the stratified and homogeneously mixed H + He structures or direct spectroscopic evidence for the presence of He.

2.2.3   Stellar coronae

While the optical emission from the majority of normal, non-degenerate stars is dominated by the photosphere, observations in the UV, EUV and soft X-ray are sampling the hot, optically thin plasma associated with the chromospheric, transition and coronal regions. Hence, the dominant emission process for EUV observations of such objects is through the recombination of ions and electrons that have been collisionally excited and ionised in the hot gas. Unlike the white dwarfs, where a detailed physical model can be constructed that can, in principle, predict what will be observed throughout the electro-magnetic spectrum, the situation is more complex for coronal/chromospheric sources. There is no existing physical model that can link the temperature and density conditions in the different discrete regions of the outer atmosphere (chromosphere, transition region and corona). Hence, any measurements made in different spectral regions are largely decoupled from each other, providing only limited information that can be of direct assistance in other wavebands.

The approach to analysing broadband X-ray and EUV data has centred on the use of computer codes that take into account the emission line formation processes in hot plasma. These incorporate the relevant atomic data and attempt to take account of all possible transitions. The emissivity of the plasma is then calculated for a particular temperature. Comparison of the predicted fluxes with the observations then allows the temperature of the source to be constrained by the data. For a practical model, the important free (variable) parameters are the plasma temperature, the hydrogen density, the element abundances and a normalisation constant. The normalisation constant is related to the volume emission measure of the source and its distance.

The temperature range covered by a typical UV-derived emission measure distribution runs over nearly two orders of magnitude from a few 10^4 K to $\approx 3 \times 10^5$ K, sampling the chromospheric and transition regions of the star. The importance of X-ray and EUV observations lies in their probing the higher temperature plasma of the corona. However, until recently, only the availability of data from ASCA have provided access to resolved lines in stars other than the Sun and even then only for the brightest stellar X-ray sources. Accordingly, many earlier analyses have considered only broadband fluxes or low-resolution spectra. In matching these to a spectral model, it is necessary to make assumptions about the temperature distribution of the plasma. The simplest is to treat the coronal source as having a single temperature. Indeed, if a single-temperature model gives an adequate match to the data, there is no *a priori* need to consider a more complex situation. All the same, it is true, for example, that high signal to noise data from the ROSAT PSPC usually require at least two temperature components to give a good match. However, this is not the same as saying that the gas really does have a single, or even several, discrete temperature components.

In reality, as can be seen from the UV observations of transition region and chromospheric plasma, a stellar corona is best represented by a range of temperatures. The characteristic temperatures associated with X-ray analyses probably arise from a combination of instrument bandpass and the location (in temperature) of the peak (or peaks) of any emission measure distribution. This effect has been discussed in a detailed review of Einstein IPC spectra by Schmitt (1988 and references therein), who shows that best-fit temperatures cluster around discrete values.

As an alternative to single or multiple temperature components a natural description is to consider a power law differential emission measure distribution, as commonly observed directly (from emission lines) on the Sun (Raymond and Doyle 1981). This is more physically reasonable as it can be explained by a plasma whose temperature structure is determined by balancing the radiative and conductive energy losses (Antiochos *et al.* 1986). The distribution can be defined as

$$Q(T) \approx (n_e^2 / T)(ds/d \ln T) = A(T/T_{max})^\alpha$$

where T_{max} is the upper limit of the range over which the distribution is non-zero and A is a normalisation constant derived from the data. The power law index (α) is related to the radiative cooling function and has a value ≈ 1 for the Sun. Drake (1999) has considered the response of the EUVE scanner and deep survey telescope to radiative loss models for coronal plasma, providing tables of the unit count rate flux calibration as a function of temperature and interstellar column density.

2.3 Grazing incidence mirror technology

The key to high-sensitivity astronomical observations in both X-ray and EUV bands is the availability of focusing telescopes capable of producing an image. Since photon reflection efficiencies near normal incidence are very low in the EUV and X-ray bands, the conventional mirror systems used in optical and UV astronomy have only limited application shortward of 300 Å. Furthermore, it is self-evident that reflecting telescopes of some description are the only solution since glass and other lens media are opaque to photons of these energies. Reflection efficiencies for high-energy photons are much greater at shallow angles, the so-called 'grazing incidence'. More recently, normal incidence mirrors have been developed for the soft X-ray and EUV using an interference technique for reflection with multilayer coatings of the optical surface. However, the bandpass over which significant efficiency can be achieved is very narrow, typically 10% of the peak wavelength. Hence, grazing incidence telescopes represent the best solution for imaging in EUV and soft X-ray astronomy, if a broad spectral coverage is required.

The construction of grazing incidence telescopes was briefly touched upon in section 1.6 in the description of Berkeley's first imaging EUV telescope. Initial designs used simple conical mirrors or the plane mirror arrangement of Kirkpatrick and Baez (1948). However, these optical systems introduce considerable optical aberration, since the focal point is not the same for all points on the mirror surface. To achieve correct focusing, the reflecting surface must be curved to match a conic section, such as a parabola or hyperbola. These principles were incorporated into the designs of Wolter (1952a; see figure 1.9), which have now been adopted as the main workhorses of X-ray and EUV astronomy. In fact, to yield a properly confocal image a grazing incidence telescope must contain an even number of reflections. Therefore, each Wolter telescope design has two reflecting surfaces. The best combinations with the least aberration have a parabolic leading surface and hyperbolic trailing reflector.

One major limitation of grazing, compared to normal incidence telescopes, is their collecting area. Apart from a small central region obscured by a secondary mirror or detector system, normal incidence telescopes utilise the entire diameter of the mirror for light collection whereas grazing incidence systems can only collect radiation in the annulus subtended by the leading reflector in the plane perpendicular to the optical axis. For example, a 30 cm diameter normal incidence mirror has a geometric area of about 700 cm^2, whereas a grazing incidence unit of the same dimensions has about one tenth the area. One solution to this problem is to build a nest of mirror shells, stacking several of decreasing diameter one inside another. However, this is only possible with the Type I design (figure 2.2).

While the use of Wolter's original designs was a major step forward in high-energy astrophysics, they still suffer from significant aberration when the imaged source is not aligned with the optical axis. In particular, the size of the focused image increases as the source moves further and further off-axis and often becomes distorted showing the aberration known as coma (figure 2.3). These effects can be minimised, although not completely eliminated (Wolter 1952b) by modifying the pure parabolic and hyperbolic figures with small corrections as described by Schwarzchild (1905) for normal incidence reflectors. The modified designs are known as Wolter–Schwarzchild telescopes and eliminate most of the coma near the optical axis but the image size still degrades towards the edge of the field (figure 2.3). The magnitude of the aberrations is a function of the field of view of any individual telescope and therefore depends upon the size of the grazing angles and focal length. The reflection efficiency at grazing incidence is a strong function of photon energy, and as energy increases the optimum angle of incidence decreases dramatically. For example, in the EUV, grazing angles $\approx 10°$ lead

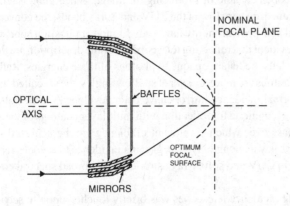

Fig. 2.2. Schematic diagram of a nest of Wolter Type I mirror shells.

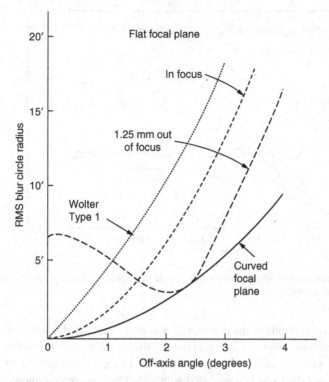

Fig. 2.3. The dependence of optical aberrations on off-axis angle in a grazing incidence telescope.

to the use of relatively short focal lengths. In this regime the Wolter–Schwarzchild design is the most appropriate. In the X-ray region, high efficiency can only be achieved with grazing angles of one or two degrees, requiring long focal lengths and limiting the allowed field of view to narrow angles. With these instruments, the effect of the Schwarzchild modifications is rather small. Hence, the basic Wolter designs are usually adopted for X-ray work.

Actual performance of real telescopes depends on the ability to machine the correct figure on a suitable substrate. The optical surface must also be smooth to tolerances close to the wavelength of the radiation to be studied, to avoid scattering of radiation outside the focused image. Large scattering effects can degrade both the image and the reflection efficiency by completely removing photons from the image. It is thus machining and polishing technology for mirror manufacture that has driven the pace of astronomical applications.

2.4   Applications of grazing incidence technology in space

Several manufacturing techniques have been developed for telescope production. The most widely used to date is the direct machining of glass or metal shells followed by polishing and gold plating. In cosmic X-ray astronomy the first two-dimensional (2-D) focusing collector was used by Gorenstein *et al.* (1975), based on the Kirkpatrick and Baez (1948) design, constructed from chromium-coated bent glass plates. As applied to Wolter Type I metal mirrors, the technique was pioneered in the mid 1970s for a sounding rocket-borne soft X-ray telescope in a joint programme with Leicester University (UK) and Massachusetts Institute

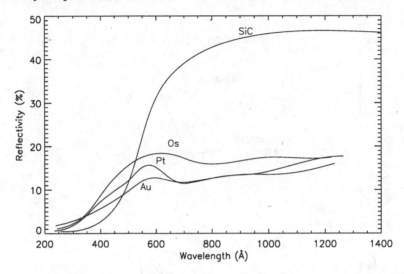

Fig. 2.4. The normal incidence reflectivities of silicon carbide, osmium, platinum and gold as a function of wavelength.

of Technology. The basic aluminium mirror structure was rough machined and then diamond turned to produce the required Wolter I figure. However, the necessary smooth surface finish could not be achieved with aluminium alone, which tends to have a pitted surface structure. Consequently, the figured shell was electroplated with nickel before overcoating with gold. Each reflecting surface was machined separately and mounted together when completed. Flown with an imaging proportional counter at the focus, this telescope represents the first use of grazing incidence mirrors in X-ray or EUV astronomy and recorded the first X-ray images of the Cygnus loop, Puppis A and IC433 supernova remnants (Cygnus, Rappaport *et al.* 1979; Puppis A and IC433, Levine *et al.* 1979).

At EUV and X-ray wavelengths, the reflectivity of the basic metal or glass mirror substrate is, in general, poor and needs to be improved by overcoating with a suitable material. The theoretical reflective properties of several suitable heavy elements, including nickel, rhodium, ruthenium, tungsten, rhenium, iridium, osmium, platinum and gold are summarised by Zombeck (1990). The normal incidence EUV and far-UV performance of some of these is shown in figure 2.4 as a function of wavelength. While the best reflectivity may be obtained with an exotic coating such as platinum, gold has a comparable reflectivity throughout most of the EUV range. Furthermore, the reflectivity curve is relatively featureless. Easier to obtain than the other materials, gold is less expensive and, having the lowest melting point, is straightforward to deposit through standard evaporation techniques. Consequently, gold has become the coating of choice for X-ray and EUV grazing incidence optics.

The Einstein and EXOSAT observatories represent the first applications of the grazing incidence telescope technology on cosmic X-ray or EUV satellite missions. However, each adopted a rather different solution to the production of the telescope optics. Einstein was the second of the series of three NASA High Energy Astronomical Observatories (HEAOs). Launched in 1978, it was designed to carry a large Wolter Type I imaging telescope system. This comprised four individual fused quartz mirror shells, each coated with a chrome–nickel alloy to enhance the X-ray reflectivity (Giacconi *et al.* 1979; figure 2.5). Spanning

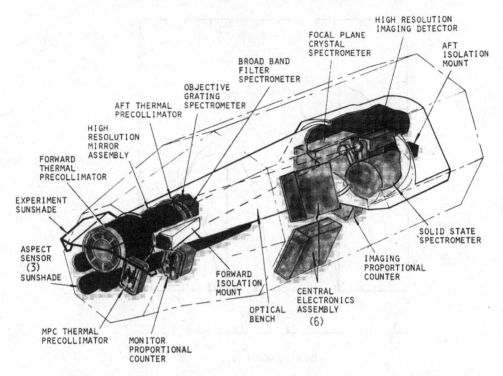

Fig. 2.5. The layout of the grazing incidence telescope and focal plane optical bench in the Einstein observatory (from Giacconi *et al.* 1979).

an energy range of ≈0.1–4 keV (3–120 Å), the mirror nest had a peak geometric area of 400 cm² at 0.25 keV (50 Å), corresponding to an effective area of 20 cm² with the High Resolution Imager (HRI) (figure 2.6) and 100 cm² with the PSPC (figure 2.7). With a focal length of 3.4 m, the useful field of view of this system was ≈1° and the on-axis resolution 2″ (rms).

In contrast the EXOSAT low energy imaging telescopes (LEITs; de Korte *et al.* 1981a,b; Paerels *et al.* 1990, figure 2.8) were manufactured using an epoxy replication technique, due to the severe mass constraint of only 7 kg allowed for the complete mirror assembly. A solid BK7 glass master was polished, on the *outside* surface, to the shape, accuracy and surface finish required for the reflecting surface of the mirror shells. Both paraboloid and hyperboloid surfaces were formed on a single unit. A non-adherent gold layer (≈900 Å) was evaporated onto the master. The mirror substrates were machined out of isopressed blocks of beryllium, with an accuracy of several microns, to ensure a uniform epoxy layer thickness. The replication process then transfers the gold layer from the master onto the beryllium substrate by means of an intermediate ≈30 μm epoxy layer. In this way, the figure and surface finish of the master can be transferred to the substrate without degradation. Two identical Wolter I telescope mirrors were manufactured and mounted coaligned in the EXOSAT spacecraft. Average grazing angles (1.8° and 1.5° for inner and outer mirror elements respectively) were chosen to give a short-wavelength cutoff at 6 Å, a factor of two longer than Einstein. The long-wavelength cutoff of ≈400 Å extends well into the EUV and was defined by broadband thin film filters. The telescope field of view was 2° with an on-axis angular resolution of

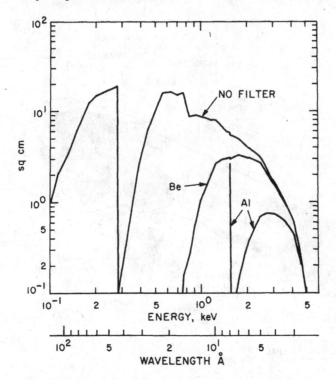

Fig. 2.6. Effective area of the Einstein grazing incidence telescope combined with the HRI detector (from Giacconi *et al.* 1979).

24 arcsec. A number of different filters gave coarse spectral information with a peak effective area of $10\,\text{cm}^2$ when combined with the channel multiplier array (or CMA, an alternative name for MCP detector), see figure 2.9. In addition, two transmission grating spectrometers (TGSs) could be deployed in the telescopes to provide spectral data for EUV-bright sources. In one telescope the grating had 500 lines mm^{-1} and covered the 8–400 Å band, while in the second a 1000 line mm^{-1} grating spanned the range 8–200 Å. Their resolutions were approximately 4 Å and 2 Å respectively. Figure 2.10 shows the efficiency of the 500 and 1000 line mm^{-1} gratings. These figures need to be multiplied by the effective areas in figure 2.9 to find the net effective areas of the telescope, grating, detector and filter combinations. This instrument covered the important region of the He II Lyman absorption line series between 304 Å and the series limit at 228 Å.

EXOSAT was launched on May 26 1983 and operated until April 9 1986 during which time it occupied a highly eccentric orbit with a 90.6 h period. The scientific instruments were operated for $\approx 76\,\text{h}$ per orbit when the satellite was outside the Earth's radiation belts. This operational strategy was very useful for monitoring variable sources and reduced the contribution of geocoronal emission to the instrument background significantly. The LEITs suffered a number of failures, restricting their use for EUV studies. One grating deployment mechanism failed after four months and shortly afterward the detector in the telescope containing the remaining grating ceased working, thus preventing any further spectral observations. However, the remaining CMA continued operation for broadband photometry until the end of the mission.

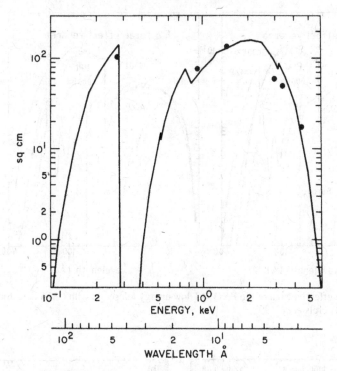

Fig. 2.7. Effective area of the Einstein grazing incidence telescope combined with the IPC detector (from Giacconi *et al.* 1979).

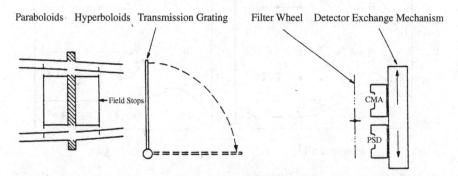

Fig. 2.8. The optical layout of the grazing incidence telescope system in the EXOSAT observatory (from Paerels *et al.* 1990).

2.5   Detector technology for space missions

The development of two-dimensional imaging detectors has had to complement the work on mirror technology. Progress with position sensitive proportional counters has run largely in parallel with that on MCP detectors. Each technique has some advantages and disadvantages associated with it. In the EUV, the need to provide a leak-tight container for the gas poses a problem. The transmission of the plastic entrance windows normally used decreases steeply into the EUV, limiting the sensitivity of the counter to the shorter wavelengths (less

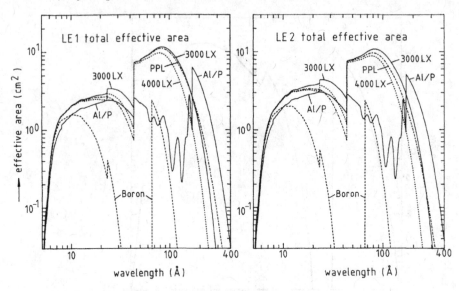

Fig. 2.9. The effective area of the EXOSAT low energy telescope with CMA detector and filters (from Paerels *et al.* 1990).

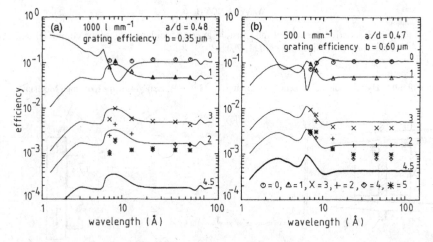

Fig. 2.10. The efficiency of the EXOSAT low energy telescope and transmission gratings (from Paerels *et al.* 1990).

than 150–200 Å). However, provided they are used below these wavelengths, they are very sensitive with typical quantum efficiencies (the fraction of incident photons detected) ranging from 10–50% in the 50–100 Å range. In addition, the proportionality between the energy of the detected photon and the electronic signal provides some spectral information, although this is very modest at long wavelengths (low energies). MCP detectors, in contrast, can operate throughout the EUV range, although they have lower quantum efficiencies than proportional counters in the wavelength range where they both operate. These detectors only have very crude energy discrimination properties at X-ray wavelengths and none at all in the EUV (see

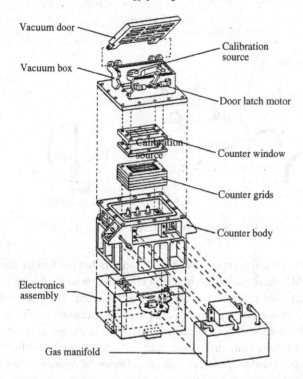

Vacuum door

Calibration source

Vacuum box

Door latch motor

Calibration source

Counter window

Counter grids

Counter body

Electronics assembly

Gas manifold

Fig. 2.11. Exploded diagram of the Einstein IPC detector (Humphrey *et al.* 1978).

Fraser and Pearson 1984; Fraser *et al.* 1985) but usually have better spatial resolution than proportional counters.

The Einstein telescope had focal plane detectors of both types to take advantage of the superior qualities of both the MCP-based HRI (Kellogg *et al.* 1976) and the IPC (Humphrey *et al.* 1978). The IPC (figure 2.11) was filled with 84% argon, 6% xenon and 10% CO_2 and had a 1.8 μm polypropylene window, incorporating 0.4 μm Lexan and 0.2 μm carbon for suppression of the UV flux from target stars and the Sun-lit Earth. Orthogonal Z-wound wire cathodes were used to record the X-ray arrival positions, yielding a spatial resolution of 1.5 mm (fwhm) at energies above 1.5 keV (8.3 Å). With a telescope plate scale of 1' per mm, the IPC resolution corresponds to 1.5' (fwhm) on the sky. The counter energy resolution, $\Delta E/E$, was 100% above 1.5 keV.

The HRI (figure 2.12) had relatively low quantum efficiency, when compared to the IPC, and no energy resolution, but gave much better spatial resolution. Utilising a pair of MCPs arranged in a chevron configuration, with 12.5 μm channel diameter and channel length-to-diameter ratio (L/D) of 80:1, the image readout was provided with a crossed grid charge divider terminated with a 16-fold partitioned RC line with 17 amplifiers per axis. This system yielded a spatial resolution of 33 μm (fwhm), corresponding to an angular resolution of 2" at the telescope focus. To compensate for the absence of any intrinsic energy resolution of the MgF_2 photocathode and MCP stack, crude spectral information was obtained by the use of broadband Al/parylene N and Be/parylene N filters. Either or both could be placed in the focused beam to give three bands. However, this system was not extensively used due to the low filter throughput and fears of jammed deployment mechanisms.

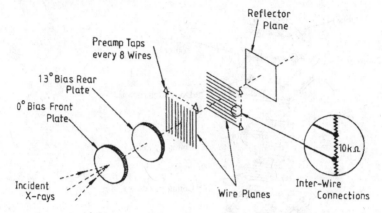

Fig. 2.12. Exploded diagram of the Einstein HRI detector (Kellogg *et al.* 1976).

EXOSAT also carried proportional counters and MCP detectors but the former failed to operate successfully. The CMA detectors were similar in construction to the Einstein HRI, comprising a pair of microchannel plates in the chevron configuration. The photocathode used was also MgF_2. However, the readout system adopted was electronically somewhat simpler than that of the HRI, comprising a circular resistive anode charge collector with four point electrodes connected to the circumference, each 90° apart (figure 2.13d). Photon positions were then calculated from the ratios of charge collected at opposite electrodes. The resistive and capacitive characteristics of the cermet anode, together with its shape, introduce considerable distortion into the output image. This was remedied by use of a 'look-up table' to provide a mapping between the apparent output coordinates and their true positions. The process of restoring the distorted image to the original shape and relative scale is often termed 'linearisation'.

2.5.1 Position encoding systems for detectors

The EXOSAT resistive anode design was far from optimum in terms of its spatial linearity, i.e. how closely the output image matches the input coordinates, and sensitivity to temperature. Investigations of anode design by Fraser and Mathieson (1981a,b) led to considerable improvements arising from adjustments to the shape and resistance of the anode, capacitance of the substrate, and the position and extent of the electrodes. Electrodes should not be point contacts but spread along a significant fraction of the square side or, alternatively, around the vertices (figure 2.13).

Apart from the need to correct for any residual distortion in the images obtained with the resistive anode technique, there are also limitations arising from resistive noise and capacitive coupling which affect the resolution and count rate capability respectively. In principle, superior resolution could be achieved utilising a conductive readout system. Low-resistivity systems have small time-constants and can cope with larger count rates than a resistive anode, although high photon fluxes do not usually pose a problem in EUV astronomy. An orthogonal wire grid readout scheme, similar to methods used in proportional counters, was adopted for the Einstein HRI (Kellogg *et al.* 1976). Terminated by a 16-fold partitioned RC line, every eighth wire of an individual grid was connected to a preamplifier, requiring 17 preamplifiers to span the 25 mm active diameter of the detector. The multi-anode microchannel

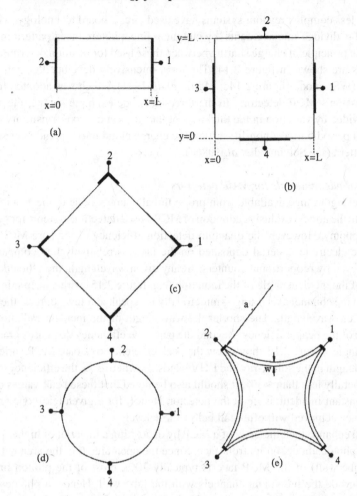

Fig. 2.13. Uniform resistive anode designs (from Fraser 1989). The numbered electrodes
are assumed to be connected, through charge-sensitive preamplifiers, to shaping amplifier of
time constant T_a. (a) 1-D resistive strip, sensitive in the x-direction only. (b) Square resistive
anode with charge collecting electrodes central to its sides. Amplitude ratio position signals
$Q_x = V_1/(V_1 + V_3)$; $Q_y = V_2/(V_2 + V_4)$, where $V_j (j = 1, \ldots, 4)$ are the peak electrode
output voltages. (c) Square resistive anode with electrodes at its vertices. Position signals
as in (b). (d) Circular anode with point contact electrodes. Position signals as in (b).
(e) Resistive anode of Gear's (1969) design. A sheet of uniform resistance is bounded
by circular arcs of equal radii a. Each arc should ideally be lined by a linear resistance.
Amplitude ratio position signals $Q_x = (V_1 + V_4)/(V_1 + V_2 + V_3 = V_4)$;
$Q_y = (V_1 + V_2)/(V_1 + V_2 + V_3 = V_4)$.

array (MAMA) detectors developed by the Center for Space Sciences and Astrophysics
at Stanford (Timothy 1986; Timothy and Bybee 1986) utilised a coincidence anode array,
comprising interleaved coarse and fine electrodes. This system can achieve pixel sizes down to
$25\,\mu m \times 25\,\mu m$ with high photon count rates; developments of the design have been used in
the Space Telescope Imaging Spectrometer (STIS) instrument (Kimble *et al.* 1999) and the
Far Ultraviolet Spectroscopic Explorer (FUSE; Seigmund *et al.* 1997).

Electronically less complex readout systems have used circuit board technology, where a conductive plane is divided into segments, by etching a fine, photoreduced pattern into the conductor, and the principle of charge sharing between these used for position determination. Several examples are shown in figure 2.14. The most intensively developed design is the wedge-and-strip (WS) anode (figure 2.14c; Martin *et al.* 1981), which was adopted for use in the EUVE mission's MCP detectors. In the three-electrode example shown, the x-axis sensitivity is provided by variation in the thickness of the strips and y-axis sensitivity by the wedges. To avoid periodic image non-linearities, the charge cloud must spread over several periods of the pattern (see Seigmund *et al.* 1986c).

2.5.2   *Photocathode materials for MCP detectors*

The limited wavelength range available from proportional counters, due to the window absorption, has led to the almost exclusive adoption of MCP-based detection systems for general use in EUV astronomy. However, the quantum detection efficiency of the bare MCP glass and the nichrome electrode material deposited on the faces is relatively low compared to what can be achieved by proportional counters. At any given wavelength, the efficiency is a strong function of the incident angle of the radiation (e.g. figure 2.15) with a deep minimum at zero degrees to the channel axis rising symmetrically to a peak at a few degrees, the exact angle depending on wavelength. The shorter the wavelength of the incident radiation, the smaller the angle of the peak efficiency. Beyond the peak, the efficiency decreases gradually with increasing angle. Figure 2.16 shows how the peak efficiency of a bare MCP varies as a function of wavelength in the soft X-ray and EUV bands. Nowhere does the efficiency exceed 20% and it is generally less than $\approx 5\%$. It should also be noted that these peak values do not correspond to constant incident angle at the detector. Hence, for a given telescope grazing angle, the efficiency achieved will often fall below these levels.

It is possible to enhance the efficiency of a MCP by depositing a material of higher photo-electron yield, a photocathode, on its front face. Since the open area (i.e. the area not filled with the glass tube wall) of the MCP face is typically 70%, most of the photon interactions take place inside the individual channels with the tube wall. Hence, a photocathode must be deposited such that it extends a suitable distance along the inside of each channel, determined by the expected angle of incidence in the telescope system. The choice of useful photocathode materials for the EUV is a careful trade-off between several properties. A detailed discussion of the physically important characteristics of photocathode materials and how these contribute to the ultimate quantum efficiency can be found in Fraser (1989, chapter 3). Briefly, the most important effects are: the primary electron yield – the efficiency of converting photons into photoelectrons; the secondary electron yield – how many further electrons are then created by interactions of the primary electrons within the photocathode; the length of the escape path of the secondary electrons from the channel wall. The basic choice is then to decide on a material with high electron yield, which can be deposited without compromising the MCP operation (usually by vacuum evaporation), and its optimum thickness.

Many experimental and theoretical studies have been carried out in search of the ideal photocathode material (e.g. Fraser 1983; Seigmund *et al.* 1987; Jelinsky *et al.* 1996), achieving a concensus that the most useful group of materials are the alkali-halides, including particularly MgF_2, CsI and KBr. A summary of measured efficiencies drawn from two different sources shows that no single material is ideal over the entire range (figure 2.17). However, it can be

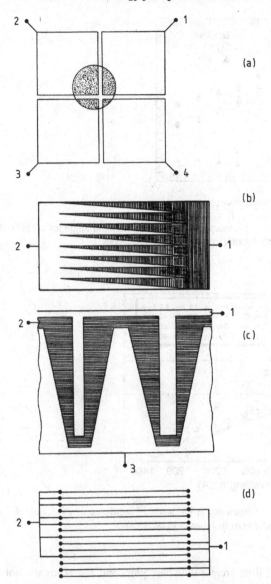

Fig. 2.14. Charge sharing (progressive geometry) encoders (from Fraser 1989). Each numbered conductor is assumed connected to a charge-sensitive preamplifier and shaping filter. $V_j (j = 1, \ldots, 4)$ are the peak electrode output voltages. (a) Four quadrant anode. Quadrant conducting planes intercept the charge cloud (shaded). Position signals: $Q_x = (V_1 + V_4 - V_2 - V_3)/(V_1 + V_2 + V_3 = V_4)$; $Q_y = (V_1 + V_2 - V_3 - V_4)/(V_1 + V_2 + V_3 = V_4)$; or $Q_x = V_1/(V_1 + V_3)$; $Q_y = V_2/(V_2 + V_4)$. (b) Planar backgammon anode. Charge divides between the interleaved conductors. Position signal $Q_x = V_1/(V_1 + V_2)$. (c) Wedge-and-strip anode. One period of the pattern is shown. Electrode 1 = strip; electrode 2 (shaded); electrode 3 (wedge). Position signals: $Q_x = 2V_1/(V_1 + V_2 + V_3)$; $Q_y = 2V_3/(V_1 + V_2 + V_3)$. (d) Graded density wire plane, with wires connected in two groups. Position signal: $Q_x = V_1/(V_1 + V_2)$.

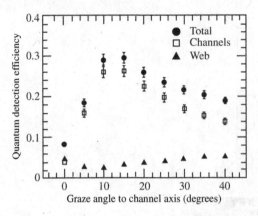

Fig. 2.15. Quantum detection efficiency as a function of incident angle for a 15 000 Å thick photocathode using 256 Å radiation (from Siegmund *et al.* 1987).

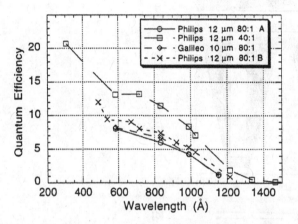

Fig. 2.16. Quantum detection efficiency at 13° angle of incidence as a function of wavelength angle for bare glass (from Jelinsky *et al.* 1996).

seen that the efficiency of MgF_2 is little greater than bare glass and, therefore, did not confer much benefit in the Einstein and EXOSAT detectors. CsI delivers the best performance over most of the EUV but it is hygroscopic (as are many alkali halides) and exposure to water vapour will impair its performance (e.g. Whiteley *et al.* 1984). KBr is much less sensitive to water vapour and can, as a result, be handled more easily; however, CsI can be used effectively with sensible handling procedures in an inert atmosphere such as dry N_2 (e.g. Whiteley *et al.* 1984).

2.6   Thin film filters

Thin film filters serve two purposes in EUV instrumentation – to define useful bandpasses for broadband photometry and to block unwanted background radiation. Their effectiveness in these tasks depends on suitable choices of material and this imposes several basic

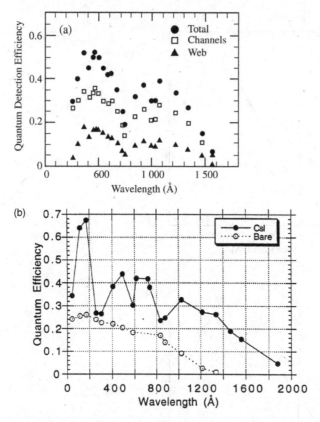

Fig. 2.17. Quantum detection efficiency as a function of wavelength for (a) a KBr photocathode, showing the total and the contribution from the channels and the inter-channel (from Siegmund *et al.* 1987) and (b) a CsI photocathode, compared to bare glass (from Jelinsky *et al.* 1996).

requirements. The ideal filter material would have a useful wavelength window of low absorption coefficient but high opacity elsewhere. It also must be thin enough to yield high efficiency in the transmission window and free of pinholes which might leak unwanted background. Finally, it must be strong enough to cover the area required without falling apart. In practice, few materials can fulfill all these requirements on their own and real filters often have to be composites of two or more different materials.

Filter materials can be divided into two categories, plastic films and metal foils. Plastic films are hydocarbon polymers and include such materials as polypropylene, parylene and Lexan, which were all used in the EXOSAT LEIT. The *Einstein* IPC window was also made of Lexan, with a conductive carbon coating added. The transmission of these is determined largely by the absorption cross-sections of the component atoms. For example, the carbon edge at 44.7 Å puts a step in the absorption coefficient function, which has a minimum just above the edge beyond which the coefficients increase towards longer wavelengths. Absorption edges like this can be used to define a cutoff in a filter transmission function and several elements, including boron, beryllium, aluminium, tin, indium and antimony, have absorption edges that are usefully placed within the EUV. For wavelengths up to ≈ 300 Å

(a)

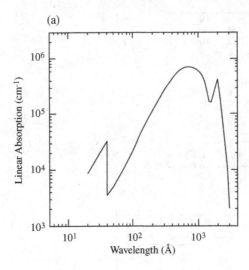

(b)

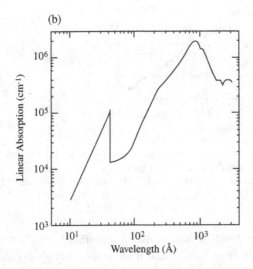

Fig. 2.18. Linear absorption coefficients for useful filter materials. (a) Lexan. (b) Carbon.

plastic films can be used to support the more brittle metal foils but beyond this the foils must be freestanding. To support large material areas or fragile metal foils the filters are often attached to wire support grids. Pinholes can be avoided if filters are constructed by sandwiching multiple layers of the materials chosen to build up the required thickness (e.g. Barstow *et al.* 1987).

Figure 2.6 shows the measured linear absorption coefficients (μ_l cm^{-1}) of a number of useful filter materials from the soft X-ray through to the far-UV. The fractional transmission (T) of a filter of known thickness (t) can be calculated as follows

$$T = \exp(-t\mu_l)$$ (2.7)

(c)

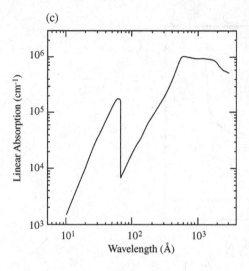

(d)

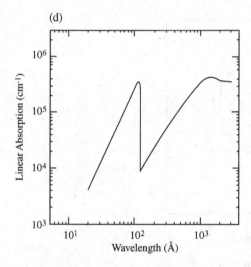

Fig. 2.18 (*cont.*). (c) Boron. (d) Beryllium.

For a composite filter consisting of several different materials, the total transmission is simply the product of the values calculated separately for each element, provided interference effects can be ignored, as they usually can in the EUV.

2.7 Selected scientific results from *Einstein* and EXOSAT

Although the *Einstein* and EXOSAT observatories were primarily X-ray astronomy missions, the longest wavelengths covered by the instruments on board extended substantially into the EUV. Furthermore, many of the observations made by the satellites in the soft X-ray régime provided scientific insights of importance for the effective exploitation of future, dedicated

(e)

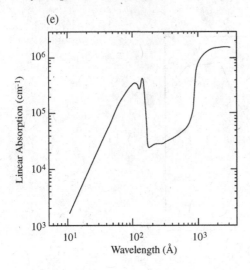

(f)

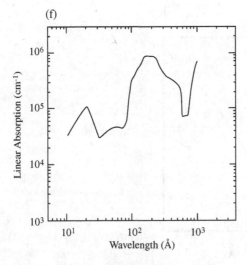

Fig. 2.18 (*cont.*). (e) Aluminium. (f) Tin.

EUV missions. Consequently, we discuss here relevant scientific results from both *Einstein* and EXOSAT, emphasising their contribution to understanding the astrophysics of the types of object concerned and setting the scene for later developments.

2.7.1   *Photometry and spectroscopy of white dwarfs*

White dwarfs are the endpoint of stellar evolution for most stars. All stars with initial masses below about eight times that of the Sun will pass through one or more red giant phases before losing most of their original mass to form a planetary nebula, leaving behind the degenerate stellar core – the white dwarf. A major problem in the study of white dwarfs has been their division into two distinct categories, having either hydrogen-rich (designated DA

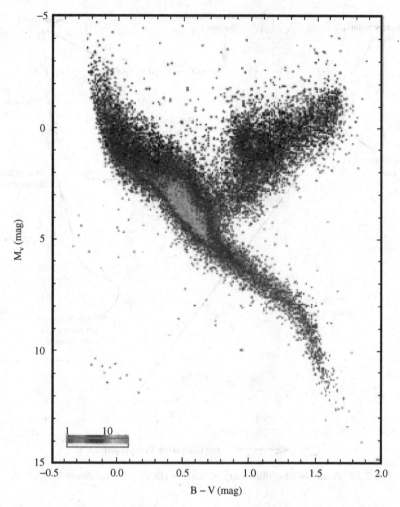

Fig. 2.19. Hertzsprung–Russell diagram showing the location of the white dwarfs in the bottom left corner (ESA, 1997).

white dwarfs) or helium-dominated atmospheres (designated DO for $T_{eff} > 45\,000$ K or DB for T_{eff} below this), and the routes by which they evolve from their possible progenitors, the diverse types of central stars of planetary nebulae (CSPN, see e.g. Sion 1986, Weidemann 1990 for reviews). Figure 2.19 shows the location of white dwarfs in the Hertzsprung–Russell (HR) diagram with their relationship to proposed evolutionary paths in figure 2.20. Studies of the white dwarf luminosity function show that the very hottest DA white dwarfs outnumber He-rich DOs by a factor of seven (Fleming *et al.* 1986); the relative number of H- and He-rich CSPN is only about 3:1. In addition, there is an apparent absence of He-rich stars in the temperature range 30 000–45 000 K, the so-called DB gap, suggesting that H- and He-dominated groups are not entirely distinct. Several competing processes can affect the composition of a white dwarf atmosphere. He and heavier elements tend to sink out of the photosphere under the influence of gravity but this can be counteracted by radiation pressure.

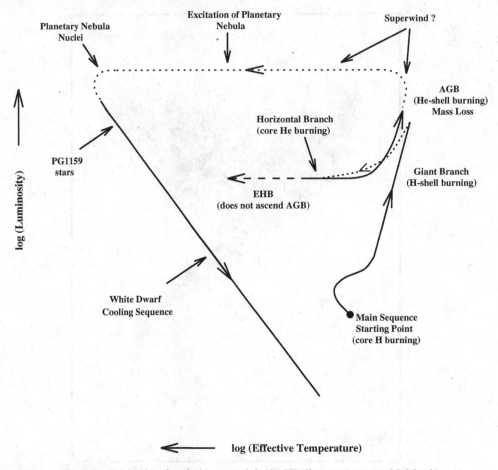

Fig. 2.20. White dwarf evolutionary path in the HR diagram (courtesy M. Marsh).

Convective mixing, accretion or mass-loss via a weak wind may also play a signficant role. These ideas are summarised in a schematic diagram of possible hot white dwarf evolution (figure 2.21).

An aim of many studies of white dwarf photospheric composition has been to understand the relationships between H- and He-rich white dwarfs. In particular, measuring the helium content in hot DA stars should lead to an understanding of how they could evolve from the DO group. The sensitivity of predicted soft X-ray and EUV fluxes to the presence of trace elements in the photosphere of a star makes observations in these wavebands of particular importance. Even in initial studies of white dwarfs with Einstein it was apparent that their soft X-ray fluxes were significantly lower than those predicted (Kahn *et al.* 1984; Petre *et al.* 1986). These deficits were interpreted in terms of absorption by helium homogeneously mixed in the stellar photosphere. From a small sample of eight detected stars, there was weak evidence for a correlation between increasing He abundance and increasing stellar temperature. Around 20 white dwarfs were observed photometrically with the EXOSAT CMA. Having several filters available, compared to the single Einstein pass band, gave potentially better discrimination between temperature and abundance effects in the stars observed. With this larger

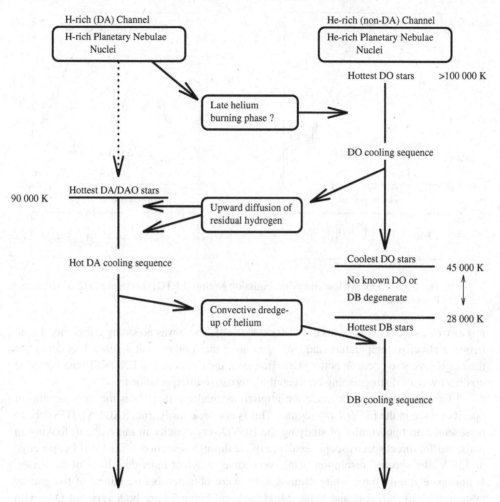

Fig. 2.21. Schematic diagram of white dwarf evolution and the relationships between H-rich and He-rich groups.

sample, the results of Jordan *et al.* (1987) and Paerels and Heise (1989) confirmed the initial Einstein conclusion. Their inferred helium abundances (He/H by number) ranged from $<10^{-5}$ to $\approx 10^{-3}$.

However, detailed theoretical studies of the diffusion of helium in H-rich white dwarf envelopes showed that radiation pressure cannot prevent He from sinking rapidly out of the atmosphere under the influence of gravity (Vennes *et al.* 1988). Consequently, the apparent abundances of helium measured by Einstein and EXOSAT for the hottest white dwarfs were two or more orders of magnitude greater than predicted. An alternative, more physically realistic interpretation of the data assumes that the atmosphere has a stratified rather than a homogeneous structure with a thin layer of H overlying a predominantly He envelope. In this situation, the helium opacity is provided by the amount that diffuses naturally across the H/He boundary and how much is seen depends on the thickness of the H layer. In stars with thinner H layers, more He would block the EUV flux. These stratified models fit the data

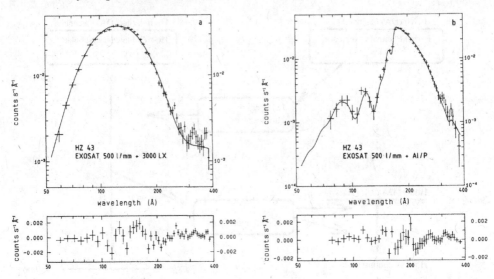

Fig. 2.22. EXOSAT low energy transmission grating (LETG) spectrum of HZ 43 (from Paerels *et al.* 1986a).

just as well as the homogeneous ones (Koester 1989). There was no strong correlation found between effective temperature and hydrogen layer mass other than a general tendency for thinner H layers to occur in hotter stars. However, the Einstein and EXOSAT data provided no direct way of distinguishing between the alternative interpretations.

All the photometric work made the implicit assumption that He is the only significant opacity source in the EUV/X-ray régime. This is not necessarily true. EXOSAT TGS spectra represented an opportunity of studying the EUV/X-ray opacity in more detail, looking in particular for direct spectroscopic evidence of He through detection of the He II Lyman edge at 228 Å, the spectral resolution being too coarse to detect individual lines of the series. Unfortunately, only three white dwarf spectra were obtained before failure of the grating system – HZ 43, Sirius B and Feige 24. HZ 43 and Feige 24 are both very hot DAs with temperatures in the range 50 000–55 000 K, while Sirius B is much cooler at $\approx$25 000 K. The spectra of both HZ 43 and Sirius B (figures 2.22 and 2.23) can be matched by the predictions of pure hydrogen model atmospheres. There is no evidence for the presence of any He or other excess opacity in either star. From detailed fits to the data, Paerels *et al.* (1986a) estimated the effective temperature of HZ 43 to be in the range 59 000 K to 62 000 K and place an upper limit (90% confidence) of 6×10^{-6} on the abundance of He. In a recent reanalysis of the EXOSAT data using fully blanketed model atmospheres, Barstow *et al.* (1995a) obtain a value of 2×10^{-6} for the He upper limit. A somewhat higher He upper limit of 1×10^{-5} was obtained for Sirius B (Paerels *et al.* 1986b) but a more important result was a refined measurement of the effective temperature (24 000 $\pm$ 500 K), which cannot be easily measured at longer wavelengths due to the glare of Sirius A. In contrast to HZ 43 and Sirius B, the spectrum of Feige 24 could not be fitted by any simple pure H or H + He model atmosphere (figure 2.24; Paerels *et al.* 1988). Although there was no evidence for any He opacity, since the He II 228 Å edge was absent from the data, the steep downturn of the flux at shorter wavelengths indicated that there must be some other absorbing elements in the photosphere. Vennes *et al.* (1989) successfully modelled the EXOSAT spectrum with a

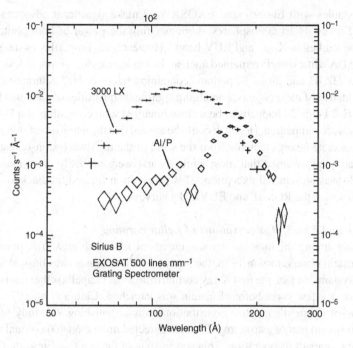

Fig. 2.23. EXOSAT LETG spectrum of Sirius B (from Paerels *et al.* 1986b).

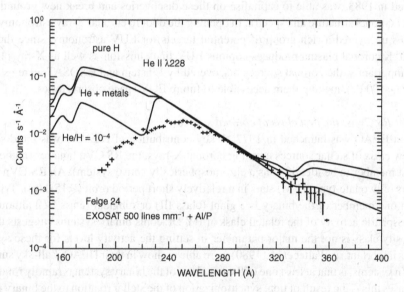

Fig. 2.24. EXOSAT LETG spectrum of Feige 24 (from Paerels *et al.* 1986b).

mixture of elements heavier than He, including C, N, O, Ne, Si, Ar and Ca, but, since there are no features in the spectrum that could be attributed to any individual element, the chosen contributors and abundances were entirely arbitrary and do not give a unique solution to the composition.

White dwarf studies with Einstein and EXOSAT did make significant advances in our understanding of white dwarf atmospheres, demonstrating the power of both photometric and spectroscopic studies in X-ray and EUV bands. However, the properties of the general population of hot DA white dwarfs remained unclear. Did most stars have more or less pure-H atmospheres, like HZ 43 and Sirius B, perhaps containing traces of He? Alternatively, was Feige 24 representative of most objects, containing significant quantities of elements heavier than either H or He? Feige 24 happens to be a close binary system consisting of a DA white dwarf with a red dwarf companion. Hence, it could be argued that the white dwarf was unusual because it could accrete heavy elements from the wind of the red dwarf. Consequently, the general picture adopted assumed that most white dwarfs were probably more like HZ 43 and that Feige 24 was an unusual exception. This proposition formed the background for subsequent studies with the ROSAT and EUVE sky surveys.

2.7.2 *Einstein and EXOSAT observations of stellar coronae*

Stellar coronae are among the most numerous sources of EUV radiation. Yet, prior to the launch of the Einstein observatory in 1978, the study of stellar coronae was limited to about 20 active binary systems. In fact, the first X-ray corona from a star (Capella) other than our Sun was discovered only a few years before Einstein was launched (Catura *et al.* 1975; Mewe *et al.* 1975). One of Einstein's greatest contributions was establishing the study of stellar coronae as a mainstream part of astronomy; Einstein detected more than 400 coronal sources within the first few years of its operation. This was in spite of the fact that Einstein, far from being an all-sky survey, covered only a few percent of the sky during its lifetime. EXOSAT, launched in 1983, was able to capitalise on these discoveries and break new ground in the study of coronal structure and spectra. In opening the door to coronal X-ray astronomy, these satellites uncovered a rich group of potential targets for EUV astronomy, since the same 10^6–10^7 K coronal plasma produces copious EUV line emission as well as X-ray photons. Most importantly, the coronal sources discovered by Einstein and EXOSAT were relatively nearby (<100 pc), making them accessible to future EUV space observatories.

2.7.2.1 *RS CVns: the first class of coronal sources*

The first HEAO was launched in 1977. A key contribution of HEAO-1 was the discovery of a new *class* of stellar sources emitting coronal X-rays, the RS CVn binaries (named after the first member to be identified as a chromospherically active system). An RS CVn binary consists of two late-type (F–M) stars in a relatively short period orbit (<15 days). Typically, at least one member of the binary is a giant (class III) or subgiant (class IV), although the chromospheric activity of the related class of BY Draconis binary systems suggests that the luminosity class is not the major parameter in setting the activity levels for these systems. What is important, as Walter *et al.* (1980) were able to show in their HEAO-1 all-sky survey of RS CVn systems, is that at least one of the members of the binary system is rapidly rotating. In most cases this is the result of tidal synchronization of the stellar rotation to the binary period. In turn, this rapid rotation is thought to enhance magnetic heating of the corona in a process not yet fully understood, but very likely related to the dynamo process which produces solar and stellar magnetic fields. Since the Sun rotates only once every 25 days or so, it is perhaps not surprising that the RS CVn systems, some with orbital (and rotational) periods less than 1 day, show much higher levels of X-ray (and EUV) emission than the Sun, and thus were the first class of coronal sources to be discovered. However, only with the much more

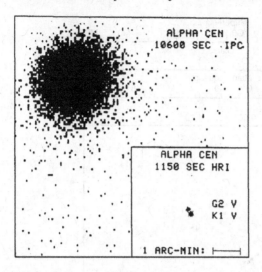

Fig. 2.25. Einstein HRI observation of α Cen (from Golub *et al.* 1982).

sensitive Einstein and EXOSAT missions were coronae from single stars detected in large numbers.

2.7.2.2 *Coronal surveys and rotation*

Pointed surveys with Einstein were able to establish that stellar coronae existed throughout most regions of the HR diagram and that the level of coronal emission varied with stellar rotation (and, indirectly, age) at a given spectral type. This was likely to be the result of magnetic activity. Hence, stellar X-ray flares dramatically more powerful than those seen on the Sun were possible and pre-main sequence stars would expected to be powerful and variable sources of stellar X-ray emission.

The key to Einstein's success was, of course, its imaging capabilities, as described earlier in section 2.5. A good example of these capabilities was the separation of the binary system α Cen into its G and K dwarf components by the Einstein HRI (figure 2.25). Surprisingly, the lower mass K star was found to be the stronger X-ray emitter, at least in the HRI bandpass (Golub *et al.* 1982).

Another important early series of observations was the pointed coronal survey of Vaiana *et al.* (1981). In this survey, a wide variety of stars of diverse spectral types and luminosities were observed in a search for coronal emission, and about 100 cool stars were detected as X-ray sources (figure 2.26). The X-ray luminosities of these stars ranged over five orders of magnitude, from about 10^{26}–10^{31} erg s^{-1}. The Vaiana *et al.* survey was instrumental in establishing magnetic activity as the only viable explanation for cool star coronae. In particular, the strength of coronal emission along the main sequence was found to vary not so much with spectral type, but with stellar rotation (or age). Among stars of a given spectral type, up to several orders of magnitude spread in stellar X-ray luminosity was found. Pallavicini *et al.* (1981) explicitly determined a correlation of X-ray luminosity with rotation, with a steeper dependence ($L_X \sim (v \sin i)^2$) than the linear relationship found by Walter *et al.* (1980) for orbital periods of RS CVn systems with HEAO-1.

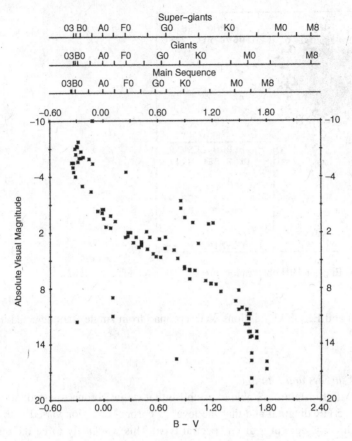

Fig. 2.26. X-ray HR diagram (from Vaiana *et al.* 1981).

This strong dependence upon stellar rotation was suggestive of dynamo-induced magnetic activity, and effectively eliminated non-magnetic acoustic heating theories, in which coronal X-ray luminosities were predicted to be a function of stellar spectral type and luminosity only. Another important result was that coronae occurred for the most part in stars of spectral type F0 and later, indicating that an outer convective zone was necessary for magnetic heating to be present. A number of very early O and B stars were also detected in X-rays, but this emission was attributable to the presence of radiatively driven shocked stellar winds. Thus a coronal 'HR' diagram does *not* have a 1:1 correspondence with the classical HR diagram.

Some interesting anomalies in the coronal 'HR' diagram were apparent, especially in stars not on the main sequence. In particular, Linsky and Haisch (1979) found evidence for a coronal 'dividing line' in the HR diagram: typically, giants cooler than about K0 did not, as a rule, possess coronae, whereas earlier giants did. Although there have been a number of explanations for this phenomenon, ranging from the onset of cool winds to the effect of stellar surface gravity, none is universally accepted. An initially curious phenomenon, termed 'saturation', was also found, most notably in the M dwarf systems (Vilhu 1984): except during flares, no star was found to have a coronal X-ray luminosity greater than about 10^{-3} of its bolometric luminosity. Although some have interpreted this phenomenon as '100%' surface coverage with stellar active regions, the saturation phenomenon is not yet completely understood.

2.7.2.3 Stellar clusters

Star clusters hold the keys to understanding how individual stars evolve. Because of their known age, distance and chemical composition, clusters are also important in understanding the evolution of stellar activity. The HEAO-1 all-sky survey lacked sufficient sensitivity for the study of nearby stellar clusters. Thus Einstein was the first satellite to perform significant studies of open star clusters.

One of the first clusters studied by Einstein was also the nearest: the Hyades cluster in Taurus (Stern *et al.* 1981). At a distance of about 45 pc, and an age of about 600 My, the Hyades provided one of the best samples of solar-like stars accessible to Einstein. Observations of the Hyades demonstrated that, for G stars around one tenth of the age of the Sun, X-rays were emitted at a rate 50–100 times that of the solar corona. The clue to such apparently startling behaviour lies in the much more rapid rotation of the Hyades G stars compared to the Sun, roughly a factor of five. Because of the phenomenon of 'saturation' alluded to earlier, however, a simple relation (e.g. a power law or exponential) between rotation and X-ray emission does not hold for even younger clusters such as the 70 My old Pleiades (Caillault and Helfand 1985).

2.7.2.4 Pre-main sequence stars

The detection of X-rays from pre-main sequence (PMS) stars was another highlight of the Einstein observations. Unfortunately, the high column densities to most star forming regions (10^{20} cm^{-2} and greater) meant that such systems could be studied only in the shortest EUV wavelengths; hence PMS stars remain one field of coronal astrophysics best investigated through X-ray, UV, optical and IR astronomy.

2.7.2.5 Binary systems

Continuing work on RS CVn systems that were first detected as a class of objects with HEAO-1, Walter and Bowyer (1981) surveyed 39 RS CVn systems for evidence of X-ray emission with the Einstein IPC, and detected all of them, with a range of X-ray luminosities from 10^{29} to 10^{31} erg s^{-1}. Algol systems, semi-detached binaries consisting of a massive unevolved primary (a main sequence star) and a low-mass evolved secondary (a subgiant or giant), were also detected with X-ray luminosities comparable to those of the RS CVn systems (White and Marshall 1983).

A key contribution of EXOSAT in this area of research was the use of eclipsing RS CVn binary systems to investigate coronal structure. EXOSAT, unlike Einstein, was ideally suited to such studies because of its highly eccentric 96 h orbit, which allowed for long pointed observations of X-ray bright RS CVn systems. Swank and White (1980) and Walter *et al.* (1983) did detect stellar eclipses in AR Lac with Einstein, despite coverage gaps in low Earth orbit. However, with EXOSAT, White *et al.* (1986, 1990) and Culhane *et al.* (1990) were able to obtain extensive coverage of AR Lac, TY Pyx, and Algol, suggesting that large ($\approx 1 R_*$), high-temperature coronal structures exist around the most active binary systems, as evidenced by their lack of eclipses in the 2–10 keV energy band. This conclusion, however, was not universally accepted, and the existence of 10^7 K plasma in large structures surrounding RS CVn systems remains controversial.

2.7.2.6 Stellar flares

Until the advent of Einstein and EXOSAT, astronomical knowledge of stellar flares was confined to optical studies of the so-called 'flare stars', typically low-mass M dwarfs that were observed to rapidly brighten for minutes or longer in the U band of optical astronomy. However, all of that changed with X-ray observations: it became clear that some of the most powerful flares were associated with the RS CVns or other binary systems (e.g. Stern 1992; Stern *et al.* 1992). With EXOSAT, White *et al.* (1986) observed a particularly large flare in the Algol system, reaching temperatures of $>60 \times 10^6$ K, and peak X-ray luminosities $>10^4 \times$ the largest solar X-ray flares. Classical M dwarf flare stars such as Proxima Centauri were also seen to produce X-ray flares (see references in Pallavicini 1989). Even young solar-type stars such as π^1 UMa were seen to flare at a level ten times that of the largest solar flares (Landini *et al.* 1986).

2.7.2.7 XUV spectroscopy

Einstein and EXOSAT significantly advanced the fledgling field of stellar coronal astronomy through imaging observations; these two missions also provided a glimpse into the complexities of coronal spectroscopy for a few bright coronal sources. Swank *et al.* (1981), using the Einstein Solid State Spectrometer (SSS) demonstrated that simple, isothermal coronae were not characteristic of RS CVn systems such as UX Ari, Capella, and others. Their two-temperature models provided a surprisingly good representation of the SSS data, and despite further advances in X-ray and EUV spectrometers, are still used by many researchers. With the TGS of EXOSAT, line complexes in the 10–200 Å range were seen for the first time in Capella, Procyon and σ^2 CrB (Lemen *et al.* 1989). Again, at least two components were required to fit the data, suggesting the presence of multitemperature plasma with temperatures in excess of 20×10^6 K. The controversy over the nature of multithermal plasma in stellar coronae exists to this day, even with the advanced spectrometers on EUVE and the ASCA satellite.

2.7.3   EUV photometric observations of CVs with Einstein and EXOSAT

Cataclysmic variables (CVs) represent several classes of binary stellar systems in which a cool late-type star, typically an M dwarf, in a close orbit about a white dwarf, episodically transfers mass to the white dwarf resulting in large changes in the luminosity of the system. These systems are widely studied over a range of wavelengths for what they can reveal about the processes of mass transfer and accretion; an extensive general review of CV systems is contained in Warner (1995). A fundamental distinction among CV systems is the presence or absence of a strong magnetic field associated with the white dwarf which is capable of confining the accreting material into columns which impact the white dwarf surface near the magnetic poles. Such systems are referred to as polars if they possess field strengths above 10 Mgauss (10^4 gauss $= 1$ T) or intermediate polars if they have lower fields between 1 to 10 Mgauss and less well organized accretion columns. Systems that do not possess strong magnetic fields form accretion discs about the white dwarfs through which material diffuses on to the surface of the white dwarf in the orbital plane of these systems. Such systems are commonly referred to as dwarf novae, classical novae and nova-like variables, depending on the nature and frequency of their outbursts as observed in the optical.

CV systems, particularly during outbursts, are frequently observed as strong sources of EUV and soft X-ray radiation. Common to magnetic and non-magnetic systems alike is that the final stages of accretion on to the white dwarf result in strong heating of the accreted material which produces intense localized sources of X-ray, soft X-ray and EUV radiation. Because these accretion zones tend to be restricted to small fractions of the white dwarf surface, orbital and rotational motions can lead to large short time scale modulations of these fluxes at characteristic frequencies, while the frequency, intensity and duration of the outbursts themselves are dominated by the overall rate of mass transfer between the two stars. Thus, the study of CV systems in the EUV focuses primarily on the details of how matter is finally transferred to the white dwarf surface and what the short-period temporal behaviour in the EUV can reveal about the geometry of the binary system and/or its accretion disc or accretion column. A more through discussion of the EUV properties of CV systems is given in chapter 9. Here we are concerned primarily with the pioneering EUV photometric observations of CV systems obtained by Einstein and EXOSAT satellites.

Einstein made pointed observations of 63 non-magnetic CV systems of which 43 were detected in the IPC soft X-ray band (van der Woerd 1987). Cordova and Mason (1984) summarised the results for 31 known and suspected CV systems, including seven observed during outburst. These observations included both the HRI and IPC instruments. In total, Cordova and Mason detected 18 of these objects with the IPC; however, only U Gem, in outburst, exhibited an enhanced soft X-ray flux in the IPC low energy (LE) channels. The U Gem observations are reported in Cordova *et al.* (1984). To the Cordova and Mason summary can be added Einstein IPC observations of 12 CV systems reported by Becker (1981). Although both Cordova and Mason and Becker report short-term flux variations, none could be directly linked to the orbital or rotational periods of these systems, in part due to the interruptions of observing sequences by the 90 min orbital period of Einstein.

EXOSAT possessed several capabilities which lent themselves particularly well to the study of CV systems. These included a four-day orbit which permitted long uninterrupted observations of CV light curves, the relatively high EUV soft X-ray sensitivity of the LE telescopes, and their relatively large field of view. All told, EXOSAT devoted 10% of its observing time to observations of CV systems.

A total of 15 non-magnetic CV systems were monitored extensively with EXOSAT, including the bright soft X-ray sources SS Cyg, U Gem, VW Hyi and OY Car (van der Woerd 1987). Notable discoveries include the demonstration that the soft component in SS Cyg can be modelled with a temperature of 0.6 keV and the EUV emission is dominated by an extremely soft flux at wavelengths longer than 170 Å (van der Woerd *et al.* 1986). Extensive monitoring of VW Hyi revealed that the spectral shape of the soft X-ray flux remained unchanged, while the total flux increased by a factor of 100 during outburst (van der Woerd and Heise 1987).

EXOSAT observations of magnetic CV systems are reviewed in Osborne (1987, 1988). One of the major achievements with EXOSAT was the elucidation of the extent, the shape and the excitation mechanisms responsible for the EUV emission from CV systems. Observations of the self-eclipsed EUV light curves of the prototype polar system AM Her (Heise *et al.* 1985) indicated that more than one EUV emitting region was involved and also provided evidence that the EUV emitting regions possessed a significant vertical extent. Observations of the polar system E2003 + 225 (Osborne *et al.* 1987) supported the view of small, localised, optically thick EUV regions which are heated by an optically thin shock-heated region and which emit the hard X-ray flux, although in some systems the EUV flux is greater than the hard X-ray flux, indicating that additional heating is required. A comparison of the medium-energy X-ray and LE fluxes clearly indicated that the EUV emitting regions were more localised than the regions responsible for the harder X-rays. Simple blackbody models of these EUV emitting regions suggested the fractional surface areas of these emission regions were quite small: of the order 10^{-2} to 10^{-3}% of the white dwarf surface area. In addition, asymmetric peaks in the EUV light curves indicated that simple circular spots on the white dwarf were insufficient to model the observations. Flickering on time scales of 4–11 min was also observed at EUV wavelengths in some systems such as E2003 + 225 (Osborne *et al.* 1987). Finally, EXOSAT, with its relatively large field of view, was successful in discovering two new CV systems: EXO 023432−5232.3 (Beuermann *et al.* 1987) and EXO 033319−2554.2 (Giommi *et al.* 1987), both AM Her type polar systems.

2.8   Far-UV spectroscopy with IUE

Launched in 1978 and operated successfully for 18 years, the International Ultraviolet Explorer (IUE) provided astronomers with reliable access to UV wavelengths between 1200 and 3000 Å. It was operated as an international observatory with an active guest observer programme which produced in excess of 100 000 spectra during its lifetime. IUE contained a 45 cm UV telescope and two pairs of UV sensitive cameras, one (SWP) covering the short wavelength region between 1200 and 1900 Å and the other (LWR) covering 1900 to 3000 Å. Each of these cameras could be operated in a low dispersion mode with a spectral resolution of $\lambda/\Delta\lambda = 200$ and a high dispersion mode with a spectral resolution of $\lambda/\Delta\lambda = 10\,000$. Of the many types of scientific investigations conducted with IUE several are of particular importance to EUV astronomy.

An area where IUE has proved especially valuable has been the discovery and study of hot white dwarf components in binary systems where the optical luminosity of the primary star obscures the white dwarf. A number of such systems were discovered during the ROSAT and EUVE all-sky surveys (see chapters 3 and 4). However, the exact nature of these sources was often difficult to determine since the EUV emission could conceivably originate in the corona of the main sequence star. A brief IUE low dispersion spectrum covering the 1200 to 1900 Å region was often all that was necessary to identify the presence of a hot white dwarf. An example of such a discovery is the white dwarf associated with the F4 V star 14 Aur C. 14 Aur C was located within the 90% confidence error circle of the WFC ROSAT source REJ0515 + 324, which is among the ten brightest EUV sources in the sky in the 100 to 200 Å band. Hodgkin *et al.* (1993) and Webbink *et al.* (1992) independently used IUE to reveal the steeply rising UV continuum of the hot white dwarf at wavelengths below 2000 Å. Figure 2.27 shows a recently constructed SWP spectrum which exhibits the composite nature of REJ0515 + 324. The characteristic broad Lyman α profile of the white dwarf at 1216 Å is clearly evident as is the strong UV continuum which begins to dominate that of the F4 primary shortward of 1700 Å. This and similar composite stellar systems are discussed in more detail in section 3.7. IUE was not only able to identify the hidden white dwarfs in such systems but it also provided valuable information on the effective temperatures and surface gravities of these stars.

A second area where IUE was instrumental in advancing EUV astronomy was the identification and subsequent quantification of the actual sources of EUV opacity in white dwarf atmospheres. As early as 1981, IUE high-dispersion observations (Bruhweiler and Kondo 1981) showed the presence of highly ionised species such as N V, C IV and Si IV in the spectra of the hot white dwarf G191-B2B. Although a link between these ions and white

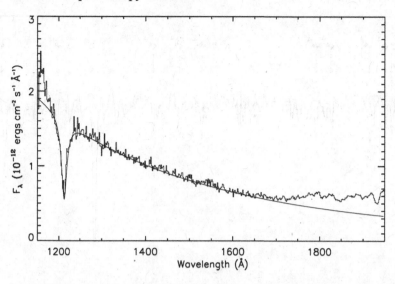

Fig. 2.27. The coadded IUE SWP low dispersion spectrum of the white dwarf in 14 Aur C compared with a model atmosphere (solid line) defined by $T_{\mathrm{eff}} = 44\,000\,\mathrm{K}$, $\log g = 8.01$ and normalised in the visual to $V = 14.05$. The coadded spectrum was constructed from the five existing low dispersion IUE spectra of 14 Aur C. Note the upturn near 1700 Å due to the F4 V primary.

dwarfs with high EUV opacity was suspected for some time, that hypothesis was not widely accepted until Barstow *et al.* (1993) used ROSAT WFC data to show that He could not be the source of this opacity (see section 3.6.2). By that time many white dwarfs were known to contain a host of heavy elements in their photospheres including C, N, O, Si, Fe and Ni. Ions of the latter two elements are particularly potent EUV opacity sources in hot DA stars. An example of the large numbers of Fe and Ni lines present in some of the hottest DA stars such as REJ2214 − 491 is shown in figure 2.28. Accurate heavy element abundances obtained from the UV absorption lines have been essential in understanding the nature of the EUV opacity in hot white dwarfs. The important link between the source and nature of the EUV opacity in white dwarfs and the high-resolution UV spectra of these stars is discussed in sections 3.6.2 and 8.4. A comprehensive review of the IUE echelle spectroscopy of hot white dwarfs is contained in Holberg *et al.* (1998).

A third area where IUE has strongly complemented EUV astronomy has been in the study of the stellar chromospheres, transition regions and coronae in late-type stars of spectral type late-A to mid-M. The wavelength range covered by IUE contains a number of important atomic transitions in ions such as C II, C III, C IV, N V, Si II, Si III and Si IV which give rise to strong emission lines in optically thin plasmas at temperatures between 2×10^4 and 10^5 K. Although well below the range of temperatures responsible for EUV emission in stellar coronae, the low-temperature plasmas probed by IUE help to provide a more complete picture of non-thermal emission occurring in all of the outer atmospheric layers of late-type stars, from the chromosphere to the corona. An example of the emission line spectra obtained by IUE is shown in figure 2.29 which displays the low dispersion SWP spectra for a set of five M stars from Linsky *et al.* (1982). In its low dispersion mode, IUE provided a spectral resolution similar to that of EUVE. Such low dispersion spectra provide accurate line strength

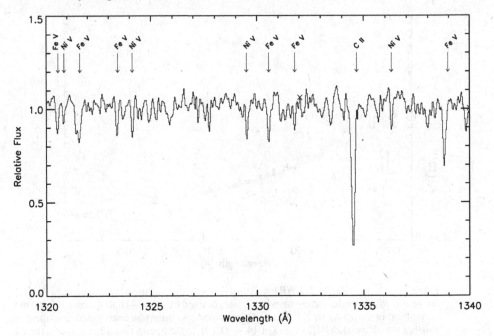

Fig. 2.28. Coadded NEWSIPS spectrum from Holberg *et al.* (1998) showing the 1320 to 1340 Å region of the metal-rich hot DA star REJ2214−491. The arrows indicate the presence of photospheric lines due to Fe V and Ni V, as well as interstellar C II.

measurements which are important as plasma diagnostics to help determine the temperature and electron densities of the emitting regions. At high dispersion, IUE observations have led to important insights into the role of turbulence and mass plasma motions in the atmospheres of late-type stars. IUE studies have also helped to establish links between the chromospheric and coronal activity observed in the Sun and stars of earlier and later spectral types than the Sun. In particular, during its long lifetime, IUE observed a large variety of late-type stars, helping to chart the occurrence of chromospheric and transition region emissions over much of the HR diagram. Studies of variations, in particular long-term activity cycles, have also been important. Discussion of EUV emission from late-type stars and the complementary role played by UV observations is included in section 3.8 and chapter 6.

The IUE satellite was finally turned off in September 1996 after a series of problems which had dramatically reduced its effectiveness. Presently, the only remaining source of UV spectroscopy is the Hubble Space Telescope (HST). The original complement of HST instruments included the Goddard High Resolution Spectrometer (GHRS) which provided high and ultra-high resolution spectroscopy of stellar sources and the Faint Object Spectrometer (FOS) which provided low resolution spectral coverage of sources throughout the UV. Although the spectral resolution of the GHRS exceeded that of IUE, it was only able to observe relatively narrow wavelength bands during any single observation. The GHRS was replaced with the STIS instrument during the January 1997 HST servicing mission. STIS observations, which are now making their impact felt, have achieved significant improvements over the GHRS for the important wavelength region from 1200 to 2000 Å where many important heavy elements have strong transitions. The launch of the Far Ultraviolet Spectroscopic

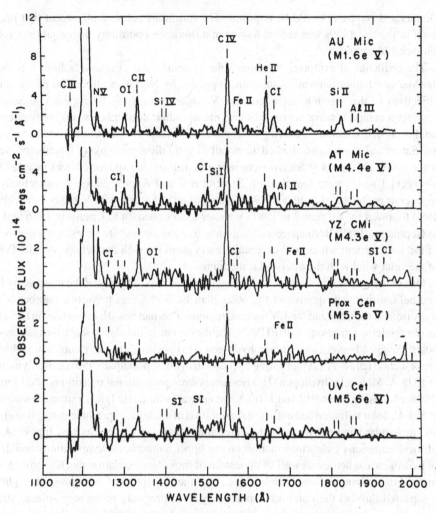

Fig. 2.29. Low dispersion IUE SWP spectra showing emission lines observed in a set of active M stars, from Linsky *et al.* (1982).

Explorer (FUSE) in 1999 has delivered spectroscopic capabilities in the little explored 900 to 1200 Å region.

2.9 EUV and far-UV spectroscopy with Voyager

During the 1960s it was realised that the outer planets – Jupiter, Saturn, Uranus and Neptune – would be orbitally aligned in such a way that a spacecraft launched from Earth in the late 1970s could fly by each planet in sequence. Moreover, such a multiplanet trajectory, if designed carefully, could use the close approach of one planet to gain the necessary velocity to proceed to the next. This favourable planetary geometry presented a unique opportunity to conduct a thorough reconnaissance of each of the giant planets using only a few spacecraft, with minimal launch energies, and in a reasonably short time. Without such gravity assist trajectories many, far less capable, missions with travel times extending over decades, especially for distant

Uranus and Neptune, would be required. The fortuitous celestial alignment that presented itself in the late 1970s was indeed a once in a lifetime opportunity that would not recur for another 176 years.

The challenge of exploring the outer solar system in this dramatic fashion was met with the Mariner Jupiter/Saturn '77 mission, approved by NASA in 1972. The prime scientific objectives of this mission, later renamed Voyager, were to fly by Jupiter and Saturn and conduct a comprehensive survey of these planets, their atmospheres, magnetospheres, and satellite systems, including the rings of Saturn, using just two spacecraft. The Voyager Jupiter and Saturn trajectories also allowed the possibility of follow-on flybys of Uranus and Neptune, if the prime objectives at Saturn were met and at least one spacecraft was in good health. Voyagers 1 and 2 were launched on September 1 and August 20 1977 respectively. Both spacecraft encountered Jupiter in 1979 and went on to Saturn, with Voyager 1 reaching it in 1980 followed by Voyager 2 in 1981. Voyager 2 continued on to Uranus in 1986 and finally to Neptune in 1989. Both spacecraft continue to function and are on escape trajectories out of the solar system which will eventually carry them through the heliopause, the boundary of the solar wind with the interstellar medium.

One of the major objectives of the Voyager mission was the study of the massive hydrogen- and helium-rich atmospheres of the outer planets. Both gases possess a number of strong transitions in the EUV and far-UV spectral regions. This fact was the major reason for selecting an ultraviolet spectrometer with EUV capabilities as one of the eleven scientific instruments on board Voyager. In particular, atomic hydrogen (H I) possesses a strong continuum absorption cross section below 912 Å and strong Lyman series lines, principally the intense Lyman α line at 1216 Å. Molecular hydrogen (H_2) possesses strong vibrational transitions, the Lyman and Werner bands, between 912 and 1100 Å and neutral helium (He I) has a strong resonance line at 584 Å. Solar radiation and collisional excitation of these gases produce strong line emission and molecular band emissions in the outer atmospheres of planets such as Jupiter. Analysis of these emissions yields information on the upper atmospheric temperatures and densities of hydrogen and helium as well as the nature of important excitation mechanisms. A second method of studying the outer atmospheres uses the absorption of sunlight or starlight which has passed through the outer atmospheric layers to precisely probe atmospheric structure. These measurements rely on the occultation of the Sun or a star by the planet's atmosphere as the spacecraft flies by.

The instrument ultimately selected by NASA in 1972 to conduct ultraviolet spectrographic investigations on Voyager was the Ultraviolet Spectrometer (UVS). It was designed to both observe faint 'airglow' emissions from the upper levels of planetary atmospheres and to perform critical occultation experiments during the flybys. The airglow observations re-quired a sensitive instrument with a large field of view, while the occultation experiments demanded a robust instrument with a large dynamic range and high time resolution. All of these features had to be combined into a small, low weight, low power instrument. The original design accepted by NASA called for an instrument similar to the UV spec-trometer successfully flown on the 1972 Mariner 10 Mercury/Venus mission which used ten fixed channeltron detectors to monitor strategic wavelengths. Such an instrument is not a true spectrometer but a multichannel photometer. During the development phase of the UVS, however, microchannel plates (MCPs) became available and were incorpo-rated in the final flight instrument and constituted the first scientific use of MCP detectors in space.

The Voyager 1 and 2 UVS instruments are compact, normal incidence, objective grating spectrometers covering the wavelength range 500 to 1700 Å. Collimation of the incoming UV radiation is accomplished with a series of 13 precision mechanical baffles which define the primary $0.10° × 0.87°$ (fwhm) field of view. A single normal incidence reflection from a concave diffraction grating then focuses the dispersed spectral image onto a photon counting detector. The grating is a platinum-coated replica, ruled at 540 lines mm^{-1}, blazed at 800 Å; dispersion in the image plane is 93 Å mm^{-1}. In addition to the primary field of view, each UVS has a small 20° off-axis port which allows direct viewing of the Sun during occultations, at decreased detector gain. The open, photon counting detector consists of a dual MCP and a 128-element linear self-scanned readout array. The MCP is normally operated at a gain of 10^6 and its output is proximity focused onto a linear array of aluminium anodes. Each of the anodes is coupled to a charge sensitive amplifier whose output is digitally converted to a 16-bit word and summed in an internal register memory. For most astronomical observations, the two prime data rates produce complete spectra every 3.84 s or every 240 s. Detailed descriptions of the Voyager instruments are contained in Broadfoot *et al.* (1977). In flight performance of the UVS from launch through 1990 is contained in Holberg (1990a) and Linick and Holberg (1991).

The primary observational objective of the UVS was the observation of atmospheric phenomena associated with the outer planets and their moons. The discoveries associated with these observations include the Io Torus at Jupiter, and the strong H_2 dayglow emission present in the atmospheres of each of the outer planets. In spite of their small size and location in the outer solar system, these instruments also possessed significant astronomical capabilities. In particular, for many years they represented the only means of obtaining observations in the EUV band between 500 and 900 Å and in the far-UV band between 900 and 1200 Å. During the long periods of interplanetary cruise it was possible to use the UVS to observe a wide variety of astronomical targets for the first time in these critical bands.

Some of the first Voyager astronomical observations dealt with the detection of EUV sources (Holberg 1990b). In 1980, shortly after the Jupiter encounter, the Voyager 2 UVS was targeted at the hot white dwarf HZ 43, which five years earlier had been detected by Apollo–Soyuz as the first extra-solar EUV source (Lampton *et al.* 1976a). The Voyager observations of HZ 43 reported in Holberg *et al.* (1980) measured the EUV spectrum of HZ 43 from 500 to 900 Å as well as the never before seen Lyman series continuum region between 900 and 1200 Å. These observations confirmed the low interstellar H I column in the direction of HZ 43. Following the success of the HZ 43 observations, a number of other known hot white dwarfs were observed with Voyager in the hope of detecting additional EUV sources in the 500 to 900 Å band. The existence of such sources would demonstrate that HZ 43 was not unique in possessing a low interstellar H I column and indicate other directions where EUV sources might be detected. At the time Voyager was obtaining these observations, the only extra-solar system EUV sources known were those detected during the brief Apollo–Soyuz mission and, of these, only HZ 43 was detected longward of 500 Å. Ultimately Voyager detected four additional white dwarfs with significant long wavelength EUV emission. Two, G191–B2B and GD153, were reported by Holberg (1984). G191–B2B is one of the brightest and hottest white dwarfs in the sky and is located in a direction quite different from HZ 43, its detection indicated that the H I column to this star was 10^{18} cm^{-2}. The second star, GD153, is only 7° from HZ 43 and, therefore, it apparently shares a similar ISM line-of-sight. No other white dwarfs were observed to have any detectable EUV emission in the 500 Å band, which indicated that they all must possess interstellar H I columns in excess of 10^{18} cm^{-2}. The ROSAT all-sky survey yielded a number of new hot white dwarf candidates. Observation of about a dozen white dwarfs, with significant ROSAT S2 band flux turned up two additional bright 500 Å EUV sources, REJ0457−281 and REJ0503−289. As the nomenclature indicates, these two stars lie very near one another in the Southern Hemisphere sharing a line-of-sight with a low column density. The observations of these stars are reported by Barstow *et al.* (1994d,e). It is now known from the EUVE all-sky survey that the total number of stellar sources in the 500 Å band numbers less than ten. The brightest of these are the B stars Epsilon and Beta CMa. Voyager had observed these stars but failed to detect the strong EUV fluxes because the scattering of the strong stellar far-UV continua into the EUV band overwhelmed the intrinsic EUV flux.

The vast majority of Voyager astronomical observations have utilised the far-UV band between 900 and 1200 Å. Along with the EUV, this spectral region has only been explored in the most preliminary fashion. Voyager observations also achieved a number of discoveries in the far-UV. These included the observation of outbursts from several CVs which helped to indicate that the outbursts must occur first at the inner edge of the accretion disc and move outward. They also helped to disprove outburst models which placed the bulk of the emitted flux in the EUV band between 200 and 900 Å. The distinct flattening of the energy distribution between 900 and 1200 Å, together with the lack of any EUV flux, were critical in this regard. Other key observations included helping to establish the absolute far-UV flux scale in the 900 to 1200 Å region. Absolute stellar fluxes are critical to much of astrophysics but are often very difficult to measure. With Voyager it was possible to compare the work of various groups who had used sounding rockets to measure the absolute far-UV fluxes of hot stars. A number of discrepancies both among the various observations and with model atmospheres were noted. It proved possible, using a subset of the sounding rocket data and white dwarf model atmospheres, to define a consistent absolute flux scale in this spectral region (Holberg *et al.* 1982, 1991). Subsequent observations with the Hopkins Ultraviolet Telescope (HUT) and Orbiting Retrievable Far and Extreme Ultraviolet Spectrometers (ORFEUS) have now largely confirmed these Voyager results.

In addition to the detection of five hot white dwarfs in the EUV and a number of important upper limits for others, Voyager also established important limits on the EUV diffuse background. Holberg (1986) discussed a deep sky background spectrum obtained by Voyager 2 near the north galactic pole. In this spectrum, which involved a total integration time of 1.5×10^6 s, it was determined that the only detected photon signal consisted of resonantly scattered solar lines from interplanetary H I and He I. After accounting for this emission, it was possible to place significant upper limits on background due to both diffuse line and continuum emission in the bands between 500 and 1100 Å. In the EUV, these limits corresponded to 2×10^4 photons cm^{-2} s^{-1} for diffuse lines and 2×10^2 photons cm^{-2} s^{-1} Å^{-1} for diffuse continuum flux.

3

Roentgen Satellit: the first EUV sky survey

3.1 Introduction

Even before the launches of the Einstein and EXOSAT observatories, it was clear that the next major step forward in EUV astronomy should be a survey of the entire sky, along the lines of the X-ray sky surveys of the 1970s. Such a survey was necessary to map out the positions of all sources of EUV radiation and determine the best directions in which to observe. Indeed, the groups at University of California, Berkeley had been selected by NASA to fly such an experiment on the Orbiting Solar Observatory (OSO) J satellite but, unfortunately, the OSO series was cancelled after the flight of OSO-I. Following the success of the Apollo–Soyuz mission in 1975, scientific interest was revived in the survey concept and the Extreme Ultraviolet Explorer (EUVE) mission was subsequently approved in 1976. Interest in the EUV waveband was also growing in Europe. Having successfully flown a series of imaging X-ray astronomy experiments between 1976 and 1978, the Massachussetts Institute of Technology (MIT)/University of Leicester collaboration sought a new direction of research, with the imminent launch of Einstein, and began development of a new imaging telescope operating in the EUV. This was seen as a direct extension of the mirror technology already refined in the soft X-ray and was combined with the MCP detector expertise acquired from work on the Einstein HRI. In 1979 the EXUV concept for an imaging combined soft X-ray and EUV sky survey was proposed to the European Space Agency (ESA), along with several other mission ideas, but was not eventually approved for flight.

All early X-ray sky surveys had been conducted with non-imaging experiments. With the development of imaging optics for X-ray astronomy, it was recognised that a new survey, taking advantage of the dramatic improvement in sensitivity made available would significantly extend the number of known X-ray sources, placing the subject on a par with optical and radio astronomy in its ability to study large samples of objects. In 1975, the Max-Planck Institüt für Extraterrestriche Physik (MPE) in Garching proposed the idea of a low cost imaging all-sky survey mission to the West German science ministry, gaining formal approval in 1979. Following the usual convention within ESA, MPE then invited proposals from other Member States for additional experiments to fly alongside their X-ray telescope. In collaboration with MIT, the University of Leicester proposed to fly an enlarged version of their EUV sounding rocket payload to complement the X-ray telescope, extending the overall coverage of the mission to longer wavelengths. This was perceived as an attractive idea by MPE and the Wide Field Camera (WFC) was selected for the Roentgen Satellit (ROSAT) mission. Formal approval for UK participation on ROSAT was subsequently obtained from the Science Research Council. Unfortunately, the MIT team were unable to find support from

NASA and had to withdraw from the programme. UK involvement in the WFC was then enlarged into a consortium including Leicester University, Birmingham University, Imperial College of Science Technology and Medicine, Mullard Space Science Laboratory and the Rutherford-Appleton Laboratory.

Following financial approval from German and UK government agencies, detailed development work began on the spacecraft and payloads, following usual patterns with construction of several models of each hardware unit for electrical and mechanical evaluation. In parallel, negotiations were conducted with NASA regarding a possible launch opportunity. Eventual agreement led to NASA providing a Space Shuttle launch in return for a share of the data from the pointed phase of the mission, following the initial sky survey. NASA was also to supply a high resolution imaging camera, similar to the Einstein HRI, for the X-ray telescope. Many scientific payloads were scheduled to be launched on board the Space Shuttle, including ROSAT and EUVE, but these programmes were thrown into crisis with the Challenger accident in January 1986 which led to a two-year grounding of the Shuttle fleet. At this stage, ROSAT had been within about one year of launch.

Retrenchment of the Shuttle programme, with a reduced number of launches per year on the return to flight, promised substantial delays in all scientific programmes. For example, an early manifest scheduled the ROSAT launch for 1994, with EUVE following a few months later! These potential problems led to re-examination of alternatives using expendable vehicle technology, which had been gradually marginalised by the Shuttle programme. A number of crash development programmes were instigated to improve the capabilities of well tried and tested vehicles such as the Delta and Atlas series. Eventually, ROSAT was offered a Delta II launch. The switch from Shuttle to Delta II involved a number of significant hardware changes. The large area of the Shuttle payload bay would have allowed ROSAT to be launched with its solar panels deployed but these had to be folded up for ROSAT to fit inside the Delta II fairing. Hence, new mechanisms had to be added and tested before the mission could fly. Furthermore, a new, specially designed fairing was required for the Delta launch vehicle to enclose the payload.

Eventually, ROSAT (figure 3.1) was launched on June 1 1990 to begin a new era in both EUV and X-ray astronomy. The mission ended after almost nine years when, on February 12 1999, the satellite control system was unable to maintain a stable attitude. In this chapter we describe the satellite and its payload, discussing technical developments crucial to the success of the mission, presenting some of the early EUV survey highlights together with a more comprehensive analysis of the properties of the survey sources. An important scientific bonus of the ROSAT mission was having coaligned X-ray and EUV telescopes observing the sky simultaneously. Therefore, we describe both instruments in detail and include results from the X-ray telescope where appropriate.

3.2   The ROSAT mission

ROSAT was a three-axis stabilised satellite designed to perform both scanning and pointed observations. The two scientific experiments (X-ray telescope and WFC) were coaligned to give simultaneous X-ray and EUV observations of cosmic sources. The spacecraft had two star trackers for position sensing and attitude determination, while the WFC had an independent star sensor to provide its own attitude solution.

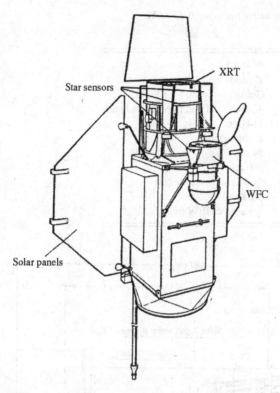

Fig. 3.1. Schematic diagram of ROSAT.

3.2.1 The X-ray telescope

The ROSAT basic X-ray telescope (XRT) design was similar to that of Einstein in concept but made use of significant enhancements in mirror and detector technology during the intervening years. The mirror assembly (XMA) utilised a Wolter I mirror design consisting of four nested shells, giving a total geometric collecting area of $1141 \, cm^2$, manufactured from a Zerodur substrate and coated in 50 nm gold to enhance the reflectivity (figure 3.2). A short-wavelength cutoff of 6 Å was determined by the grazing angle of $2.5°$, itself arising from the need to cover a large ($2°$) field of view for the purpose of the all-sky survey (Aschenbach 1988). An important advance beyond Einstein was the improved smoothness of the mirror surface, reducing scattering in the wings of the focused image and, as a result, improving the achievable signal-to-noise. These mirrors were among the smoothest optical surfaces ever manufactured with a residual scattering power less than 4% of the source flux, at energies below 1 keV within a detection cell radius of 100 arcsec. The angular resolution of the mirror array had a half energy width ≈ 4.8 arcsec (measured on the ground, pre-launch) on-axis increasing to an rms radius ≈ 4 arcmin, due to the intrinsic optical aberrations of the Wolter I design, at the edge of the field of view. The characteristics of the mirror assembly are summarised in table 3.1.

ROSAT had two different types of detector in the focal plane of the XRT. A gas-filled position sensitive proportional counter (PSPC) was chosen as the prime instrument for the sky survey. This was similar to the IPC on Einstein but of superior efficiency and energy

Table 3.1. *ROSAT X-ray mirror assembly (XMA) characteristics.*

Mirror type	Wolter I
Mirror material	Zerodur
Reflective coating	Gold
Number of shells	4 paraboloid and hyperboloid pairs
Geometric area	1141 cm^2
Field of view	2° diameter
Aperture diameter	84 cm
Focal length	240 cm
Focal plane scale factor	11.64 μm per arcsec
High energy cutoff	≈2 keV
Angular resolution	≈5 arcsec HEW on-axis

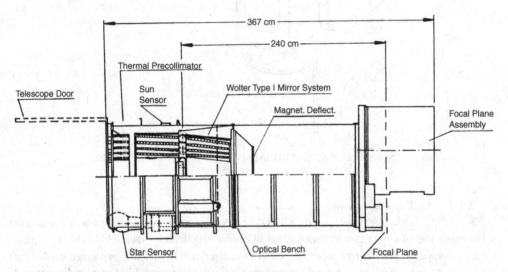

Fig. 3.2. The ROSAT X-ray mirror assembly.

resolution. Two PSPC units were mounted on a carousel to provide redundancy in the event of a failure. Both were operated from the same gas system for purging and filling. The gas supply was expected to last for several years and the last PSPC observations were made at the end of 1994.

The PSPC detectors were multiwire proportional counters each with a cathode strip readout scheme for position determination. Both spectral and spatial resolution increased with increasing X-ray energy. X-rays entered the counter through a polypropylene entrance window, which was coated with carbon and Lexan to decrease the UV transmission. A four-position filter wheel was mounted in front of the detectors. Open and closed positions were used for 'standard' observations and for monitoring the particle background; the third position was used for spectral calibration and the fourth position contained a boron filter. Insertion of the boron filter into the optical path allowed an increase in spectral resolution at lower energies.

Table 3.2. *ROSAT PSPC characteristics.*

Window size	8 cm diameter
Field of view	2° diameter
Gas mixture	65% argon, 15% methane, 20% xenon
Operating pressure	1.466 bar at 22°C
Energy resolution	43% at 0.93 keV
Spatial resolution	300 μm ($\approx$25 arcsec) at 1 keV
Entrance window	1 μm polypropylene
Support grid transmission	72% (on average)
Temporal resolution	$\approx$130 μs
Dead time	3% (100 cts s^{-1}) to 12% (400 cts s^{-1})

Table 3.3. *ROSAT HRI characteristics.*

Window size	8 cm diameter
Field of view	38 arcmin square
Spatial resolution	1.7 arcsec (fwhm)
Quantum efficiency	30% at 1 keV
Window transmission	75% at 1 keV
Background	3.8 $\times$ 10^{-3} cts arcmin^{-2}s^{-1} (typical total)
Temporal resolution	61 μs
Dead time	0.36 to 1.35 ms

The quantum efficiency of the PSPCs was determined by the entrance window (1 μm polypropylene coated with 50 μg cm^{-2} carbon and 40 μg cm^{-2} Lexan) as the absorption of the counter gas was near 100%. Energy resolution was close to the ideal of a single wire proportional counter ($\Delta E/E = 0.43(E/0.9)^{-0.5}$ fwhm) over its entire sensitive area and the position resolution was 300 μm. At the centre of the field, the position resolution was energy dependent, while at large off-axis angles (>20 arcmin) resolution of the complete XMA + PSPC system was dominated by the telescope and was, therefore, energy independent. Table 3.2 summarises the PSPC performance.

In addition to the PSPC units, constructed by MPE, the Center for Astrophysics at Harvard provided a MCP based detector, the HRI, for use in the pointed observation programme (see table 3.3). Apart from the use of a CsI photocathode in place of MgF$_2$, the HRI was identical to the instrument used on Einstein. A fixed UV blocking filter was mounted in front of the HRI to prevent contamination of the X-ray observations with flux from longer wavelengths to which the CsI is sensitive. In the case of the PSPC, the boron carbide/polypropylene window of the counter acted to block the UV. In addition, a boron filter could be deployed in the central part of the field of view to provide additional energy discrimination beyond that from the counter itself.

The HRI utilised 36 mm diameter MCPs, delivering an open aperture of 30 mm when mounted in the detector structure. The readout system had a working area of 27 mm so that the net field of view was given by the intersection of this with the circular MCP aperture. Since the CsI-coated MCP was sensitive to UV radiation as well as X-rays, it was protected from the geocoronal UV and hot stellar UV continua by two aluminised plastic windows.

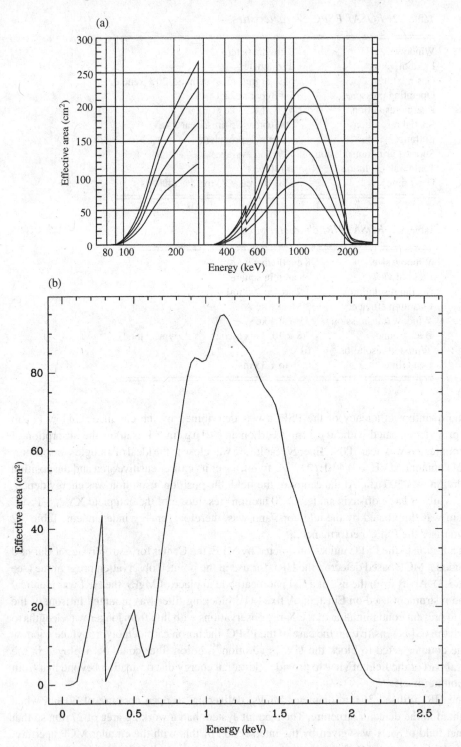

Fig. 3.3. Effective area of the ROSAT XRT with (a) the PSPC detector and (b) the HRI detector.

These also served the purpose of blocking the flux of positive ions, which would otherwise have been accelerated into the detector aperture by the high negative potential of the input face of the MCP stack. The measured spatial resolution of the detector was 8.6 ± 0.2 μm and 8.2 ± 0.1 μm (rms) in X and Y axes respectively, correponding to a fwhm angular resolution of 1.7 arcsec for the 2.4 m focal length of the XMA.

Figure 3.3 shows the total effective area of the XRT with both PSPC and HRI detectors. The most important band from the point of view of joint studies with the WFC was the region between the carbon K edge at 44.7 Å and the roll-off of the proportional counter window (or HRI blocking filter).

3.3 The ROSAT Wide Field Camera

The WFC was designed to cover the EUV range during the ROSAT mission, principally from 60 to 200 Å during the survey itself, but with a capability of observing at longer wavelengths, to 600 Å, during the pointed phase of the mission. The instrument consisted of a grazing incidence telescope comprising a set of three Wolter–Schwarzchild Type I gold-coated aluminium mirrors (Willingale 1988), with a MCP detector in the focal plane (Barstow *et al.* 1985; Fraser *et al.* 1988; Barstow and Sansom 1990). A focal plane turret mechanism allowed selection between two identical detector assemblies. An aperture wheel mounted in front of the focal plane contained eight thin film filters to define the wavelength pass bands, suppress geocoronal background (which would otherwise saturate the detector count rate), and prevent UV radiation from hot O and B stars from being imaged indistinguishably from the EUV. A baffle assembly mounted in front of the mirrors excluded scattered solar radiation from the mirrors and provided thermal decoupling between the mirrors and space. A UV calibration system (Adams *et al.* 1987) was mounted on the mirror support, to give in flight monitoring of the detector gain, electronic stability and mirror alignment. A Geiger–Muller tube and CEM measured charged particle background and generated a signal for switch-off of the detector during passages through the high background regions of the South Atlantic Anomaly (SAA) and Auroral Zones. Background electrons were swept away from the detector aperture by a magnetic diverter system mounted just behind the mirrors (Sumner *et al.* 1989). Figure 3.4 shows a cutaway diagram of the WFC and MCP detector assembly, illustrating the major subsystems, while figure 3.5 is a schematic diagram showing the main optical components. A complete description of the WFC can be found in Sims *et al.* (1990) but the most important information is summarised in table 3.4.

3.3.1 WFC mirrors

The geometric collecting area of the WFC mirror nest was 456 cm^2 and, with a mean grazing angle of 7.5°, it had a 5° field of view (see Willingale 1988). The plate scale at the detector was 160 μm = 1 arcsec. The spatial resolution of the WFC, dominated by the mirror performance, was roughly constant at 20 arcsec over the central $\approx 1°$ radius of the field of view but increased rapidly towards the edge (figure 3.6). An important feature of the Wolter–Schwarzchild design is the curvature of the optimum focal surface which exaggerates the optical aberrations for a flat detector and which dominates at the edge of the field. The on-axis focused image of these mirrors has a weak dependence on wavelength as increased scattering due to the intrinsic surface roughness enlarges the point response function at shorter wavelengths. Averaged over the survey, the point spread function was very similar to the value measured at a position

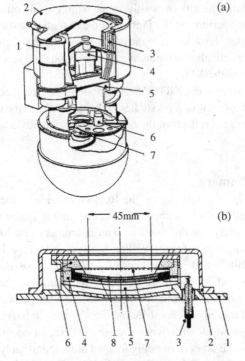

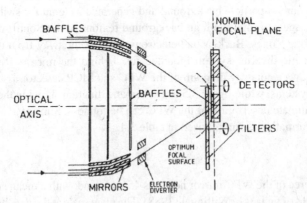

Fig. 3.4. Cutaway diagram of the ROSAT (a) WFC and (b) MCP detector assembly.
(a): 1, Startracker; 2, telescope door; 3, baffles; 4, calibration and protection system; 5,
electron diverter; 6, filter wheel; 7, MCP detector and turret. (b): 1, 2, stainless steel housing;
3, HV feed-through; 4, 6, ceramic structure; 5, resistive anode; 7, repeller grid; 8, MCPs.

Fig. 3.5. Schematic diagram of the optical systems of the ROSAT WFC.

2° off-axis, with a fwhm of 2 arcmin. The main characteristics of the WFC mirror nest are
summarised in table 3.5.

3.3.2   MCP detectors for the WFC

The MCP detectors for the WFC (see table 3.6) consisted of a chevron pair with a resistive
anode readout system (see Barstow *et al.* 1985; Barstow and Sansom 1990). A novel development in the design was to curve the individual MCPs and anode with a radius of 165 mm to

Table 3.4. *Performance characteristics of the ROSAT WFC.*

Field of view	5° diameter (2.5° with P1 and P2 filters)
Spatial resolution:	
survey	2 arcmin (fwhm)
pointed	1 arcmin (fwhm) within 1° of optical axis
Source location accuracy:	(error circle 90% confidence radius)
survey	≈60 arcsec for a weak (5σ) source
	≈25 arcsec for a strong ($>15\sigma$) source
pointed	≈45 arcsec for a weak (5σ) source
	≈25 arcsec for a strong ($>15\sigma$) source
Time resolution	40 ms

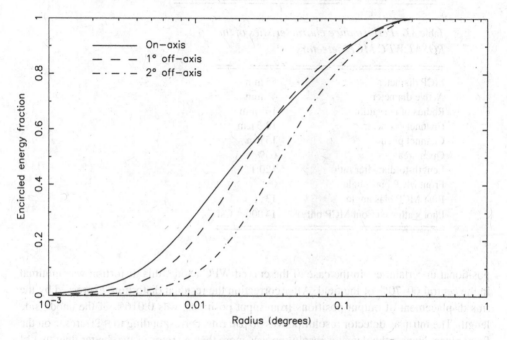

Fig. 3.6. Off-axis imaging performance of the ROSAT WFC mirror nest.

match the curvature of the telescope focal plane, thus avoiding the degradation in the image that would occur with a flat detector. This approach allowed the maximum possible sensitivity to be achieved during the survey as the signal to background is improved with a better point response function. A CsI photocathode was used to give the best quantum efficiency across the EUV wavelength range for the 30° angle of the radiation incident at the detector. A further enhancement to the quantum efficiency was obtained by the use of a repeller grid in front of the detector. This high transparency (fractional transmission 0.92) grid was biased such that photoelectons created by photon interactions with the glass matrix of the channel plate were directed into the nearest channel.

A known property of resistive anode readout systems is a distortion introduced into detector images, for which correction must be made. Unquantified effects can also contribute to

Table 3.5. *Performance characteristics of the ROSAT WFC mirrors.*

Optical design	Wolter–Schwarzschild Type I
Material	Nickel-plated aluminium
Reflective coating	Au
Number of shells	3
Field of view	5° diameter
Geometric area	456 cm²
Aperture diameter	576 mm
Focal length	525 mm
Focal plane scale factor	160 μm = 1 arcmin
High energy cutoff	0.21 keV (10% of peak)
On-axis resolution	1.7 arcmin (HEW) at 0.04 keV

Table 3.6. *Performance characteristics of the ROSAT WFC MCP detectors.*

MCP diameter	55 mm
Active diameter	45 mm
Radius of curvature	165 mm
Channel diameter	12.5 μm
Channel pitch	15 μm
Open area	63%
Length-to-diameter ratio	120:1
Front MCP bias angle	0°
Rear MCP bias angle	13°
Photocathode (front MCP only)	14 000 Å CsI

positional uncertainties. In the case of the curved WFC anode, this distortion was minimal in the central 60–70% of the field. After correction the residual distortion, expressed by the rms displacement of output positions from input positions, was 0.016% of the anode side length. The intrinsic detector resolution was 50 μm rms, corresponding to ≈20 arcsec on the focal plane. Hence, the detector resolution was more than a factor of two better than that of the mirror nest and, as a result, did not contribute significantly to the overall performance of the WFC.

3.3.3   Thin film filters

The survey and pointed observation bandpasses were defined by a series of thin film filters installed in the filter wheel and rotated into position as required (see Kent *et al.* 1990). Survey filters covered the full field of view of the instrument and were 80 mm in diameter. Each of these was a composite metal foil with a Lexan support structure. The short wavelength (S1) band was constructed in two forms – boron carbide + carbon + Lexan (S1a) and carbon + Lexan (S1b), each in a multilayer sandwich structure to improve the strength and minimise the effects of any pinholes. The boron carbide was added to the S1a filter to prevent possible erosion from atomic oxygen bombardment of the carbon foil and carbon-based

Table 3.7. *Nominal wavelength ranges defined by thin film filters in the ROSAT WFC.*

Filter type		Survey (S) Pointed (P)	FOV diameter (degrees)	Bandpass (Å)	Thickness (μm) (10% peak efficiency)
S1a:	C/Lexan/B$_4$C	S+P	5	60–140	0.2/0.3/0.1
S1b:	C/Lexan	S+P	5	44–140	0.3/0.3
S2:	Be/Lexan (×2)	S+P	5	112–200	0.3/0.2
P1:	Al/Lexan	P	2.5	150–220	0.2/0.2
P2:	Sn/Al	P	2.5	530–720	0.2/0.2
OPQ:	Lexan/Al	S+P	2.5		Opaque to EUV

plastic. However, it had the additional effect of suppressing the sensitivity of the WFC to relatively hard or strongly absorbed sources by reducing the soft X-ray 'leak' below 44 Å, and better defined the EUV transmission band of the filter. The remaining filters, containing metal foils were expected to produce stable oxide layers. The longer wavelength survey filters (S2) consisted of beryllium + Lexan and were identical in thickness and construction. For the pointed phase additional filters were available, covering half the 5° field of view and 40 mm in diameter, consisting of aluminium + Lexan (P1) and tin + aluminium (P2). In addition, a carbon + Lexan + aluminium filter was installed which was designed to be opaque to the cosmic background EUV radiation and used to monitor the particle background reaching the detector. All filters, except the Sn/Al, were mounted on 50 μm thick etched stainless steel support grids, having 70% transmission, designed to prevent damage during the launch and operation. Unlike the other filters, the Sn/Al was a free standing foil without support from a plastic substrate which would be too opaque at very long wavelengths. The foil was mounted on a mesh by the manufacturer (Luxel Corporation) but, to cover the large area required by the WFC, was made up of six triangular segments. Filter pass bands are summarised in table 3.7 and their effective areas, when combined with the detector and mirror responses, are shown in figure 3.7.

3.3.4 *ROSAT sky survey strategy*

ROSAT was placed in a circular orbit at an altitude of approximately 580 km, giving an orbital period of ≈96 min. The orbital inclination was 53°. The sky survey began on July 30 1990, after about two months of instrument and satellite checkout and was planned to last for six months. However, it was cut short by about two weeks on Janury 25 1991 when the spacecraft lost attitude control for a time. This created a small gap, comprising about 5% of the sky, which was filled in during August 1991. Use was also made of the 'mini-survey' carried out between July 11 and 16 1990 as part of the mission performance and verification mission (Pounds *et al.* 1991). During the survey, the sky was scanned in a series of great circles passing through the ecliptic poles and crossing the ecliptic plane at a fixed angle to the Sun, nominally within 12° of 90°. To maintain the angle of the solar panels within this attitude to the Sun the scan path was advanced by ≈1° per day, covering the whole sky in a period of approximately six months. The scan path was synchronised with the orbital motion so that the telescopes always pointed away from the Earth. In the WFC, sources on the scan

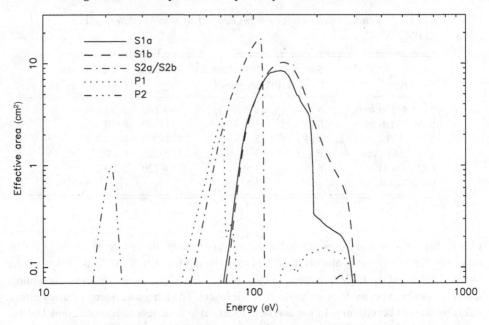

Fig. 3.7. Effective area of the ROSAT WFC in each of the survey and pointed filter
bandpasses.

path were viewed for up to 80 s per orbit. Those located near the ecliptic plane were scanned
each orbit for five successive days, the coverage increasing towards the ecliptic poles which
were scanned throughout the entire six months. With a smaller (2°) field of view the exposure
time in the XRT was only ≈30 s and the coverage for an individual source on the ecliptic
plane two days.

To conduct the EUV survey in two bands the S1a and S2 filters were exchanged in the
field of view once per day. Allowing for unavoidable losses of data during satellite passage
through the Auroral Zones and SAA useful data were collected for about 74% of each day. An
additional average loss of about 10% occurred throughout the survey from minor spacecraft
and WFC hardware problems and spacecraft maintenance time. The final effective exposure
in each survey waveband ranged from ≈1600 s near the ecliptic equator to ≈70 000 s near
the poles.

3.4   Highlights from the WFC EUV sky survey

Ultimately a full appreciation of the importance of the WFC to EUV astronomy and the
detailed astrophysical knowledge revealed by the mission requires a comprehensive review
of the data, dealing with each class of EUV source in turn (see section 3.5). Nevertheless,
whenever a new waveband is opened up with a first comprehensive sky survey, some of the
first scientific results stand out as milestones in the field. In particular, when only a fraction of
the sky had been observed and the data subsequently processed, studies during the early part
of the mission tended to concentrate on individual objects of special interest. Furthermore,
a thorough understanding of the general population of sources would take several years to
come to fruition. We present here some selected early highlights from the WFC sky survey.

Fig. 3.8. First light image recorded through the S1a filter of the WFC, revealing a point source of EUV radiation.

3.4.1   A new source at 'first light'

One of the most significant (and stressful) phases at the beginning of the lifetime of a satellite is the testing of instrumentation, to verify that everything is operating correctly, leading up to the the moment of truth – a first observation of the sky, 'first light'. Planned observations of known sources are then used to establish that the actual instrument performance matches that expected. On ROSAT, testing of the XRT and WFC were carried out in an overlapping sequence. For the first WFC observation of the sky, on June 17 1990, ROSAT was pointed at the A2199 cluster of galaxies for XRT calibration, a source which was not expected to be detected in the EUV. However, a source of EUV radiation was immediately apparent within the field of view of the WFC during real-time data acquisition, with an event rate ≈ 1 count s^{-1}. The first light image (figure 3.8) revealed a point source of radiation approximately 2 degrees from the optical axis, which could not be associated with the cluster. It was possible to position the source within a 90% confidence contour of radius ≈ 1 arcmin centred at (equinox J2000) RA $= 16$ h 29 min 29 s, Dec $= +78°$ 04' 31". Of six objects lying near or within the error circle, one blue star, five magnitudes brighter than the rest seemed the most plausible candidate. Subsequent optical observations obtained with the William Herschel Telescope showed a blue spectrum characteristic of a DA white dwarf but the red region shows clear evidence of TiO absorption, suggesting that the white dwarf has a cool companion of spectral class M (figure 3.9; Cooke *et al.* 1992). This new source, designated REJ1629 + 781 (RE = ROSAT EUV) from its location in the sky, is a rare DA + dM binary system, similar in nature

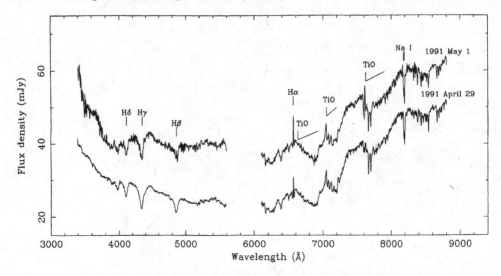

Fig. 3.9. Optical spectrum of the DA+dM binary REJ1629+781, showing the Balmer lines characteristic of the white dwarf and the TiO bands of the M dwarf (from Catalan *et al.* 1995).

to Feige 24 (see chapter 2). However, it appears to have different characteristics. At 35 000 K, the REJ1629 white dwarf is cooler and appears to have a more or less pure H composition (see Catalan *et al.* 1995; Sion *et al.* 1995). Searches for radial velocity variations reveal no detectable changes indicating that either the orbital period is very much larger than that of Feige 24 (4.23 d) or that the inclination of the system is close to 90°.

3.4.2   *First detection of a stellar flare in the EUV*

One consequence of the strategy adopted by ROSAT to cover the whole sky during the survey was the repeated observation of sources once per satellite orbit (96 min) for the ≈1 min duration of their passage through the WFC telescope field of view. The actual exposure time depended upon the distance of the source path from the optical axis, with a maximum value of 80 s when this cuts across the full diameter of the field, passing through the optical axis. The rate of drift of the scan path then determined the visibility of individual objects. Those at higher galactic latitudes received longer coverage than those near the equator, which were typically monitored for ≈5 d. The double lined spectroscopic binary system BY Draconis was observed for a period of 19 d, from September 23 to October 11 1990. UV observations were also obtained with the IUE satellite coinciding with the optimum WFC exposures, between September 29 and October 4. BY Dra has an orbital period of 6 d and an optical photometric period of 3.8 d, which is assumed to be the result of rotational modulation by spotted regions on the more active component, although both stars have been seen to flare.

The mean source count rates were 0.051 ± 0.004 counts s^{-1} and 0.057 ± 0.004 counts s^{-1} in the S1 and S2 bands respectively, Since these rates are identical, within experimental error, it was possible to merge the separate S1 and S2 times series (the filters were exchanged once per day in the field of view) to produce a single composite light curve (figure 3.10). Two distinct features can be seen in the EUV data. First there is a clear and possibly periodic

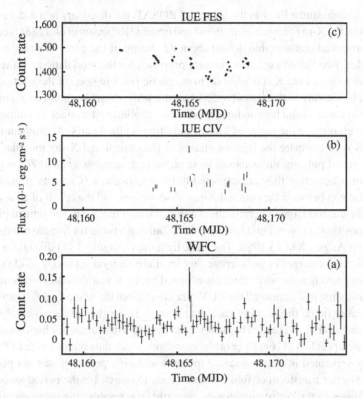

Fig. 3.10. Fifteen days of EUV observations of BY Dra (a) recorded during the ROSAT all-sky survey. Simultaneous far-UV (b) and optical (c) data from IUE span the central six days of the survey coverage (from Barstow *et al.* 1991).

modulation of the flux. Second, there is a strong flare, occurring at a minimum in the quiescent flux, between Modified Julian Date (MJD) 48165 and 48166. The peak count rate is a factor 3.5 higher than the prior quiescent level. The flare can also be seen in the far-UV and optical bands, which are also shown in figure 3.10.

The observed flare was the first to be unambiguously detected in the EUV on any star other than the Sun. EUV emission had only once before been reported from a dMe flare star, Proxima Centauri (Haisch *et al.* 1977). Because in that study the star was only detected on one of two brief observations, it was assumed that a flare had occurred, but there was no other evidence to support this assertion. The long decay time of the BY Dra flare indicates that it was probably similar in nature to a two-ribbon solar flare. With a total energy radiated in the EUV of $\approx 7 \times 10^{32}$ erg, some 50% of the typical flare energies seen in X-ray observations of BY Dra (e.g. Pallavicini *et al.* 1990), this result emphasised that radiation in the EUV carries away a substantial fraction of the total energy released.

3.4.3 Optical, EUV and X-ray oscillations in V471 Tauri

V471 Tauri is short period (0.52 d) eclipsing binary system consisting of a DA white dwarf and a K2V star. Lying in the Hyades cluster its distance (49 ± 1 pc) and age (5×10^8 y) are well determined and, as an example of a post common-envelope binary system it has been the subject of considerable study. Prior to the launch of ROSAT, the discovery of a 555 s regular modulation in both the X-ray (Jensen *et al.* 1986) and optical (Robinson *et al.* 1988) flux from V471 Tau had provoked considerable debate about the nature of the components, the white dwarf in particular. Two viable explanations were proposed for the modulations. Either they arise from the accretion of the K star wind on to magnetic polar regions of the white dwarf, which must then be rotating with a period of 555 s, or the white dwarf is a non-radial pulsator. In the latter case, the star would have to lie in some new instability strip, since its temperature ($\approx$35 000 K) is well outside the range of known pulsating white dwarfs. A stringent test of these hypotheses is to compare the relative phases of the optical and X-ray modulation. If the star is a non-radial pulsator these should be in phase (e.g. Barstow *et al.* 1986), whereas if the mechanism is accretion they are expected to be in anti-phase (Clemens *et al.* 1992). However, the time gap between the original X-ray observations and the epoch of an accurate optical ephemeris was too large to permit direct comparison of the X-ray and optical phases.

Occupying a position at low ecliptic latitude, V471 Tau was visible for five days during the WFC survey from August 8 to 13 1990. The EUV light curve (figure 3.11) displays a range of variability. Discrete changes in the average flux level are seen once per day as the survey filters are exchanged. When the white dwarf is eclipsed by the K star the observed count rate drops to zero, indicating that almost all the EUV flux arises from the white dwarf. In addition, the remaining scatter in the flux level of each data point is larger than expected purely from the photon counting statistics. This can be attributed to the 555 s modulation, which is sampled at a different phase at each WFC once per orbit 'snapshot'. Since data points in the EUV time series are widely separated in time compared to the modulation period, it was not possible to calculate the Fourier transform or fold the time series to search for the period associated with the modulations in the conventional way. Nevertheless, because the mean duration of the time slots is a fraction ($\approx$1/10) of the period, it was possible to isolate the phase of each sample quite accurately. Although the optical data obtained in 1990 and 1991 were not contemporaneous with the ROSAT observations, the accuracy of the pulsation ephemeris was sufficient to restrict the cycle count error to less than 10% in phase. Hence, not only could the data be folded by the optical period to determine the EUV pulse shape, but the absolute optical

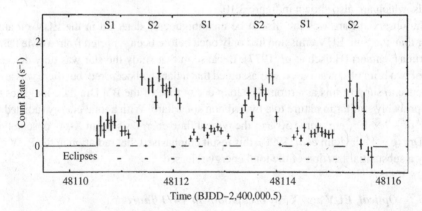

Fig. 3.11. ROSAT EUV light curve of V471 Tauri. The horizontal bars below the zero count rate line in the lower panel mark the predicted times of the white dwarf eclipse (from Barstow *et al.* 1992b).

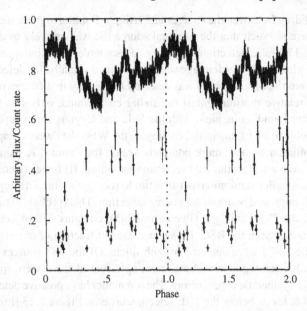

Fig. 3.12. EUV (bottom, circles), soft X-ray (centre, stars) and optical (top) pulse phases calculated from the optical pulsation ephemeris (Clemens *et al.* 1992). Phase errors are not included for clarity but amount to ≈ 0.05, ≈ 0.015 and ≈ 0.09 for EUV, soft X-ray and optical observations respectively (from Barstow *et al.* 1992b).

phase could be recovered with sufficient accuracy to allow direct comparison with the ROSAT data (figure 3.12; Barstow *et al.* 1992b). The phase of the optical maximum matches the point of EUV/X-ray minimum demonstrating that the white dwarf is not a non-radial pulsator and avoiding the need to invoke the existence of a new instability strip. Accretion from the wind of the K star onto the magnetic pole(s) of the white dwarf creates a local hotspot that is optically bright but dark in the EUV due to the opacity of the accumulated material. The modulation then arises from the rotation of the white dwarf. Direct observational confirmation of the accretion has recently been obtained with the Hubble Space Telescope, where Sion *et al.* (1998) observed a modulation of the strength of the Si III absorption line and were able to estimate the size of the opaque polar region and the accretion rate.

3.5 The WFC EUV catalogues and the source population

While many interesting results arose from studies of known objects, such as V471 Tauri and BY Draconis, which might have been expected to be among the EUV source detections, the most significant results of any sky survey emerge from detailed study of the source population as a whole. Two EUV source catalogues have been published from analysis of the WFC data and the objects included in these have formed the basis for subsequent studies of individual groups of objects.

3.5.1 The catalogues

The Bright Source Catalogue (BSC; Pounds *et al.* 1993) contains 383 EUV detections from the first analysis of the survey. A rather conservative automated source search technique was used, passing a circular sliding box over the images and using a Poisson significance test. Significance thresholds were set such that the eventual source list was unlikely to contain more than about three (i.e. 1%) false detections. In many cases sources were independently detected in both S1 and S2 filter bands. Each image was also scanned visually, for defects and extended sources. Apart from the Moon, which was detected fortnightly in different regions of the sky, because of its relative motion against the stellar background, only two bright, extended EUV sources were found, coincident with the Vela and Cygnus Loop supernova remnants. Following the publication of this initial catalogue, the WFC data were completely reprocessed before performing a second, more complete search for sources. A number of significant improvements were made in the data reduction including: (i) better rejection of periods of poor aspect and smaller random errors in attitude reconstruction; (ii) improved background screening; (iii) improved methods for source detection. This (probably) final list of sources from the WFC, the 2RE catalogue (Pye *et al.* 1995), contains 479 objects. The majority of these are also included in the BSC. However, a greater fraction of objects (80% compared with 60% for the BSC) were detected in both filters. Of the 383 sources in the BSC, 359 are also in the 2RE catalogue but 24 are 'missing'. Two of these were spurious detections arising from large systematic aspect errors. The remainder have positive detections in the reprocessed data but at levels below the 2RE selection criteria. Figure 3.13 shows the distribution of the 2RE sources on the sky.

3.5.2   *Identification of EUV sources*

If the nature of EUV sources remains unknown, then it is difficult to carry out any detailed analysis of the source population. Hence, to fully exploit the contents of the catalogues, it is important to associate the individual sources with optically identified objects. The ability to do this depends on the accuracy of point source location, the source density and the optical brightness of the likely counterparts. With an effective spatial resolution of about 3 arcmin fwhm, averaged over the field of view, the WFC was able to locate sources to an accuracy typically better than 1 arcmin. Initial source identifications were found by cross-comparing the WFC positions with a number of optical catalogues (see Pye *et al.* 1995 for details), including for example the McCook and Sion (1987) (note: a revised version of this catalogue

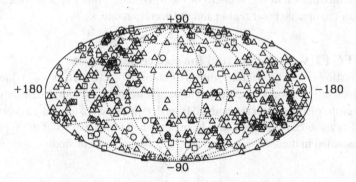

Fig. 3.13. Map of the sky in galactic coordinates, showing the locations of the EUV sources in the 2RE catalogue. The projection is Hammer equal area. Sources detected in both S1 and S2 wavebands are indicated by triangles, while those detected only in S1 or S2 are marked with circles and squares respectively.

is now available as McCook and Sion 1999) list of spectroscopically identified white dwarfs, the Gliese and Jahreiss (1991) catalogue of nearby stars and the SIMBAD database. In the BSC, 73% (279) of the listed sources were found to have a probable identification, based on the existing knowledge of those objects believed to be potentially strong EUV emitters, although, in 92 cases, the catalogue search yielded more than one possible counterpart.

The optical catalogues do not necessarily represent spatially complete samples. For example, a significant fraction of the white dwarfs in McCook and Sion (1987) originated from the Palomar Green (PG) survey (Green *et al.* 1986b). This was a magnitude limited optical survey of UV bright objects covering approximately 25% of the celestial sphere and was predominantly located in the Northern Hemisphere. This survey was very effective at identifying white dwarfs with T_{eff} above $\approx 10\,000\,\text{K}$, but did not cover the whole sky observed by ROSAT. Hence, a follow-up programme of optical spectroscopy was undertaken, to attempt to identify those EUV sources without an obvious counterpart, to resolve the ambiguity of multiple counterparts and to check that the proposed identifications of a sample of objects are correct. These observations were carried out in both Northern and Southern Hemispheres using the Isaac Newton 2.5 m and South African Astronomical Observatory 1.9 m telescopes respectively. A total of 195 sources was identified in this programme, comprising 69 previously uncatalogued white dwarfs, 114 active stars, seven new magnetic cataclysmic variables and five active galaxies (Mason *et al.* 1995). As a result of this work, the total of BSC identifications was increased to 337 objects (87%). Many of the fainter sources not in the BSC but ultimately included in the 2RE catalogue were also in the optical identification programme. A total of 444 (93%) of the 2RE sources have proposed identifications.

Figure 3.14 illustrates the distribution, by optical type, of all the 2RE sources for which an identification exists. The largest single group of optical counterparts is active stars, of spectral type F to M. The numbers of F, G, K and RS CVn stars predicted before the survey agreed remarkably well with the detections (Pounds *et al.* 1993). However, a factor 4–10 fewer M stars were seen than expected. Hot white dwarf stars ($T_{\text{eff}} > 2 \times 10^4\,\text{K}$) form the second major group. Many of the brightest sources seen fall into this class. Furthermore, these objects also have the highest EUV/optical luminosity ratio, as can be seen from figure 3.15, which compares the EUV source count rates with the visual magnitudes of the counterpart. Nevertheless, the total number of white dwarfs detected (125) is significantly less than the 1000–2000 expected pre-launch (Barstow and Pounds 1988; Barstow 1989). Several hot central stars of planetary nebulae are included in this group. Apart from the ≈ 20 cataclysmic variables detected, all the remaining categories – active galactic nuclei, X-ray binaries, early-type stars and supernovae remnants – include only a handful of objects.

The distribution of sources on the sky (figure 3.13) and also of the count rates can provide important information regarding the effects of interstellar absorption in the EUV and the location of the mainly galactic source population. Since the accumulated exposure on the sky was not uniform, due to the survey strategy and data losses caused by high background or temporary malfunction, the sky survey sensitivity varied from location to location on the sky. After estimating the minimum detectable point source count rate at a grid of sky locations, 'coverage corrected' source counts can be constructed. The resulting number/count rate ($\log N - \log S$) cumulative distributions for each waveband are shown in figure 3.16, displaying the whole 2RE catalogue with white dwarf and late-type star subgroups. The flatness of the white dwarf distribution dramatically shows the effect of ISM absorption while the late-type stars are consistent with a 'Euclidean' distribution and, hence, are little affected by the ISM.

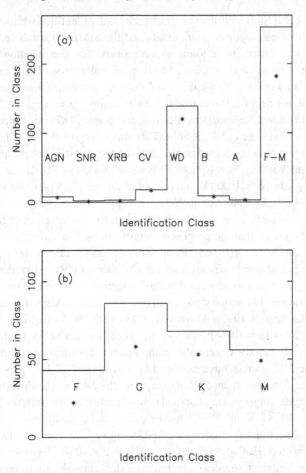

Fig. 3.14. (a) Distribution of optical counterparts in the 2RE catalogue. The star symbols show the same information, but for the BSC. (b) Distribution of identifications in the main subgroup of late-type stars (from Pye *et al.* 1995).

This indicates that the white dwarf sample lies on average at a greater distance than the active stars, which is consistent with their greater EUV luminosity.

3.5.2.1 WFC source names

The naming convention for sources in the WFC catalogues uses a suitably truncated RA and Dec, as laid down by the International Astronomical Union (IAU), combined with a catalogue designation and an indication of the epoch (J2000). The BSC catalogue was given the code RE (ROSAT EUV) and 2RE is used for the subsequent catalogue of that name. Hence, a source at J2000 coordinates RA = 00 h 03 m 55.17 s and Dec = +43 d 35′ 41.2″ will carry the name 2REJ0003 + 436, the last digit of the Dec being the decimal fraction of a degree corresponding to 35 arcmin. In general, objects are usually referred to by the names they acquire on discovery. Hence, EUV sources with pre-existing catalogued counterparts should refer to the optical catalogue name, while newly identified sources in the BSC carry the 'REJ' appelation. Only the additional sources in the 2RE catalogue should use the '2REJ' form.

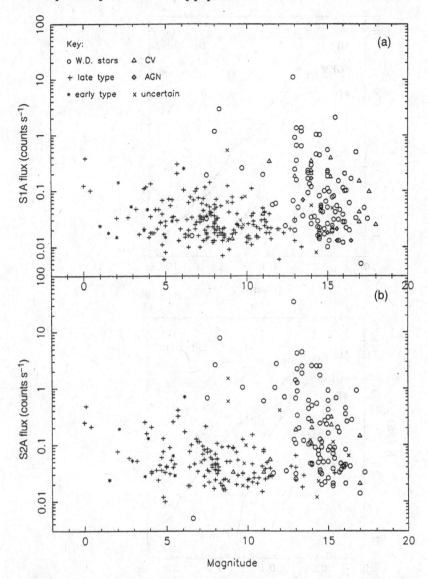

Fig. 3.15. EUV count rate versus optical magnitude for the main classes of identified object (from Pounds *et al.* 1993).

3.6 Properties of the white dwarf population

3.6.1 Composition and luminosity of DA white dwarfs in the EUV

As already noted, hot white dwarfs constitute the second largest group of EUV sources, amounting to some 125 objects. Not surprisingly, the vast majority of these are DA white dwarfs, with only a handful of He-rich DOs or PG1159 stars included. However, the actual number detected was considerably below (by at least a factor of ten) that expected, based on the existing knowledge of the white dwarf population and distribution of absorbing gas in the LISM. In predicting the likely number of white dwarf detections the most important

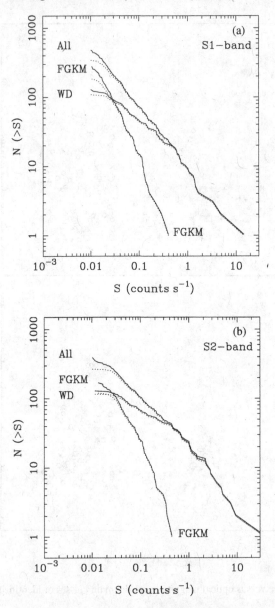

Fig. 3.16. Raw (dotted line) and corrected (solid line) source count distributions for the full 2RE catalogue (All), white dwarfs (WD) and late-type stars (FGKM), in (a) the S1 band and (b) the S2 band (from Pye *et al.* 1995).

assumptions included are a knowledge of their space density and EUV luminosity. The space density of hot white dwarfs is well determined, from the magnitude limited PG optical survey (see Fleming *et al.* 1986), revealing a homogeneous distribution of stars in the galactic disc with a scale height ≈200 pc. Conversely, the expected EUV luminosity depends strongly on both the stellar temperature and photospheric composition. While the work of Fleming *et al.* (1986) gives an indication of the relative numbers of 'hot' white dwarfs, with temperatures in the range 20 000–40 000 K and 40 000–80 000 K, optical data can provide little information on the possible variance in composition. However, as demonstrated by work with *Einstein* and EXOSAT traces of He in otherwise pure H atmospheres will lower the EUV luminosity at a given temperature. The presence of significant quantities of He (and, indeed, heavier elements) will effectively decrease the volume within which white dwarfs can be detected and increase the temperature above which they become significant EUV sources from the $\approx$20 000 K threshold for a pure H envelope. With the limited data available from the small numbers of white dwarfs studied by *Einstein* and EXOSAT, He was assumed to be the dominant source of opacity with a canonical photospheric abundance of 10^{-4} (He/H by number; see Barstow and Pounds 1988; Barstow 1989).

A first indication of the possible explanation for the shortfall in white dwarf detections comes from an examination of the fraction of optically selected PG survey white dwarfs detected at X-ray wavelengths (figure 3.17), where the effects of interstellar absorption are reduced. All DAs hotter than 30 000 K (roughly corresponding to $M_V = 10$) should have been detected but, as can be seen, this is not the case (Fleming *et al.* 1993a). In fact, although the fraction of detections increases with decreasing absolute magnitude until $M_V \approx 9$, a steep decline sets in thereafter. The most striking feature of figure 3.17 is the absence of any detections among the 6 PG stars at $M_V < 8$ ($T_{\mathrm{eff}} > 60\,000$ K). This result indicates that the assumed composition, and, as a result, the implied EUV luminosity of the general population of DA white dwarfs is incorrect.

The second, and most important result arises from a detailed analysis of the X-ray and EUV photometric fluxes for the white dwarfs actually detected in the survey. Such a study relies on the comparison between observed count rates and those predicted for theoretical white dwarf model atmospheres, when the synthetic spectra generated from these are folded through the instrument response functions. A simple H+He atmosphere is specified by four parameters – effective temperature, surface gravity, helium abundance and a distance-radius

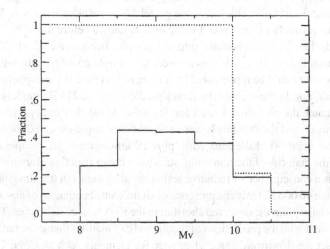

Fig. 3.17. The fraction of optically selected PG survey DA white dwarfs detected at X-ray wavelengths by the ROSAT PSPC, as a function of absolute magnitude (i.e. effective temperature). The dotted histogram shows what would be expected if all hot DAs had a homogeneous atmosphere with He:H = 10^{-4} while the solid line shows the fraction actually observed.

related normalisation constant. Two different atmospheric structures can be considered, either homogeneously mixed H and He or a stratified model where a thin layer of hydrogen overlies a predominantly He envelope. The latter case is more physically realistic, since any He present in the envelope of a DA white dwarf is expected to sink rapidly out of the atmosphere in the strong gravitational field (Vennes *et al.* 1988). In this case, the free abundance related parameter is then the H layer mass and any opacity arising from He is related to the diffusion across the H/He boundary. In addition, the effects of interstellar absorption must be taken into account using an appropriate model. The main contributions to interstellar opacity in the EUV are neutral hydrogen (H I), neutral helium (He I) and singly ionised helium (He II). When dealing with broadband photometric data, it is convenient to assume that there is negligible ionisation along the line-of-sight and that the ratio of He I to H I matches the cosmic He/H abundance of 0.1. Hence, the interstellar absorption can be expressed in terms of a single variable, the H I column density (see section 1.2).

A minimum of five free parameters must be specified to match predicted and observed count rates. However, although the combination of the ROSAT X-ray and EUV surveys improved on the data available from *Einstein* and EXOSAT, with narrower filter bands, only three data points were available. Hence, it is necessary to know the value of at least two variables in the complete model spectrum to arrive at a unique solution to the question of which model yields the best fit. Since, uniquely among EUV sources, the entire spectrum of a white dwarf can be specified by the physical parameters of its atmosphere, measurements from wavebands other than in the EUV can be used to contrain these. For example, temperature and surface gravity can be determined from the Balmer line profiles in the optical (see e.g. Holberg *et al.* 1986; Kidder 1991; Bergeron *et al.* 1992). In addition, a visual magnitude measurement (usually the V band) provides a convenient normalisation point for the absolute flux level of the predicted spectrum. For full exploitation of the EUV/soft X-ray data this information needs to be available for all the white dwarfs detected.

Initially, Barstow *et al.* (1993b) were able to carry out an investigation of a group of 30 well studied DA white dwarfs, for which T_{eff}, log g and M_V were already known. These stars consisted almost exclusively of catalogued white dwarfs known before the ROSAT sky survey and represented, therefore, an optically selected sample. Irrespective of whether or not homogeneous or stratified H+He models were used, the sample could be split into two broad categories. Many stars could be represented by a more or less pure H atmosphere but a substantial fraction could not. In these cases the fluxes predicted by the H+He models were completely unable to match the photometric data for any value of the He opacity (either He abundance or H layer mass) and the flux levels were always below those expected for a pure H atmosphere. Furthermore, most of the stars with 'pure H' atmospheres had temperatures below 40 000 K, while the majority of the remaining stars were hotter than this. Barstow *et al.* interpreted this result as a consequence of radiative levitation of elements in the atmospheres of the DA stars hotter than 40 000 K. Only the presence of significant abundances of absorbing material other than He can explain the observed shortfall in the EUV/soft X-ray fluxes. This is also qualitatively consistent with the predictions of theoretical calculations that show radiative levitation can counteract the downward force of gravity for elements such as C, N, O and Si at temperatures above $\approx$40 000 K (e.g. Morvan *et al.* 1986; Vauclair 1989; Chayer *et al.* 1995a,b). Abundances of Fe and Ni are expected to be significant at temperatures above 50 000 K (Chayer *et al.* 1995a,b), and these elements have a dramatic effect on EUV opacity.

One outstanding question was whether or not the observed properties applied only to an optically selected sample. Several surveys based on ROSAT pointed observations confirmed the results of Barstow *et al.* (S. Jordan *et al.* 1994; Wolff *et al.* 1995) but the selected targets were still dominated by white dwarfs identified primarily through optical studies. Following the identification programme for the ROSAT WFC sources, a complete sample of stars was available for study based on selection by their EUV flux. From a total of 120 DA white dwarfs detected by the WFC, Marsh *et al.* (1997b) assembled 89 objects for which T_{eff}, log g and m_V were measured from optical follow-up observations and where the EUV fluxes were large enough to place sensible constraints on the photospheric and interstellar opacity. A useful way of comparing the EUV fluxes from individual stars is to estimate the absolute flux in each waveband from the respective count rates, taking in to account the stellar magnitude and effective area of each waveband (see Barstow *et al.* 1993b; Marsh *et al.* 1997b). These values can only be considered to be representative of the true stellar flux if the intervening interstellar opacity is small. Figure 3.18 shows the fluxes for each of the three ROSAT survey bands – PSPC, WFC S1 and WFC S2 – as a function of stellar temperature, after excluding all stars with significant interstellar opacity. The fluxes are normalised to a V magnitude of 11.0 and the predictions of a pure H atmosphere with modest interstellar absorption.

The effect reported by Barstow *et al.* (1993b) is clearly evident, but with a larger group of objects more detail can be seen, particularly in the 40 000–50 000 K range which was sparsely populated in the earlier work. In this region, the majority of objects show flux decrements,

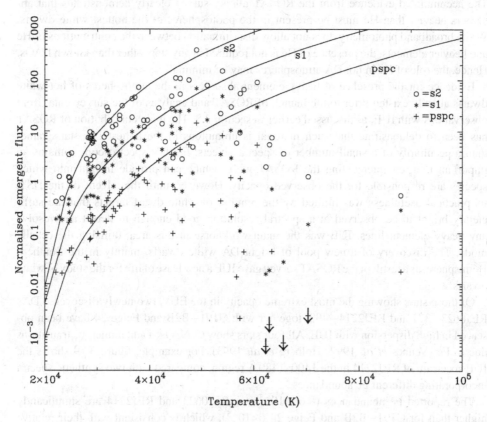

Fig. 3.18. Normalised emergent EUV/X-ray fluxes as a function of temperature for S2, S1 and PSPC bands as indicated. The curves correspond to the predictions of a pure H atmosphere and H I column of 5×10^{18} cm^{-2} for each filter/instrument combination. Arrows indicate upper limits.

compared to the pure H prediction, indicating that some source of photospheric opacity is present. However, the fluxes of several objects other than HZ 43, already known to have a pure H atmosphere, appear to be consistent with pure H models. All stars of temperature greater than 50 000 K show flux deficits in all wavebands implying that there is significant photospheric opacity in these objects. Above $\approx$55 000 K the flux deficits become very large showing very high levels of opacity. This is likely to be the result of levitation of Fe and Fe-group elements, indicating that the mechanism is no longer efficient for these species once the temperature falls below $\approx$55 000 K. However, for a typical hot DA gravity ($\approx$7.5), the theoretical calculations predict that significant Fe should be present above $\approx$40 000 K. Another difficulty for the radiative levitation theory is that there is a large spread in the observed opacity for stars of similar temperature and gravity. For example, between 50 000 K and 60 000 K, the opacity ranges from a very low level, equivalent to a pure H atmosphere, to a level yielding a flux only 10^{-4} of the level expected for pure H (figure 3.18). Differences in surface gravity within the very hot DA group are not large and cannot account for such large flux differences. Certainly, there is no obvious correlation between the observed flux and gravity.

3.6.2   *The nature of the photospheric opacity in DA white dwarfs*

The accumulated evidence from the ROSAT all-sky survey clearly demonstrates that absorbers heavier than He must be present in the photospheres of the hottest white dwarfs. While broadband photometry does not allow discrimination between the contributions of He and heavier elements, the presence of He is not required in any stars other than known DAOs. Hence, the role of He in hot DA atmospheres may be minimal.

Evidence for the presence of heavy element absorbers in the atmospheres of hot white dwarfs already existed prior to the launch of ROSAT and analysis of the survey data, from observations with IUE, as discussed earlier in section 2.8. The main contribution of ROSAT has been to demonstrate that such material is ubiquitous in the hottest DA stars, rather than a peculiarity of a small number of special objects. Far-UV spectroscopy remains an important tool, complementing the ROSAT work, which can provide insight into which species are responsible for the observed opacity. However, until the advent of the HST, its practical usefulness was limited by the number of white dwarf targets that are sufficiently bright to be observed at a spectral resolution good enough to detect and resolve any heavy element lines. IUE was the main workhorse in this area, utilising the echelle mode. The discovery of a new pool of bright DA white dwarfs, mainly in the Southern Hemisphere, as a result of the ROSAT survey gave IUE a new lease of life for the study of white dwarfs.

Of those stars showing the most extreme opacity in the EUV, two newly discovered DAs REJ0623−371 and REJ2214−492, together with G191−B2B and Feige 24 have been observed in high dispersion with IUE. All four stars show C, N, O, Si and numerous transitions due to Fe (Vennes *et al.* 1992; Holberg *et al.* 1993). For example, figure 3.19 shows the IUE spectrum of REJ2214 in the 1400–1480 Å region, compared with two synthetic spectra incorporating different Fe abundances.

The reported Fe abundances ($\approx$6 × 10^{-5}) for REJ0623 and REJ2214 are significantly higher than for G191−B2B and Feige 24 ($\approx10^{-5}$), which is consistent with their relative EUV fluxes. More recently, Werner and Driezler (1994) and Holberg *et al.* (1994) have reported significant quantities of Ni in the UV spectra of G191−B2B and REJ2214−492.

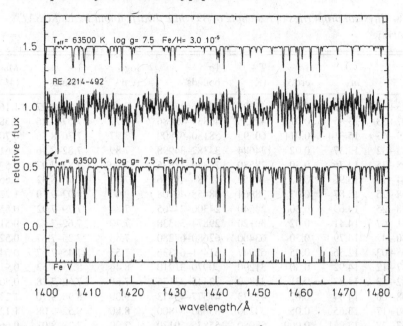

Fig. 3.19. A comparison of the observed 1400-1480 Å spectrum of REJ2214−492 (normalised at 1.0) to model spectra containing Fe abundances N(Fe)/N(H) = 3.0×10^{-5} (top) and 1.0×10^{-4} (bottom). The line at the bottom labelled 'Fe V' shows the locations and strengths of the Fe V lines expected.

The presence of all the species detected is qualitatively consistent with the predictions of radiative levitation calculations in that they should be present in stars with temperatures above 50 000 K. In the most detailed recent theoretical study, Chayer *et al.* (1995a) compared the abundances measured for these four stars with predicted values. Except for Si, which had a greater measured abundance than that expected, these theoretical predictions were as much as a factor of ten in excess of the observed values. Further improvements to the calculations (Chayer *et al.* 1995b) did not solve the problem. In addition, when synthetic EUV spectra were calculated for models incorporating the measured abundances of C, N, O, Si, Fe and Ni, the predicted flux levels were one to two orders of magnitude greater than those observed. Hence, questions also remained regarding the reliability of the model atmospheres and the resulting abundance estimates.

3.6.3 The mass distribution of the EUV selected sample of white dwarfs

The selection of white dwarfs on the basis of their EUV flux will inevitably introduce biases different to those arising from optical samples. For example, only those stars with temperatures in excess of $\approx$25 000 K, above which the EUV emission becomes significant, will be included in the list of ROSAT WFC detections. It is interesting and important to investigate any selection effects by comparing the ROSAT sample with other studies and in particular the white dwarf masses. Once $T_{\rm eff}$ and $\log g$ have been determined from the Balmer line spectra, the mass can be found from a theoretical mass–radius relation such as that of M. A. Wood (1992), which has subsequently been extended to deal with the thick ($10^{-4} M_{\odot}$) H layers appropriate for hot DA white dwarfs. Table 3.8 lists

Table 3.8. *Physical properties of a sample of 89 white dwarfs found in the ROSAT WFC survey.*

Target	m_V	1σ err.	T_{eff} (K)	1σ bounds	$\log g$ (cm s^{-2})	1σ bounds	Mass (err.) $(M_\odot)$
REJ0003+43	16.82	0.30	46205	44476–47850	8.85	8.73–8.97	1.16 (0.07)
REJ0007+33	13.85	0.01	47936	46839–49086	7.77	7.67–7.86	0.58 (0.04)
REJ0029−63	15.314	0.011	60595	58130–63800	7.97	7.76–8.16	0.70 (0.10)
REJ0053−32	13.37	0.02	34684	34382–34998	7.89	7.82–7.96	0.61 (0.03)
REJ0108−35	14.76	0.30	28580	28320–28835	7.90	7.83–7.97	0.59 (0.03)
REJ0134−16	13.96	0.03	44850	42786–48296	7.96	7.70–8.32	0.66 (0.17)
REJ0138+25	15.87	0.03	38964	38458–40708	9.00	8.92–9.10	1.22 (0.05)
REJ0148−25	14.69	0.30	24540	24300–24765	7.84	7.79–7.92	0.55 (0.02)
REJ0151+67	14.41	0.02	30120	29874–30338	7.70	7.62–7.77	0.51 (0.03)
REJ0230−47	14.79	0.30	63400	62680–67280	7.43	7.25–7.67	0.52 (0.07)
REJ0235+03	12.56	0.05	62947	61481–64336	7.53	7.45–7.62	0.54 (0.03)
REJ0237−12	14.92	0.30	31290	30970–31710	8.44	8.35–8.53	0.91 (0.05)
REJ0322−53	14.83	0.3	32860	32445–33135	7.66	7.62–7.78	0.50 (0.05)
REJ0348−00	14.04	0.02	42373	41577–43473	9.00	8.90–9.10	1.23 (0.05)
REJ0350+17	13.65	0.05	34200	33600–34800	8.80	8.52–9.08	1.12 (0.17)
REJ0457−28	13.951	0.009	58080	55875–60170	7.90	7.78–8.07	0.66 (0.07)
REJ0505+52	11.73	0.01	57340	56040–58670	7.48	7.34–7.58	0.51 (0.04)
REJ0512−41	17.257	0.013	53960	51400–58435	7.62	7.33–7.86	0.55 (0.10)
REJ0521−10	15.815	0.028	31770	31090–32160	8.70	8.61–8.82	1.07 (0.06)
REJ0550+00	14.773	0.024	45748	44360–47405	7.79	7.66–7.90	0.59 (0.05)
REJ0550−24	16.371	0.012	51870	48750–56400	7.29	7.02–7.59	0.44 (0.07)
REJ0552+15	13.032	0.009	32008	31962–32287	7.70	7.63–7.73	0.51 (0.03)
REJ0558−37	14.369	0.006	70275	66400–78780	7.37	7.13–7.50	0.52 (0.07)
REJ0605−48	15.825	0.017	33040	32260–34370	7.80	7.57–8.26	0.56 (0.20)
REJ0623−37	12.089	0.001	62280	60590–64140	7.22	7.13–7.40	0.46 (0.05)
REJ0632−05	15.537	0.015	41765	40330–44420	8.51	8.35–8.65	0.97 (0.07)
REJ0645−16	8.35	0.1	24700	23700–25700	8.65	8.35–8.95	1.03 (0.20)
REJ0654−02	14.82	0.03	32280	31740–32840	8.34	8.23–8.45	0.85 (0.07)
REJ0715−70	14.178	0.015	44300	43310–45445	7.69	7.62–7.80	0.54 (0.04)
REJ0720−31	14.87	0.04	53630	52400–54525	7.64	7.54–7.74	0.55 (0.03)
REJ0723−27	14.60	0.30	37120	36390–37860	7.75	7.65–7.85	0.55 (0.04)
REJ0827+28	14.22	0.03	51934	48457–56007	8.00	7.73–8.22	0.70 (0.13)
REJ0831−53	14.65	0.30	29330	28970–29590	7.79	7.75–7.89	0.54 (0.05)
REJ0841+03	14.475	0.028	36605	36180–37015	7.69	7.64–7.76	0.52 (0.03)
REJ0902−04	13.190	0.020	23310	22560–24120	7.74	7.64–7.83	0.50 (0.04)
REJ0907+50	16.54	0.2	33459	32882–34239	7.86	7.70–7.99	0.59 (0.08)
REJ1016−05	14.21	0.05	53827	52629–55931	8.08	7.98–8.17	0.65 (0.05)
REJ1019−14	14.93	0.30	31340	30895–31680	7.79	7.70–7.89	0.55 (0.04)
REJ1024−30	16.67	0.30	36610	35780–38360	8.69	8.50–8.80	1.06 (0.10)
REJ1029+45	16.13	0.03	34224	33889–34740	7.85	7.76–7.93	0.58 (0.04)
REJ1032+53	14.455	0.020	44980	44210–45310	7.68	7.64–7.77	0.54 (0.03)
REJ1033−11	13.012	0.009	22790	22570–23030	7.57	7.53–7.60	0.43 (0.01)
REJ1036+46	14.34	0.05	28766	28349–29236	7.92	7.79–8.06	0.61 (0.07)
REJ1043+49	16.234	0.028	47560	46550–48630	7.62	7.52–7.70	0.53 (0.03)
REJ1044+57	14.64	0.05	29016	28750–29304	7.79	7.71–7.87	0.54 (0.04)
REJ1058−38	13.78	0.30	27970	27680–28200	7.88	7.81–7.95	0.58 (0.04)

Table 3.8. (*cont.*)

Target	m_V	1σ err.	T_{eff} (K)	1σ bounds	$\log g$ (cm s^{-2})	1σ bounds	Mass (err.) ($M_\odot$)
REJ1100+71	14.68	0.2	39555	38747–40503	7.66	7.54–7.78	0.52 (0.05)
REJ1112+24	15.773	0.028	38970	38620–39780	7.91	7.81–7.96	0.63 (0.05)
REJ1126+18	14.127	0.028	55640	54770–56940	7.62	7.54–7.69	0.55 (0.03)
REJ1128−02	15.73	0.30	31280	30840–31770	8.11	8.00–8.21	0.71 (0.06)
REJ1148+18	14.33	0.30	25107	24764–25489	7.81	7.73–7.87	0.54 (0.04)
REJ1236+47	14.38	0.05	55570	54540–56720	7.57	7.49–7.64	0.53 (0.03)
REJ1257+22	13.38	0.02	37880	37209–38584	7.69	7.59–7.79	0.53 (0.04)
REJ1316+29	12.99	0.05	49000	47000–51000	7.70	7.50–7.90	0.56 (0.07)
REJ1340+60	16.941	0.020	42970	42030–44240	7.68	7.57–7.84	0.54 (0.06)
REJ1431+37	15.277	0.016	34419	33582–35135	7.66	7.52–7.82	0.50 (0.06)
REJ1529+48	15.083	0.028	46230	45540–47000	7.70	7.62–7.75	0.55 (0.03)
REJ1614−08	14.013	0.028	38500	38160–38890	7.85	7.79–7.90	0.60 (0.03)
REJ1623−39	11.00	0.01	24760	24333–24805	7.92	7.88–8.09	0.60 (0.09)
REJ1638+35	14.83	0.05	36056	35479–36662	7.71	7.61–7.80	0.53 (0.04)
REJ1650+40	15.831	0.028	37850	36800–38430	7.95	7.81–8.12	0.64 (0.09)
REJ1726+58	15.45	0.05	54550	49616–58835	8.49	8.25–8.77	0.97 (0.16)
REJ1738+66	14.606	0.028	88010	85400–90400	7.79	7.70–7.93	0.65 (0.21)
REJ1746−70	16.60	0.30	51050	48410–53930	8.84	8.68–9.10	1.16 (0.12)
REJ1800+68	14.74	0.04	43701	42700–44759	7.80	7.68–7.92	0.59 (0.05)
REJ1820+58	13.949	0.028	45330	44600–46290	7.73	7.68–7.81	0.56 (0.03)
REJ1847+01	12.95	0.05	28744	28485–29017	7.72	7.65–7.79	0.51 (0.03)
REJ1847−22	13.720	0.028	31920	31540–32260	8.00	7.91–8.09	0.65 (0.05)
REJ1943+50	14.62	0.30	33500	33330–33690	7.86	7.82–7.90	0.59 (0.02)
REJ2004−56	15.05	0.30	44456	43381–45327	7.54	7.43–7.66	0.49 (0.04)
REJ2009−60	13.59	0.30	44200	43650–44800	8.14	8.03–8.22	0.76 (0.06)
REJ2013+40	14.6	0.60	47057	45680–48230	7.74	7.49–8.04	0.57 (0.12)
REJ2018−57	13.61	0.30	26579	26034–26892	7.78	7.72–7.85	0.53 (0.03)
REJ2024−42	14.74	0.30	28597	28230–28988	8.54	8.38–8.67	0.97 (0.10)
REJ2024+20	16.59	0.30	50564	49332–51810	7.96	7.83–8.06	0.67 (0.06)
REJ2029+39	13.38	0.01	22153	21749–22560	7.79	7.72–7.87	0.52 (0.04)
REJ2112+50	13.08	0.02	38866	38139–39634	7.84	7.74–7.94	0.59 (0.05)
REJ2127−22	14.66	0.30	48297	46455–49934	7.69	7.56–7.82	0.55 (0.05)
REJ2154−30	14.17	0.30	28741	28363–29152	8.18	8.08–8.26	0.75 (0.06)
REJ2156−41	15.38	0.30	49764	46513–53131	7.75	7.57–7.92	0.58 (0.07)
REJ2156−54	14.44	0.30	45860	44600–47200	7.74	7.64–7.83	0.57 (0.04)
REJ2207+25	14.58	0.30	24610	24490–24690	8.16	8.14–8.20	0.73 (0.02)
REJ2210−30	14.79	0.30	28268	27823–28640	7.60	7.53–7.72	0.46 (0.04)
REJ2214−49	11.708	0.007	65600	63790–68510	7.42	7.24–7.57	0.52 (0.05)
REJ2244−32	15.66	0.30	31692	31312–32188	8.07	7.96–8.17	0.69 (0.06)
REJ2312+10	13.09	0.01	57380	56190–58670	7.86	7.78–7.93	0.64 (0.03)
REJ2324−54	15.197	0.017	45860	44490–47320	7.73	7.61–7.82	0.56 (0.05)
REJ2334−47	13.441	0.007	56682	54883–59500	7.64	7.57–7.78	0.56 (0.05)
REJ2353−24	15.444	0.014	29120	28720–29620	8.14	8.01–8.21	0.73 (0.08)

the physical properties of the 89 brightest white dwarfs in the WFC survey (Marsh *et al.* 1997a).

Currently, the largest optically selected sample of white dwarfs is that studied by Bergeron *et al.* (1992, BSL), containing 129 objects. Recently, their spectroscopic masses have been redetermined using the latest evolutionary models of Wood (1992), yielding a mean of $0.590 \pm 0.134 M_\odot$ (Bergeron *et al.* 1995). By comparison the mean mass of $0.628 \pm 0.162 M_\odot$ for the ROSAT stars is somewhat greater. A detailed comparison of the mass distributions (figure 3.20), conducted by Marsh *et al.* (1997a), shows some striking differences which are hidden in the modest discrepancy between the mean masses. It can be seen that the distributions have similar well defined peaks with more than 50% of stars lying in the range 0.5–$0.6 M_\odot$. The ROSAT distribution has an interesting but modest enhancement in the number of stars with masses in the range 0.65–$0.75 M_\odot$ by a factor 4. At larger

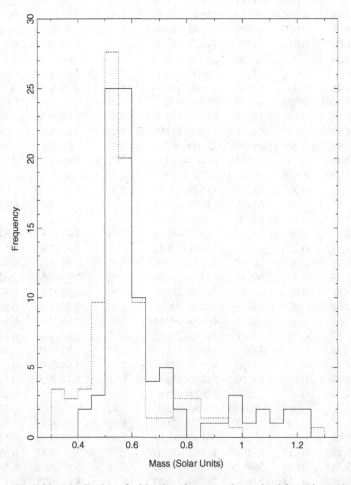

Fig. 3.20. Distribution of white dwarf masses, determined from the evolutionary models of Wood (1992), for the 89 stars in the EUV selected sample (solid histogram) and the 129 objects in that of BSL (dotted histogram). The BSL data have been scaled by the ratio of the sample sizes for ease of comparison.

masses, $\approx 1\ M_\odot$, there is a large excess of stars compared to BSL at a significance level of 99.7%.

Flux limited optical samples are biased against high mass white dwarfs because such stars have smaller radii and, therefore, occupy a smaller volume of interstellar space than less massive white dwarfs of the same temperature. If the EUV luminosity were mass independent then the same bias would also apply, in addition to the obvious EUV sampling bias due to the increase of ISM opacity with distance. Consequently, the excess of high mass white dwarfs in the ROSAT sample implies that high mass white dwarfs are significantly more EUV luminous than low mass white dwarfs of the same temperature. It would appear that the deficit in flux associated with the smaller radii of the high mass objects is more than offset by an increase in their intrinsic EUV luminosity. Hence, the higher mass, higher surface gravity stars are more luminous than their less massive counterparts. This is consistent with the idea that lower abundances of heavy elements can be supported by radiative levitation.

The high mass tail of a white dwarf mass distribution derived from single star evolution is expected to be monotonically decreasing, if it is the outcome of a reasonably constant star formation rate. In the ROSAT sample the high mass tail is clearly not monotonic and a secondary peak near $1.0 M_\odot$ is evident. Indeed, one of the stars in the high mass peak, GD50, has already been suggested as a possible merged double degenerate system (Vennes *et al.* 1996a) on the basis of its large mass and suggested high rotation rate. It seems reasonable to suggest that the ROSAT peak may result from a population of double degenerate mergers but there are presently no detailed calculations predicting the expected number or mean mass of white dwarfs that might arise from double degenerate mergers. Finley *et al.* (1997) examined a larger sample drawn from both the WFC and EUVE sky surveys, together with a significant group of optically selected objects, and they also found a high mass excess. However, they propose that this arises from a relative deficit of stars with more typical masses, due to interstellar absorption eliminating many of the more distant EUV sources, rather than an anomalously high number of more massive objects.

3.7   Hidden white dwarfs in binary systems

The overwhelming majority of known white dwarfs are isolated stars. This is perhaps not surprising, as they have mostly been identified through optical blue star colour and proper motion surveys (e.g. Green *et al.* 1986b). Thus there exists an inherent bias against the detection of white dwarfs in unresolved binary systems, particularly if there is no active mass transfer which might otherwise draw the observer's attention, as in the cataclysmic variables. For example, the spectrum of any companion star of type K or earlier will completely dominate that of the white dwarf, rendering it undetectable at optical wavelengths. Even if the companion is an M star, the composite spectrum is likely to redden the colours, thereby preventing the white dwarf appearing as a blue object. Indeed, were it not for its proximity to the Earth (2.64 pc), the Sirius binary system could not be resolved, and the visible light from Sirius B, the white dwarf component, would then be completely hidden. Several unresolved binaries with hot white dwarf companions have been discovered serendipitously as a result of detection of eclipses (e.g. V471 Tau; Nelson and Young 1970) and the spectral signatures of others have been found in the IUE spectra of a number of stars – ζ Cap (Böhm-Vitense 1980), 56 Peg (Schindler *et al.* 1982; Stencel *et al.* 1984), $4o^1$ Ori (Johnson and Ake 1986) and HD27483 (Böhm-Vitense 1993). However, a systematic search of the then and existing IUE archive by Shipman and Geczi (1989) revealed no further white dwarf companions.

It was evident even in the very early days of the ROSAT survey that a number of EUV and X-ray sources associated with apparently normal main sequence stars were anomalously luminous compared to what might have been expected from the known emission mechanisms. This was particularly striking in the case of a number of A star detections, since A stars were not known to be significant emitters of X-rays and were not expected to be EUV sources either. Comparison of the EUV and X-ray properties of white dwarfs with normal stars shows that the former also have very distinctive soft spectra, when the interstellar column density is relatively low. Many of these 'anomalous' sources showed this same characteristic, indicating that the emission was probably from a hidden white dwarf companion. However, confirmation of the nature of the object requires further spectroscopic evidence. For white dwarfs with companions later than K, their binary nature is evident in composite optical spectra, showing the broad H Balmer series typical of a DA white dwarf and TiO bands from the M star companion (e.g. figure 3.9). For companions of earlier spectral type, the white dwarf component cannot be discerned in the optical spectrum. However it remains possible to discriminate between the two stars using far-UV spectra (e.g. figure 3.21). Several binary systems with white dwarf components were quickly identified in this way, using IUE

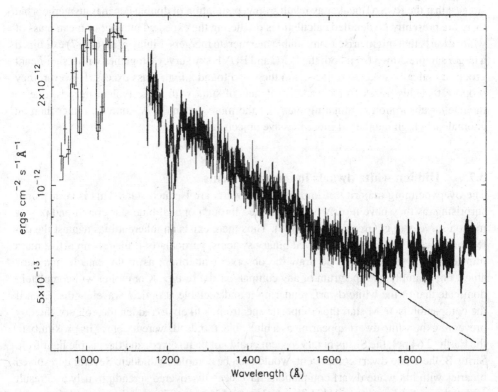

Fig. 3.21. Far-UV spectrum of the F star + DA white dwarf binary HD33959C recorded with IUE and Voyager 2. The strong blue continuum and Lyman aborption lines of the white dwarf dominate at short wavelengths but the F star component can be clearly seen longward of 1600 Å.

including $-\beta$ Crt (Fleming *et al.* 1991), HD33959C (Hodgkin *et al.* 1993), HD15638, HR1608 and HR8210 (variously Barstow *et al.* 1993a; Landsman *et al.* 1993; Wonnacott *et al.* 1993).

A first systematic search for binary systems containing white dwarfs was conducted by Barstow *et al.* (1994a), concentrating just on systems with A–K companions. This revealed a rather disparate group of nine systems with a range of properties, which are included in the more extensive compilation of Burleigh (1999; see table 3.9). Binary periods range from a few days to several years. Magnetic activity in the primary (inferred from the detection of coronal X-ray emission), undetectable in some systems, is reminiscent of young, rapid rotators in others. Furthermore, some of these binaries are likely to have had an episode of common-envelope evolution in the past. The fact that the active primaries are associated with white dwarfs places a lower limit on their ages which makes it unlikely that they are young ($<10^9$ y). They may, therefore, be examples of older stars whose rotational velocities have been reinforced by tidal interaction with their companions and it is tempting to speculate that some of these systems might have been RS CVn binaries, whose primaries have now evolved into white dwarfs.

Tidal effects will only be significant in short period (a few days or less) binary systems, yet a few wider systems also show unusually high levels of coronal activity. Jeffries and Stevens (1996) have proposed that in these, the main sequence primary has been spun up through the accretion of the wind of the white dwarf progenitor during its red giant phase. BD+08°102 is the prototype of this group, christened WIRRing stars (Wind-accretion Induced Rapid Rotators).

The initial work of Barstow *et al.* (1994a) dealt with the most obvious potential binaries in the EUV survey data, chosen mainly on the basis of their X-ray and EUV colours. Since approximately 30% of the isolated white dwarfs have similar colours a simple scaling of this population of new binaries indicates that a further 10–20 white dwarfs with A–K companions might be present in the EUV source catalogue, but these will not necessarily have immediately remarkable properties to attract attention. Indeed, individual systems have continued to be discovered intermittently – HD18131 (Vennes *et al.* 1996), REJ1027+322 (Genova *et al.* 1995), REJ0357+283 (Burleigh *et al.* 1997) and MS0354.6–3650 (D. J. Christian *et al.* 1996). Burleigh *et al.* (1997) considered the problem in a systematic way, examining 11 new candidate systems (including HD18131, REJ1027 and REJ0357) but white dwarfs were not found in six objects observed, indicating that not many binaries now remain to be discovered. Burleigh (1999) reviews the properties of the complete sample of EUV-selected binary systems with primary spectral types earlier than M (see table 3.9), while Vennes *et al.* (1998) present follow-up data dealing with the physical parameters of the white dwarf components and the orbital parameters of the binaries.

If WD+dM systems are considered in conjunction with those white dwarfs with A–K companions there are ≈30 white dwarfs residing in non-interacting binary systems in the EUV sample, corresponding to $\approx25\%$ of the total number of white dwarfs detected. Hence, these objects represent an important, previously unaccounted for, contribution to the total space density of white dwarfs. Even so, as between 50 and 80% of all stars are believed to lie in binary or multiple systems, it seems probable that there might be more binaries. Vennes and Thorstensen (1994) suggest that faint, low mass red dwarf companions of spectral type M6 or later will not be seen with current optical techniques. Perhaps five to ten such binaries with a hot ($T_{eff} > 40\,000$ K) white dwarf component may await discovery in the presently known EUV source population.

Table 3.9. *Sirius-type binaries discovered with ROSAT/EUVE and IUE.*

Name	Primary	D (pc)	$\log g$	T_{eff} (K)	Mass $M_\odot$	Sep.[b] ('')	Sep.[b] (AU)	Comment
β Crt	AIII	77–87[a]	~7.5	~31 000	≤0.44			Unlikely to be a close binary
14 Aur C	F4V	74–93[a]	~8.0	~46 000	0.53–0.69	2.01	149–187	Third, wide member of a triple system
HD15638	F3–6IV	172–237[a]	7.0–7.5	44–48 000	>0.44			Period = 903 ± 5 days. Active primary
HR1608	K0IV	52–58[a]	8.0–8.1	27–29 000	0.51–0.67			Close binary ($P = 21.7$ d)
IK Peg	A8V	44–48[a]	~9.0	~35 600	>1.17			Min M_{WD} (astrometric) = 1.17 $M_\odot$
HD223816	G0V	115	8.2–8.5	~70 000	≥0.60	0.58	33–91	EUVE data suggest heavy elements
HD217411	G5V	87	≥8.2	≥35 500	≥0.68			
RE J1925−56	G5–8V	100–120	8.1–8.2	~51 000	0.70–0.76	0.21	21–25	Rapid rotator primary (WIRRing star) Barium giant progenitor system?
BD+08°102	K1–3V	50–60	~8.5	~29 000	~0.90			Active primary
HD18131	K0IV	92–120[a]	<7.8	<34 000	~0.50			Active primary
HD2133	F7/8V	121–165[a]	~7.75	~27 000	~0.50			
HD27483	F6V	44–48[a]	~8.5	~22 000	~0.95	0.6	73–99	Hyades triple system (F6V+F6V+WD)
RE J1027+32	G0–4V	380–550	≤7.5	29–35 000	~0.45			Rapid rotator primary (WIRRing star) Barium giant progenitor system?
RE J0357+28	K2V	>107	≤7.9	≤31 000	≤0.60			Rapidly rotating primary
BD+27°1888	A8–F2V	185–218	~7.25	~34 000	~0.40			
MS 0354−36	G2V	320–410	~8.0	~52 000	~0.70			Active primary
RE J0702+12	K0IV/V	91–145	7.5–8.5	32–38 000	0.4–0.9			Lies near β CMa hole in LISM
RE J0500−36	F6/7V	500–1000	~7.25	~38 000	~0.40			First confirmed B star/WD binary
y Pup	B5V	157–187[a]	≥8.5	≥43 000	0.93±0.15			B5V star M=6.7±1.8 $M_\odot$
θ Hya	B9.5V	38–41[a]	≥8.5	~28 000	≥0.62±00.09			
16 Dra	B9.5V	115–131[a]	≥8.0	29–35 000	≥0.69			PM companion 17 Dra is B9V

[a] Distance derived from *Hipparcos* parallax.
[b] Resolved through HST/WFPC2 imaging.
Source: Main references: Barstow *et al.* 1994a; Burleigh 1999; Vennes *et al.* 1998.

3.8   EUV emission from late-type stars

Late-type stars constitute the largest single class of objects in the WFC EUV source catalogue. With EUV luminosities in the range $\approx 10^{27} - 10^{28}$ ergs s^{-1} (the EUV luminosity of the Sun is $\approx 10^{27}$ ergs s^{-1}), one to two orders of magnitude below those of typical white dwarfs, this source population is expected to occupy a smaller volume within 10 to 25 pc of the Sun. Hence, most of these objects should be found in the third catalogue of nearby stars (CNS3; Gleise and Jahreiss 1991), the most complete listing of stars within 25 pc of the Sun. This is illustrated in the work of Warwick *et al.* (1993), who show the number of detections of white dwarfs and late-type stars, the latter subdivided by class, as a function of distance (see figure 3.22). The peak distance for white dwarfs lies at 60–80 pc, while most late-type star classes have maxima at ≈ 20 pc. G stars appear to be visible out to greater distances (50 pc) probably because of the population of RS CVn binaries, which are more active than isolated stars and which generally have a component of this spectral type.

In sampling the EUV emission from mainly nearby stars, the WFC survey is examining a population with relatively well determined distances and for which observed count rates can be simply converted into luminosities. The EUV luminosity of an individual star is a measure of its level of coronal activity. Hence, the WFC sample then provides the information which defines the activity levels in specific spectral classes and for examining evolution of this activity and, as a consequence, the rotation/convection driven magnetic dynamo which is thought to power the coronae.

3.8.1   EUV luminosity functions and the nearby stellar sample

Two complementary approaches have been used to study the general properties and luminosity of the EUV selected nearby stars. B. E. Wood *et al.* (1994) considered a volume limited sample, within 10 pc. By comparison Hodgkin and Pye (1994) used the CNS3 as a reference and studied the EUV sources associated with each spectral type out to the distances for which CNS3 is believed to be complete, encompassing a larger volume than that of Wood *et al.* (1994).

Of the 220 known star systems within 10 pc, Wood *et al.* find that only $\approx 25\%$ (41 objects) are detected by the WFC, but the number is large enough to construct EUV luminosity functions for stars of different spectral types and compare these with X-ray luminosity functions derived from *Einstein* IPC data. The M stars appear to be less luminous in the EUV, compared to the FGK spectral types, than is the case in X-rays. Late M stars are found to be significantly less luminous in the EUV than other late-type stars. The explanation proposed by Wood *et al.* is that of coronal saturation, that later-type stars can emit only a limited fraction of their total luminosity in the EUV or X-ray bands. Hence, stars with very low bolometric luminosities must also have very low EUV and X-ray luminosities.

Both EUV and X-ray luminosities are well correlated with stellar rotational velocity (figure 3.23). From this, it also seems that most stars in binary systems do not emit EUV radiation at higher levels than single stars. Those few binary systems (e.g. AT Mic, ζ UMa, FL Aqr, Wolf 630, ζ Boo and EQ Peg) which do have high EUV luminosity have a larger ratio of X-ray to EUV emission, indicating that they have hotter coronae.

Extending the sample out to the nominal 25 pc distance limit for the CNS3 catalogue (10 pc for M dwarfs) increases the number of EUV sources that may be included in a study of the luminosity functions. Hodgkin and Pye (1994) have measured count rates and determined upper limits for all the 3803 CNS3 catalogue entries, but using a detection threshold lower than

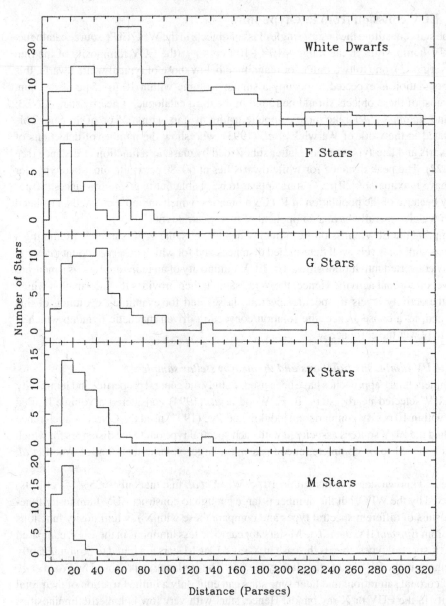

Fig. 3.22. The distribution of known distances for white dwarfs and active late-type stars in the Bright Source Catalogue. The late-type stars have been subdivided by spectral class (F, G, K and M; figure from Warwick *et al.* 1993).

used for the Bright Source Catalogue. Hence, a factor of two more CNS3 stars (159 compared with 78) were detected in this analysis, but this still represents only a small fraction of the complete catalogue. Table 3.10 lists the detections from the CNS3 catalogue where: (1) star name (Gl = Gliese (1969), Wo = Woolley *et al.* (1970) number, NN = star not yet assigned a number in CNS3, otherwise an additional identifier is used if known); (2) component (for

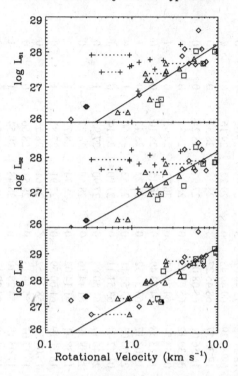

Fig. 3.23. Activity–rotation relations for the WFC S1 and S2 EUV luminosities together with the Einstein IPC luminosity. The boxes, triangles and diamonds represent FG stars, K stars and M stars, respectively. The plus signs represent upper limits. For binaries the luminosity was divided equally between the two components and the stars are connected by a dashed line in the figure. The solid lines represent a linear least squares fit to the data (figure from Wood *et al.* 1994).

multiple systems); (3) distance (in pc) obtained from resultant parallax; (4) spectral type; (5) apparent V magnitude; (6) $B-V$ colour index (in some cases $B-V=0.0$ indicates that the colour is not available); (7–8) equatorial coordinates (equinox J2000, epoch 1990.0); (9) spectroscopic-binary flag (3 = SB, 2 = SB?, 1 = variable radial velocity, 0 = no information); (10–11) EUV luminosity for the S1 and S2 bandpasses; (12–13) detection flags for S1 and S2 bands (0, detection; –1, upper limit).

Cumulative EUV luminosity functions were determined, utilising both detections and upper limits, employing a 'survival analysis' technique. The total sample was divided into subgroups of F, G, K and M spectral types and further into single stars and binaries (figures 3.24 and 3.25 respectively). The subsequent luminosity functions can be described by simple power laws. The spectroscopic binaries typically have flatter slope and higher mean EUV luminosity than the single stars. A high luminosity tail is apparent in the G star function, which can be attributed almost entirely to spectroscopic binaries, most of which are RS CVn systems. While the mean and median EUV luminosities of the F, G and K stars broadly match the X-ray luminosity functions obtained by *Einstein*, the M star luminosities are about a factor of ten lower. This probably arises from the influence of a high luminosity M dwarf tail in the X-ray function, with more recent studies (Barbera *et al.* 1993) giving better agreement.

Table 3.10. *ROSAT EUV detected nearby stars reproduced from Hodgkin and Pye (1994) but without the WFC count rates, which are reproduced in Table A.2 of the Appendix. Detailed notes on these stars can be found in Hodgkin and Pye (1994).*

| | | | | | | | | | L_{EUV} | | Det | |
| | | | | | | | | | S1 | S2 | | |
Name (1)	Cmpnt (2)	Distance (pc) (3)	Spectral type (4)	m_V (5)	$B-V$ (6)	α_{J2000} (7)	δ_{J2000} (8)	SB (9)	(10)	(11)	(12)	(13)
GJ 1255	AB	24.75	K0 V	8.00	0.86	20 37 21	75 35 59	3	28.94	29.11	0	0
GJ 1255	C	24.75		10.40	0.00	20 37 21	75 35 59	0	28.94	29.11	−1	−1
Gl 909	A	10.36	K3 V	6.40	0.98	23 52 24	75 32 37	3	27.94	28.02	0	0
Gl 909	B	10.36	M2	11.70	0.00	23 52 24	75 32 37	0	27.94	28.02	−1	−1
Gl 420	A	12.26	dK5	7.68	1.06	11 15 13	73 28 31	0	28.40	27.85	−1	0
Gl 420	B	12.26	M2	11.40	0.00	11 15 13	73 28 31	0	28.40	27.85	−1	0
Gl 713	AB	7.54	F7 V	3.57	0.49	18 21 02	72 44 02	3	27.68	27.63	−1	0
Gl 355.1		24.75	G4 III	4.56	0.77	09 34 29	69 49 48	0	29.22	29.16	0	0
BD+68 949		19.61	F5 V	4.80	0.43	17 36 57	68 45 27	3	28.38	28.31	0	0
Gl 648		16.95	F6 V	4.89	0.48	16 56 01	65 08 08	3	27.97	28.24	0	0
Gl 311		13.85	G1 V	5.64	0.62	08 39 12	65 01 12	0	28.34	28.31	0	0
BD+64 458		25.00	G0	7.74	0.00	04 33 54	64 38 01	0	28.71	28.93	0	−1
Gl 708.1		19.88	F5 V	5.03	0.38	18 13 53	64 23 49	0	28.15	28.10	0	0
Steph 1676		25.00	M0	10.60	1.42	19 03 17	63 59 32	0	28.22	28.45	0	−1
Wo 9809		22.73	dM0	10.82	1.42	23 06 03	63 55 11	0	28.52	28.60	−1	0
Gl 75		10.29	K0 V	5.63	0.81	01 47 44	63 51 11	0	27.90	28.14	0	−1
Gl 807		13.14	K0 IVe	3.43	0.92	20 45 17	61 50 09	0	27.90	27.93	−1	0
AC+61:27026		24.39	dK8	10.28	1.23	17 38 39	61 14 16	0	28.69	28.77	−1	0
GSC 036B-821		24.39		14.70	0.00	17 38 41	61 14 02	0	28.74	28.79	−1	−1
G227−028		12.80	DC:9	15.65	0.97	18 21 19	61 00 59	0	27.91	27.83	−1	0
Gl 609.1		20.53	F8 IV-V	4.01	0.52	16 01 54	58 33 52	3	28.61	28.55	0	0
Gl 616.2		21.05	dM1.5e	9.96	1.48	16 17 07	55 16 13	0	28.51	28.66	0	0
Wo 9584	C	23.26	M3 :	13.50	0.00	17 05 19	54 28 10	0	28.48	28.54	−1	−1
Wo 9584	A	23.26	F7 V	5.63	0.47	17 05 19	54 28 10	0	28.48	28.54	0	0
Wo 9584	B	23.25	F7 V	5.73	0.00	17 05 19	54 28 10	3	28.48	28.54	−1	−1

Name			Sp									
Gl 212		11.61	M1	9.75	1.48	05 41 31	53 29 28	0	28.04	28.31	−1	−1
Gl 211		12.09	K1 Ve	6.23	0.84	05 41 21	53 28 58	0	28.13	28.39	0	−1
BD+54 1499		22.73	K2	8.03	0.88	12 11 28	53 25 20	0	28.31	28.49	0	−1
LTT 13390		22.73	K3	8.23	0.00	12 11 27	53 25 08	0	28.33	28.58	0	−1
Gl 549	A	12.35	F7 V	4.06	0.50	14 25 12	51 51 08	0	28.49	28.41	0	0
Gl 549	B	12.35	M3	11.50	1.50	14 25 11	51 49 56	0	28.49	28.44	−1	−1
Gl 719		17.24	K6 Ve	8.10	1.22	18 33 56	51 43 12	3	28.95	28.95	0	0
Gl 765	A	18.52	F4 V	4.48	0.38	19 36 27	50 13 15	0	28.37	28.58	0	−1
Gl 765	B	18.52		13.00	0.00	19 36 27	50 13 15	0	28.37	28.58	−1	−1
Steph 598		19.23	M0	11.09	1.50	06 31 01	50 02 53	0	28.58	28.70	0	−1
Gl 380		4.69	K2 Ve	6.59	1.36	10 11 23	49 27 21	0	27.17	27.44	0	−1
Gl 275.2	A	11.56	sdM5	13.56	1.71	07 30 44	48 12 14	0	27.93	28.02	0	−1
Gl 575	A	11.75	F9 V n	5.19	0.65	15 03 48	47 39 16	0	28.80	28.82	0	0
Gl 575	B	11.75	dG2	5.96	0.00	15 03 48	47 39 16	0	28.80	28.82	−1	0
Wo 9832		20.79	G8 III	3.82	1.01	23 37 33	46 27 32	3	29.44	29.56	0	0
Gl 368		13.28	G0.5 V	5.09	0.62	09 48 35	46 01 16	3	28.21	28.41	−1	0
Gl 194	B	12.66	G0 III	0.96	0.00	05 16 41	45 59 59	0	29.58	29.60	−1	−1
Gl 194	A	12.66	G5 III	0.71	0.80	05 16 41	45 59 59	3	29.58	29.60	−1	−1
Gl 873		5.08	dM4.5e	10.26	1.61	22 46 50	44 20 07	3	28.17	28.30	0	0
G190−028		14.88	M2	11.87	1.52	23 29 23	41 28 25	0	27.92	28.62	0	0
G190−027		14.81	M3 :	12.44	1.61	23 29 22	41 28 13	0	27.92	28.65	0	−1
GJ 1124		18.48	K2 V	7.63	0.99	09 22 26	40 12 07	3	28.91	29.18	0	−1
Gl 815	A	15.13	dM3 e	10.34	1.52	21 00 05	40 04 17	3	28.30	28.49	0	0
Gl 815	B	15.13		11.90	0.00	21 00 05	40 04 17	0	28.30	28.49	−1	−1
Gl 820	B	3.46	K7 Ve	6.03	1.37	21 06 51	38 44 14	0	26.53	27.21	0	−1
Gl 820	A	3.46	K5 Ve	5.21	1.18	21 06 51	38 44 14	0	26.53	27.21	0	−1
Gl 161.1		23.26	F7 V	5.52	0.46	04 08 36	38 02 22	0	28.80	29.05	0	−1
Gl 277	B	12.05	dM4.5e	11.78	1.52	07 31 57	36 13 48	0	28.04	28.05	0	0
Gl 277	A	12.05	dM3.5e	10.58	1.47	07 31 58	36 13 12	3	28.01	28.05	0	0

Table 3.10. (cont.)

Name (1)	Cmpnt (2)	Distance (pc) (3)	Spectral type (4)	m_V (5)	B–V (6)	α_{J2000} (7)	δ_{J2000} (8)	SB (9)	L_{EUV} S1 (10)	S2 (11)	Det (12)	Det (13)
BD+36 2975		20.00	G5	7.84	0.00	17 55 25	36 11 18	3	28.79	28.91	0	0
Gl 29.1		21.19	dM0 e	10.52	1.50	00 42 48	35 32 53	3	28.77	28.91	0	−1
Gl 490	A	21.10	M0 Ve	10.68	1.44	12 57 40	35 13 32	2	28.29	28.84	0	−1
Gl 490	B	21.10	dM4 e	13.20	1.64	12 57 39	35 13 20	0	28.27	28.84	0	−1
BD+33 4828		20.41	G2 V	6.11	0.62	00 04 53	34 39 34	1	28.54	28.85	0	−1
Gl 615.2	A	22.52	F8 V	5.64	0.51	16 14 41	33 51 29	3	29.82	29.98	0	0
Gl 615.2	B	22.52	G1 V	6.72	0.00	16 14 41	33 51 29	0	29.82	29.98	−1	−1
G119–062		12.20	dM4	12.38	0.00	11 11 52	33 32 09	0	28.20	28.23	0	−1
GJ 1108	A	17.24	dM0.5e	10.05	1.35	08 08 56	32 49 15	0	28.16	28.37	0	0
GJ 1108	B	17.24	dM3 e	12.12	1.53	08 08 55	32 49 06	0	28.16	28.34	0	0
BD+32 1398		21.23	K3 V	8.77	0.96	06 46 05	32 33 20	0	28.07	28.63	0	−1
G087–004		21.23	M0.5	12.17	1.53	06 46 07	32 33 15	0	28.07	28.62	0	−1
Gl 856	B	10.99		11.60	0.00	22 23 29	32 27 33	0	28.13	28.16	−1	−1
Gl 856	A	10.99	dM0 e	11.41	1.57	22 23 29	32 27 33	0	28.13	28.16	0	0
Gl 278	A	14.64	A1 V	1.94	0.04	07 34 36	31 53 20	3	28.86	28.97	−1	−1
Gl 278	B	14.64	A m	2.85	0.00	07 34 36	31 53 20	3	28.86	28.97	−1	−1
Gl 278	C	14.64	M0.5Ve	9.07	1.49	07 34 37	31 52 08	3	28.87	28.96	0	0
Gl 274	B	16.92		12.50	0.00	07 29 07	31 46 60	0	28.31	28.53	−1	−1
Gl 274	A	16.92	F0 V	4.18	0.32	07 29 07	31 46 60	2	28.31	28.53	0	0
Gl 423	B	10.42	G0 Ve	4.80	0.00	11 18 11	31 31 48	3	28.62	28.63	−1	−1
Gl 423	A	10.42	G0 Ve	4.33	0.59	11 18 11	31 31 48	3	28.62	28.63	0	0
Gl 113.1		12.82	G9 e	6.76	0.96	02 48 43	31 06 56	3	29.09	29.08	0	0
Gl 455		20.28	M3	12.84	1.75	12 02 19	28 35 17	0	28.37	28.57	0	−1
G188–038		8.93	m	12.01	1.63	22 01 13	28 18 23	0	27.83	27.85	−1	0
Gl 171.2	B	16.37	DC8	15.80	0.65	04 36 45	27 09 51	0	28.91	29.09	−1	−1
Gl 171.2	A	16.37	dK5 ep	8.42	1.12	04 36 48	27 07 56	3	28.91	29.09	0	0

Name												
Gl 354.1	A	18.87	dG9	7.01	0.77	09 32 44	26 59 20	0	28.57	28.62	−1	0
BD+25 2191		20.83	K0 V	7.90	0.88	10 00 02	24 33 10	3	28.98	28.97	0	0
Gl 140	A	16.67	dM0	10.64	1.51	03 24 06	23 47 06	0	28.42	28.55	0	0
Gl 140	B	16.67		12.00	0.00	03 24 06	23 47 06	0	28.42	28.55	−1	−1
Gl 140	C	16.67	m	11.89	1.50	03 24 13	23 46 24	0	28.38	28.56	−1	−1
Wo 9638		24.45	K2	8.09	0.91	18 55 53	23 33 28	3	29.08	29.15	0	0
BD+22 4409		24.57	dK8	9.25	1.05	21 31 02	23 20 08	0	28.81	28.95	−1	−1
BD+21 1764		16.95	K5	9.80	1.38	08 08 13	21 06 18	0	28.05	28.69	0	−1
NN		16.95	m	11.00	0.00	08 08 13	21 06 12	0	28.05	28.71	0	−1
GJ 2079		16.95	dM0 e	10.20	1.36	10 14 20	21 04 39	0	28.42	28.73	0	−1
Wo 9520		10.92	dM0 e	10.11	1.51	15 21 52	20 58 41	0	28.25	28.32	0	−1
Gl 174		14.05	K3 V	8.00	1.10	04 41 19	20 54 09	0	28.75	28.90	0	0
Gl 222	AB	9.70	G0 V	4.40	0.59	05 54 23	20 16 33	3	28.42	28.24	0	0
Gl 896	B	6.58	dM6 e	12.40	1.65	23 31 52	19 56 15	0	28.15	28.03	−1	0
Gl 896	A	6.58	dM4 e	10.38	1.54	23 31 52	19 56 15	0	28.15	28.03	−1	0
Gl 388		4.90	M4.5Ve	9.40	1.54	10 19 37	19 52 11	0	27.77	27.87	0	0
GJ 2069	B	8.77		13.32	0.00	08 31 38	19 23 57	0	27.60	28.23	0	−1
GJ 2069	A	8.77	M5 e	11.89	0.00	08 31 38	19 23 45	0	27.61	28.21	0	−1
Wo 9652	A	20.88	M3	11.55	0.00	19 14 40	19 18 57	0	28.43	28.78	0	−1
Wo 9652	B	20.88	M3.5	13.27	0.00	19 14 40	19 18 15	0	28.47	28.72	0	−1
Gl 567		11.51	K2 V	6.02	0.84	14 53 24	19 09 09	3	27.95	28.09	0	0
Gl 566	A	6.71	G8 Ve	4.70	0.73	14 51 23	19 06 02	0	27.67	27.85	0	0
Gl 566	B	6.71	K4 Ve	6.97	1.16	14 51 23	19 06 02	0	27.67	27.85	0	0
Gl 233	AB	15.01	K2 V e	6.76	0.94	06 26 10	18 45 23	3	28.54	28.70	0	0
Gl 501	B	19.34	F5 V	5.17	0.00	13 09 59	17 31 45	0	28.79	28.60	−1	−1
Gl 501	A	19.34	F5 V	4.98	0.45	13 09 59	17 31 45	3	28.79	28.60	−1	−1
Gl 202		15.34	F8 Ve	4.99	0.53	05 24 25	17 22 59	0	28.56	28.55	0	0
Gl 505	B	11.91	M1 V	9.60	0.00	13 16 51	17 01 02	0	27.97	28.08	−1	−1
Gl 505	A	11.91	K1 V	6.59	0.94	13 16 51	17 01 02	0	27.97	28.08	0	−1

Table 3.10. (cont.)

Name (1)	Cmpnt (2)	Distance (pc) (3)	Spectral type (4)	m_V (5)	B–V (6)	α_{J2000} (7)	δ_{J2000} (8)	SB (9)	L_{EUV} S1 (10)	L_{EUV} S2 (11)	Det (12)	Det (13)
Gl 494		11.12	dM1.5e	9.75	1.47	13 00 47	12 22 33	0	28.31	28.29	0	0
Wo 9207		23.04	F5 IV-V	5.04	0.42	06 16 26	12 16 20	0	28.47	29.05	0	-1
Gl 208		11.62	dM0	8.80	1.40	05 36 31	11 19 40	0	28.00	28.20	0	0
LP 476-207		7.04	dM3	11.47	1.52	05 01 59	09 59 00	0	27.82	27.99	0	0
Gl 206		14.11	dM4 e	11.52	1.63	05 32 15	09 49 18	3	28.42	28.52	0	-1
Gl 504		13.48	G0 V	5.20	0.58	13 16 47	09 25 26	0	28.47	28.59	-1	0
G041-014		4.46	k	10.89	1.67	08 58 56	08 28 30	0	27.54	27.68	0	0
Gl 735		11.29	dM3 e	10.11	1.53	18 55 28	08 24 10	3	28.03	28.34	0	-1
BD+ 6 211		15.15	G5	7.33	0.00	01 22 57	07 25 07	3	29.02	29.02	0	0
Gl 406		2.39	M6	13.45	2.00	10 56 32	07 01 21	0	26.77	26.90	-1	0
Gl 178		7.51	F6 V	3.19	0.45	04 49 50	06 57 38	0	28.17	28.27	0	0
Gl 35		4.33	DZ7	12.38	0.55	00 49 09	05 23 44	0	27.05	27.31	0	-1
Gl 280	B	3.50	DA	10.70	0.00	07 39 18	05 13 41	0	27.91	28.11	-1	-1
Gl 280	A	3.50	F5 IV-V	0.38	0.42	07 39 18	05 13 41	3	27.91	28.11	0	0
Wolf 1494		16.39	m	14.34	0.00	13 48 49	04 06 05	0	28.14	28.18	0	-1
Gl 285		6.21	dM4.5e	11.20	1.60	07 44 40	03 33 13	0	27.88	28.02	0	0
Gl 106.1	C	23.81	K5	10.16	1.36	02 42 30	03 22 32	0	29.16	29.04	0	-1
Gl 137		9.57	G5 Ve	4.82	0.68	03 19 21	03 22 13	2	28.27	28.45	0	0
Gl 748		10.06	dM4	11.10	1.51	19 12 13	02 53 22	0	27.94	28.10	0	0
G099-049		5.37	M4	11.33	1.68	06 00 04	02 42 21	0	27.29	27.53	0	0
Gl 702	A	5.02	K0 Ve	4.21	0.86	18 05 28	02 30 10	3	28.02	27.63	-1	0
Gl 702	B	5.02	K5 Ve	6.00	1.15	18 05 28	02 30 10	3	28.02	27.63	-1	0
Gl 207.1		15.08	dM2.5e	11.53	1.57	05 33 44	01 56 41	0	28.22	28.25	0	-1
Gl 182		16.39	dM0.5	10.09	1.39	04 59 35	01 46 56	0	28.32	28.86	0	-1
Gl 449		10.12	F9 V	3.61	0.55	11 50 41	01 45 56	2	27.83	28.18	0	-1
G045-011		16.67	m	13.85	0.00	10 52 04	00 32 40	0	28.30	28.56	-1	0

Name			Sp			RA	Dec					
Gl 402.1		21.37	dK8	10.20	0.90	10 52 03	00 09 37	0	28.43	28.56	0	−1
Gl 159		18.12	F6 V	5.37	0.50	04 02 37	00 16 06	1	28.93	28.98	0	0
Gl 157	A	14.71	K4 V	8.04	1.11	03 57 29	−01 09 31	0	28.35	28.57	−1	−1
Gl 157	B	14.71	dM3 e	11.61	1.47	03 57 29	−01 09 31	3	28.35	28.57	0	−1
Gl 482	A	10.13	F0 V	3.46	0.36	12 41 40	−01 26 56	1	27.93	28.18	0	0
Gl 482	B	10.13	F0 V	3.52	0.00	12 41 40	−01 26 56	0	27.93	28.18	0	0
Steph 497		14.71	M1	10.59	1.45	04 37 37	−02 29 32	0	28.40	28.88	0	−1
Gl 234	A	4.13	M4.5	11.13	1.71	06 29 23	−02 48 43	0	27.29	27.57	0	−1
Gl 234	B	4.13		14.60	0.00	06 29 23	−02 48 43	0	27.29	27.57	−1	−1
Gl 23	B	16.29	G1 V	6.40	0.00	00 35 14	−03 35 36	3	28.72	28.20	−1	0
Gl 23	A	16.29	F6 V	5.65	0.57	00 35 14	−03 35 36	3	28.72	28.20	−1	0
Wo 9721	A	22.42	dM2	9.44	1.13	21 08 45	−04 25 36	0	28.52	28.83	0	−1
Wo 9721	B	22.42		13.40	1.63	21 08 45	−04 25 36	0	28.52	28.83	−1	−1
G159−003		10.00	m	14.60	0.00	01 51 04	−06 07 08	0	28.01	27.95	0	−1
Gl 166	A	4.83	K1 Ve	4.43	0.82	04 15 18	−07 38 38	0	27.59	27.52	−1	−1
Gl 166	B	4.83	DA4	9.52	0.03	04 15 24	−07 38 57	0	27.52	27.69	−1	−1
Gl 166	C	4.83	dM4.5e	11.17	1.67	04 15 24	−07 38 57	0	27.52	27.69	−1	−1
BD− 8 801		23.81	G2 IV-V	7.04	0.00	04 09 41	−07 53 37	3	29.46	29.60	0	0
Gl 643		5.82	sdM4	11.80	1.69	16 55 26	−08 19 13	3	28.18	28.37	−1	−1
Gl 644	B	6.50		9.90	0.00	16 55 29	−08 20 01	0	28.22	28.42	−1	−1
Gl 644	A	6.50	M3	9.69	1.57	16 55 29	−08 20 01	0	28.22	28.42	0	0
Gl 517		19.23	K5	9.31	1.21	13 34 44	−08 20 30	0	28.63	28.80	0	0
Gl 644	C	6.50	M7	16.78	1.99	16 55 36	−08 23 31	0	27.85	27.97	−1	−1
Gl 144		3.27	K2 V	3.73	0.88	03 32 56	−09 27 31	0	27.77	27.78	0	0
Gl 355		19.23	K0	7.82	0.92	09 32 26	−11 11 06	0	28.93	28.93	0	0
Gl 105.4	B	14.58	F5 V	5.60	0.00	02 39 34	−11 52 17	0	28.57	28.47	−1	−1
Gl 105.4	A	14.58	F5 V	5.50	0.44	02 39 34	−11 52 17	3	28.57	28.47	−1	−1
Gl 256		21.28	K4 V	9.15	1.16	06 58 26	−12 59 29	2	28.52	28.75	−1	0
Gl 838.5		21.88	F1 III	5.08	0.37	21 53 17	−13 33 08	0	29.00	29.01	−1	0

Table 3.10. (cont.)

Name (1)	Cmpnt (2)	Distance (pc) (3)	Spectral type (4)	m_V (5)	B–V (6)	α_{J2000} (7)	δ_{J2000} (8)	SB (9)	L_{EUV} S1 (10)	S2 (11)	Det (12)	(13)
Gl 225		13.62	F1 III	3.72	0.33	05 56 25	−14 10 04	0	28.22	28.22	−1	0
BD−14 3902		25.00	K4 V	10.40	1.28	14 14 22	−15 21 24	0	28.89	28.67	−1	0
Gl 837		14.68	A6 m	2.87	0.29	21 47 02	−16 07 35	3	28.75	28.73	0	0
Gl 244	B	2.63	DA2	8.44	−0.03	06 45 10	−16 42 48	0	29.11	29.49	0	0
Gl 244	A	2.63	A1 V	−1.43	0.00	06 45 10	−16 42 48	1	29.11	29.49	−1	−1
Gl 859	A	16.45	G3 V	6.21	0.62	22 26 35	−16 44 30	0	28.52	28.61	0	−1
Gl 859	B	16.45	G3 V	6.40	0.00	22 26 34	−16 44 31	0	28.52	28.62	0	−1
Gl 897	A	12.90	M3.5	10.95	1.51	23 32 46	−16 45 05	0	28.36	28.45	0	0
Gl 897	B	12.90		11.40	0.00	23 32 46	−16 45 05	0	28.36	28.45	−1	−1
Gl 898		12.90	K5/M0	8.60	1.28	23 32 49	−16 50 41	0	28.22	28.45	−1	−1
Gl 54.1		3.74	dM5 e	12.05	1.84	01 12 30	−17 00 05	0	27.04	27.32	0	−1
Gl 84		8.47	M3	10.19	1.51	02 05 03	−17 36 53	0	27.75	27.84	−1	0
Gl 65	B	2.63	dM5.5e	12.70	0.00	01 38 59	−17 57 07	0	26.64	26.50	0	0
Gl 65	A	2.63	dM5.5e	12.57	1.85	01 38 59	−17 57 07	0	26.63	26.49	0	0
Gl 31		18.35	K1 III	2.04	1.01	00 43 35	−17 59 09	0	29.04	29.10	−1	0
Gl 867	A	8.67	dM2 e	9.10	1.51	22 38 45	−20 37 12	3	28.38	28.46	0	0
Gl 867	B	8.67	dM4 e	11.45	1.60	22 38 45	−20 37 12	0	27.90	27.98	0	0
Steph 545		12.05	M3:	11.66	1.51	05 06 49	−21 35 05	0	28.50	28.53	−1	−1
BD−21 1074		12.05	M2	10.29	1.52	05 06 50	−21 35 05	0	28.50	28.55	0	0
GJ 1257		20.24	K5 V	9.70	1.11	20 41 42	−22 19 01	0	28.47	28.85	0	−1
Gl 216	B	8.01	K2 V	6.13	0.94	05 44 26	−22 25 14	0	27.58	27.89	0	−1
Gl 216	A	8.01	F6 V	3.58	0.47	05 44 28	−22 26 50	0	27.58	27.79	−1	−1
Gl 416.1		19.12	A2 III	4.48	0.03	11 11 40	−22 49 29	0	29.47	29.91	0	0
Gl 97		12.05	G1 V	5.20	0.60	02 22 32	−23 49 01	0	28.17	28.26	0	−1
CD−23 3577		16.39	G5 V	6.38	0.72	06 13 45	−23 51 45	0	28.49	28.63	0	0
Gl 455.3		14.14	F2 III	4.02	0.32	12 08 25	−24 43 43	3	28.27	28.69	0	−1

Gl 663	A	5.33	K1 Ve	5.07	0.85	17 15 21	−26 35 54	0	27.26	27.87	0	−1
Gl 663	B	5.33	K1 Ve	5.11	0.86	17 15 21	−26 36 00	0	27.25	27.88	0	−1
CD−27 6141		21.74	G5 V	6.87	0.69	08 59 43	−27 48 59	3	28.86	28.95	0	0
Gl 127	B	13.76	G7 V	6.70	0.00	03 12 04	−28 59 20	0	28.66	28.60	−1	−1
Gl 127	A	13.75	F7 IV	3.95	0.51	03 12 04	−28 59 20	0	28.66	28.60	−1	0
Gl 60	C	17.24	M2 V	10.40	0.00	01 35 01	−29 54 36	0	28.76	28.94	−1	−1
Gl 60	A	17.24	K3 V	7.78	0.92	01 35 01	−29 54 36	0	28.76	28.94	0	0
Gl 60	B	17.24	K4 V	8.00	0.00	01 35 01	−29 54 36	0	28.76	28.94	−1	−1
NN		11.36	M4	13.00	0.00	02 16 34	−30 58 10	0	27.56	28.16	0	−1
CD−31 909		11.36	M4	12.00	0.00	02 16 40	−30 59 22	0	27.59	28.11	0	−1
Gl 803		9.35	M0 Ve	8.81	1.42	20 45 09	−31 20 24	0	28.84	28.58	0	0
Gl 799	A	8.14	dM4.5e	10.99	1.57	20 41 51	−32 26 04	0	27.98	28.25	0	0
Gl 799	B	8.14	dM4.5e	11.00	0.00	20 41 51	−32 26 04	0	27.98	28.25	0	0
Gl 525.1		21.28	F3 IV	4.23	0.38	13 45 41	−33 02 36	3	28.65	28.80	0	−1
Gl 773.4		19.38	F8 V	5.66	0.49	20 00 20	−33 42 13	1	28.82	29.02	0	−1
Gl 886.2		19.92	F0 IV	5.11	0.29	23 03 30	−34 44 58	3	28.88	29.06	−1	0
Gl 755		20.83	G5 V	6.48	0.62	19 21 30	−34 58 57	0	28.81	28.84	0	−1
Gl 320		10.43	K1 V	6.56	0.93	08 43 18	−38 53 03	0	27.84	28.01	0	−1
Gl 620.1	A	12.82	G3/5 V	5.39	0.63	16 24 02	−39 11 36	0	28.57	29.06	−1	−1
Gl 620.1	B	12.82	DA2	11.00	−0.14	16 23 34	−39 13 44	0	29.33	29.76	0	0
Gl 370		9.69	K5 V	7.64	1.18	09 51 07	−43 30 06	0	27.95	27.91	0	−1
Gl 103		11.47	K7 Ve	8.85	1.39	02 34 22	−33 47 42	3	28.54	28.56	0	0
Gl 375		14.47	M3.5	11.27	1.55	09 58 34	−36 25 24	0	28.44	28.47	0	0
Gl 52.1		21.28	K1 V	8.85	0.94	01 06 47	−50 59 28	0	28.98	28.97	0	−1
Gl 841	A	14.62	M0	10.40	1.50	21 57 41	−50 59 45	3	28.51	28.69	0	0
Gl 81	A	18.25	G5 IV	3.70	0.85	01 55 57	−51 36 33	0	28.94	28.75	0	−1
Gl 81	B	18.25		10.70	0.00	01 55 57	−51 36 33	0	28.94	28.75	−1	−1
GJ 2036	B	10.53		12.15	1.60	04 53 30	−55 51 08	0	28.23	28.44	0	0
GJ 2036	A	10.53	M2 Ve	11.13	1.57	04 53 30	−55 51 08	0	28.23	28.44	0	−1

Table 3.10. (cont.)

Name (1)	Cmpnt (2)	Distance (pc) (3)	Spectral type (4)	m_V (5)	B–V (6)	α_{J2000} (7)	δ_{J2000} (8)	SB (9)	L_{EUV} S1 (10)	S2 (11)	Det (12)	(13)
Gl 189		13.70	F7 V	4.71	0.52	05 05 31	–57 28 21	0	28.53	28.60	0	0
GJ 1075		12.35	K7 V	9.02	1.40	05 05 47	–57 33 17	0	28.09	28.31	–1	–1
NN		20.41	m	13.40	0.00	18 52 02	–60 45 47	0	29.28	29.12	0	–1
Gl 559	A	1.33	G2 V	0.01	0.64	14 39 41	–60 50 14	0	26.66	26.95	0	0
Gl 559	B	1.33	K0 V	1.34	0.84	14 39 41	–60 50 14	0	27.00	27.29	0	0
CP–61 688		21.74	G0 V	6.34	0.62	06 38 00	–61 31 60	0	28.85	28.60	0	0
NN		21.74		8.30	0.00	06 38 00	–61 31 60	0	28.85	28.60	–1	–1
Gl 551		1.29	dM5 e	11.05	1.83	14 29 49	–62 40 56	0	26.55	26.62	0	0
Gl 865		11.12	k-m	11.48	1.61	22 38 28	–65 22 42	0	27.99	28.23	0	0
Wo 9791		22.94		14.00	0.00	22 38 25	–65 22 53	0	28.62	28.92	–1	–1
GJ 1280		22.73	K4 V	8.78	1.02	23 10 21	–68 50 24	0	28.58	28.84	0	–1

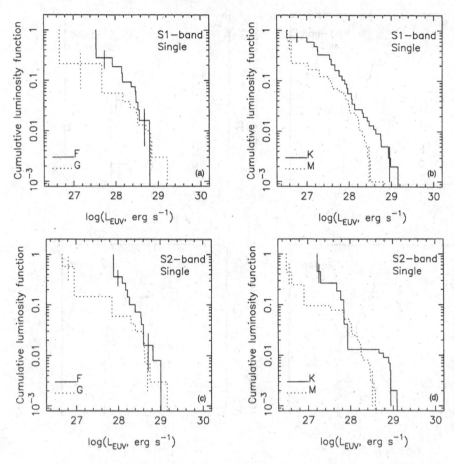

Fig. 3.24. Cumulative luminosity functions for single stars (from Hodgkin and Pye 1994).

It explains why, contrary to the pre-launch expectations of Pye and McHardy (1988) and Vedder *et al.* (1991), the late-type stellar population of EUV sources is not dominated by M dwarfs. Indeed, while the expected number of F, G and K stars agrees reasonably (within a factor of two) with those observed, the M star predictions were a factor of four to ten too high. The completeness of the WFC all-sky survey ensures that these results are statistically representative of the EUV properties of the local late-type dwarf population.

Although the Bright Source Catalogue includes part of the CNS3 sample, there are many more coronal sources that lie outside the CNS3 volume (e.g. see figure 3.22). For F, G and M stars the star counts predicted using the volume limited EUV luminosity functions are broadly in agreement with those derived for the more spatially disparate BSC sample, indicating that, at least for these spectral types, the local stellar population is representative of a flux limited sample. However, for K stars the BSC source counts lie significantly above the levels predicted by the nearby-star EUV luminosity function. This discrepancy cannot be explained by excluding the known RS CVn systems and other spectroscopic binaries. Hence, it would appear that the WFC detected a population of EUV-bright coronal sources which are not significantly represented in the solar neighbourhood. A similar excess, composed of very

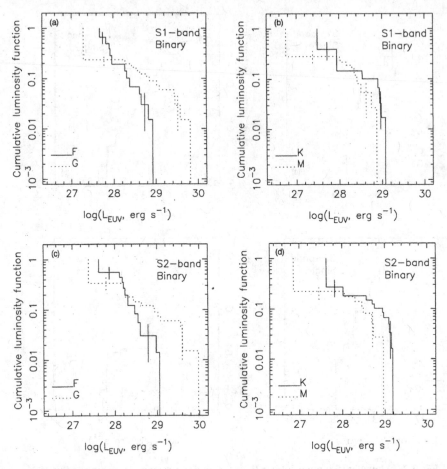

Fig. 3.25. Cumulative luminosity functions for binary stars (from Hodgkin and Pye 1994).

young active stars has also been identified in volume limited X-ray luminosity functions (Favata *et al.* 1993a,b). However, Jeffries *et al.* (1995) suggest that the WFC result is a statistical effect arising from the low space density of active binary stars.

3.8.2   *Kinematic properties of EUV selected active late-type stars*

Since the dispersion of stellar space motions increases with age, the kinematic distribution of a sample of stars can reveal information about expected levels of coronal activity. Coronal activity is related to stellar rotation rate and, therefore, age. Hence, it might be expected that kinematically young populations of late-type stars would show the greatest magnetic activity, while binary systems might remain active for a range of kinematic properties.

A second reason for examining the kinetics of active stars is to consider the coherence of their space motions. For example, in the solar vicinity there are separate groups of stars that share the same space motion as the Hyades (called the Hyades supercluster) and the U Ma cluster of stars (the Sirius supercluster), as reported by Eggen (1984) and Wilson (1990). These objects, related by birth to their associated superclusters, have ages of 0.6 Gy and

0.3 Gy respectively. Since these ages are relatively young, it might be expected that single stars belonging to these groups would be rapidly rotating. Hence they should be active, and comparatively bright EUV sources, as indeed are the stars in the Hyades. Eggen (1983) has claimed that there may be an even younger moving group associated with the space motions of the Pleiades and α Per clusters in the vicinity of the Sun.

Jeffries and Jewell (1993) have taken the sample of CNS3 stars detected by the ROSAT WFC, for which kinematical information in the form of heliocentric U, V and W velocities is available. Eliminating all objects known to be evolved, lying in close binaries or having no measured space motions yielded a group of 60 objects. They find that the velocity dispersions within the sample are significantly less than the inactive late-type stars in the solar vicinity, indicating a mean age of 1–2 Gy or possibly lower and confirming the link between activity, rotation and age. There is also evidence for a kinematic clustering in those active stars with a high EUV/bolometric luminosity ($L_{EUV}/L_{bol} > 10^{-5}$). In particular a kinematic group is coincident with the Local Association, a stream of stars found to share the properties of the young Pleiades and α Per open clusters. The membership of at least some of the identified Local Association candidates is supported by other evidence such as rotation rates, level of chromospheric and coronal activity and lithium abundance.

For individual objects, it is difficult to use the observed EUV coronal activity as a direct age diagnostic, while, in general, the Local Association candidates do have a higher level of EUV activity. For a given spectral type the range of luminosities is at least a factor of three. There is a dearth of the most coronally active F–K stars ($L_{EUV}/L_{bol} > 10^{-4}$) in this WFC sample of nearby stars, compared with the complete, unlimited volume, catalogues. It is likely that these very active stars represent the high coronal luminosity tail of a population of very young field stars, identified as Local Association kinematic group members. If their coronal activity is similar to that of the Pleiades, the numbers of Local Association F–K members within 25 pc of the Sun is unlikely to exceed 10–15. However, as many as 10^3 stars within 100 pc, detected in the ROSAT PSPC X-ray survey, will be late-type members of the Local Association, becoming an increasing fraction of the late-type stellar sources in deeper X-ray surveys.

3.8.3 Cataclysmic variables

Approximately 20 cataclysmic variables (CVs) were detected in the WFC sky survey, including a significant number (seven) of new magnetic systems. This important subclass contains an accreting magnetic white dwarf with a strong field lying in the range 1–50 MG. The field controls the accretion flow from the main sequence companion over large distances from the white dwarf companion, leading to infall of material onto one or more regions close to the magnetic pole caps. The infalling gas produces a shock above the white dwarf surface (or perhaps in the photosphere) and cools predominantly by hard X-ray radiation and optical or infrared cyclotron emission. These systems divide into two groups: the polars (from the strongly polarised optical/infrared signal), which are characterised by very strong EUV/soft X-ray components representing a significant fraction of the total system luminosity; and the intermediate polars, detected strongly in the hard X-ray band.

In the optical and IR, both types of system exhibit strong photometric modulation at the spin period of the white dwarf and strong emission line optical spectra due to photoionisation of the accretion stream. The space density of magnetic CVs is low ($\approx 10^{-6}$ pc^{-3}), so despite their distinctive characteristics, less than 30 systems had been catalogued prior to the ROSAT

Table 3.11. *Observational properties of new magnetic CVs discovered through the WFC survey (from Watson 1993).*

Object	WFC (S1)[a] (counts ks^{-1})	Mag[b]	Type[c]	Period (min)	Comments
REJ0453–42	44	19	AM	95	Not IDed in WFC prog.
REJ0531–46	16	16	AM:	135:	
REJ0751+14	20	14.5	IP	311	$P_{spin} = 13.9$ min
REJ1149+28	138	17	AM	90	
REJ1307+53	41	17–21	AM	80	
REJ1844–74	92	16	AM:	90	
REJ1938–46	387	15	AM	140	
REJ2107–05	19	15	AM	125	Eclipsing system

[a] WFC survey count rate in S1 filter.
[b] Typical *V* magnitude, taken from references in Watson (1993).
[c] AM = AM Her system. IP = Intermediate Polar system. The colon indicates that mean values/class are uncertain.

sky survey. Hence, the addition of seven new systems from the EUV source catalogue was a significant increase in the sample size. Table 3.11 lists these objects and their properties, obtained from follow-up optical observations.

Watson (1993) discusses the observed EUV to optical luminosity ratios of these seven systems and compares them with the ratios for a subset of known polars and intermediate polars drawn from the BSC, or from pointed observations. Several features are immediately apparent. The two intermediate polars, EX Hya and REJ0751+14, have much lower EUV luminosity than any of the polar systems, in keeping with earlier results that they are weak or undetectable in the soft X-ray band. The polar systems span a large range of luminosity ratios but there is a distinct tendency for the newly discovered objects to have the larger ratios. Indeed, four of the six lie well above the average value defined by the known systems, AN UMa, QQ Vul, V834 Cen and BL Hyi, which cluster near a value ≈ 1. The fact that new systems tend to have a large EUV/optical luminosity ratio should not be too surprising, since an EUV-selected sample is expected to be dominated by EUV-bright objects. Nevertheless, it does indicate that polar systems with high EUV/optical luminosity may be more common than previously anticipated.

The distribution of the orbital periods of CVs, and polar systems in particular, is an important parameter that can be used to verify evolutionary models and constrain system parameters such as the component masses. While the new CVs discovered increase the overall size of the sample, this is not enough statistically to change the orbital period distribution. Even so, the periods measured are interesting. For example, REJ1307+53 has the shortest known period (79.7 min) amongst the polars, lying very close to the theoretical minimum which is placed in the range 61.5–74.5 min (Rappaport *et al.* 1982).

Two objects have periods in the so-called 'period gap', between 2 and 3 h and well above the 'spike' in polar periods at 114 min. It is possible that REJ2107–05, with a period of 124.5 min was born with a period above the gap and has evolved across it in the usual way, but resumed accretion at a higher period because of a greater white dwarf mass (Hameury *et al.* 1988, 1991). This explanation cannot be used for REJ1938–46 because its 140 min period is too long for a system which has evolved across the gap, whatever the white dwarf mass, unless the evolutionary models are incorrect. Therefore, this object must be a rare example of a system born in the gap.

In non-magnetic cataclysmic variables, X-rays are believed to arise from a boundary layer between an accretion disc and the white dwarf, when material settles onto the surface. Provided the white dwarf is not rotating near its break-up speed, half the total accretion luminosity must be released in this transition zone. With a thin boundary layer, when the transition from the Keplerian to white dwarf rotation velocity is sudden, all the boundary layer luminosity should be emitted as X-rays. Consequently, non-magnetic systems are not promising EUV sources. Exceptions may be systems with relatively high accretion rates, such as dwarf novae in outburst or nova-like variables. In these, the boundary layer density may become sufficiently high so that it is at least partially optically thick to its own radiation. Hard X-rays are thermalised and re-radiated in the EUV. Before the launch of ROSAT, soft emission components had been observed in three systems while in outburst – all bright dwarf novae: SS Cyg, VW Hyi and U Gem (Pringle *et al.* 1987; Mason *et al.* 1988; Jones and Watson 1992). If this picture is correct, EUV emission should be significant in all high mass transfer systems with a luminosity comparable to the UV and optical emission of the disc. Van Teeseling and Verbunt (1994) show that there is a deficiency of boundary layer luminosity (the hard X-rays compared with optical and UV) which increases as the accretion rate rises, which might indicate the importance of an unobserved EUV component.

Six non-magnetic CVs were detected in the WFC survey, of which four are the dwarf novae VW Hyi, SS Cyg, Z Cam and RXJ0640−24. The remaining two are the nova-like variables IX Vel and V3885 Sgr. Fortuitously, three dwarf novae (SS Cyg, VW Hyi and Z Cam) were detected during outburst. SS Cyg was caught during a decline from maximum optical light (figure 3.26) after an ordinary outburst. Both EUV and X-ray components declined more rapidly than the optical flux. The luminosity of the EUV component was estimated to be at least an order of magnitude less than that of the disc. The hard X-ray flux remained constant throughout the EUV and optical decline (Ponman *et al.* 1995). In VW Hyi, the suppression of the hard X-ray flux during the outburst is clear in light curves recorded with both the ROSAT PSPC and Ginga, which recover from a minimum to the quiescent level at the end of the outburst (Wheatley *et al.* 1996). However, in this case the EUV light curves did not show any evidence for an EUV component replacing the hard X-rays. Again the EUV luminosity must be substantially below that of the disc. Z Cam exhibited a clear EUV enhancement at the maximum of the optical outburst with a decline more rapid than that in the optical band. Overall it appears that while the expected EUV component is present during outburst in these systems, the observed luminosity does not reach the anticipated levels.

Only the two optically brightest nova-like variables were detected during the WFC survey (Wheatley 1996). IX Vel appears to be an extremely soft source, with an S1:S2 ratio $\approx$3, similar to that found in white dwarfs. However, the PSPC data indicate that the plasma temperature is $\approx 10^7$ K (Beuermann and Thomas 1993). At this temperature the EUV flux is dominated by line emission, with the strong lines present in the S2 band mimicking the effect of a soft spectrum. A 10^7 K optically thin plasma is able to account for both WFC and PSPC observations, indicating that only a single plasma component is present. Similarly, the EUV flux of V3885 Sgr is consistent with an extrapolation of the PSPC spectrm. Hence, the WFC survey presents no evidence of a distinct EUV emitting component in these systems.

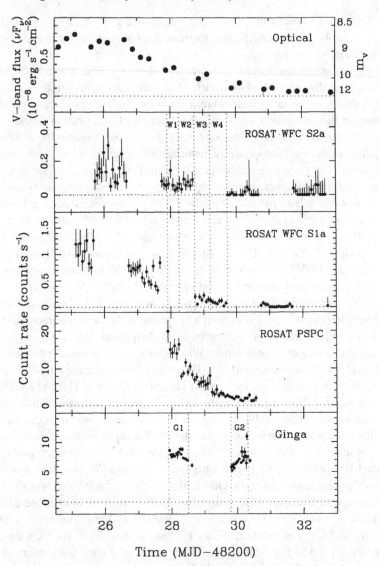

Fig. 3.26. ROSAT survey light curves of SS Cygni, including the PSPC, S1 and S2 bands. Also included are contemporaneous optical (V band) and hard X-ray (Ginga) data (from Ponman *et al.* 1995).

3.8.4   EUV emission from active galactic nuclei

It is generally accepted that the emission from active galactic nuclei (AGN) is driven principally by the accretion of material by a supermassive black hole. However, the flux seen at different wavelengths can arise from different parts of the system. At X-ray wavelengths AGN spectra can generally be modelled by a simple power law arising from regions close to the black hole. In many objects it is clear that, when longer wavelength data are considered, there is an excess flux over the the extrapolation of the power law. This so-called soft (or EUV) excess might be emitted from the inner edges of the accretion disc (e.g. Czerny and

Elvis 1987). Alternatively, the EUV emission may arise from reprocessed higher energy flux, for example, by a Comptonising atmosphere.

The possibility of detecting extragalactic sources was always expected to be limited by the opacity of the interstellar medium within our own Galaxy and the intervening intergalactic gas. With the spectral bandpass of the S1 filter limited to a lower value of $67\,\text{Å}$ by the boron component, objects with line-of-sight column densities greater than $\approx 2 \times 10^{20}\,\text{cm}^{-2}$ could not be detected. Even so, seven sources in the BSC were associated with AGN with the addition of a single further AGN and a normal galaxy in the 2RE catalogue. The AGN are split into two groups. Five are Seyfert type galaxies and the remaining three are BL Lac objects. Not surprisingly, these extragalactic sources were only detected in the S1 filter, consistent with the blocking of the S2 signal by interstellar absorption.

The Seyfert galaxies are probably detected as a result of their steep spectral components which have been found to dominate their emission at energies below 1keV (e.g. Turner and Pounds 1989). Interestingly, while X-ray observations clearly demonstrate the presence of this component in most systems, those systems detected in the EUV appear to have unusually high ratios of soft (0.15–0.5 keV) to harder (1–2 keV) fluxes, an order of magnitude larger than for a typical Seyfert galaxy. In contrast the EUV flux from the BL Lac objects appears to be consistent with that expected from an extension of the featureless power law seen over a wide spectral band. However, in the case of PKS2155−304, the EUV flux appears to be higher than predicted by an accretion disc model for the object (Gondhalekar *et al.* 1992), although the EUV and X-ray data can still be described by a single power law.

One of the Seyfert galaxies, Zwicky 159.034 (REJ1237+264), has been found to have unusual properties (Brandt *et al.* 1995). This object is a peculiar spiral galaxy and probably an outlying member of the Coma cluster of galaxies. ROSAT pointed observations taken after the survey reveal that the 0.1–2.5 keV flux had decreased by a factor of ≈ 70, similar to the amplitude of variability seen in another ultrasoft Seyfert, E1615+061. However, while the flux decreases in E1615 are associated with a decrease in the spectral slope, the spectrum of ZWG 159.034 remained steep.

3.8.5   *EUV emission from supernova remnants*

Apart from regular detections of the Moon, the only extended sources seen during the WFC survey were the Vela and Cygnus loop supernova remnants. In both cases the diffuse emission was only seen in the S1 filter, indicating that interstellar absorption is blocking the longer wavelength S2 flux. The origin of the observed flux is recombination emission lines from highly shocked regions where the supernova blast wave is encountering slow moving interstellar gas. Most of the EUV flux from Vela is confined to a D-shaped arc around a well-defined limb of radius 170 arcmin (figure 3.27). A detailed comparison between the EUV and soft X-ray emission reveals that two temperature components are required to model the observed flux, with large spatial variations in both temperature and absorption structure. For example, a typical bright knot seen in the EUV gives temperatures $\approx 10^6\,\text{K}$ and $\approx 6 \times 10^6\,\text{K}$ for the two thermal components and a typical interstellar column density of $1.8 \times 10^{20}\,\text{cm}^{-2}$, with an emission measure ratio of 20 between cooler and hotter plasma (Warwick 1994).

3.9   The interstellar medium

The role of the interstellar medium (ISM) in EUV astronomy is of great importance. Its absorbing effect on the EUV flux from astronomical sources has been extensively discussed

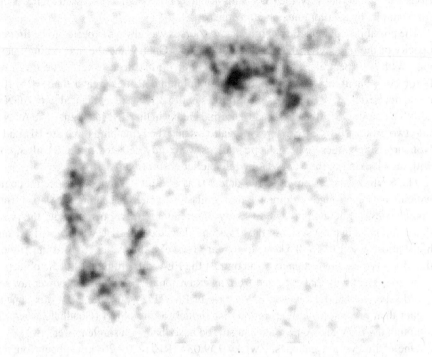

Fig. 3.27. An image of the Vela supernova remnant derived from the WFC all-sky survey data.

earlier in this book and, although this can restrict our view of these sources, considerable information on the structure of the ISM itself can be extracted. In addition, hot interstellar gas is expected to be a source of diffuse background, extending to EUV wavelengths.

3.9.1   A search for the signature of the diffuse background

In any study or search for a cosmic background component, it is necessary to understand and isolate any other possible sources of background in the instrument. Five non-cosmic components have been identified in the WFC data. A spacecraft glow is a persistent background phenomenon which shows a clear correlation with the orientation of the spacecraft and velocity vector. It is believed to be due to an interaction between the atmosphere and spacecraft surfaces and ranges from an intensity of 1 to 1000 counts s^{-1} (West *et al.* 1994). This component is well characterised and can be removed from the survey data by ignoring the affected periods. Equally intense at certain times is the background from soft ($E \approx 50\,keV$) auroral electrons, precipitated at high geomagnetic latitudes. Again, periods of contaminated data can be excluded. He II 304 Å geocoronal radiation is effectively suppressed by the WFC S1 filter. Its identification in the S2 filter shows that it is not present in the S1 band (West *et al.* 1996). Finally, the cosmic ray component can be modelled to an accuracy better than $\approx 10\%$ and the MCP dark count is constant throughout the survey and can be subtracted.

West *et al.* (1996) have reported their detailed analysis of the background components in the WFC survey data. After removal, by subtraction or discrimination, of all the spacecraft glow, auroral background and cosmic ray contributions, cleaned data covering approximately 40% of the sky coverage were available. The total exposure was $\approx 130\,\text{ks}$. The residual total count rate in the field of $2.00 \pm 0.12\,\text{counts s}^{-1}$ showed no statistically significant variation with galactic longitude or latitude and was consistent with the MCP detector background of $1.97 \pm 0.17\,\text{counts s}^{-1}$. Hence, no signal is detected from the cosmic background and a conservative upper limit of $0.5\,\text{counts s}^{-1}$ was placed on the averaged count rate in the S1 filter for the whole field of view.

Predictions of the expected count rates in the S1 filter based on the Wisconsin B-band sky survey (McCammon *et al.* 1983) for the regions of the sky covered by clean data periods, indicate that large scale structures should have been visible in the WFC survey data. Indeed, the detection of the low surface brightness structures of the Vela and Cygnus loop supernova remnants is good evidence that the WFC should be able to detect the predicted ISM structures. Hence, the WFC and Wisconsin data are clearly inconsistent. These predictions were made on the assumption that the interstellar gas is represented by a hot ($\approx 10^6$ K) optically thin plasma in thermal equilibrium with normal metallic abundances and that the foreground column of absorbing neutral material is that solely in what is called the 'local fluff', residing in the interstellar cloud surrounding the Sun ($N_\text{H} < 10^{18}\,\text{cm}^{-2}$). Reconciliation of the results may require non-equilibrium conditions or depletion of heavy elements in the hot gas.

3.9.2   Structure of the local ISM

Two different, but complementary, approaches have been taken in studies of the structure of the local ISM (LISM) based on the WFC survey. Warwick *et al.* (1993) used the differential source counts to delineate the boundaries of the so-called 'local bubble' of low density gas, while Diamond *et al.* (1995) used detected count rates and upper limits to estimate the line-of-sight column densities for white dwarfs and late-type stars.

As the two largest categories of EUV source, white dwarfs and late-type stars provide the most useful probes of the LISM. The WFC sees late-type stars ranging in distance from a few parsecs up to $\approx 100\,\text{pc}$, whereas the white dwarf detections extend out to $300\,\text{pc}$ or so. The initial analysis of the BSC sample showed that the slope of the number counts of white dwarfs was very flat, consistent with the expected level of interstellar absorption (Pounds *et al.* 1993). The more detailed study of Warwick *et al.* (1993) shows that the differential source counts for both classes of object are flatter than the uniform Euclidean form. Assuming a uniform distribution of absorbing gas to match the observations requires an average gas density of $\approx 0.3\,\text{atoms cm}^{-3}$ for the white dwarfs compared with only $\approx 0.05\,\text{atoms cm}^{-3}$ for the late-type stars, demonstrating that the assumption of a uniform distribution of absorbing gas is invalid.

Modelling the source counts in terms of a local bubble within which the gas density is relatively low but which is bounded by a medium of much higher density gives far better agreement. The WFC data are consistent with an interior gas density of $0.05\,\text{atoms cm}^{-3}$, average bubble radius of $80\,\text{pc}$ and a five-fold increase in gas density at the bubble boundary. The spatial distribution of the EUV sources helps identify regions where the EUV transmission of the LISM departs from the global average value. This is most clear in the general direction of the galactic centre (figure 3.28, which can be interpreted as a relatively nearby (10 pc) wall of absorption, which is consistent with the optical/UV absorption measurements of Frisch and York (1983) and Paresce (1984)). An excess of late-type stars in the direction of $l \approx 210°$, $b \approx -30°$ is consistent with an ultra-low gas density.

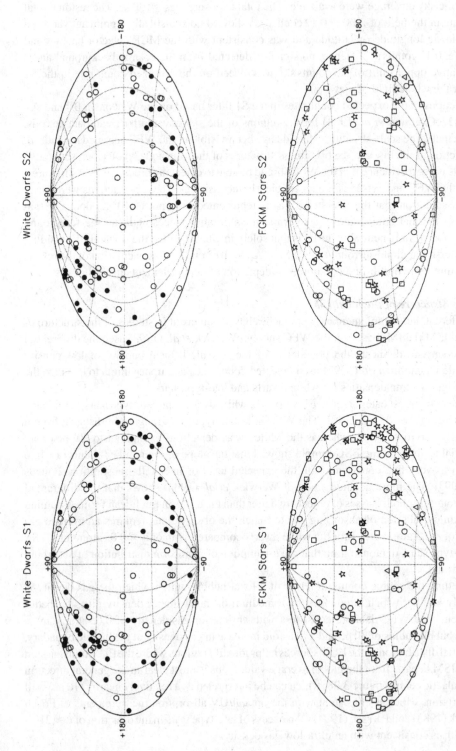

Fig. 3.28. The spatial distribution of white dwarfs and active late-type stars plotted in galactic coordinates. The plots for the late-type stars show objects at distances >100 pc as filled circles and <100 pc as open circles. These are also distinguished between spectral types: F stars, triangles; G stars, circles; K stars, squares; M stars, star symbols (from Warwick *et al.* 1993).

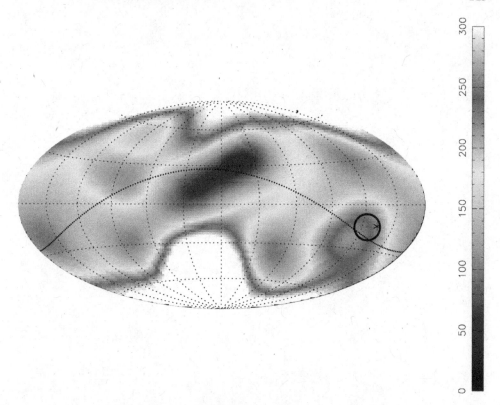

Fig. 3.29. Grey-scale representation of the extent of the local bubble based on the distance to white dwarfs with 'hard' EUV colours detected in the WFC survey. The distance scaling is from <50 pc (black) to ≈150 pc (white) (courtesy R. S. Warwick).

The extent of the local bubble can be studied using the distances of the detected white dwarfs which have relatively hard spectra (i.e. S2/S1 count rate <2.5). In general, these objects are probably embedded in the walls of the local cavity and, consequently, are more absorbed in the S2 filter than in the S1 passband. Warwick *et al.* (1993) determined the mean distance of these 'hard' white dwarfs as a function of position on the sky shown as a grey scale representation in figure 3.29.

The detailed study of the line-of-sight column densities carried out by Diamond *et al.* (1995) gives results that are in close agreement with those of Warwick *et al.* (1993). They find that the column density within the local bubble is less than 10^{19} cm^{-2}, within a region 70 pc from the Sun, and does not rise to greater than 10^{20} cm^{-2} within 100 pc. The bubble appears to be offset from the Sun with its centre at a distance of 30 pc in the direction $l \approx 135°$.

It is interesting to note that more recent work by Hutchinson *et al.* (2001), based on studies of a much larger number of sources in the ROSAT X-ray catalogue, has refined the original interpretation of the EUV results. In this case, the simple spherical model of the local bubble is pinched along the plane of Gould's Belt, with two lobes of low density material extending north and south of this plane.

4

The Extreme Ultraviolet Explorer and ALEXIS sky surveys

4.1 The Extreme Ultraviolet Explorer

S. Bowyer and his team at the University of California, Berkeley, the pioneers of EUV astronomy, originally proposed the Extreme Ultraviolet Explorer (EUVE) mission to NASA in 1975. Selected for development in 1976, it was eventually launched in June 1992, almost exactly two years after ROSAT. The science payload was developed and built by the Space Sciences Laboratory and Center for EUV Astrophysics of the University of California at Berkeley. Like the WFC, a principal aim of EUVE was to survey the sky at EUV wavelengths and to produce a catalogue of sources. However, the two missions differed in several respects. While the WFC had a filter complement allowing it to observe at the longer wavelengths of the EUV band (P1 and P2 filters), the survey was only conducted at the shorter wavelengths from 60 Å to ≈200 Å. By contrast, the EUVE survey was carried out in four separate wavelength ranges, extending out to ≈800 Å. In addition, EUVE carried on board a spectrometer for pointed observations following the survey phase of the mission. The spectrometer will be discussed in detail in chapter 6, but half the effective area of its telescope was utilised for a deep survey imager, giving exposure times significantly larger than either the WFC or EUVE all-sky surveys but over a restricted region of sky. The following sections include a detailed description of the payload components drawn from several papers in the scientific literature (see e.g. Bowyer and Malina 1991a,b). A schematic diagram of the complete EUVE payload is given in figure 4.1. After a highly successful in-orbit lifetime of almost nine years, EUVE was switched off in February 2001.

4.1.1 The telescopes

The all-sky survey section of the EUVE payload had a complement of three single grazing incidence mirrors, two of an identical Wolter–Schwarzchild Type I design (see sections 2.3 and 2.4) and one Wolter–Schwarzchild Type II. The deep survey/spectrometer telescope is also a Wolter–Schwarzchild Type II telescope. Schematic drawings of these telescopes are shown in figure 4.2 while their design parameters and performance are discussed in detail by Finley *et al.* (1986b) and J. Green *et al.* (1986) and summarised in table 4.1.

In a standard Wolter–Schwarzchild Type I design the short wavelength limit of the telescope bandpass is determined by the grazing angles. Hence, X-ray mirrors utilise smaller grazing angles than those used in the EUV. At short wavelengths in the EUV, the mirror cutoff lies just below the typical bandpasses of the filters. However, at longer wavelengths, in the 400–800 Å range, no filter design can provided a bandpass without also having contaminating throughput below 100 Å, a soft X-ray 'leak'. A novel Wolter–Schwarzchild Type II design has been developed to solve this problem, which has sufficiently large grazing angles to

Table 4.1. *Parameters of the EUVE telescope mirrors.*

	Short wavelength scanners	Long wavelength scanner	Deep survey/ spectroscopy
Focal length (cm)	56	70	136
f/ratio	1.41	1.75	3.3
Geometric area (cm^{-2})	138	407	453
Field of view (diameter, deg)	5	4	2
Primary/Secondary ave. graze angles (deg)	5/5	29/21	9/5.5
Surface coating (Å)	Au, 1000	Ni, 13 000	Au, 350
fwhm (optical″)	1–2.4	2.3	1.6
HEW (optical″)	8–32	18	12

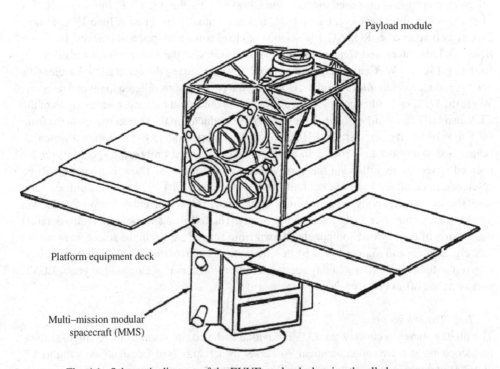

Payload module

Platform equipment deck

Multi–mission modular spacecraft (MMS)

Fig. 4.1. Schematic diagram of the EUVE payload, showing the all-sky survey scanners and the deep survey/spectrometer telescope pointing at right-angles to these (courtesy S. Bowyer).

prevent transmission below 100 Å but still allows high throughput above 400 Å (see figure 4.2b; Finley *et al.* 1986b).

Unlike the WFC mirrors which were incorporated into a three-fold nest to produce a single telescope, the EUVE mirrors each formed part of a separate telescope unit. However, their individual geometric collecting areas of 140 cm^2 for the Type I scanner and 390 cm^2 for the Type II design, gave a total of 670 cm^2, about 25% more than that obtained with the WFC nest. Their 5° fields of view were identical to that of the WFC, while that of the deep survey

Scanner A/B $f/1.31$ 58–234 Å
Wolter–Schwarzschild Type I

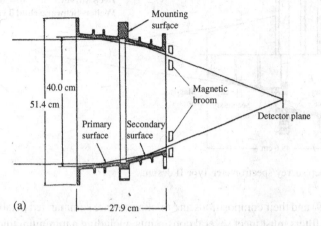

(a)

Fig. 4.2. Schematic diagram of the three Wolter–Schwarzschild telescope designs used by EUVE (from Bowyer and Malina 1991a). (a) Shorter wavelength survey configuration.

Scanner C $f/1.65$ 368–742 Å
Wolter–Schwarzschild Type II

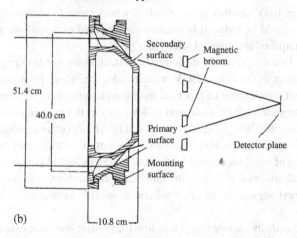

(b)

Fig. 4.2. (b) Longer wavelength survey Type II telescope.

telescope was 2°. The on-axis imaging performance of the EUVE telescopes was superior to the WFC, with an ≈8 arcsec half energy width compared to 20 arcsec. However, at large off-axis angles, the image quality was dominated by the intrinsic aberrations of the Wolter–Schwarzschild optical design (see section 2.3).

4.1.2 The thin film bandpass filters

As outlined earlier (section 3.3.3), combinations of plastic and metal films can be used to define useful bandpasses in the EUV. Four filter designs were included in the EUVE complement giving bandpasses covering the total wavelength from 60 to 800 Å. A detailed discussion of the properties of these filters is given in several papers (Vallerga *et al.* 1986,

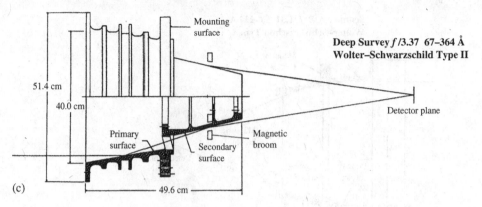

Fig. 4.2. (c) Deep survey/spectrometer Type II design.

1992; Vedder *et al.* 1989) and their compositions and thicknesses are summarised in table 4.2. The design of thin film filters must meet several constraints, including a minimum thickness for mechanical strength, ensuring that the total throughput falls within the limits of the telemetry rate and low sensitivity in the far-UV to ensure that bright early-type stars do not contaminate the data. The deep survey scanner, mounted at right angles to the all-sky survey telescopes, always pointed in the anti-Sun direction, along the Earth/Sun line and, as a result, was exposed to a much lower EUV (mainly geocoronal) sky background. Therefore, these filter designs (see table 4.2.) could be reduced in thickness, compared to the all-sky survey units, within limits of their required mechanical strength.

From a mechanical point of view, a particular difficulty is the need to use free standing metal foils to give useful transmission at the longer EUV wavelengths, i.e. where plastic support films are completely opaque. Development of practical long wavelength filters was the result of considerable innovation from the EUVE team and followed from the examination of the properties of many different materials (e.g. Jelinsky *et al.* 1983). An important consideration has been the long term stability of various combinations. For example, metals may become oxidised and the performance of indium/tin combinations was seen to degrade by an electrochemical interaction (Chakrabarti *et al.* 1982). In the latter case, the problem was ultimately solved by introducing an inert separating layer of silicon monoxide between the tin and indium components.

The bombardment of materials by atomic oxygen in low Earth orbit has been extensively studied, as oxidation may alter performance and affect structural integrity through erosion of polymeric materials (e.g. Leger *et al.* 1984; Slemp *et al.* 1985; Visentine *et al.* 1985). In general, the metal foils used for EUV filters, such as Be or Al, form stable oxide layers ≈ 100–200 Å thick which then prevent further changes. Furthermore, as these layers already exist following the filter manufacturing process, no in-orbit changes would be expected. However, the oxidation of carbon produces carbon monoxide or carbon dioxide which are both gases. Hence, atomic oxygen bombardment will continually erode a carbon foil or plastic film, since the latter is also largely composed of carbon. A solution proposed by the Berkeley team, and also adopted for the WFC (see section 3.3.3), was to replace at least some of the carbon with a protective boron film.

The EUVE filter arrangement followed a different philosophy to the ROSAT WFC, which used a rotating wheel to deploy a single filter in the field of view at any one time. Each

Table 4.2. *Configuration of the filters used in the EUVE sky survey.*

Quadrants	Material	Bandpass (Å)
0,1 (Scan A)/2,3 (Scan B)	Lexan/boron 2400 Å/1000 Å	44–240
2,3 (Scan A)/0,1 (Scan B)	Al/Ti/C 2400 Å/250 Å/600 Å	140–360
0,1 (Scan C)	Sn/SiO 3484 Å/100 Å	520–750
2,3 (Scan C)	Ti/Sb/Al 374 Å/686 Å/1256 Å	400–600
DS Edges	Al/C 2039 Å/399 Å	157–364
DS Centre	Lexan/B 1498 Å/1172 Å	67–178

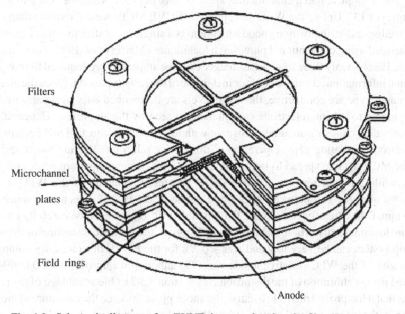

Fig. 4.3. Schematic diagram of an EUVE detector showing the filter arrangement, the MCP stack and wedge-and-strip image readout (from Bowyer and Malina 1991a).

EUVE detector had a single fixed filter unit, which was divided into several segments. For the survey scanners there was a four-fold division (figure 4.3), with the Type I units containing pairs of Lexan/boron and aluminium/carbon filters with the identical compositions mounted in opposite quadrants (table 4.2, figure 4.3). The Type II scanner had the titanium/antimony (dagwood) and indium/silicon monoxide/tin filters located in a similar way. In contrast, the circular field of the deep survey detector is divided into three parallel filter strips comprising aluminium/carbon (outer strips) and Lexan/boron (inner strip).

4.1.3   The detectors

All the detectors developed for the EUVE mission were identical in their basic design, utilising MCP arrays combined with a wedge-and-strip anode for position encoding (figure 4.3). Magnesium fluoride photocathodes were used on the detectors for the short wavelength Type I scanners and the deep survey telescope, while no photocathode was used for the

Type II, long wavelength, scanner. All the spectrometer channel detectors were coated with potassium bromide. In parallel with similar work on the WFC detector design, the EUVE development programme attempted to optimise the photoelectric yield of the photocathodes obtaining the best results with thick deposits of $\approx 15\,000\,\text{Å}$ of cathode material (Seigmund *et al.* 1986b, 1987, 1988). To further enhance the quantum detection efficiency of the MCP detector, a mesh grid was used to generate an electric field to redirect those photoelectrons produced in the interchannel glass matrix into adjacent channels, as originally suggested by Taylor *et al.* (1983).

The EUVE detector used a 'Z stack' of three MCPs to generate high gain (a few $\times 10^7$), with a narrow pulse height distribution and low levels of ion feedback (e.g. Seigmund *et al.* 1985). The front plate had a channel bias angle of zero degrees, while the rear pair both had bias angles of $13°$. Unlike the WFC detectors, the EUVE MCPs were flat rather than curved.

The wedge-and-strip position encoding system is a single layer structure that can be easily manufactured using conventional photo-etch techniques (Martin *et al.* 1981; Seigmund *et al.* 1983a,b, 1986b). Only three low-noise charge sensitive amplifiers are required to provide 2-D positional information, compared to four in the case of the resistive anode. Since the electrodes of this anode type are conductive, the noise levels are determined only by a capacitive load, corresponding to the interelectrode capacitances as seen by the amplifiers. Hence the noise is minimised by choosing an anode substrate with low dielectric constant and optimising the interelectrode insulating gaps. Operation of this device relies on allowing the charge cloud from the MCP stack to spread to larger dimensions than the anode pattern repetition period. Hence, while the spacing between the MCPs and resistive anode of the WFC detectors is only ≈ 1 mm, the gap between MCPs and anode in the EUVE detectors is much larger at ≈ 15 mm. Consequently, the electric field in this 'drift' region must be carefully controlled to ensure low image distortion. In an optimised system, the intrinsic distortion of the wedge-and-strip system can be very low and was $<0.3\%$ for the EUVE detectors. By comparison, the linearity of the WFC resistive anode was 2%, although it improved to $<0.016\%$ when corrected using calibration of the distortion (see section 3.3.2). One advantage of the resistive anode is that it has proved possible to curve the anode plane to match the curvature of the MCPs and, hence, the optimum focal surface of the WFC telescope mirrors. A similar geometry has not been attempted for the wedge-and-strip readout.

4.1.4   *The overall performance of the EUVE imaging telescopes*

Figure 4.4 shows the effective area of each of the all-sky survey bands and the deep survey telescope after taking into account the MCP detector efficiency, filter transmission, mirror reflectivity and geometric area. Comparing these with the similar set of curves for the ROSAT WFC (figure 3.7) does reveal significant differences. In particular, as the filters are thinner, the deep survey bands have a factor two to three times the effective area of both the WFC and EUVE all-sky bands at the same wavelengths. However, where the WFC and EUVE all-sky bands overlap, the effective areas are not too different.

Assessing the relative sensitivities of the ROSAT WFC and EUVE sky surveys is difficult, since this depends to a large extent on the magnitude of the various sources of sky background and their behaviour in time and space at the different epochs of the surveys. Nevertheless, some key factors concerned with the survey strategy also have a large influence. How these determine the total exposure time achieved is important. In the first instance, the WFC survey was only conducted in two bands and the filters used exchanged once per day in the field

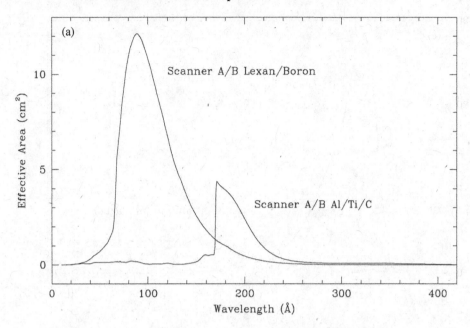

Fig. 4.4. (a) The effective area of the 100 Å and 200 Å filter bandpasses, summed over the A and B EUVE scanner telescopes, taking into account the mirror, detector and filter performance (from EUVE 1993).

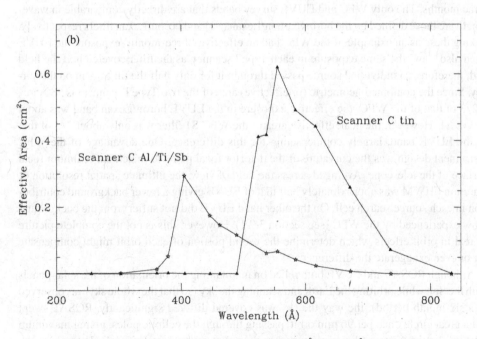

Fig. 4.4. (b) The effective area of the scanner C 400 Å and 600 Å filter bandpasses, taking into account the mirror, detector and filter performance (from EUVE 1993).

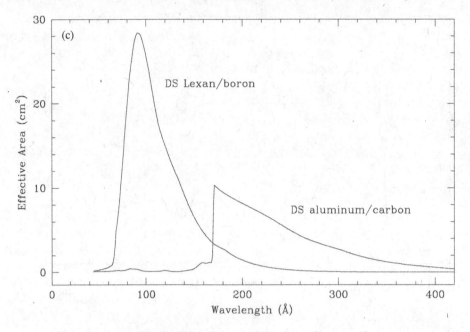

Fig. 4.4. (c) The effective area of the deep survey 100 Å and 200 Å filter bandpasses, taking into account the mirror, detector and filter performance (from EUVE 1993).

of view. Hence, for the planned six-month survey period, each band was covered for three months. The only WFC and EUVE survey bands that are directly comparable in wavelength are those defined by the boron/carbon/Lexan (S1) and boron/Lexan filters respectively. Taking these as an example, if the WFC had an effective 'three-month' exposure the EUVE band also 'saw' the same exposure in each Type I scanner, as the filter covered half the field and, therefore, an individual source passed through it for only half the time spent on the survey. Since the combined geometric (or effective) area of the two Type I scanners is 280 cm², ≈2/3 of that of the WFC, the effective exposure in the EUVE boron/Lexan band was about 2/3 of S1. However, the peak effective area of the WFC S1 filter was only about 2/3 of that of the EUVE band, largely compensating for this difference. One advantage of the WFC instrument design was the curvature of the detector focal plane to match the optimum focal surface of the telescope. Averaged across the field of view, the ultimate spatial resolution of 3 arcmin FHWM was approximately half that of EUVE giving a lower background contribution in each source search cell. On the other hand EUVE did not suffer from the background glow experienced by the WFC (see section 3.9.1). However, this is not the complete picture as real in-orbit effects which determine the useful portion of each orbit might compensate for or even exaggerate the difference.

Although ROSAT and EUVE both relied on maintaining the orientation of the solar panels with the Sun while orbiting the Earth and scanning the sky (so that the whole sky was observed in a six-month period), the way the sky was covered differed significantly. ROSAT swept out a great circle once per 96 min orbit, passing through the ecliptic poles, giving maximum exposure at the poles and a minimum near the equator (see section 3.3.4). In contrast the EUVE spacecraft had a lower inclination orbit (20° cf. 53°) and rotated three times in each

96 min period, giving a more even sky coverage. For the boron/Lexan band, a maximum exposure of $\approx 40\,000\,s$ was achieved at the ecliptic poles down to $\approx 1\,000\,s$ at the equator. These values can be compared with the corresponding ones for the WFC S1 band of 70 000 s and 1 600 s respectively.

From the above figures, bearing in mind the uncertainties in this analysis, it would appear that the ROSAT and EUVE all-sky surveys have a roughly comparable sensitivity. Ultimately, the most direct comparison of the two hinges on the number of sources detected in the various wavebands.

4.2 The EUVE all-sky survey

While having separate origins in earlier programmes, the development of the ROSAT WFC and EUVE missions took place in parallel over approximately ten years. With very similar scientific goals, a certain level of rivalry was inevitable. Even so information was freely exchanged between instrument teams for the greater good of the science as a whole, promoting a spirit of friendly cooperation. Which mission would arrive in orbit first remained unclear for a long time and was ultimately determined by the political lottery of funding programmes. Once a first sky survey has been completed, questions are often raised about the value of a second. However, the history of X-ray astronomy has already demonstrated that new and different scientific results are obtained by multiple sky surveys. While certain results from a first look at the sky in a new waveband can only be 'discovered' once, there remain important things still to be found. In this context, the WFC and EUVE surveys should be treated as complementary rather than competitive. The WFC survey was only conducted at wavelengths below $\approx 180\,\mathring{A}$, whereas EUVE covered the entire EUV spectrum in four colours. Separated by two years, comparison of detections between each survey were expected to provide important information on source variability on long time scales, of particular interest when the majority of EUV sources are late-type stars and not expected to be constant. Finally, EUVE added two further dimensions to EUV astronomy which the WFC did not possess – the deep survey and spectroscopy (see chapters 5–10).

The EUVE mission conducted its sky survey between July 1992 and January 1993. During this period, 97% of the sky was surveyed at some level with approximately 80% receiving an exposure of 500 s or more. The gaps in the data were due mainly to calibration periods when the survey was briefly paused at the time an appropriate source was visible within the satellite attitude constraints. A preliminary bright source list was published by Malina *et al.* (1994), followed by the first EUVE catalogue (Bowyer *et al.* 1994). In detail, the technical organisation of the EUVE survey analysis differs from the approach of Pounds *et al.* (1993) and Pye *et al.* (1995) for the WFC, but yields a similar result – a catalogue of point sources in the four survey bands. The larger effective area of the Lexan/boron filter compared to the WFC S1 band delivers higher count rates – 34 counts s^{-1} compared to 14 counts s^{-1} – but there is a window of transmission in the UV (Vallerga *et al.* 1992) which might introduce spurious sources. Hence, any 100 Å band sources associated with known stars of spectral class O or B were rejected.

Bowyer *et al.* (1994) published a catalogue of 410 sources, comprising 287 objects detected in the all-sky survey, 35 deep survey detections and a further 88 drawn from the Malina *et al.* (1994) bright source list which were identified through a variety of special processing techniques but which did not meet all the rigorous and uniform criteria of Bowyer *et al.* (1994). As might be expected the majority of sources were detected in the shortest wavelength 100 Å band, where interstellar absorption is least. Progressively fewer sources were detected at longer wavelengths: less than 100 in the 200 Å band; 10–20 at the very longest wavelengths. This makes an interesting contrast to the WFC survey, where roughly equal numbers were detected in the S1 and S2 bands, the latter lying in between the EUVE 100 Å and 200 Å filter responses.

The Bowyer *et al.* (1994) catalogue had a number of limitations, mainly because several regions of sky remained poorly or completely unexposed due to interruptions in the survey due to calibrations. These were subsequently filled in during the spectroscopic phase of the mission between February 1993 and July 1993, with 100% of the sky receiving some exposure and 98% more than 500 s. At the same time, this 'gap filling' extended the coverage of the deep survey by providing views of regions of the ecliptic outside the nominal 270° to 90° range of longitudes covered during the primary six-month survey. The source search algorithms used on the deep survey data were also improved to take account of the variation in telescope imaging performance that occurs for sources at different off-axis locations. In addition, during spectroscopy observations, when the EUVE spacecraft is placed in an inertially stabilised attitude, the scanner telescopes are fixed on a region of sky perpendicular to the spectroscopy target. Fields examined in this way achieve exposure times in the range 10^4 to 10^5 s. Such observations are highly sensitive probes of the fainter population of EUV objects (McDonald *et al.* 1994).

4.2.1 The Right Angle Program

After completion of the all-sky survey portion of its mission, EUVE was used primarily in its spectroscopic mode. However, when the spectrometers and deep survey (DS) telescope were pointed at a particular primary target it was often the case that the spacecraft roll axis could be selected in a manner which allowed the EUVE scanner telescopes to simultaneously observe a secondary target, if it lay at 90° to the telescope axis and satisfied other operational constraints. This allowed EUVE to pursue a novel and highly useful set of observations under a programme known as the EUVE Right Angle Program or RAP.

There were several classes of scientific objectives that could be effectively pursued with the RAP. These include the following.

- Detection of new EUV sources too faint to have been detected in the EUV all-sky surveys. This is primarily a result of the longer exposure times afforded by spectroscopic observations.
- Obtaining deeper and more accurate photometric data on known sources. Such observations could confirm weak sources and improve locations, which aids in source identification.
- Simultaneous photometric coverage in all four of the scanner bands centred on 100, 200, 400 and 600 Å. This was particularly important for sources where low interstellar column densities allow observation of long wavelength fluxes.
- Time resolved photometric coverage for the study of variable and periodic sources such as cataclysmic variables and active late-type stars.

Initially the RAP utilised the random directions in which the survey telescopes were pointed during a spectroscopic observation to obtain deep images of the sky. In its guest observer mode, investigators proposed lists of targets for inclusion in the RAP along with a brief scientific justification of the observations. When a prime spectroscopic or DS target was scheduled for observations the list of approved RAP targets was then searched for those lying on a great circle 90° from the spectrometer/DS axis. If such a target existed and it met other observational constraints then the spacecraft roll axis could be selected to observe the target.

A description of the RAP is given in McDonald *et al.* (1994), where they report the serendipitous detection of 99 new sources during the first year of the RAP. The longer times available during a typical RAP observation made it possible to detect sources approximately a factor of ten below the detection thresholds of the survey. The new sources reported by McDonald *et al.* included a B star and three extragalactic sources.

An excellent example of the type of detailed investigation of variable sources that were pursued to great advantage are the observations of Warren *et al.* (1995) of UZ For. UZ For is an AM Her type magnetic cataclysmic variable system with an orbital period of 125.6 min and the second brightest CV system in the EUVE all-sky survey. Warren *et al.* considered two datasets, one in which UZ For was observed with the scanner telescopes while the primary target was λ And and another in which UZ For was fortuitously located in the DS telescope field while the spectrometer was observing the planetary nebulae NGC 1360. Together, these two observations totalled over 150 orbits of the UZ For system. The data, which were all photometric in nature, allowed the construction of highly time resolved and highly detailed EUV light curves for UZ For. By folding the data about the established orbital period, Warren *et al.* were able to model some geometric properties of the accretion stream with remarkable detail. For example, they were able to locate the stagnation point at which the ballistic portion of the accretion stream meets the magnetically controlled portion. Among other things, they were also able to specify the extent of the accretion spot on the white dwarf ($<5°$) and show that it must extend vertically by 0.03 to 0.07 white dwarf radii above the surface.

All the EUV sources arising from the three different exposure strategies – the all-sky survey, deep survey and so-called RAP – are included in the second EUVE source catalogue (Bowyer *et al.* 1996). This catalogue was constructed using the same analysis strategy and source selection criteria as the first catalogue. Figure 4.5 shows the distribution of EUVE source detections on the sky in galactic coordinates. A total of 514 objects were detected during the all-sky survey. Thirty five sources were found in the deep survey, of which three were also detected in the all-sky survey. The combination of the RAP and long exposures with the DS instrument outside the survey yielded 188 new sources, bringing the total of catalogued sources detected by EUVE to 734. Of these, 211 sources were not included in any of the earlier EUV catalogues. As for the WFC and earlier EUVE catalogues, potential counterparts for the sources were identified by comparing their positions with a range of optical, UV and X-ray catalogues and source categories are indicated in figure 4.5. Approximately 65% of the catalogue had plausible optical counterparts, while a programme of optical spectroscopy worked on the remaining identifications (e.g. Mathioudakis *et al.* 1995; Craig *et al.* 1996).

4.2.2 Comparison of the WFC and EUVE sky surveys

While the EUVE mission added further objects to the list of known EUV sources, the heterogeneous nature of the observing conditions and exposure times associated with these makes any meaningful analysis of this source population and comparison of the total number of sources in the second EUVE catalogue with earlier work impossible. However, the results of the EUVE shorter wavelength filters are directly comparable with the WFC all-sky survey because they overlap with the S1 and S2 bands (Barber *et al.* 1995). For example, Bowyer *et al.* (1996) plot the WFC S1 count rate against the EUVE Lexan/boron count rate for

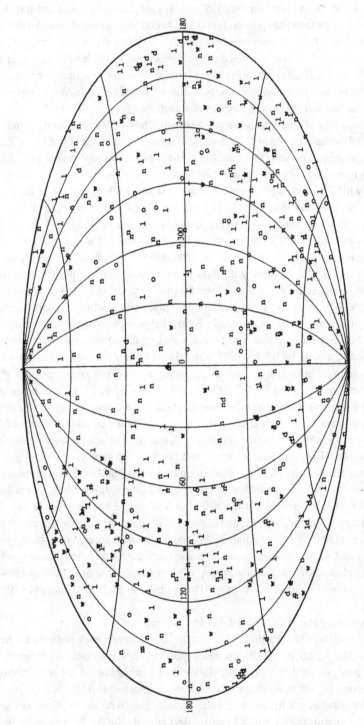

Fig. 4.5. Distribution of EUV sources, in galactic coordinates, detected by the EUVE all-sky survey and deep survey. All filter bands are included. The map is centred on $l = 0$. Symbols: w = white dwarf; l = late-type star; n = not identified; o = other; d = deep survey detection (from Bowyer et al. 1996).

Table 4.3. *The overlap of source detections in the first EUVE source catalogue with the ROSAT WFC all-sky survey (from Barber et al. 1995).*

EUVE filter	Number of sources	WFC S1 detections	WFC S2 detections	WFC S1 and S2 detections	WFC S1 or S2 detections
100 Å	283	223	218	203	238
200 Å	93	78	85	77	86

Table 4.4. *The overlap of source detections in the first EUVE source catalogue with the ROSAT WFC all-sky survey (from Barber et al. 1995).*

Source type	EUVE 100 Å sources	EUVE 100 Å and ROSAT WFC S1 detections	EUVE 200 Å sources	EUVE 200 Å and ROSAT WFC S2 detections
WD stars	95	84	50	49
FGKM stars	125	106	30	30
CVs	12	8	4	1
OB stars	13	7	3	2
Other IDs	11	6	1	1
Unidentified	27	12	5	2
Total	283	223	93	85

sources detected in both filters, finding that the majority of sources follow the relationship S1:Lexan/boron = 0.5. Barber *et al.* (1995) take the comparison a stage further, using the all-sky survey data from the first EUVE source catalogue as a primary input list for an analysis of the WFC data. For each source detected in either the 100 or 200 Å filters they searched the all-sky survey data for a WFC counterpart. Tables 4.3 and 4.4 give the details of the total number of EUVE sources detected in the WFC, in various categories.

Over 80% of the EUVE 100 Å sources are detected by the WFC with an even higher ($\approx$90%) overlap between the WFC and the EUVE 200 Å band. This degree of overlap is remarkable given the complexity of the way the respective survey threshold sensitivity varies across the sky. Secondly, this result also indicates that the typical amplitude of variability of the EUV source population on time scales of two years (the time gap between the surveys) is not large, less than a factor of two. The fraction of EUVE sources detected by the WFC is very high for the dominant populations of the white dwarfs and late-type stars but is lower for other source categories. This may be due to the effects of variability, for example, half of those CVs listed in the Bright Source Catalogue (Pounds *et al.* 1993) are not included in the EUVE list.

A measurement of the relative sensitivities of the two surveys can be made by comparing the fraction of detections associated with optical catalogues representing the two main types of EUV source, white dwarfs and late-type stars. Approximately 23% of the sources in the EUVE sky survey are in the *Third Catalogue of Nearby Stars* (Gleise and Jahreiss 1991) while in the WFC the corresponding fraction is 22%. Similarly, 13% of the EUVE sources are in the McCook and Sion 1987 white dwarf catalogue (note this catalogue has subsequently been updated to include new WFC and EUVE discoveries (McCook and Sion (1999)), as are 12% of the WFC objects. Hence, both surveys appear to be probing the nearby star and white dwarf distribution to the same relative depth.

4.3   Key EUVE survey results
4.3.1   *Science with the long wavelength filters*

Although few EUV sources were detected in the 400 and 600 Å bands, they are all important as they must lie along viewing directions with the very lowest column densities. Of particular note are the giant B stars β CMa (B1 II–III) and ϵ CMa (B2 II), which were, surprisingly, the brightest sources in these wavelength ranges. Indeed, with a 600 Å band count rate of 78 counts s^{-1}, ϵ CMa is now known to be the brightest EUV source in the sky at any wavelength. There are two reasons for this. First, the stars lie in a tunnel through the ISM which was known to have very low column density ($N_{H\,I} \approx 1$–2×10^{18} cm^{-2}) even before the launch of EUVE (see references in Welsh *et al.* (1999)), even though they lie at distances greater than $\approx$200 pc. Second, their photospheric fluxes exceed the pre-launch predictions from model atmospheres that adequately fit the UV and visual energy distributions (e.g. Vallerga *et al.* 1993).

The β CMa variables (also sometimes referred to as the β Cephei variables) constitute an important instability strip which runs through the HR diagram in the region of the early B stars from B0.5 to B2 and luminosity classes II, III and IV. The β CMa phenomenon is characterised by non-radial pulsation modes having periods of 3 to 7 h. In the optical, these pulsations have low (typically <0.1 mag) photometric amplitudes and exhibit corresponding radial velocity variations. At shorter wavelengths, particularly in the UV where the bulk of the luminosity is radiated, photometric variations can be significantly larger. There are roughly 100 such stars known. The pulsational instability mechanism is believed to originate with an opacity driven instability associated with the abundance of iron deep in the stellar atmosphere.

Actual EUV flux has only been detected from two B stars, ϵ CMa and β CMa. Hoare *et al.* (1993) obtained pointed ROSAT observations of these two stars using the Sn/Al P2 filter ($520 < \lambda < 730$). Both stars were strongly detected and β CMa exhibited evidence of 'regular' photometric variations. The observations, which consisted of nine samples with average integrations of 1270 s during each orbital night, showed marginal evidence of variability with an amplitude of ∼0.04 magnitude. The allowed range of periods for these variations is one to three hours not the ∼6 h modes observed in the optical. The observations of Cassineli *et al.* (1995) obtained with the EUVE long wavelength (LW) spectrometer (see section 6.3), which did detect a blend of the main pulsational modes of β CMa, provide a more definitive view of the pulsational behaviour of this star at EUV wavelengths.

The B star λ Sco was also detected in both the ROSAT and EUVE all-sky surveys. As with the B stars β Dra, θ Hya and y Pup (see section 3.6), the EUV excess in λ Sco is perhaps best explained by the presence of a hot DA white dwarf companion (Berghöfer *et al.* 2000). Nevertheless, Berghöfer *et al.* noted that the EUV flux from λ Sco was not constant and showed the possible presence (at the 2σ level) of pulsations near the 4.66 h β CMa period of the B star.

Prior to the launch of EUVE, observations with the Voyager ultraviolet spectrometers revealed several directions where the line-of-sight column density was sufficiently low to allow detection of objects at wavelengths longer than the instrument cutoff at 500 Å. These were the white dwarfs HZ 43, GD 153 and G191–B2B (see section 2.9). Subsequently, Voyager

Table 4.5. *H I column densities for HZ 43, GD153, REJ0457–281 and REJ0503–285 (from Vennes et al. 1994).*

Object	T_{eff} (K)	log $N_{H\,I}$
REJ0457–281	59–63	17.80–17.90
REJ0503–285	60–70	17.75–18.00
GD153	38–41	17.70–17.90
HZ 43	50–54	17.70–17.90

studies of new EUV sources detected by the ROSAT WFC detected REJ0457–281 and REJ0503–285 (Barstow *et al.* 1994a,b), a DA and DO white dwarf lying just 1° apart on the sky. However, these studies followed a rather piecemeal approach and the EUVE LW observations represent the first systematic search for very low column density lines-of-sight, from which a statistical distribution of these LISM structures and their relationship to the local cloud can be studied. Vennes *et al.* (1994) list nine sources which are detected beyond the 504 Å photoionisation edge of neutral helium including the B stars and white dwarfs detailed above with the addition of the DA stars Feige 24 and GD50. Vennes *et al.* (1994) report a detailed study of four stars – HZ 43, GD153, REJ0457–281 and REJ0503–285 – combining published estimates of the stellar effective temperatures with the 400 Å and 600 Å band count rates to determine the interstellar H I column densities (table 4.5).

These values are significantly larger than those obtained with Voyager for REJ0457 and REJ0503 (see Barstow *et al.* 1994d,e) but are in good agreement with the Voyager studies of GD153 and HZ 43. It may be that second order contributions to the Voyager flux may be more significant for the two hotter stars, giving erroneous measurements of the column density.

The area of the celestial sphere defined by β and ϵ CMa near the galactic plane and by REJ0457 with REJ0503, which are approximately 23° away, is seen to be essentially devoid of neutral hydrogen and helium beyond the local cloud. Studies by Frisch and York (1983) and Paresce (1984) had earlier identified this region as having extremely low column density. The β CMa and ϵ CMa neutral hydrogen column densities are approximately twice those of the nearer white dwarfs: $1 - 2 \times 10^{18}$ cm^2 (Gry *et al.* 1985) and $1 - 1.5 \times 10^{18}$ cm^2 respectively (Hoare *et al.* 1993; Vallerga *et al.* 1993). These stars are approximately twice the distance of REJ0457 and REJ0503, indicating that the local cloud is the only predominantly neutral structure along this line-of-sight. If all the observed neutral hydrogen does lie in the local cloud in these directions and has a representative density of 0.1 atoms cm^{-3}, the heliocentric path length through the local cloud is $\approx 1.7 - 2.6$ pc.

Welsh *et al.* (1999) have studied the galactic distribution of 450 sources detected by EUVE which have reliable distance estimates, comparing this with the neutral absorption boundary of the local bubble determined from NaI observations by Sfeir *et al.* (1991). Their results are broadly similar to those of Warwick *et al.* (1993), with late-type stars dominating the nearby source distribution and the majority of more distant detections being white dwarfs. The vast majority of these sources are found within the boundary delineated by the NaI measurements which mark the wall of the local bubble inferred by Warwick *et al.* (1993) from the ROSAT WFC data. The number of extragalactic detections at high galactic latitudes suggests that the local bubble is open ended in the galactic halo forming what Welsh *et al.* call the 'Local Chimney'. This is found to be perpendicular to the Gould Belt and opens towards regions of enhanced 0.25 keV soft X-ray background. A detailed examination of the soft X-ray distribution from the ROSAT survey by Hutchinson *et al.* (2001) yields a similar orientation for the local bubble, which in this case is 'pinched' along the axis of Gould's Belt but is not open ended at the poles in their model.

4.3.2   Hot white dwarfs

Given the large overlap in the sample of white dwarfs detected by the WFC and EUVE surveys, it is inevitable that the broad properties determined from the WFC survey data and follow-up optical observations should be largely matched by the EUVE sample. For example, a review of the EUVE white dwarf population by Vennes (1996) shows a similar mass distribution to that reported by Marsh *et al.* (1997a) for the ROSAT WFC sample, revealing an excess of objects with high masses, above $1 M_\odot$. In addition, the EUVE photometric data show a flux dependency similar to the variation with effective temperature seen in the WFC (see Barstow *et al.* 1993b; Marsh *et al.* 1997b; Finley 1996).

Finley (1996) has combined the white dwarf observations of both the WFC and EUVE to quantify the metallicity of each star in the sample. Since individual element abundances cannot be determined from photometric data alone, a canonical set of heavy element abundances was determined by matching the EUVE spectrum of G191–B2B with a local thermodynamic equilibrium (LTE) model incorporating just bound-free opacities. All the abundances were then scaled by a common factor relative to the G191–B2B 'compositions', keeping the relative abundances of individual heavy elements constant. Stellar effective temperatures and gravities were held within the optically determined limits. No white dwarfs with temperatures lower than 48 000 K or with log g > 7.9 require the presence of heavy elements, with the exception of GD394. Estimating the observed abundances of each element using the scale factor shows a good correlation with the predicted values of Chayer *et al.* (1995a; see figure 4.6). However, the observed abundances vary more rapidly with $T_{\rm eff}$ and log g than predicted. Significant

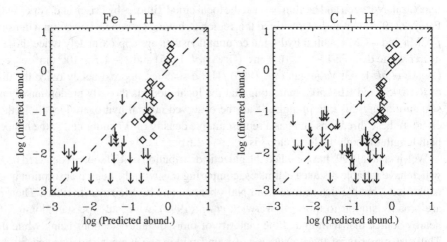

Fig. 4.6. Relative abundance measurements (diamonds) and upper limits (arrows) as a function of the values of Fe and C predicted by Chayer, Fontaine and Wesemael (1995a). The dashed line has a slope of unity (from Finley 1996).

discrepancies between predicted and observed metallicity are found for four objects – HZ 43, GD2, REJ1043+492 and REJ1032+532. All of these stars have observational metallicity upper limits an order of magnitude below the predicted abundances. It is notable that these stars all lie near the temperature dividing line between effectively pure H atmospheres and the stars rich in heavy elements.

Follow-up studies of white dwarfs detected in the EUVE survey have in many cases centred on those in binary systems. Vennes and Thorstensen (1994) had already noted the presence of a class of short period binaries that have emerged from a common envelope phase (de Kool and Ritter 1993). A substantial fraction of the white dwarfs detected in the EUV reside in binary systems, as noted by Barstow *et al.* (1994a) and Burleigh *et al.* (1997, 1998). The study of these systems may be able to place constraints on models of binary evolution and, as Vennes (1996) has pointed out, a well-defined sample may help resolve the issue of mass correlation in binaries.

Several new binary systems have been discovered through their detection as EUV sources by EUVE. The source EUVE J0254–053 has been identified with the K0IV star HD18131. An IUE low dispersion spectrum of the object reveals the presence of a DA white dwarf companion with an effective temperature ≈30 000 K (Vennes *et al.* 1995a). This system shares some properties with the DA+K0 binary HR1608 (Landsman *et al.* 1993). Both components of these systems are evolved, comprising a young (i.e. hot) white dwarf and a subgiant. Because both components were formed together, they share a similar evolutionary age and, consequently, a similar initial mass. Those binaries which consist of one or two evolved components offer important insights into the formation and evolution of stars. In particular, for cases like HD18131 and HR1608, the mass distribution in binaries may favour components of similar mass. This mass correlation in turn allows observational constraints to be placed on the initial–final mass relations for white dwarfs. Vennes *et al.* (1998) have recently published an important survey of the properties of all the binary systens (excluding those with M dwarf companions) discovered by both WFC and EUVE surveys, drawing together most of the known properties of these objects and including measurements of their radial velocities. However, apart from those systems where the periods were already known (HR8210 and 14 Aur C) none appear to be close binaries.

The close binaries which have white dwarf+M dwarf components are an important subclass of EUV sources which belong to the pre-cataclysmic binaries, i.e. binary systems that may eventually evolve into cataclysmic variables. Several DA+dM systems are EUV sources and are common to both WFC and EUVE survey catalogues. However, a determination of whether they are pre-CV systems depends upon knowledge of their orbital elements. Of four systems intensively observed three are found to have short periods – REJ0720–317 (= EUVE J0720–317; Vennes and Thorstensen 1994, 1995; Barstow *et al.* 1995a); REJ1016–053 (= EUVE J1016–053; Tweedy *et al.* 1993; Vennes and Thorstensen 1995); and REJ2013+400 (= EUVE J2013+400; Thorstensen *et al.* 1994; Barstow *et al.* 1995d). Table 4.6 summarises their properties. The fourth system, REJ1629+781, the first light source discovered by the WFC, exhibits no apparent radial velocity variations (Catalan *et al.* 1995; Sion *et al.* 1995).

4.3.3 Observations of the Moon
EUV studies can potentially contribute significantly to the understanding of the physical and chemical properties of the Sun, planets and their satellites. Reflected solar EUV radiation from the Moon was detected at regular intervals during both the WFC and EUVE

Table 4.6. *Orbital parameters of the pre-cataclysmic binaries observed/discovered in the ROSAT WFC and EUVE sky surveys (from Vennes and Thorstensen 1995, with updates from more recent papers referenced in the text).*

Name	Types	P (d)	M_1/M_2
Feige 24	DA+dM1–2	4.2316	1.59
V471 Tau	DA+dK2	0.5212	0.97
REJ0720–317	DAO+dM0–2	1.26245 ± 0.00004	1.59
REJ1016–053	DAO+dM1–3	0.7893	1.8
REJ1629+781	DA+dM4	?	3.1
REJ2013+400	DA+dM1–4	0.7055 ± 0.063	1.3

surveys. Prior to this the only EUV observations of the Moon had been made by the UV spectrometer on board Mariner 10, which obtained albedos in the 550–1250 Å wavelength range (Wu and Broadfoot 1977). Gladstone *et al.* (1994) have reported on a detailed analysis of the EUVE observations.

In its all-sky survey mode, EUVE rotated three times during each 96 min orbit. Since the scan path was perpendicular to the Earth–Sun axis, the Moon was scanned each month at first and last quarter and was visible in the 5° field of view for approximately 20–25 s or less per orbit and for only around five or six orbits in a row, before passing beyond the EUVE scan path. The signal-to-noise ratio achieved with each lunar apparition varied according to observing conditions, some being limited by being unfavourably positioned with respect to the South Atlantic Anomaly. Hence, Gladstone *et al.* selected two first quarter dates (August 5 1992 and October 3 1992) and two last quarter dates (September 19 1992 and November 17 1992) for study (e.g. figure 4.7). Geometric albedos were obtained by folding a model solar EUV spectrum through the EUVE filter responses and comparing the observed count rates with the predictions, yielding average values of 0.15% ($\pm0.02\%$), 3.1% ($\pm0.2\%$) and 3.5% ($\pm0.2\%$) over wavelength intervals of 150–240 Å, 400–580 Å and 550–650 Å respectively. These are shown in figure 4.8 together with UV measurements and an X-ray albedo derived from the data of Schmitt *et al.* (1991). The data are well matched by the scaled reflectivities of SiO_2 glass and Al_2O_3. Hence, the reflectivity is dominated by surface reflection rather than volume scattering. It appears that the EUV brightness of the Moon is consistent with reflected sunlight rather than X-ray fluorescence.

4.3.4 *Isolated neutron stars*

Despite considerable efforts to search for neutron stars in the WFC survey database, none were found (Manning *et al.* 1996). Many of the possible candidates proposed were ultimately identified with white dwarfs in binary systems. One reason is that detection of the EUV emission from neutron stars requires long integration times since most of these objects are 0.5 to 10 kpc away. High column densities will attenuate any EUV emission and theoretical models of neutron star cooling predict that after a few million years surface temperatures will fall below 10^5 K, leading to very low or non-existent EUV fluxes. In contrast, the number of known radio pulsars continues to increase. One of the recently discovered millisecond pulsars, PSR J0437–4715 (Johnston *et al.* 1993) is the closest known pulsar, at a distance of

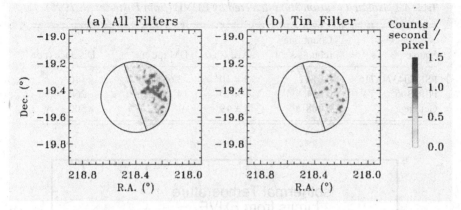

Fig. 4.7. The first quarter Moon of August 5 1992 as seen through (a) all the filters of the EUVE survey scanners and (b) only the tin filter of the long wavelength scanner. The Sun is located to the right in both images. Each resolution element represents ≈100 km on the Moon's surface (from Gladstone *et al.* 1994).

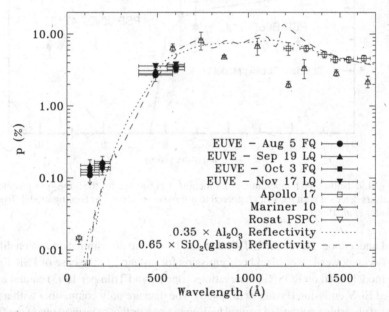

Fig. 4.8. The geometric albedo of the Moon at EUV and far-UV wavelengths. EUVE measurements are plotted as filled symbols, data from other sources, including Apollo 17, Mariner 10 and ROSAT are plotted as open symbols. Scaled reflectivity curves for Al_2O_3 and SiO_2 are plotted for comparison (from Gladstone *et al.* 1994).

140 pc. Edelstein *et al.* (1995) and Shemi (1995) suggest that a number of the closest neutron stars might be EUV sources.

Three pulsars have been detected with EUVE – PSR B0656+14, Geminga and the milli-second pulsar PSR J0437–4715. All of these are spin-powered pulsars and their deep survey 100 Å band count rates, spin down age, dispersion measure (DM) and distance are listed in table 4.7 (Foster *et al.* 1996). The mechanisms giving rise to the EUV emission from pulsars

Table 4.7. *Isolated neutron stars detected by* EUVE *(from Foster et al. 1996).*

Source	Count rate (counts s^{-1})	Age (y)	DM (pc cm^{-3})	Distance (pc)
PSR J0437–4715	0.0143	5×10^9	3.65	140
PSR B0656+14	0.024	1×10^5	14	260
Geminga	0.008	5×19^5	—	<500

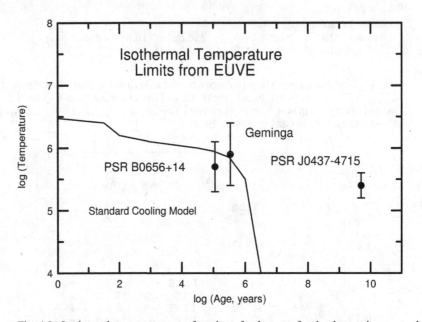

Fig. 4.9. Isothermal temperature as a function of pulsar age for the three spin-powered pulsars detected with EUVE, compared to a standard neutron star cooling model (from Foster *et al.* 1996).

are not well understood. Hence the EUVE data can be used to discriminate between different models, between power law or blackbody emission for instance. In the case of PSR J0437, a power law model based on ROSAT observations (Becker and Trümper 1993) cannot explain the observed EUV emission (Foster *et al.* 1996). The data are only compatible with a power law if the total absorbing column of neutral hydrogen exceeds that measured out of the Galaxy in this direction.

An alternative explanation is that the observed flux has a thermal origin. Edelstein *et al.* (1995) propose that both X-ray and EUV components are consistent with a single blackbody at a temperature $\approx 5.7 \times 10^5$ K from an emitting region of 3 km^2 and galactic absorbing column density of $\approx 5 \times 10^{19}$ cm^{-2}. If the emission is assumed to be from the entire surface of the neutron star (10 km radius, at 140 pc) a lower limit to the surface temperature is 1.6×10^5 K. Applying the additional constraint of the known column density within the galaxy gives an upper limit to the temperature of 4.0×10^5 K. Applying a similar isothermal model to each of the two other neutron star detections, it is possible to derive temperature limits as a function of spin-down age (figure 4.9). The two younger neutron stars (Geminga and PSR B0655+14) fit the theoretical cooling curve very well but the 5×10^9 y millisecond pulsar shows an excess of emission compared to that expected for its age.

If the EUV emission is thermal in origin, some form of surface reheating must be going on in PSR J0437–4715. Edelstein *et al.* (1995) considered five possible mechanisms, including crust–core friction from spin-down, Bondi–Hoyle accretion from the ISM, accretion from the white dwarf companion star, a particle wind nebula generated from the neutron star and magnetic monopoles. All are difficult to justify observationally.

4.3.5   EUV emission from classical novae

The accretion driven dwarf novae have been established as a class of EUV sources since the detection of SS Cygni during the ASTP mission (Margon *et al.* 1978). Several other examples have since been observed and are included in the ROSAT WFC and EUVE source catalogues. In constrast, classical novae, occurring when the white dwarf in a binary system accretes sufficient hydrogen-rich material from a non-degenerate companion to trigger thermonuclear runaway, have not been detected. Three were seen by EXOSAT in the soft X-ray waveband, including Nova GQ Muscae 1983, Nova PW Vulpeculae 1984 and Nova QU Vulpeculae 1984 (Ögelman *et al.* 1984, 1987). Subsequently, Nova Herculis 1991 (Lloyd *et al.* 1992) and Nova Cygni 1992 (Krautter *et al.* 1992) were detected by the ROSAT X-ray telescope.

After the optical decline of a nova outburst, the stellar remnants evolve at constant, near-Eddington, bolometric luminosity to effective temperatures $>2 \times 10^5$ K before turning off. In principle, X-ray and EUV observations can be used to determine the mass of the underlying white dwarf, place limits on the rate of mass loss by stellar winds and the rate of mass gain due to accretion from the stellar companion, through the determination of the maximum effective temperature reached by the remnant before turn-off (see MacDonald (1996) for a review). This measurement requires appropriate stellar atmosphere models for comparison with the EUV and soft X-ray data, since blackbody fits can grossly overestimate (by as much as a factor of two) the effective temperature.

Model calculations (e.g. MacDonald and Vennes 1991) for novae remnants indicate that these should be bright EUV as well as soft X-ray sources. The absence, until recently, of any EUV detections can probably be ascribed to a combination of their relative rarity (few occur when observable by space borne instruments) and their high luminosity. Their consequent great distance implies that typical line-of-sight neutral hydrogen columns will be rather high, absorbing any EUV emission present. Nevertheless, Nova Cygni 1992 (V1974 Cyg) was detected by EUVE during the all-sky survey after the nova had entered its nebula phase, about 285 days after maximum visual brightness (Stringfellow and Bowyer 1993). The survey observation spanned the period November 27 1992 to December 8 1992.

Based on the models of MacDonald and Vennes (1991), the observed EUVE count rates imply a white dwarf temperature lying between 3×10^5 and 5×10^5 K with a high column density, in excess of 10^{21} cm^{-2}. Hence, essentially all the detected emission from the nova arises from supersoft X-rays at wavelengths below 44 Å. While this lies outside the nominal pass band of the 100 Å and 200 Å filters in which the source was detected, the instrument effective area is still sufficiently high to allow detection of very bright soft X-ray sources such as Sco X-1 (Stringfellow and Bowyer 1996).

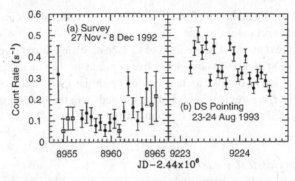

Fig. 4.10. Light curves of V1974 Cyg observed by EUVE. (a) Lexan/boron filter observation from the all-sky survey. (b) Pointed observation with the deep survey detector. Each survey bin represents the accumulated exposure ($\approx 100 \pm 50$ s) during a 12 h interval, whereas the deep survey data correspond to a continuous pointed exposure (about 31 min) per EUVE orbit. (From Stringfellow and Bowyer 1996.)

Analysis of the EUVE survey light curve and a subsequent deep survey observation of V1974 Cyg show that the source is significantly variable (figure 4.10). Even more significant variations are seen in ROSAT PSPC data (Stringfellow and Bowyer 1996). This implies that the source of the supersoft X-rays is a compact region and that accretion instabilities are present. Hence, the concept of stable thermonuclear burning on the surface of the white dwarf needs to be re-examined.

4.3.6   Activity in late-type stars

In their comparison of the WFC and EUVE all-sky surveys, Barber *et al.* (1995) found levels of variability around a factor of two in late-type star fluxes at the two survey epochs. Apart from the differences in measured fluxes, another indication is the number of stellar detections that are not common to both surveys because their EUV fluxes were not above the sensitivity threshold of one instrument at the time. As in the WFC survey, the time coverage of the EUVE strategy was well suited to studying stellar variability on time scales of several days. The deep survey was particularly well adapted to variability studies due to its higher sensitivity and more continuous coverage of an individual source, since the telescope pointed down the Earth shadow along the spacecraft's rotational axis. The deeper exposures also allowed lower levels of activity to be probed than with the all-sky surveys. Observations made in parallel with spectroscopic studies, either in the survey scanners, through the RAP, or in the DS telescope often provided very long exposures.

The RS CVn binary V711 Tauri (HR1099) was studied twice by EUVE during 1992 for periods lasting several days, once during the all-sky survey and secondly during a pointed calibration observation. The survey coverage spanned 1.74 cycles of the 2.84 d orbital period, while the DS telescope exposure totalled $\approx 260\,000$ s and 1.05 orbital periods. Both light curves show distinct modulation, as already hinted at by an earlier WFC observation (Barstow *et al.* 1992a), as illustrated in figure 4.11 (Drake *et al.* 1994b). There are no obvious flares in the survey data but a small one, with a decay time of around 1 h, seems to have occurred near MJD = 48918.2. The modulation appears to be phase dependent, with similar gross features in both datasets, characterised by a minimum EUV flux centred at phase 0.5 and a maximum near phase 1.0.

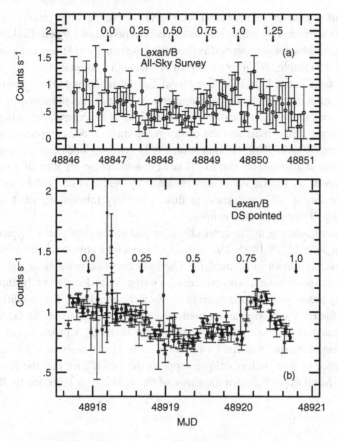

Fig. 4.11. V711 Tau count rates in the Lexan/boron filter of (a) the all-sky survey telescope, binned every spacecraft orbit and (b) the deep survey telescope, binned every 1000 s, except during the period near MJD-48918.2 (200 s), where a small flare may have occurred (from Drake *et al.* 1994).

A joint analysis of the EUVE photometric data and the published PSPC observation from the ROSAT X-ray survey suggests that most (75%) of the EUV emission from V711 Tau originates in a hot coronal 10^7 K plasma. Since the modulation amplitude is much larger than 25%, it is very likely that the variability arises from obscuration of the 10^7 K plasma on the most active of the two stars in the system (Drake *et al.* 1994b). This would imply a small scale height for any structures relative to the stellar radius. Doppler imaging of the system (Donati *et al.* 1992) indicates the presence of an equatorial spot that is rotationally occulted, which is consistent with the EUV light curve. However, optical photometry, contemporaneous with the deep survey data have neither a direct correlation nor anti-correlation with the EUV light curves, minimum optical emission appearing to lead the maximum EUV flux by $\approx 90°$ in phase.

Even though V711 Tau is not an eclipsing RS CVn system, modulation of the EUV flux is still present. In principle, observation of eclipsing binary systems provides a more powerful tool for studying coronal structures through occultation effects associated with each star. In particular, it is generally possible to unambiguously associate active regions of X-ray or EUV emission with a particular star. The eclipsing system AR Lac was observed during calibration observations of the all-sky survey scanners and deep survey telescope. The system consists of G2IV and K0IV stars separated by $9.2 R_{\odot}$ and has an orbital period of 1.983 d. Earlier X-ray observations have made substantial progress in elucidating the structure of the corona, but with conflicting results. For example, White *et al.* (1990) saw no sign of orbital modulation of the X-ray flux with EXOSAT in the 1–10 keV X-ray band, but did see strong orbital modulation in soft X-rays, suggesting that the two distinct temperature components proposed by Swank *et al.* (1981) were associated with physically different regions in the corona, the hotter being extended beyond several stellar radii and the cooler being much more compact. However, in contrast, more recent ROSAT data (Ottman *et al.* 1993), covering one orbital cycle, show a clear primary eclipse at all energies (up to $\approx$4 keV) contradicting the idea of a spatially extended hot component. A similar result was obtained by White *et al.* (1994), using the ASCA satellite, revealing a 50% reduction in flux at primary minimum with a shallow minimum associated with the secondary eclipse.

AR Lac was observed during the in-orbit checkout and calibration phase immediately following launch, on June 27/28 1992. The scanner observations covered a single primary and secondary eclipse, as shown in figure 4.12. The light curve shows a clear rise from a minimum at primary eclipse but there is no evidence for a significant decrease in the flux at the secondary eclipse, a pattern similar to that seen in the X-ray observations. This modulation is direct evidence that the EUV emission arises predominantly from regions on the G star with spatial dimensions less than or comparable to the size of the K star (D. Christian *et al.* 1996). The minimum fraction of the flux that can be attributed to the G star is about 2/3 of the total count rate. The absence of a secondary eclipse supports the earlier idea that the K star has an extended corona. Similarities between the shape of the light curves recorded by ROSAT

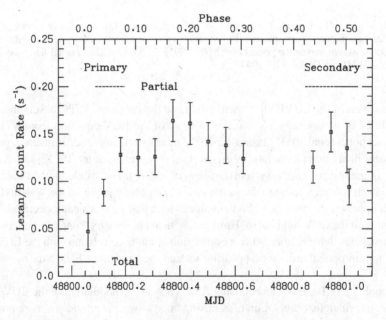

Fig. 4.12. The 100 Å scanner band count rate for the eclipsing RS CVn binary AR Lac. The orbital phase associated with the light curve is indicated together with the times of primary and secondary eclipse, including the partial phases (from D. Christian *et al.* 1996).

(Ottmann *et al.* 1993) and the EUVE data suggest that the associated coronal features may be stable on time scales of years.

An important example of the sensitivity of the deep survey telescope with a long observation time is the detection of quiescent emission from the nearby (6.4 pc) very low mass dwarf VB8 (Drake *et al.* 1996). This star has a spectral type M7eV and a mass of $\approx 0.8 M_\odot$, placing it well below the theoretical fully convective mass limit and near the hydrogen burning limit. Detection of this and other similar late-type stars at X-ray wavelengths by EXOSAT (Tagliaferi *et al.* 1990), Einstein (Johnson 1981) and ROSAT (Barbera *et al.* 1993; Fleming *et al.* 1993b) have confirmed that such late M dwarfs do indeed have hot X-ray emitting coronae. All the EXOSAT flux was associated with a flare, a strong indication that the corona of VB8 is heated by magnetic processes. Study of very low mass stars like VB8 is important for an understanding of how magnetic field generation and coronal heating might work.

VB8 was detected serendipitously in the deep survey imager during a spectroscopic observation of Wolf 630 (Drake *et al.* 1996). The light curve (figure 4.13) shows some structure, suggestive of variability, which is not consistent with a constant count rate with a flare-like event near MJD49572, increasing from the quiescent count rate by a factor of ten. The decay time scale of the flare is 1–2 h. For an assumed flare temperature of 10^7 K, the total energy in the flare is $\approx 5 \times 10^{31}$ erg. Using an optically thin plasma radiative loss model (e.g. Landini and Monsignori-Fossi 1990), the observed X-ray and EUV count rates can be used to determine the emission measure of the plasma as a function of temperature, assuming that the plasma is isothermal (figure 4.14). The formal best estimate of the temperature based on all the EUVE, EXOSAT and Einstein loci is $\log T = 6.6$, with the EUVE observations excluding temperatures below $\log T \approx 6.5$.

To sustain an active relatively constant corona (over time scales ≈ 10 y) at a temperature of several million degrees, dynamo processes must be at work in VB8. Drake *et al.* (1996)

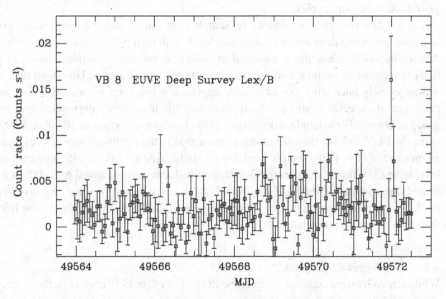

Fig. 4.13. Deep survey telescope light curve of VB8. Each time bin corresponds to one orbit of EUVE, about 5540s of elapsed time and between 500s and 200s of exposure (from Drake *et al.* 1996).

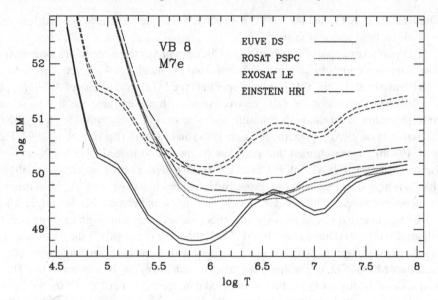

Fig. 4.14. Logarithmic emission measure/temperature loci for the combined X-ray and EUV observatons of VB8. Each pair of curves spans the uncertainties in the observed count rates (from Drake *et al.* 1996).

interpret this as providing evidence that a turbulent dynamo mechanism, rather than a solar-like large scale field shell dynamo, dominates the magnetic activity of fully convective M dwarfs. This then implies that the same mechanism also dominates the young active stars with convective cores (see Stern *et al.* 1995b). As they then spin down with age their activity will change from a régime dominated by a field dynamo to more solar-like behaviour, giving rise to activity cycles.

The EUVE catalogues were based on search algorithms looking for significant excesses of counts above the determined background level, with no *a priori* information on the likely sources included. While this is essential to provide a statistically uniform survey, it potentially ignores weak sources, that fall below the sensitivity threshold, but which may still be astrophysically interesting. An alternative approach is to search for sources using an input catalogue of expected source positions, to extract the maximum information for a specific group of objects. For example, Mitrou *et al.* (1997) performed a study of 104 RS CVn systems using the EUVE all-sky data, detecting 11 more than in the published survey. This represents an increase of 40% in the sample available for study. Mitrou *et al.* identify a general trend of increasing EUV flux (Lex/B band, 50–180 Å) with decreasing rotational period (figure 4.15). In addition, comparing the EUV fluxes with the X-ray output shows that radiative losses in the EUV are a smaller fraction of the total losses from the stellar coronae of the RS CVns than they are in the active late-type stars.

4.3.7   *Extragalactic objects*

While few AGN were detected in either the ROSAT WFC or EUVE surveys, these detections are significant since they indicate the low column density lines-of-sight through the galaxy. In addition, the measured EUV fluxes can provide important constraints on the emission

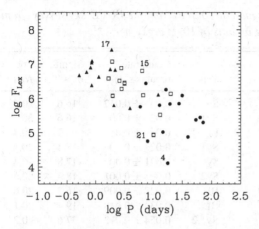

Fig. 4.15. The log–log diagram of surface Lex/B flux (ergs cm^{-2} s^{-1}) against photometric period (days). Circles, systems comprising giants; squares, subgiants; triangles, dwarfs (from Mitrou *et al.* 1997).

processes in these systems (see section 3.8.4). The uniform detection criteria applied to the analysis of the survey data revealed nine and eleven AGN in the WFC and EUVE surveys respectively. The fact that the samples, although similar in size, are not identical in content is an indication of different survey coverage and also possible variability in the objects themselves (see e.g. Barber *et al.* 1995). As discussed in section 4.3.6 for the RS CVn binaries, additional detections can be generated using an input catalogue of objects by measuring the count rates at their positions in the survey database. The *a priori* knowledge that a possible source is located at a particular position increases the confidence of any detection. All the AGN detected in the all-sky surveys share the common property that they are also X-ray sources, which is not surprising since the EUVE 100 Å and WFC S1 bands partially overlap the X-ray range. Hence, Fruscioni (1996a,b) has used a catalogue of $\approx$2500 extragalactic X-ray sources to search the EUVE all-sky survey data archive.

This programme found 21 extragalactic objects as likely sources of EUV radiation. In all but one case, the X-ray galaxy was the only catalogued object within 3 arcsec of the EUV source position. As the white dwarf GD336 is close to the position of EXO1429.9+3717 and probably contributes most of the EUV flux, the former object was excluded from the final list of 20 summarised in table 4.8. Eight of these were not detected during the normal survey analysis and all previously reported detections were recovered in this analysis. Figure 4.16 shows the distribution of these objects on the sky together with the galactic distribution of H I from Dickey and Lockman (1990). As expected the column densities of all the detected EUV AGN are below $\approx$3 × 10^{20} cm^{-2} in all cases except Mrk 507 and PKS2005-489. However, the exposure of Mkn 507, which lies at high ecliptic latitude, was more than 35 000 s, which could explain its detection through a higher column density.

Few diffuse sources have been detected in either WFC or EUVE sky surveys and all the AGN in the catalogues and the work of Fruscione (1996b) are point sources. However, a deep ROSAT pointing of the galaxy cluster A2199 revealed an extended source of EUV emission associated with the cluster core, suggesting that similar structure may be detectable in other clusters in favourable circumstances. An EUVE deep survey telescope observation of the central galaxy of the Virgo cluster M87 reveals a source position consistent with being

Table 4.8. *X-ray selected EUV extragalactic sources from the EUVE sky survey (from Fruscione 1996b). Source fluxes are in units of* $10^{-12}\ ergs\ cm^{-2}\ s^{-1}$.

Source name	Other name	Type	Count rate (counts s^{-1})	Signif $\sigma > 2$	Log $N_{H\,I}$	Flux
MS0037.7–0156		Sy	0.035 ± 0.012	16.6	20.5	4.33
1H1023+513C	Mrk142	Sy1	0.026 ± 0.008	16.6	20.1	1.36
1WGA118.8+1306	M65	LI	0.049 ± 0.015	23.7	20.3	3.84
EXO1128.1+6906		Sy1	0.022 ± 0.006	17.4	20.1	1.15
1H1226+128	M87	Sy	0.031 ± 0.009	17.8	20.4	3.06
2E1748.8+6842	Mrk507	Sy2	0.005 ± 0.001	18.3	20.6	0.76
RXJ20094–4849	PKS2005–489	BL	0.027 ± 0.009	16.3	20.6	4.08
1H2351–315.A	H2356–309	BL	0.034 ± 0.010	19.4	20.1	1.78
WGA0057–2222	TONS180	Sy1.2	0.054 ± 0.012	37.6	20.2	3.42
1H0419–577	1eS0425–573	Sy	0.036 ± 0.006	72.0	20.4	3.55
REJ1034+393	X12325	Sy	0.023 ± 0.007	19.4	20.2	1.46
1H1104+382	Mrk421	BL	0.052 ± 0.010	50.2	20.2	3.29
H1226+023	3C273	QSO	0.050 ± 0.015	20.0	20.3	3.92
1H1350+696	Mrk279	Sy1	0.024 ± 0.005	44.4	20.2	1.52
1H1415+255	NGC5548	Sy1.5	0.029 ± 0.008	24.1	20.2	1.83
1H1430+423	WGAJ1428+4240	BL	0.029 ± 0.007	29.2	20.1	1.51
1H1429+370	Mrk478	Sy1	0.061 ± 0.010	83.6	20.0	2.71
1H1651+398	Mrk501	BL	0.038 ± 0.007	56.6	20.2	2.40
1H2156–304	PKS2155–304	BL	0.269 ± 0.021	571.7	20.1	14.1
1H2209–4470	NGC7213	Sy1	0.032 ± 0.009	20.0	20.4	3.16

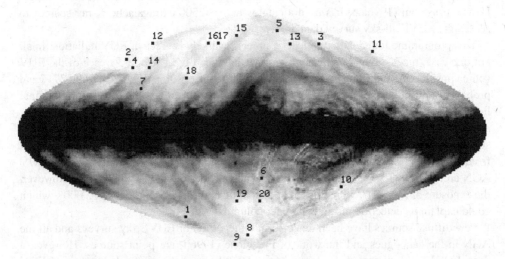

Fig. 4.16. Galactic distribution of the X-ray selected EUV extragalactic sources from the EUVE sky survey, superimposed on the neutral hydrogen sky map of Dickey and Lockman (1990). The map (from Fruscione 1996a) is centred on the galactic centre with longitude increasing to the left. Numbers correspond to the sources listed in order in table 4.8.

centred on M87 and a halo of diffuse emission extending to a radius of about 20 arcmin (Lieu *et al.* 1996). A detailed survey of H I in the region towards M87 indicates that the line-of-sight H I column density is between 1.8 and 2.1 $\times 10^{20}$ cm^{-2}.

The EUV data were compared with ROSAT PSPC data from the all-sky survey, after subtracting the effects of the diffuse soft X-ray background. This is mainly galactic, since the extragalactic component is only $\approx 2 \times 10^{-4}$ counts s^{-1} arcmin^{-2}, accounting for less than 1/5 of the surface brightness within 20 arcmin from M87. Comparing the EUVE deep survey and ROSAT PSPC data simultaneously with the Mewe optically thin thermal plasma model, no satisfactory agreement can be found if only a single temperature is assumed. The main reason for a poor fit is the presence of count rate excesses in the deep survey data when the PSPC data are adequately matched by the model. This excess cannot be explained by adjusting the H I absorption, depleting the abundances of interstellar C and O or considering non-cosmic abundances for the emitting gas. A significantly improved fit can be obtained by using a two temperature plasma model. To account for the soft excess, the additional plasma component has a considerably lower temperature, in the range 5–10×10^5 K. The cooling time of the cool plasma component ($\approx 2 \times 10^8$ y) is much less than the age of the cluster. Hence, a mechanism for sustaining the gas is necessary. The 'mass accretion rate' of $\approx 340 M_\odot$ per year is much higher than the $\approx 10 M_\odot$ per year for an implied cooling flow. Therefore, the gas is probably maintained by heating which results from a shock in the central region of the inflow.

The difficulty in explaining the presence of the excess EUV emission has prompted several authors to examine its reality. For example, Arabadjis and Bregman (1999) suggest that the effect may arise from varying absorption cross-sections of H and He, due to different ionisation states of the intervening galactic ISM. However, the EUVE results cannot be explained in this way. Bowyer *et al.* (1999) reanalyse the EUVE data for Abel 2199 and Abell 1795, concluding that the existence of an excess is very sensitive to the characterisation of the DS telescope. They find no evidence for an excess in either of these clusters but do find extended EUV emission in the Coma cluster. However, Lieu *et al.* (1999) counter this result with a reobservation of A2199, determining the *in situ* background with an offset DS pointing, which appears to show that the excess is present.

Further work by Bowyer *et al.* (1999), using the same data reduction techniques, also reveals EUV emission from the Virgo cluster and the jet of M87. In general the observed emission from clusters is not compatible with a thermal plasma origin. It has been suggested that the EUV flux observed in clusters arises from inverse Compton scattering of relativistic electrons again the 3 K blackbody background. Bowyer *et al.* show, at least for the Virgo cluster, that the EUV emission cannot be produced by extrapolation to lower energies of the observed synchroton radio emitting electrons and that an additional component of low energy relativistic electrons is needed.

4.3.8 *Shadowing of the EUVE background*
There is little evidence of a diffuse EUV background component in the ROSAT WFC all-sky survey (West *et al.* 1996). Detection of any such emission was made more difficult by the dominance of other sources of background such as spacecraft glow, geocoronal radiation and high energy electrons but no cosmic component was visible in the clean data periods. The EUVE DS telescope had increased sensitivity through higher filter transmission, leading to a larger effective area, coupled with a deep exposure along the ecliptic plane. Bowyer *et al.*

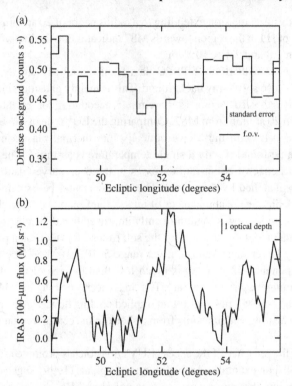

Fig. 4.17. (a) The diffuse background as measured by the EUVE deep survey telescope using the Lexan/boron filter as a function of ecliptic longitude showing an absorption feature centred at 52°. The dashed line is the local background level derived from data either side of the feature. The 1σ errors in the count rates and the exposure averaged deep survey field of view are indicated. (b) The IRAS 100 μm continuum subtracted flux averaged over the deep survey field of view as a function of ecliptic longitude. The flux level corresponding to an optical depth of EUV absorption is marked (from Bowyer *et al.* 1995).

(1995) report a shadow in the EUV background measured by the deep survey telescope. Figure 4.17a shows the diffuse background count rate, integrated over the 1.64° telescope field of view in the ecliptic longitude range from 47° to 57°. The boundaries of this EUV depleted region are aligned with an enhancement in the infrared flux observed by the Infrared Astronomical Satellite (IRAS) at that position (figure 4.17b). An estimate of the H I absorbing column associated with the IRAS cloud is 6×10^{19} cm^{-2}, corresponding to about three optical depths at the mean wavelength of the DS telescope. Hence, the EUV flux from behind the cloud is completely blocked and the observed count rate in this direction due entirely to foreground emission.

Bowyer *et al.* (1995) determined the distance to the absorbing cloud using Strömgren photometry, measuring the reddening and absolute magnitude of stars in this region. Calculating the distances to each star from their absolute magnitudes and considering the spatial correlation of the degree of reddening indicates that the cloud lies within a distance of 40 pc. Using a plasma emissivity code to interpret the data and combining the observed count rate in the direction of the cloud with the estimated temperature of the diffuse flux (7×10^7 K;

Lieu *et al.* 1993) gives an emission measure of $0.0077\,\mathrm{cm}^{-6}\,\mathrm{pc}$ for the hot gas present. Its pressure can be estimated using the relation

$$P = 1.92kT(EM/L)^{0.5}$$

giving $P/k = 19\,000\,\mathrm{cm}^{-3}\,\mathrm{K}$. The pressure of the warm local cloud surrounding the Sun is considerably lower than this ($730 \pm 20\,\mathrm{cm}^{-2}\,\mathrm{K}$), disproving the widely held assumption that the ISM should be in pressure equilibrium. However, if warm and hot phases are not in pressure equilibrium, cooler clouds should not persist. One possibility is that the gas is confined by magnetic fields (e.g. Cox and Snowden 1986). Cox (1996) points out that this pressure imbalance has been known for about ten years. However, in the absence of a viable explanation, the McKee and Ostriker (1977) thermal pressure equilibrium model of the ISM has continued to be the main point of reference. The work of Bowyer *et al.* (1995) adds further impetus to the need for development of a new model.

4.4 The ALEXIS mission

The ROSAT WFC and EUVE payloads have provided high sensitivity broad band surveys of the sky aimed at detecting point sources of EUV radiation with high sensitivity and probing the cosmic EUV background. A third survey of the sky has been conducted covering a series of narrow bands using the Los Alamos pathfinding small space mission ALEXIS (Array of Low Energy X-ray Imaging Sensors). A prime consideration, in comparison with the relatively large, high cost observatory class missions, is to achieve low cost and rapid development time for the technology while still providing a unique and useful scientific return.

The ALEXIS small satellite contained an ultrasoft X-ray/EUV monitoring experiment and a very high frequency (VHF) broadband ionospheric survey experiment called BLACK-BEARD. The whole project was led by the US Los Alamos National Laboratory (LANL). Payload data processors were provided by the Sandia National Laboratory, detectors and high voltage supplies by the University of California at Berkeley Space Sciences Laboratory and the custom, low cost miniature satellite bus by AeroAstro Inc. ALEXIS was launched on a Pegasus booster into a 400–450 nautical mile orbit on April 25 1993. The satellite and experiments were controlled from a small ground station at LANL.

In contrast to the WFC and EUVE missions, which utilise grazing incidence telescope technology to provide the focusing optical system (see sections 3.3.1 and 4.1.1), the ALEXIS experiment used normal incidence reflecting telescopes with multilayer coatings to yield high reflectivity at these angles. Normal incidence optics have an advantage in offering a larger geometric collecting area compared to grazing incidence units for a given mirror diameter. This translates into smaller, less massive and, consequently, lower cost payloads to achieve a given throughput. The major disadvantage of these systems is the need to tune the multilayer coatings to a specific narrow band $\approx 30 - 40\,\mathrm{\mathring{A}}$ wide, outside of which the reflectivity will be very small. Consequently, these optical systems are only useful for scientific applications that can utilise narrow bands and are not suitable for missions requiring broad and continuous spectral coverage, such as the EUVE spectroscopy telescope. ALEXIS provided a narrow band sky survey and monitoring mission which was well-matched to the normal incidence technology.

4.4.1   *Normal incidence mirror technology for EUV astronomy*

The relatively poor reflectances of materials at EUV wavelengths places strong constraints on instrument designs which employ normal incidence reflections. Nevertheless, it is possible to design effective normal incidence telescopes and spectrometers for use in the EUV, particularly at wavelengths longer than 400 Å. There are three basic classes of materials which have been used in space applications. Heavy metal reflective coatings, in particular gold, platinum and osmium have relatively good reflectivities, maintaining reflectances of 10% to 20% down to 400 Å. A very different type of coating is represented by compounds such as silicon carbide (SiC) and boron carbide (BC), which are twice as reflective as gold or platinum over these same wavelengths (Keski-Kuha *et al.* 1997). Figure 2.4 shows typical reflectivities for gold, platinum, osmium and SiC over the wavelength range 200 to 1200 Å. Finally, it is also possible to construct multilayer thin film mirrors which employ interference to optimize reflectance over a relatively narrow band of EUV wavelengths. In general, factors such as the surface roughness of the substrate, the surface preparation and the method of deposition of the coating can play a significant role in achieving high levels of reflectivity. In orbit, direct exposure to the residual atmospheric stream of atomic oxygen in the ram direction can rapidly degrade the reflectivity of any coating.

Because of the superior performance of SiC in the EUV and the far-UV, this material is preferred for normal incidence reflective surfaces between 400 and 1050 Å. At still longer wavelengths, coatings such as LiF and MgF_2 are preferred. The subject of reflective coatings in the EUV is reviewed in Samson (1967) and Hunter (2000) and the measured EUV reflectivities of a wide range of materials are presented in Windt *et al.* (1988a,b).

There are several examples of the instruments that use different normal incidence optical designs. The Voyager ultraviolet spectrometers (see section 2.9) employed a collimated objective grating spectrograph using a platinum coated concave grating. The ORFEUS and HUT instruments used large primary mirrors and concave gratings which focused the spectra on the Rowland circle. It is interesting to note that first flight of HUT in 1990 used osmium coated optics, while the second flight in 1995 achieved much higher throughput with the use of SiC coated optics. The *FUSE*, which operates in the 900 to 1170 Å band, is designed with four separate apertures which are grouped into two parallel 'sides' each consisting of a pair of channels with a SiC coated primary mirror and concave grating and a LiF coated primary and grating. In this design a reasonable effective area is achieved across the entire far-UV band. The use of multilayer mirrors has primarily been restricted to EUV imaging applications. Examples of such instruments are the ALEXIS cameras (see section 4.4) and the Extreme Ultraviolet Imager (EUV) on the IMAGE mission (Sandel *et al.* 2000). The latter instrument uses a multilayer mirror consisting of a uranium oxide layer followed by six and a half bilayers of uranium/silicon that image the Earth's plasmasphere in the resonantly scattered He II 304 Å emission line. The multilayer mirror produces a band pass which peaked near 304 Å, while effectively rejecting the stronger He I 584 Å line.

4.4.2   *The ALEXIS scientific payload*

The X-ray/EUV monitor experiment consists of six compact normal incidence telescopes operating in narrow bands centred on 130, 176 and 186 Å (Priedhorsky *et al.* 1988). Although normal incidence multilayer X-ray/EUV telescopes have been used for solar observations, ALEXIS represented their first successful use for non-solar cosmic studies. The ALEXIS telescope design used a simple $f/1$ optical system consisting of an annular entrance aperture,

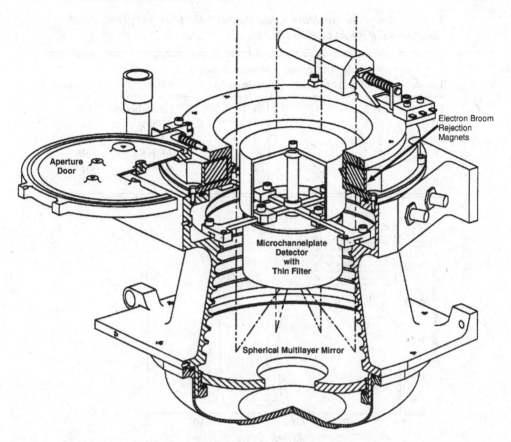

Fig. 4.18. Cross-sectional view of one of the six ALEXIS telescopes. Each telescope has an identical design (courtesy J. J. Bloch).

a spherical mirror, optical/UV blocking filter and a curved MCP detector (figure 4.18). The MCP detector had a convex curved front to match the curvature of the focal surface of the spherical mirrors so that the spatial resolution was approximately constant over the entire 33° field of view. Spherical aberration restricted the overall spatial resolution to 0.25°.

The ALEXIS mirrors were coated with molybdenum and silicon layers and their spacing is the main factor determining the telescope's spectral response (Smith *et al.* 1989, 1990). These multilayers also employ a 'wave-trap' feature to significantly reduce the reflectivity at 304 Å to reduce contamination from the He II geocoronal radiation which would otherwise be a significant source of background. The individual telescope bandpasses, filter and photocathode complement are listed in table 4.9 and the on-axis effective areas are shown as a function of photon energy in figure 4.19.

The six EUV telescopes were coaligned in pairs and covered three overlapping 33° fields of view (figure 4.20), angled as listed in table 4.9. The satellite was spin stabilised, rotating once every 50 s, during which time the telescopes scan most of the anti-solar hemisphere. The geometric collecting area of each telescope was about 25 cm^2 and peak on-axis effective areas ranged from 0.25 to 0.05 cm^2 (figure 4.19). In one 12 h data collection period, the brightest EUV sources in the sky (e.g. HZ 43 and GD153) can be detected.

Table 4.9. *Telescope bandpasses, together with the filter and photocathode complement of the ALEXIS payload.*

Telescope	View direction offset from spin axis (deg)	Multilayer mirror bandpass (Å)	Filter	Photocathode
1A	87.5	130	Lexan/boron	MgF_2
1B	87.5	172	Al/Si/C	NaBr
2A	56	130	Lexan/boron	MgF_2
2B	56	186	Al/Si/C	NaBr
3A	31.5	172	Al/Si/C	NaBr
3B	31.5	186	Al/Si/C	NaBr

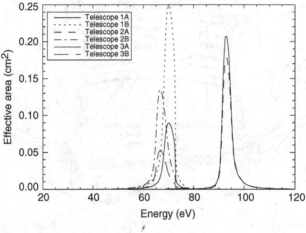

Fig. 4.19. Pre-flight on-axis effective area for each ALEXIS telescope (from Bloch 1996).

4.4.3   ALEXIS scientific goals

With its wide fields of view (33°) and narrow wavelength bands, ALEXIS complemented the surveys of the ROSAT WFC and EUVE which cover a broad spectral range with comparatively narrow fields of view ($\approx 5°$) and broad spectral coverage. The 172 and 186 Å bandpasses are tuned to the Fe IX–XII emission line complex which is characteristic of million degree optically thin plasmas which exist in stellar coronae, are believed to occupy a large fraction of interstellar space, and generate the soft X-ray background. Although the effective areas of each telescope were small compared to the WFC and EUVE, because of the large solid angle covered by the ALEXIS units the experiment was well-suited to studies of the diffuse background. In addition, since large areas of sky were regularly monitored by the observing strategy and telescope design, the EUV sky could be effectively monitored for transient events.

4.4.4   ALEXIS in-orbit loss and recovery

Following its launch into a 884×749 km orbit, with an inclination of 70°, initial attempts to contact ALEXIS were unsuccessful. Apparently, one of the solar paddles had deployed prematurely. During the weeks after launch, it became clear that, after arrival in-orbit, ALEXIS had switched on long enough to deploy the three undamaged solar panels. On June 30 1993

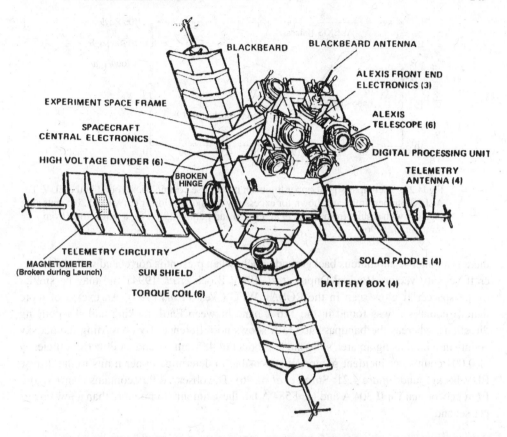

Fig. 4.20. The in-orbit configuration of the ALEXIS satellite showing the deployment of the six telescopes. The solar panel and magnetometer damaged during launch are indicated (from Bloch 1996).

ALEXIS transmitted a strong signal, remaining in contact for nearly 4 min. During this period, telemetry data showed that ALEXIS was spinning about an axis nearly 90° from the Sun. All systems were found to be functional, except the magnetometer. Following this, regular contact was established with ALEXIS, the satellite was brought under control and scientific operations began at the end of July 1993 (see Bloch (1996) for a more detailed summary). Initially, telescope data were collected blindly until it was possible to derive an attitude solution algorithm that took into account the modified moment of inertia of the spinning satellite, the absence of magnetometer data and the possible motion of the broken solar panel. Several developments have led to imaging of the Moon and the white dwarf HZ 43 (Bloch *et al.* 1994). Refinement of the software produces solutions with an accuracy better than 0.5° and the ability to recover data generated during spacecraft manoeuvres.

4.4.5 *Scientific results*

4.4.5.1 *The diffuse EUV background*

Although the telescopes were designed to reduce or eliminate the main sources of non-cosmic background, several unwanted components are present in the ALEXIS data. In particular,

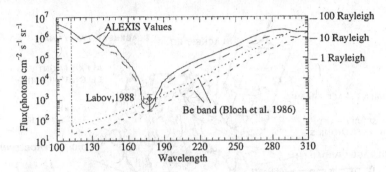

Fig. 4.21. A comparison of upper limits on the diffuse cosmic background in the EUV. The ALEXIS upper limits are shown for excess sky rates of 10 (lower curve) and 20 counts s⁻¹ together with results from Bloch *et al.* (1986) and Labov (1988) for a single line (from Smith *et al.* 1996).

there is an intense anomalous background which, when present is correlated with the spacecraft velocity vector (Roussel-Dupré *et al.* 1996; Bloch *et al.* 1994) and may be similar to the spacecraft glow seen in the ROSAT WFC (West *et al.* 1994). An excess of more than 10 counts s⁻¹ was found in the $176\,\text{Å}$ band between Earth-looking and sky-pointing directions, whereas the bandpass at $133\,\text{Å}$ shows no difference. By converting the net sky counts into flux, using an area solid angle product of $4.22\,\text{cm}^2$ sr and an effective efficiency of 0.004 counts per incident photon, it is possible to determine upper limits to the diffuse EUV background (figure 4.21; Smith *et al.* 1996). The observed flux contains contributions from geocoronal He II $304\,\text{Å}$ and He I $584\,\text{Å}$ but these amount to no more than a few counts per second.

4.4.5.2 The narrow band source survey
More than 18 steady EUV sources were detected in an analysis of a year of ALEXIS survey data (Roussel-Dupré *et al.* 1996). The majority of these are hot white dwarfs and were already detected in the WFC and EUVE sky surveys (table 4.10). Since all the objects are bright enough for spectroscopic observations, there is probably little that ALEXIS could add to what is known about these objects. However, they provided a useful check of the in-orbit sensitivity of the ALEXIS telescopes and cross-calibration with other EUV instruments (table 4.11).

4.4.5.3 Transient phenomena
Detections of bright transient EUV sources is the most significant contribution that ALEXIS has made to EUV astronomy. Interest in this area was heightened by the serendipitous discovery of a transient source (REJ1255+266) in the WFC data during a ROSAT observation of the Coma cluster of galaxies (Dahlem *et al.* 1995). This object brightened by more than a factor of 4000 from its quiescent state, decaying back to quiescence over a period ≈ 14 d. REJ1255+266 has subsequently been identified with a close binary system comprising a white dwarf and late-type star but the mechanism for the outburst remains uncertain (Watson *et al.* 1996).

Three methods were employed to search for transients in the ALEXIS data: (i) archival search; (ii) daily automated sky map searches; and (iii) manual inspection of the daily skymaps (see Roussel-Dupré *et al.* 1996). Five transient/variable sources have been reported

Table 4.10. *Previously identified sources detected by ALEXIS (from Roussel-Dupré et al. 1996).*

Name	RA (deg)	DEC (deg)	Spectral type
WD0050-332 (GD659)	12.74	−33.29	DA1
Feige 24	38.44	3.58	DA0
V471 Tau	57.54	17.18	DA2+dK
VW Hyi	62.60	−71.05	CV
MCT0455−2812	73.97	−28.05	DA
G191−B2B	76.41	52.94	DA0
WD0549+158 (GD71)	88.21	15.67	DA1
WD0642−166 (Sirius B)	101.12	−16.86	DA2
U Gem	118.05	22.00	CV
REJ1032+532	157.80	53.41	DA
WD1254+223 (GD153)	194.27	22.01	DA1
HZ 43	199.23	29.19	DA1
WD1501+55	222.54	66.39	DZ0
WD2111+498 (GD394)	318.04	49.94	DA2
REJ2156−543	329.62	−54.35	DA
REJ2214−491	333.64	−49.19	DA
WD2309+105 (GD246)	348.04	10.98	DA0

Table 4.11. *Comparison of pre-flight ALEXIS count rate predictions with observed values (from Roussel-Dupré et al. 1996).*

	Count rate (count s^{-1})		
Telescope	Pre-flight	Observed (Jan 94)	Observed (Mar 95)
HZ 43			
1A	0.12	0.02 ± 0.015^a	0.025 ± 0.015
1B	0.49	0.46 ± 0.05	0.55 ± 0.05
3A	0.18		0.26 ± 0.03
3B	0.13		0.18 ± 0.02
G191−B2B			
1B	0.057		0.05 ± 0.015

^{a} Mirror response much narrower than initially predicted.

(table 4.12). Three of these are CVs, two of which (VW Hyi and U Gem) have been previously well studied. VW Hyi was seen in a superoutburst from May 30 to June 6 1994. ALEXIS J1139–685 and ALEXIS J1644–032 both have similar time signatures, having outbursts lasting for 24–36 h. EUVE target of opportunity observations also failed to detected them, despite starting within 24–36 h after maximum light. These have a different signature compared to REJ1255+266 and are potentially a new type of transient. Searches of the positional error boxes have failed to produce good candidates for counterparts to these systems.

Table 4.12. *Transient EUV sources detected by ALEXIS (from Roussel-Dupré et al. 1996).*

Name	RA (deg)	Dec (deg)	Spectral type	Telescope
VW Hyi	61.60	−71.05	CV	(1B 172 Å)
U Gem	118.05	22.00	CV	(2A 130 Å)
ALEXIS J1114+430	168.66	42.73	CV	(1B 172 Å)
(=AR UMa)				(1A/2A 130 Å)
				(2B 186 Å)
ALEXIS J1139–685	174.51	−69.9	unknown	(1B 172 Å)
ALEXIS J1644–302	251.23	−3.21	unknown	(2B 186 Å)

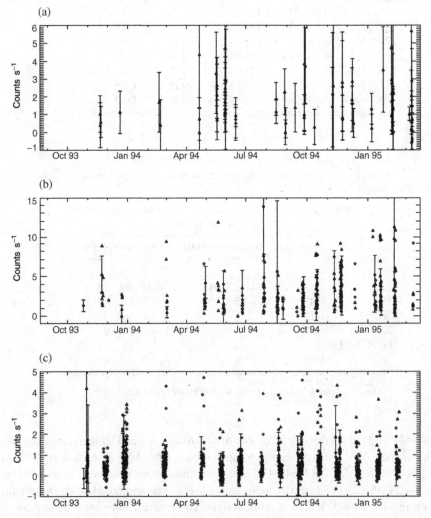

Fig. 4.22. Lunar count rates observed by ALEXIS. (a) Telescope pair 1, (b) pair 2 and (c) pair 3. In all plots the diamonds are for the A telescopes and triangles for the B telescopes (see table 4.9; data from Edwards *et al.* 1996).

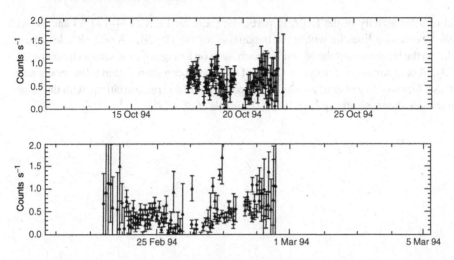

Fig. 4.23. Lunar count rates observed by ALEXIS for several representative periods. Diamonds represent the 3A telescope and triangles the 3B telescope (from Edwards *et al.* 1996).

4.4.5.4 *Gamma ray bursts*

The detection of EUV counterparts to gamma ray burst sources would be of tremendous importance in understanding the nature of these enigmatic objects. Early arguments, prior to 1997, centred on whether the burst sources were cosmological distances or comprised a nearby population. Detection of X-ray and optical counterparts to a number of bursts now shows that many, if not all, bursts are associated with very distant galaxies. However, there is still the remote possibility that more than one population of burst sources may exist.

Before the first identification of gamma ray burst counterparts, the detection of their EUV emission was seen and indicated that a burst source would have to be within a few hundred parsecs of the Earth, due to the opacity of the ISM. Owens *et al.* (1993) searched the ROSAT WFC all-sky survey for short duration (<50 s) bursts but found no evidence for any such events in the 171 day dataset. This is not too surprising, because the instantaneous 5 degree field results in a low probability that any single burst detected by the Compton Gamma Ray Observatory (CGRO) could have fallen within the area of sky observed by the WFC. In contrast ALEXIS scanned a much larger area of sky with every 50 s rotation of the satellite, giving a significant chance that a gamma ray burst detected by the CGRO might also have been observed by ALEXIS. In an examination of 39 gamma bursts reported by Bloch (1996), ten were not in any area surveyed by ALEXIS, 12 occurred behind the Earth from ALEXIS' viewpoint and three occurred during satellite shutdown periods. The remaining 14 bursts are candidates to search for pre- or post-burst EUV emission and four are good candidates for near simultaneous observations. That is, ALEXIS may have observed the error box at the exact moment the burst occurred.

4.4.5.5 *Obervations of the Moon*

ALEXIS semi-continually monitored the Moon for two weeks around the full moon of every lunar cycle. Since the lunar emission is primarily reflected solar radiation (see section 4.3.3; Edwards *et al.* 1991; Gladstone *et al.* 1994), the brightness of the Moon is an indication of the level of solar activity in the EUV. In particular the solar Fe X (174–177 Å) and Fe XI (180–190 Å) emission lines lie within the bandpasses of the 1B, 2B, 3A and 3B telescopes. Variability in the brightness of the Moon is clearly seen on a range of time scales (figures 4.22 and 4.23). Flux changes greater than a factor of two have been seen within a few hours and as much as a factor of five over a few days. Comparison of these rate variations with the solar 10.7 cm indices shows no strong correlation (Edwards *et al.* 1996).

5

Spectroscopic instrumentation and analysis techniques

5.1 The limitations of photometric techniques

The photometric all-sky surveys conducted in the EUV by the ROSAT WFC and EUVE have been sources of important information concerning the general properties of groups of objects contained in the EUV source population, including late-type stars, white dwarfs, cataclysmic variables and active galactic nuclei. However, when considering individual objects in detail, the amount of information that can be extracted from three or four such data points is limited. For example, if heavy elements are present in the atmosphere of a hot white dwarf, the survey data can only give an indication of the level of opacity and are unable to distinguish between the possible species responsible and especially whether or not helium is present. Similarly, in studying the emission from stellar coronae, only crude estimates of conditions in the plasma can be made and usually only when simplifying assumptions such as the existence of a single temperature component are incorporated into the analysis. The overwhelming advantage of spectroscopic observations lies in the ability to study individual spectral features or blends of features, giving a more detailed picture of the underlying physical processes responsible for the EUV emission.

5.2 The Extreme Ultraviolet Explorer spectrometer

The main components of EUVE have been described in detail in chapter 4 with the exception of the spectrometer. This instrument made use of part of the converging beam of the Wolter type II deep survey telescope, intercepting this with three variable line space reflection gratings (Hettrick et al. 1985). Such gratings are mechanically ruled plane gratings with a line density that varies continuously from one end of the grating to another to correct first order aberration arising from the varying incident angle at the grating. Each grating used one sixth of the telescope aperture and was focused onto a separate MCP detector/filter unit. The filter attenuated the higher order spectra and emission outside the main bandpass arising from possible grating scatter. The individual gratings were tuned to different wavelength ranges, two of which were susceptible to geocoronal helium resonance lines at 304 Å and 584 Å. These two channels were precollimated by wire grid collimators to limit the background emission from diffuse sources. Figure 5.1 shows an exploded view of the instrument and the light path for the deep survey channel. Several papers have been published describing the spectrometer with a useful overview and details of the instrument calibration being presented by Welsh et al. (1989) and Abbott et al. (1996).

The combination of the three gratings provided moderate resolution ($\lambda/\delta\lambda \approx 200$) spectroscopy in the 80 to 760 Å band. This coverage is divided into three channels, the short wave (SW, 70–190 Å), medium wave (MW, 140–380 Å) and long wave (LW, 280–760 Å).

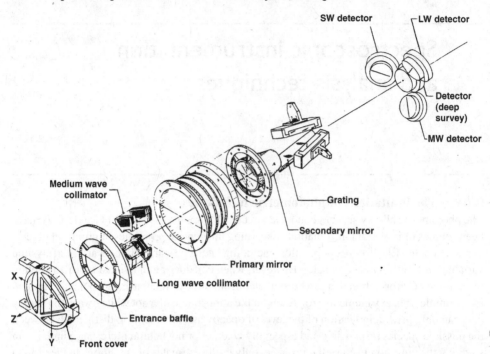

Fig. 5.1. Exploded view of the EUVE deep survey/spectrometer (DSS) telescope. The axes drawn at the front of the instrument represent the DSS coordinate system (from Hettrick *et al.* 1985).

Contributions from each element in the optical system, including grating aberrations, line spacing errors, on/off-axis mirror aberrations detector resolution, alignment errors and in-orbit aspect uncertainty, determined the overall spectral resolution. Figures 5.2 and 5.3 show the spectral resolution, measured in-orbit (Abbott *et al.* 1996), and effective area as a function of wavelength. As can be seen in figure 5.2, the spectral resolution varied by about a factor of two across each band, with typical average values of 0.5 Å, 1.0 Å and 2.0 Å for SW, MW and LW ranges respectively. A detailed description of the performance can be found in the references noted above and the basic parameters are summarised in table 5.1. The MW and LW channels had wire grid collimators placed in the aperture before the mirrors in order to limit the field of view and exclude some of the sky background, which is dominated by geocoronal emission lines. There are no geocoronal lines in the SW wavelength range, hence a collimator was not needed in that case. Each collimator had a peak transmission of $\approx 60\%$ on-axis and consisted of 15 etched molybdenum grids, spaced exponentially and held in a thermally stable molybdenum claw.

The MW and LW spectrometer channels suffer from contamination by the higher spectral orders, which must be taken into account in any data analysis and whose effective areas are included in figure 5.3. As a result, it is not possible to produce a flux calibrated spectrum simply by dividing the counts by the instrument effective area. When the source spectrum mainly arises from emission lines, with minimal continuum at the shorter wavelengths, this problem is not too severe. The second and third order features can be readily identified at $2\times$ and $3\times$ the line wavelengths. However, if the source has a strong short wavelength continuum, as is the case for a white dwarf, the second and third order components are

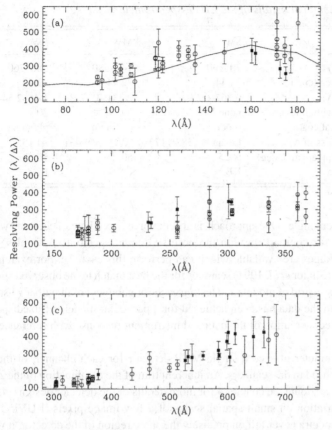

Fig. 5.2. Spectral resolution of each spectrometer band as a function of wavelength (from Abbott *et al.* 1996).

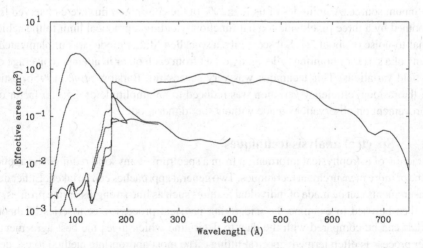

Fig. 5.3. Effective area of the EUVE spectrometer as a function of wavelength. The upper curves are the first order values for SW, MW and LW bands while the lower curves are the respective second and third order effective areas.

Table 5.1. *Technical details of the EUVE spectrometers.*

	SW	MW	LW
Bandpass (Å)	70–190	140–380	280–760
Geometric area (cm^2)	72.44	72.44	72.44
Resolution (Å)	0.367–0.636	0.731–1.27	1.46–2.54
Collimator fwhm (arcmin)	none	20	20
Grating optical coating	rhodium	platinum	platinum
Filter (thickness in Å)	Lexan/B 1588/1172	Al/C 1566/431	Al 1553
Detector diameter (mm, deg)	50, 5	50, 5	50, 5
Photocathode	KBr	KBr	KBr

more difficult to disentangle. One approach is to convolve a predicted spectrum with the instrument response for comparison with the observed count spectrum. Several so-called 'spectral fitting' packages are available which can perform this exercise. For example, the programme XSPEC (Shafer *et al.* 1991) searches for the best match to the observed spectrum by interpolating across a grid of models and testing each for goodness of fit with a χ^2 statistic. Once a good match to the data has been achieved for a particular model, a fluxed spectrum can be obtained by deconvolution of the data and instrument response with reference to the model.

The EUVE spectrometer utilised a single MCP detector for each channel, of the same basic design as those used in the scanners. An inherent trait of the combination of the Z-stack and wedge-and-strip readout (see chapter 4 for more details of the detector design) is a fixed pattern efficiency variation on small spatial scales of a few image pixels (EUVE 1993b). Since the dispersed spectra never fell on precisely the same region of the detector, it was not possible to correct directly for this effect by flat-field calibrations. Hence, spurious absorption and emission features were incorporated into the observed spectrum of an otherwise smooth continuum source. Amplitudes of up to $\pm20\%$ of the continuum flux were observed for data smoothed by a three pixel wide top hat function, yielding a practical limit to the achievable signal-to-noise of about 5:1. Subsequently, a so-called 'dither mode' was implemented, consisting of a series of pointings all slightly offset from each other in attempt to average out the flat-field variations. This technique was very successful. Barstow *et al.* (1995b) estimated that the residual efficiency variation was reduced to an amplitude of $\pm5\%$, a factor of four improvement on observations made without the 'dither mode'.

5.3   Spectral analysis techniques

Extraction of astrophysical information from a spectrum of any kind requires the application of one or more measurement techniques. Two general approaches can be taken. Either detailed measurements can be made of individual features such as line strengths or line profiles, which can then be used to examine the underlying physical processes, or a series of theoretical models can be compared with the data to determine which gives the best agreement. This latter process is often termed 'spectral fitting'. The most appropriate method to use depends on the complexity of the particular spectrum to be analysed and the availability of theoretical models. To illustrate these ideas, we outline examples of the study of continuum spectra from hot white dwarfs and emission lines from stellar coronae.

This 'spectral fitting' technique has been widely used in X-ray astronomy and examples have been described in detail in many papers. However, until very recently, the spectral resolution typical of X-ray instruments was inferior to that of EUVE. Hence, the number of independent resolution elements is much larger in EUVE than has been dealt with routinely. The application of the technique to EUVE data has been reported in the work of Barstow *et al.* (1994b,c). However, it is useful to discuss this here, paying particular attention to how errors are estimated for the measured parameters.

Two approaches to spectral fitting can be taken. Either a flux calibrated spectrum can be compared directly with a model or a model can be folded through a function that describes the response of the instrument, taking into account spectral resolution and effective area, and then matched to the observed spectrum in instrumental counts. The second approach is more appropriate for the analysis of EUVE data, as noted in section 5.2, because there is considerable overlap of the higher spectral orders with the prime wavelength range, as a result of the difficulty in finding suitable order separation filters in the EUV. In general, it is not possible to provide a flux calibration for an EUVE spectrum simply by dividing the counts by the instrument effective area function. In either approach, the quality of the match, or 'goodness of fit', can be described by a standard χ^2 statistic (a good description can be found in Press *et al.* 1992), which quantifies the difference between model prediction and data on a point-by-point basis. It assumes that the statistical errors of the data points are independent of one another and distributed normally. Several software packages are available which can search a grid of models and automatically minimise the value of χ^2 to find the best 'fit' to the data. For example, a widely used, publicly available program is XSPEC (Shafer *et al.* 1991).

For EUVE the folding operation must calculate the count spectra for each order of an individual grating and then coadd them to model the effective spectrometer response. Once a good fit to the data has been achieved with a suitable model, a fluxed spectrum can be obtained by deconvolution of data and instrument response with reference to the model. For this analysis, the instrument response function can be drawn from the EUVE reference calibration data archive, which is available in the public domain and is supplied with the Guest Observer data, and takes into account the effective areas of the overlapping orders and the wavelength variation of the spectral resolution. However, Dupuis *et al.* (1995) examined the effective areas of the overlapping spectral orders of the LW spectrometer, finding that the second, third and fourth order responses were all over estimated. They recommended that a grey correction factor be applied to each, dividing them by 1.4, 1.2 and 1.1 respectively.

Estimates of the uncertainty in any fitted parameter of the model can be determined by considering the variation of χ^2 as the parameter value is stepped in small increments away from the best fit value. All other parameters must be allowed to vary so that the new value of χ^2 is a minimum for the fixed value of the parameter being considered. The difference in two values of χ^2, $\delta\chi^2$, then gives an indication of the probability that the true value of the parameter lies between the best fit value and the new one. Stepping the parameter to yield a predetermined $\delta\chi^2$, corresponding to a particular confidence level, gives the estimate in the possible uncertainty in the best fit value. For a given confidence level, the appropriate value of $\delta\chi^2$ depends on the number of 'interesting' parameters, or the number of variables in the model that can significantly effect the quality of the fit. For example, with only one such parameter a 1σ (68%) confidence limit is given by $\delta\chi^2 = 1.0$. With two, three, four and five

Table 5.2. $\delta\chi^2$ *as a function of confidence level (p) and degrees of freedom (v).*

	ν					
p (%)	1	2	3	4	5	6
68.3 (1σ)	1.00	2.30	3.53	4.72	5.89	7.04
90	2.71	4.61	6.25	7.78	9.24	10.6
95.4 (2σ)	4.00	6.17	8.02	9.70	11.3	12.8
99	6.63	9.21	11.3	13.3	15.1	16.8
99.73 (3σ)	9.00	11.8	14.2	16.3	18.2	20.1
99.99	15.1	18.4	21.1	23.5	25.7	27.8

variables the corresponding $\delta\chi^2$s are 2.3, 3.8, 4.6 and 5.9 respectively (see Lampton *et al.* 1976b, Press *et al.* 1992 for a detailed discussion). Standard criteria for quoting confidence limits on estimated parameters are 1σ (68%), 2σ (95%), 3σ (99%) or 90%.

5.4 Theoretical spectral models

The practicality of studying any observed spectrum with a fitting technique, as described above, depends upon the availability of suitable theoretical spectra to compare with the observation. While the 'goodness of fit' can be determined for any individual model, a search for the best fit requires a grid of models indexed by all the important parameters and spanning an appropriate range. For example, a simple white dwarf model composed of just hydrogen and helium is specified by three parameters – effective temperature (T_{eff}), log surface gravity and helium abundance (or H layer mass for a stratified model). Combining these with the detailed ISM model of Rumph *et al.* (1994) adds a further three parameters – H I column density, He I column density and He II column density, to which must be added a distance/radius related normalisation constant, giving seven variables in all. The problem can be simplified by using information from other wavebands to constrain some parameters. For white dwarfs, Balmer line fits to the optical data give estimates of T_{eff} and log g within narrow bounds. Furthermore, a measurement of V magnitude, from optical photometry, can be used to determine the normalisation. Hence, only four parameters may be completely unknown and an estimate of the 1σ error for each requires $\delta\chi^2$ to be 5.9. Table 5.2. lists values of $\delta\chi^2$ for a range of significance values and degrees of freedom.

Inclusion of other elements in the model atmosphere calculations will increase the potential number of free parameters in the analysis. However, since, in general, individual lines cannot always be resolved in EUVE spectra, because several features may be blended, the increase in complexity may be difficult to handle and a spectral fitting technique no longer appropriate. Even so, using the χ^2 parameter to evaluate the quality of agreement can still be useful. Common ways of dealing with large numbers of heavy elements are to identify one or two which are expected to dominate the EUV opacity or to work with a single abundance parameter and fixed scaling between the abundances of individual elements. For example, relative abundances could be held at solar values scaled to the predicted abundances of radiative levitation theory.

5.4.1 Computation of white dwarf model spectra

While a range of simple models, such as a blackbody or power law, can be used to parameterise an observed spectrum, these are usually unable to provide much insight into the physics of the objects under study. Calculation of model spectra for comparison with observational data is essentially a computational effort, particularly when trying to include realistic input physics as a means of understanding the underlying physical processes.

The two groups of objects where most progress has been made in producing detailed models for the interpretation of EUVE data are the white dwarfs and coronal sources. In the former, detailed stellar atmosphere models have existed for a long time. The most simple structures considered have been the homogeneous pure H and H+He models (e.g. Wesemael *et al.* 1980). Many workers have since carried out similar calculations. An important improvement was made by Bergeron *et al.* (1992), applying the occupation probability formalism of Hummer and Mihalas (1988), who showed that the treatment of the upper levels of the H atom affected the model structure and, consequently, the shape of the Balmer lines. Stratified H+He models were also developed by Jordan and Koester (1986) and Vennes *et al.* (1988), dealing with a thin H envelope overlying a predominantly He atmosphere.

These comparatively simple models have typically been computed under the assumption that the ion and level populations can be specified by the temperature and density of the gas: local thermodynamic equilbrium (LTE). This seems to be adequate for stars with very high atmospheric density, although there may be some differences in detailed line strengths between LTE and non-LTE, which might give a different abundance (or limit) for the He in a star (see Napiwotzki 1995). A good example of a non-LTE effect is the narrow emission core seen in the Hα line of the hot DA white dwarf G191$-$B2B (Reid and Wegner 1988).

While the continuum fluxes of white dwarf models may be relatively insensitive to the physical validity of assuming LTE, absorption lines are more likely to reveal non-LTE effects. As line blanketing is significant in determining the atmospheric structure, departures from non-LTE may influence the outcome of model calculations through these effects. When only H and He are included, the number of lines that must be considered is only a few tens, having only a modest effect on the stellar atmosphere. However, as more and more elements are included in a model the level of line blanketing and, therefore, the non-LTE effects will become increasingly important. Indeed, the Fe group has millions of lines in the EUV range and non-LTE calculations are essential to provide a physically realistic representation of the stellar atmosphere. Advanced non-LTE codes for the calculation of white dwarf models have been developed by Hubeny (1988) and Werner (1986). These make use of the technique of Accelerated Lambda Iteration (ALI) to make the solution of the statistical equilibrium calculations that specify the ion and level populations in non-LTE, coupled with radiative transfer, a tractable computational problem. Together with increasing computer performance, further improvements (e.g. by Dreizler and Werner 1993; Hubeny and Lanz 1992, 1995) now allow calculations of complex models including many elements and nearly ten million lines (Lanz *et al.* 1996).

5.4.2 Plasma codes and emission line analyses

Techniques developed initially for the analysis of solar data and also applied in the UV for the interpretation of IUE spectra can also be used with the spectral data acquired by EUVE. The emission in a single spectral line can be related to the physical conditions in the gas from which it arises. For a given line, the element and its ionisation state must be identified and the excitation cross-section of the transition known. The strength of the observed line then depends on the population of the particular species at the appropriate ionisation stage and level. This in turn is related to the temperature and density of the gas and also the abundance. In a dense, photospheric plasma, the level populations can, in principle, be specified simply by the standard Boltzmann and Saha equations for excitation and ionisation equilibrium respectively, although departures from this assumption of LTE do occur in many situations. These equations can be found in any basic astrophysics textbook. However, it is necessary to assume that the plasma is dominated by collisional rather than radiative processes. While this is often a reasonable assumption in stellar photospheres, at high temperature and low densities (such as those found in the chromospheric/coronal layers of a star) radiative transitions dominate causing large departures from the LTE approximation.

Even when the LTE approximation is no longer valid, it is possible to construct a stellar photosphere model by solving the equations of statistical equilibrium (see section 5.4.1) that take into account the radiation field to determine the level and ion populations. This is feasible because the general structure of the atmosphere is determined by the equations of hydrostatic equilibrium; in simple terms, the balance between gas and radiation pressure and gravity. However, the existence of the chromosphere and corona cannot be explained in the same manner. Indeed, the mechanism(s) which support these low density/high temperature regions are poorly understood and a solution to this is one of the great outstanding problems of astrophysics. An important goal of stellar EUV spectroscopy is to contribute to understanding the nature of these outer stellar atmospheres and the mechanisms which maintain them.

In the absence of a physical model from which to construct a synthetic spectrum to compare directly with all the available observational data, as has proved possible in the case of hot white dwarfs, for example, the aim of observations of chromospheric and coronal spectra is to reconstruct the physical structure of the atmosphere above the photosphere. From these empirical models the heating required to sustain them can be estimated and then compared to the predictions of particular non-thermal processes. The EUV lies in between two spectral regions, the UV and X-ray, for which rather different analysis techniques have evolved, for largely historical reasons.

5.4.3   The emission measure and its distribution

In the UV (far-UV in particular), the IUE mission provided access to spectra capable of resolving individual lines. As different ionisation stages are related to different temperature plasma, a study of the emission from a range of ions of individual elements provides information on the volume and density across a range of temperatures, provided the atomic data is known for the observed transitions. The integral of the electron density is often termed the 'emission measure' and defines the emissivity (ϵ_λ, $\mathrm{erg\,cm^{-3}\,s^{-1}}$) of the plasma. However, care must be taken in the interpretation of the observations, since the emission measure (and hence, the emissivity) can be defined in slightly different ways. For example, it can be the integral of either n_e^2 or $n_H n_e$ (n_e is the volume electron density and n_H is the volume hydrogen density). Of course if the hydrogen is fully ionised, n_H is equal to n_e.

When interpreting the emission measure inferred from a flux measurement in terms of the physical parameters of an emission region, geometric effects have to be considered. A particularly useful summary of the problem is included in a review of EUV stellar spectroscopy by Jordan (1996), which we follow here. The volume emission measure can be derived from a flux measured at the Earth (F)

$$F = \int \epsilon_\lambda N_e^2 \, dV / (4\pi d^2)$$

where d is the distance to the star and ϵ_λ is the line emissivity. However, this implicitly assumes that all photons escape from the emitting region without interception by the star. This might be a good approximation if the plasma extends to large stellar radii and has a volume much greater than the star. However, this may not be the case and the apparent volume emission measure is

$$Em(V)_{\mathrm{app}} = 4\pi d^2 F / \epsilon_\lambda$$

For a spherically symmetric outer envelope, the fraction of the photons not intercepted by the star is

$$G(r) = 0.5 \left[1 + \left(1 - (R_*/r)^2 \right)^{1/2} \right]$$

When dealing with regions close to the stellar photosphere, a plane-parallel atmosphere can be assumed with $r = R_*$, giving $G(r) = 0.5$. The apparent emission measure over height is then

$$Em(h)_{\mathrm{app}} = (Fd^2)/(\epsilon_\lambda R_*^2) = Em(V)_{\mathrm{app}}/4\pi R_*^2$$

The true volume emission measure is

$$Em(V) = \int N_e^2 4/3\pi r^3 \, dr$$

Hence, what is observed is

$$Em(V)_{\mathrm{app}} = \int N_e^2 4\pi r^2 G(r) \, dr$$

Consequently, the relationship between the true emission measure over the radial extent, $Em(r)$, and the apparent volume emission measure is

$$Em(r) = Em(V)_{\mathrm{app}} (R_*/r)^2 / 4\pi R_*^2 G(r)$$

Because of these additional geometrical factors the apparent emission measure distribution will not have the same gradient with temperature as the true emission measure distribution.

In a thin plasma the state of ionisation of the gas, or indeed an individual element, is determined by the balance between ionisation and recombination rates. Raymond (1988) provides a good summary of the relevant atomic physics that goes into such calculations. Arnaud and Raymond (1992) have computed these rates for Fe, an astrophysically important element, and used them to predict the ionisation equilibrium as a function of temperature.

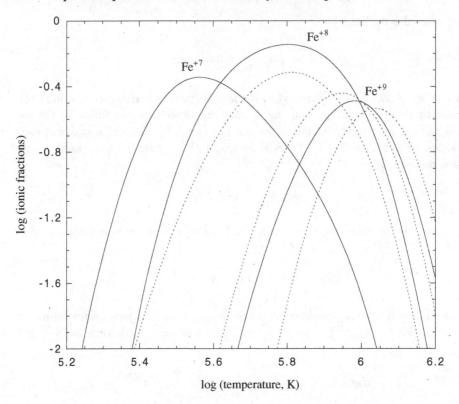

Fig. 5.4. Ionic fractions vs. temperature for Fe^{+7}, Fe^{+8} and Fe^{+9}. The plain curves are the most recent work of Arnaud and Raymond and the dashed curves are from Arnaud and Rothenflug (1985) (from Arnaud and Raymond 1992).

Figure 5.4 shows, for example, the ionic fractions for Fe VIII (7+), Fe IX (8+) and Fe X (9+) in the log temperature (K) range from 5.2 to 6.2. It can be seen that the functions each have a quite strongly peaked maximum. The difference between the Arnaud and Raymond (1992) and the Arnaud and Rothenflug (1985) results is explained by the higher dielectronic recombination rates used by the former authors.

For any line of a given ion, the emissivity of a plasma depends on both the population and the transition probability (the sum of all processes that give rise to the transition). Since a significant population of the ion in question exists over a finite temperature range (as described above) the value of emission measure derived from a line flux must depend on the assumed temperature. Thus, it is conventional to compute the emission measure, integrated over a finite region, at a range of temperatures $\log T_e = \log T_m \pm 0.15$, where T_m is the temperature at which the ion population is predicted to have a maximum. In principle, the emission measure distribution can then be built up as a function of T_e by using many lines. This is practical when considering the solar spectrum as there are a large number of lines observed which can define the mean emission measure distribution. However, for other stars only the stronger lines are observable and it is then useful to consider the emission measure as if all the lines were formed at each T_e in turn. This gives a locus of emission measure which is an upper limit to the true value as a function of T_e. As an example, the emission measure

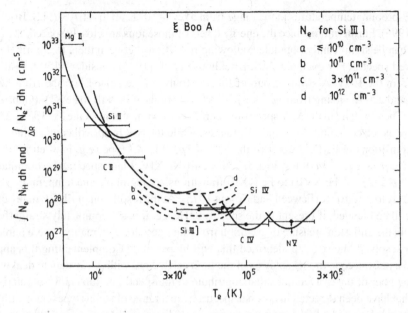

Fig. 5.5. Emission measure distribution for ξ Boo A (G8 V) (from Jordan *et al.* 1987). The horizontal bars are the mean values required to match the observed line flux centred on the temperature shown. The dashed curves illustrate the density sensitive nature of Si III.

distribution for ξ Boo A is shown in figure 5.5, derived from IUE observations of Mg II, Si II, C II, Si III, Si IV, C IV and N V lines.

If a particular transition has a high enough probability, the spontaneous decay rate dominates any other emission processes and the emission measure is sensitive only to the ion population (i.e. T_e). However, when a transition has low spontaneous transition probability, collisional de-excitation can compete with or even exceed the spontaneous decay. Then the emissivity of the plasma is also sensitive to the electron density. This is the case for the Si III 1892 Å line, the four dashed curves in figure 5.5 showing the emission measure distribution for a range of electron densities from 10^{10} to 10^{12} cm^{-3}.

5.4.4 What is needed to analyse EUV spectra?

The approach that might be adopted to analyse the data from a particular instrument depends on the quality of information provided by that instrument, which is basically determined by the spectral resolution. The best spectral resolution available for stellar spectroscopy has been that provided by EUVE but this was still limited to ≈0.5–1 Å. Corresponding to a resolving power $(\lambda/\delta\lambda) \approx 400$, this is similar to the low resolution capability of IUE in the far-UV. In principle, it should be possible to determine the emission measure distribution directly from the individual line fluxes as shown in the earlier example, for ξ Boo A (figure 5.5). However, for the EUV the situation is not completely clear cut. In IUE spectra the emission features are all attributable to resonance lines and features arising from excited levels are not seen. This is not so in the EUV band.

In the EUV, the emission line spectrum is expected to be dominated by ionised iron. Ions from Fe IX through to Fe XXIV all have lines falling within the wavelength range covered by

EUVE, probing temperatures in the range from a few $\times 10^5$ K up to $\approx 10^8$ K (N.S. Brickhouse *et al.* 1995). Furthermore, since the energy levels of these ions are closely spaced, the excited levels can be significantly populated, allowing transitions between them to contribute to the observed spectra. For example, N.S. Brickhouse *et al.* (1995) consider ≈ 150 transitions of Fe ions in the EUV. A strong group of lines is found in the solar EUV spectrum between 170 Å and 220 Å, arising from $3p^n - 3p^{n-1}3d$ transitions of Fe VIII to Fe XIV (Gabriel *et al.* 1966). These will lie in the M̂W spectrometer range. Iron transitions of the $3s^23p^n - 3s3p^{n+1}$ type are expected in both MW and LW ranges, while the SW band will contain lines arising from transitions in Fe IX–Fe XII of the $3p^n - 3p^{n-1}4s, 4d$ type (e.g. Fawcett *et al.* 1972) besides $3p^n - 3p^{n-1}3d$ transitions of Ni IX to Ni XIV. Flare spectra are dominated by $2s^22p^n - 2s2p^{n+1}$ Fe XVIII to Fe XXIII transitions arising in plasma temperatures around 6×10^6 K to 2×10^7 K (Fawcett and Cowan 1975). At the resolution of EUVE many of these lines will be blended, if present in the spectrum. Hence, it is not completely straightforward to obtain the emission measure distribution from the lines alone. In reality, where unblended, clearly resolved lines can be identified this will be possible. Complementing this approach will be a plasma code model analysis of the kind carried out with X-ray data to deal with the large number of unresolved lines that contribute to the spectrum. Table 5.3. summarises the lines that have been detected in coronal spectra from a range of stellar types, indicating their rest wavelengths and whether or not they are single lines or a blend of more than one, at the resolution of EUVE. For line identifications and line ratio diagnostics, Keenan (1996) has produced a bibliography of the most reliable emission and absorption line ratio calculations currently available in the wavelength range from 50 to 3000 Å.

5.4.5   *Plasma codes for data analysis*

Three main plasma codes are in common use in X-ray/EUV astrophysics. The earliest one of these developed and, perhaps, the most widely used is that of Raymond and Smith (1977), which is based largely on ionisation rates calculated by them. However, initially at least, this was mainly focused on the analysis and interpretation of X-ray data, with Stern *et al.* (1978) providing a complementary coverage of the EUV range. Subsequent developments have dealt with both X-ray and EUV spectral regions. The code of Mewe and colleagues (Mewe *et al.* 1985; Kaastra 1992) makes use of the ion balance calculations of Arnaud and Rothenflug (1985). Comparing spectra generated by the two codes (with Mewe *et al.* 1985, specifically), Raymond (1988) notes that they agree quite well at short wavelengths but that line strengths differ by a factor of two at the longer wavelengths. Raymond (1988) attributed this mainly to different sources for the ionisation balance calculations together with differing estimates of the Si and S collision strengths. A third effort is the Arcetri code of Landini and Monsignori-Fossi (1990). They also use the excitation–autoionisation rates from Arnaud and Rothenflug (1985) but adopt the expressions and fitting coefficients of Shull and van Steenberg (1982a,b) for direct ionisation, radiative and dielectronic recombination.

All these codes are made readily available by the authors for use by other astronomers. For example, both the Mewe *et al.* (1985) and Raymond and Smith (1977) are implemented in the XSPEC spectral fitting package (Shafer *et al.* 1991; Arnaud 1996), while the Arcetri spectral code is available on the internet (www.arcetri.astro.it). However, as for the stellar atmosphere calculations discussed earlier, these spectral models cannot be treated as 'black boxes' when used to interpret astronomical spectra. The reliability and accuracy of the spectral predictions

Table 5.3. *Emission lines detected in EUV spectra. Lines linked by '/' are blended.*

Ion	log T	λ (Å)
He II		303.78
O V	5.4	629.72
O VI	5.5	150.6, 172.94/173.08/173.09
Ne IV	5.2	421.61
Ne V	5.45	416.20
Ne VII	5.7	465.22
Mg VII	5.8	434.92
Mg VIII	5.9	436.73
Mg IX	5.95	368.07
Si VII	5.75	275.35
Si VIII	5.9	319.84
Si IX	6.0	226.99/227.30
Si X	6.1	261.27
Si XII	6.3	520.67
S VIII	5.85	198.55
S IX	6.0	224.75/225.25
S X	6.15	228.18/228.70, 259.52, 264.24
S XI	6.25	188.67, 191.26, 246.90/247.13, 291.59
S XII	6.3	227.47
S XIV	6.45	417.62
Ar XII	6.2	224.23
Ar XV	6.55	221.10
Fe VIII	5.8	167–169.5, 185.19, 186.58
Fe IX	5.95	171.07, 224.91
Fe X	6.05	174.53, 175.26 177.224, 184.54
Fe XI	6.1	178.05, 181.13, 182.16, 184.70/184.78, 187.44, 188.25
Fe XII	6.15	186.87, 192.42, 193.53, 195.14
Fe XIII	6.2	201–203, 209.62/209.92, 221.83, 246.21, 251.95, 320.8, 359.65
Fe XIV	6.25	211.33, 219.13, 264.78, 270.51, 274.20, 334.17
Fe XV	6.3	285.15, 243.79, 284.17
Fe XVI	6.35	262.98, 335.41, 360.80
Fe XVIII		93.92, 103.94
Fe XIX		91.02, 101.55, 108.37, 109.97, 111.70, 120.00
Fe XX		118.66, 121.83
Fe XXI		97.88, 102.22, 128.73, 142.27, 145.65
Fe XXII		117.17, 135.78
Fe XXIII		132.85
Fe XXIV		192.04
Ni XI	6.15	148.40
Ni XVII	6.4	249.18

depends on the completeness of the input physics and the quality (and also completeness) of the atomic data included. An aspect of this is the source of the ionisation balance calculations discussed above. Raymond (1988), in his review, provides a comprehensive discussion of the physical processes that need to be included in the calculations. The key components are the ionisation balance (ionisation rates vs. radiative and dielectronic recombination processes) and radiation (collisional excitation, two photon continuum, bremmstrahlung and recombination). To reduce the complexities of the problem a number of simplifying assumptions need to be made: the plasma is non-relativistic, has only weak electric and magnetic fields, has no significant photoionisation, no time dependence, no diffusion, no optical depth and has a Maxwellian velocity distribution. Hence, if the real plasma departs from any of these the validity of the predicted spectrum is called into question. Raymond (1988) suggests that all these can affect the emissivity of the plasma at a level similar to the uncertainties in the atomic data, which range from tens of per cent to a factor of two for the most complex species. A direct comparison of model calculations with an observed solar flare X-ray spectrum carried out by Raymond (1988) suggested typical errors $\approx 50\%$ in the predicted strengths of individual lines.

The problem of possible inaccuracies in the spectral codes has been brought into focus recently by the results of X-ray spectroscopy of coronal sources with the ASCA satellite. These observed spectra consistently require peculiar element abundances in the models, a factor of three to four less than the solar photospheric values (e.g. Antunes *et al.*1994; Singh *et al.*1996). Similar results were found in an entirely different environment, in studies of the hot plasmas that lie in clusters of galaxies (e.g. Fabian *et al.* 1994). The question raised by all this work is whether the apparent abundance deficits are real, the spectra reflecting the true composition of the plasma, or an artifact of the limitations in the input physics and the atomic data. Misleading abundance measurements might arise from incorrect transition probabilities in an individual line or group of lines for a particular ion, so that the emissivity is not correctly determined. Alternatively, if the ionisation balance calculations are not reliable the observed ion population may not reflect the actual element abundance.

N. Brickhouse *et al.* (1995) addressed this issue with a detailed comparison of the available plasma codes. Their basic conclusions are that the codes are in broad agreement, the main source of discrepancy being the adopted ionisation balance (i.e. Arnaud and Rothenflug 1985 or Arnaud and Raymond 1992). However, uncertainties in even the latest ionisation balance calculations are significant, as the best dielectronic recombination rates are uncertain at the 30% level. A particular problem was identified for the intensity ratios of the transitions between $n = 2$ to $n = 3$ and $n = 2$ to $n = 4$ levels of highly ionised systems. This led to discrepancies between the intensities of the Fe XV and Fe XVI spectral lines and Fe XVIII–XXII in the 40–85 Å and 85–125 Å ranges respectively. Similar problems are likely to exist for analogous transitions of Si, S and Mg. However, it was noted that, in general, Fe data of importance in the EUVE range were adequate, although Ni collision rates would need to be updated.

Mason (1996) provides a useful discussion of the relative merits and deficiencies of the various plasma codes, concentrating in particular on the accuracy of the atomic calculations. Good electron scattering calculations are now available for many of the ions which are abundant in coronal and transition region plasma. These include Fe IX–Fe XIV, Fe XV–Fe XVI and Fe XVIII–Fe XXIV, grouped by wavelength region. The most recent calculations highlight some deficiencies in earlier data. For example, recent predictions of Fe XV line strengths are completely different to those observed in solar spectra. However, accurate Fe XV data have only recently been published and remain to be incorporated in the spectral codes. Hence, this could lead to misinterpretation of any data incorporating these lines.

The main conclusion of Mason (1996) is that the best plasma codes are those that are able to represent the atomic rates most accurately. Therefore, improvement of these codes by incorporation of new data will be a more or less continuous process. The international collaboration known as the Iron Project (coordinated by D. Hummer; Hummer *et al.* 1993) aims to calculate accurate electron excitation data for many ions of Fe and other elements using the most accurate methods available. The results of this project are reported in a series of *Astronomy and Astrophysics* papers (see e.g. Storey *et al.* (1996) for Fe XIV or Saraph and Storey (1996) for Ar VI, K VII and Ca VIII).

5.4.6   Abundance measurements and the FIP effect

Despite the reservations concerning the limitations of the spectral codes, a large abundance deficiency of Fe observed by ASCA in the spectra of AR Lac, Algol and Capella is a significant effect that does not seem to be explained by these uncertainties. Nor can a complex temperature distribution account for what is observed (White 1996). Singh *et al.* (1996) consider the optical depth of the He-like resonance lines of Fe, Si and Mg, finding that for AR Lac that of the Fe XXV resonance line could be significant, with a favourable geometry. However, the Fe 6.7 keV line observed by ASCA is a blend of the Fe XXV resonance line and other satellite lines of Fe XXV and Fe XXIV, contributing only about one quarter of the power in the complex. Hence, even if Fe XXV were completely suppressed, the observed decrease would not match the factor of 2–3 underabundance of Fe that is observed.

Abundance anomalies are not unknown in the Sun. Pottasch (1963) first noted that the coronal abundance of some elements, most notably Si, Mg and Fe, was significantly higher than the accepted photospheric values. Meyer (1985) established that the differences are related to the first ionisation potential (FIP) of the element in question and are similar to those abundance patterns seen in cosmic rays. Elements with low FIP (<10 eV; e.g. Si, Mg, Fe) show average enhancements of a factor ≈ 4 when compared with elements of high FIP (≥ 10 eV; e.g. O, Ne, Ar). The mechanism by which the FIP effect is produced is not known, nor are the sites responsible for it identified. Interestingly, the abundance anomalies seen in this case are in the opposite sense to those observed by ASCA.

As solar EUV spectroscopy would be a complete subject for a book on its own, it is not included in this work. However, the Sun does represent an important baseline against which observations of other stars may be compared. In doing so, it is important to remember that much of the information we have about the Sun is determined from discrete features that are spatially resolved. In terms of coronal studies, these may be individual loops or active regions. In contrast, when observing a distant star, which we cannot resolve with our instrumentation, what we see are the effects of features averaged across the entire disc. Spatial structure can sometimes be studied but only if features are occulted by eclipses or stellar rotation. Consequently, it is helpful to consider how the Sun would appear if observed at great distance. Laming *et al.* (1995) have reanalysed the full disc quiet Sun spectrum of Malinovksi and Heroux (1973), using up-to-date atomic data, to determine whether the FIP effect is seen only in discrete features or on average across the whole disc. Using atomic data that are the same as those to be used in the analysis of EUVE spectra also provides internal consistency when comparing the results, avoiding any possible systematic biases that might arise from using different information. Laming *et al.* (1995) recover from the full disc analysis the results obtained from previous work on discrete solar regions, with a factor of three to four coronal abundance enhancement for low FIP elements, but only for lines formed at temperatures greater than 10^6 K. At lower temperatures, the FIP effect seems to be substantially smaller. This may suggest that the FIP effect is a function of altitude, with the lower temperature full disc emission being dominated by the supergranulation network.

5.5   EUV spectroscopy with other instruments

There exists a limited amount of spectroscopy at EUV wavelengths conducted with instruments other than EUVE. These include a sounding rocket experiment by the University of Colorado group, launched in 1992, the HUT I far-UV telescope which possessed a significant EUV spectral capability by utilising the second order response of the spectrometer, the ORFEUS II mission, and the UVSTAR (UltraViolet Spectrograph Telescope for Astronomical Research) missions; which is an Italian–US programme of Space Shuttle based far-UV and EUV telescopes.

The Extreme Ultraviolet Spectrograph (EUVS) was a sounding rocket experiment flown on January 10 1992 from White Sands Missile Range which obtained the first modest resolution EUV spectra of the white dwarf G191–B2B. This instrument employed a grazing incidence telescope coupled with a toroidal grating to yield a spectral resolution of ≈ 2 Å over the 200 to 320 Å wavelength range. The resulting spectrum (Wilkinson *et al.* 1992) was an important development in our efforts to understand the nature of the EUV opacity known to be present in the atmospheres of a number of hot hydrogen-rich white dwarfs such as G191−B2B. Most significantly, it yielded no spectroscopic evidence of an He II Lyman continuum edge in the G191−B2B. The existence of such an edge is precisely what was to be expected if He was the source of the short wavelength opacity that early EUV and soft X-ray photometric observations required. The absence of this feature helped to shift attention away from He as the primary source of white dwarf opacity and helped lead to a consideration of the role of heavier elements. Moreover, several curious edge-like features and an apparent absorption feature in the rocket spectrum were interpreted as ionisation edges and lines due to O III. At the time, the presence of O III, as well as other ions, was known from HST and IUE observations and attempts were made to interpret these results in terms of the primary EUV opacity source being due to O III. These results were not entirely successful and it was later shown from EUVE spectra that the features in the rocket spectrum were artefacts and not due to O III.

The HUT was part of a suite of astronomical telescopes flown as the Astro-I mission on the space shuttle in December 1990. Although HUT was primarily designed to provide spectroscopy in the far-UV wavelength range between 900 and 1850 Å, the spectrometer could also be used in the second order to cover the spectral coverage from 350 and 700 Å, when a thin aluminium filter was used to exclude the first order far-UV spectrum. Relatively few spectra were acquired in this mode but one significant result which stands out is the measurement of the interstellar He I 504 Å absorption edge in the spectrum of G191−B2B (Kimble *et al.* 1993a). The presence of He I is virtually impossible to detect at longer wavelengths and this result provided a valuable pre-EUVE measure of the column density of this ion and an estimate of the ratio of He I to H I along the line-of-sight to G191−B2B. Although an earlier sounding rocket observation of G191−B2B (Green *et al.* 1990) also observed the He I ionisation edge, the HUT results of Kimble *et al.* allowed significant constraints to be placed on the ionisation state of He in the local cloud. The He I to H I ratio was determined to be 11.6 ± 1.0, a result consistent with the more recent measurements of Lanz *et al.* (1996) for G191−B2B obtained with EUVE. Only a few stars have sufficient unattenuated 500 Å flux to produce an observable He I ISM absorption edge, however, as Rumph *et al.* (1994)

have shown, it is also possible to measure ISM He I columns for some white dwarfs using observations of the He I autoionisation features near 206 Å.

The ORFEUS II mission, flown on the space shuttle in late 1996, was primarily a far-UV spectroscopic instrument but also had a high resolution EUV spectroscopic capability. The ORFEUS II Berkeley EUV spectrometer, described by Hurwitz and Bowyer (1996), achieved a spectral resolution of ≈0.5 Å between the wavelengths of 520 and 665 Å. Among the observations which utilized this capability were the ε CMa observations of Cohen *et al.* (1998). As discussed in chapter 6, the bright B2 II star ε CMa is the brightest long wavelength EUV source in the sky. The ORFEUS spectra were able to resolve a number of features previously seen with EUVE. Important results included the specific identification of a number of lines predicted by non-LTE model atmospheric codes and presence of asymmetric wind broadened features due to O V and S IV with blue-edge velocities near 800 km s^{-1}.

A more recent spectroscopic capability in the EUV is available with the EUV channel on UVSTAR, which involves a pair of twin astronomical telescopes and spectrographs which have flown several times as part of the International Extreme-ultraviolet Hitchhiker (IEH) program on the space shuttle (Stalio *et al.* 1999). The EUV channel covers the wavelength band from 535 to 935 Å with a spectral resolution of approximately 1 Å. Although designed primarily to obtain spatially resolved EUV spectra of the Jupiter's Io plasma torus, the UVSTAR EUV channel also obtained unique spectra of the bright B star ε CMa at wavelengths up to and including the Lyman limit at 912 Å. UVSTAR observed ε CMa on both the STS 69 (September 1995) and STS 95 (November 1998) shuttle missions (Gregorio *et al.* 2001). Because of the differing wavelength coverage of UVSTAR and superior spectral resolution with respect to EUVE it has been possible to identify a number of photospheric lines in the UVSTAR spectrum of ε CMa. Gregorio *et al.* list a number of transitions due to C II, N II, O II, and S II, which are identified in the 595 to 765 Å wavelength band. Gregorio *et al.* also compared the UVSTAR continuum fluxes with those of EUVE and ORFEUS II in the region of mutual wavelength overlap finding significant apparent differences in absolute flux levels with respect to EUVE and ORFEUS.

6

Spectroscopy of stellar sources

6.1   Emission from B stars

Prior to the EUV sky surveys, O and B stars exhibiting strong mass-loss were expected to be a minor, but nevertheless important, group of EUV sources, the emission arising from hot, shocked gas in the stellar winds. Little thought was given to the likelihood of detecting photospheric EUV flux since photospheric helium was expected to restrict emission to the longest EUV wavelengths, most affected by interstellar attenuation. Nevertheless, the existence of the so-called β CMa tunnel of low column density, extending over distances of 200–300 pc (e.g. Welsh 1991) promoted the hope that a few such objects might be detected in this direction at wavelengths longward of 504 Å. The subsequent detection of the B2 II star ϵ CMa (Adhara, $d = 188$ pc) in the 500–740 Å (tin) filter during the EUVE sky search was not, therefore, particularly remarkable. However, the intensity of the flux recorded outshone all other non-solar sources of EUV radiation, including the well-known hot white dwarf HZ 43, previously believed to be the brightest EUV source, although this star remains the brightest object at the shortest EUV wavelengths (Vallerga et al. 1993).

The magnitude of the detected EUVE tin count rate (98 ± 10 counts s^{-1}) was a strong indication that the line-of-sight column density was even lower than the upper limit of 3×10^{18} cm^{-2} estimated indirectly by Welsh (1991) from NaI absorption line studies. Indeed a fit to the EUVE data, using a grid of LTE model atmospheres computed by Kurucz (1979, 1992) with an assumed temperature of 25 000 K and $\log g = 3.3$ (Kudritzki et al. 1991), gives a neutral hydrogen column density $N_{H\,I} = 1.05 \pm 0.05 \times 10^{18}$ cm^{-2}. However, this result is very sensitive to the model atmosphere used. For example, Vallerga et al. (1993) repeated the analysis with a non-LTE spectrum computed using the TLUSTY code of Hubeny (1988), updated by Hubeny and Lanz (1992). Using the same values for the temperature and gravity yields a value of $N_{H\,I}$ a factor 2.5 times lower (4×10^{17} cm^{-2}) than the LTE result. This difference must arise from a strong (and expected) difference between the EUV spectrum predicted by the LTE and non-LTE calculations. This problem, coupled with the brightness of ϵ CMa, made the star an important target for the EUVE spectrograph.

β CMa (B1 II–III) is the second brightest object detected by EUVE in the two longest wavelength filters during the sky survey (Bowyer et al. 1994, 1996; Malina et al. 1994). Since it has a similar temperature (23 250 K), surface gravity ($\log g = 3.5$) and distance (206 pc) to ϵ CMa, the factor 12 lower count rate in the tin filter is probably an indication of a somewhat larger neutral hydrogen column density in that direction.

6.2   ϵ **Canis Majoris**

ϵ CMa is an interesting star from a general point of view, since the mass loss rate and wind density is intermediate between the O stars and main sequence B stars. Understanding the EUV emission from B stars is of great interest for several reasons. For example, the boundary between O and B stars marks a sharp difference in the size of H II regions because of differences in the fraction of stellar radiation in the EUV range, depending strongly on stellar temperature for T_{eff} below 30 000 K. Along the main sequence, wind lines are very weak in the UV spectra of stars later than B1 as wind driving is a strong function of stellar luminosity. Furthermore, in the supergiant spectral sequence a sharp decrease in the wind speed for stars later than B2 Ia is seen as driving switches from being driven by strong EUV lines to more numerous, but weaker lines in the UV when the wind becomes optically thick in the EUV (Pauldrach and Puls 1990). Several interesting groups of stars lie near the position of ϵ CMa in the HR diagram, including the emission line Be stars and the β Cephei stars, an important class of pulsating stars that occur in the early B spectral range.

ϵ CMa was one of the first targets to be observed during the spectroscopic phase of the EUVE mission, immediately following the all-sky survey, on January 17–19 1993. Consequently, the ≈60 000 s exposure suffered from the fixed pattern efficiency variation inherent in the detector and discussed in section 5.1. Subsequently, a second ≈140 ks observation was carried out on March 8–12 1994, using the dither mode to eliminate this particular problem. Figure 6.1 shows the count spectrum from the first of these observations for the

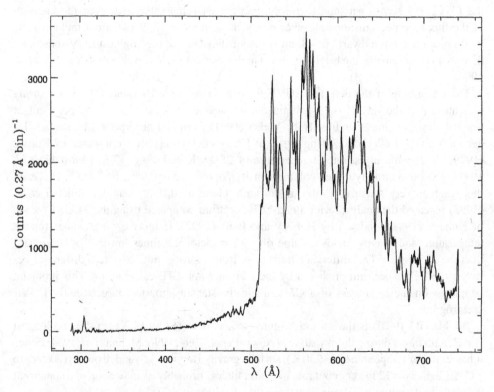

Fig. 6.1. Long wavelength EUVE spectrum of the B2 II star ϵ CMa, displayed in counts s^{-1} Å^{-1}, obtained in January 1993 prior to implementation of the dither mode (from Cassinelli *et al.* 1995).

Table 6.1. *Stellar parameters for ε CMa and β CMa.*

Parameter	ε CMa	β CMa	Reference
Spectral type	B2 II	B1 II–III	Hoffleit and Jaschek 1982
M_V	1.5	1.98	Hoffleit and Jaschek 1982
Distance (pc)	188	206	Bohlin *et al.* 1978
θ_d (mas)	0.80 ± 0.05	0.52 ± 0.03	Hanbury Brown *et al.* 1974
$N_{H\,I}$ (cm^{-2})	$0.7 - 1.2 \times 10^{18}$	$2.0 - 2.2 \times 10^{18}$	Cassineli *et al.* 1995, 1996
T_{eff} (K)	$20\,990 \pm 760$	$25\,180 \pm 1130$	Code *et al.* 1976
log g (cgs)	3.2 ± 0.15	3.4 ± 0.15	Drew *et al.* 1994
R_*	$16.2 \pm 1.2 R_\odot$	–	Bohlin *et al.* 1978; Hanbury Brown *et al.* 1974
$V \sin i$ (km s^{-1})	35	30	Uesigi and Fukuda 1982

wavelength range from 300 to 740 Å. The dominant feature is the strong EUV continuum longward of the He I 504 Å interstellar absorption edge. Considerable structure is apparent in the continuum flux which can be attributed to photospheric absorption features. Interestingly, in general, the sizes of these features are much larger than the typical 20% amplitude of the fixed pattern efficiency variation. Hence, most of the original analysis conducted using the first observation will not have been compromised by the fixed pattern problem. In addition to the dominant continuum longward of 504 Å strong emission lines of He II 304 Å and O III 374 Å are clearly seen.

ε CMa is one of the few objects for which an angular diameter has been determined, using an optical intensity interferometer (Hanbury Brown *et al.* 1974). Combining this observation with measurements of the total bolometric luminosity allows an empirical determination of effective temperature. Assuming a distance, the angular diameter measurement also leads directly to an estimate of the stellar radius. Hence, ε CMa is an ideal object for theoretical modelling and, in particular, provides an important test for stellar atmosphere calculations. Table 6.1 lists some of the known properties of ε CMa.

Cassinelli *et al.* (1995) have presented a comprehensive analysis of the EUV spectrum of ε CMa. As the angular diameter of the star is known and the complete spectral range of its emitted flux accessible to observation, the effective temperature is well constrained (see table 6.1). The unexpected EUV flux excess corresponds to less than 1% of the bolometric flux and, therefore, will not alter the temperature measurement significantly. Taking a plausible surface gravity (based on the stellar distance and the evolutionary tracks of Maeder and Meynet (1988)) with this temperature allows the predictions of stellar atmosphere calculations to be compared directly with the observed spectrum. After removing the effect of interstellar absorption ($N_{H\,I} = 1.0 \times 10^{18}$ cm^{-2}), it is clear that, while the far-UV and optical fluxes are reproduced well by a 21 000 K LTE model (calculated using the ATLAS9 code of Kurucz 1979, 1992), the observed EUV spectrum is well in excess of the predicted flux level (figure 6.2).

An important question is whether or not the discrepancy can be explained as arising from the neglect of non-LTE effects. Cassinelli *et al.* constructed a line blanketed model using the TLUSTY code of Hubeny (1988), using the same temperature and surface gravity as derived from the Kurucz model, while treating hydrogen, helium, Fe III and Fe IV (the dominant ionisation stages of Fe in this temperature range) in non-LTE. The treatment of iron is described in detail by Hubeny and Lanz (1995). A comparison of the non-LTE (TLUSTY)

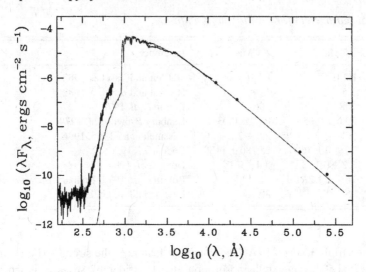

Fig. 6.2. Comparison of the energy distribution of ϵ CMa from EUV through to mid-IR wavelengths compared with a 21 000 K LTE model (from Cassinelli *et al.* 1995).

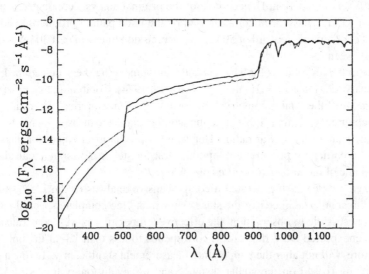

Fig. 6.3. Comparison of the 21 000 K, log $g = 3.2$ LTE (solid line) and non-LTE (dashed line) models (from Cassinelli *et al.* 1995).

and LTE (ATLAS9) models (figure 6.3) shows little difference in the absolute flux level. Indeed, in the Lyman continuum the predicted EUV flux is lower in the non-LTE case enlarging the discrepancy with the observed spectrum.

An interesting point is the similarity in the Lyman and mid-IR continua formation depth within the atmosphere. However, they arise by different processes. In particular, the dominant infrared opacity sources are free–free and bound–free absorption (the latter only involving high lying bound states); the source function is much less sensitive to non-LTE effects. Closer inspection of the 12 and 25 μm fluxes of ϵ CMa, drawn from the IRAS point source catalogue

(Beichmann *et al.* 1988) shows that they lie significantly above the model atmosphere prediction (figure 6.2). Since the mass-loss rate from ε CMa is very small (Drew *et al.* 1994), this excess cannot be attributed to free–free emission from a wind.

If the infrared excess, with respect to model predictions, resides within the stellar photosphere, it indicates that the atmospheric temperature at the depth of the continuum formation is some 16% higher with respect to the LTE model structure, i.e. $\approx 17\,000$ K compared with $\approx 15\,000$ K. As a consequence, the temperature of the Lyman continuum forming layer, at a similar depth, must also be enhanced giving rise to a larger flux than predicted by the models. Using this new temperature estimate for the Lyman continuum forming region allows the interstellar H I absorbing column to be redetermined, giving a value of about 1×10^{18} cm^{-2}. Improved atmospheric models will be necessary to understand the cause of the temperature enhancement. In addition, Hubeny and Lanz (1996) note that there is a flux discrepancy at wavelengths near 2200 Å, indicating that there must be significant missing opacity in this region. Historically, similar discrepancies between models and observation were assigned to the well-known 2200 Å interstellar feature. However, in this case the interstellar column density is too low to provide a tenable explanation. Hence, this problem must also be addressed by developments in the model atmospheres.

6.2.1 Absorption lines in the spectrum of ε CMa

As has been discussed already, the EUV region is rich in spectral lines and the EUVE observation of ε CMa shows many absorption and a number of emission features. However, interpretation of the origin of the observed features is complicated by the fact that some will arise in the stellar photosphere, while others might be attributed to a wind. The use of a synthetic spectrum, generated from a model atmosphere calculation, provides an identification for the photospheric lines. Significant lines not coincident with the model predictions will be candidates for wind lines, particularly if they are identified with higher ionisation stages. Figure 6.4 shows the long wavelength section of the EUV spectrum, from 450–735 Å, together with the positions of the predicted lines, listing their wavelengths. An asterisk (*) is shown if a line is a resonance transition or originates from an excited level <3 eV above the ground state. It is clear that, at the resolution of EUVE, many of the visible absorption features are blends of lines arising from several transitions, some from different elements and ionisation stages. The high density of lines in this EUV spectrum makes it difficult to establish the level of the continuum flux. Consequently, great care must be taken in measuring the equivalent widths of absorption lines in EUVE data.

Evidence for a stellar wind in ε CMa already exists from UV observations made with the Copernicus and IUE satellites. In non-B supergiant stars, the wind lines do not typically show a distinct shortward edge to the P Cygni absorption trough but a gradual rise towards the continuum. This leads to uncertain estimates in terminal wind speed. However, estimates of this velocity, derived from UV observations, range from 600 to 820 km s^{-1} (see Cassinelli *et al.* 1995) and lie below the ≈ 1200 km s^{-1} resolution of EUVE in the LW range. Hence, it is not possible to distinguish between photospheric and wind features by line width or velocity. Since the ionisation level in a wind is higher than in the photosphere (because of the X-rays that are present in the wind), the best discriminant is the ionisation potential of the ion producing the line. For example, the feature at 629 Å is coincident with the location of

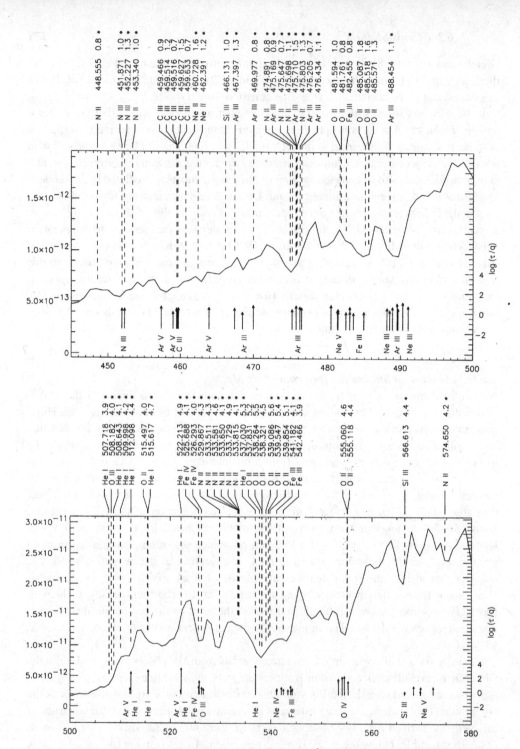

Fig. 6.4. Line spectrum of ε CMa from 450 to 7350 Å in cgs flux units at the Earth. Along the upper portion of each figure are the line identifications and indicatory line strengths in units of log r, where r is the ratio of the line centre opacity to that in the continuum, which are predicted by a non-LTE model atmosphere with $T_{eff} = 21\,000$ K. Along the lower portion of the figure are lines which are expected to be strong in the stellar wind, based on ionisation calculations. The values of the wind line strength can be read from the scale on the right of each panel (from Cassinelli *et al.* 1995).

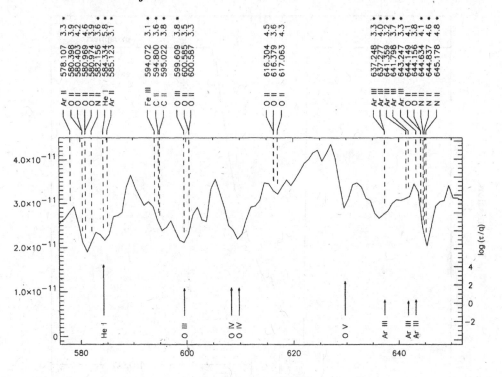

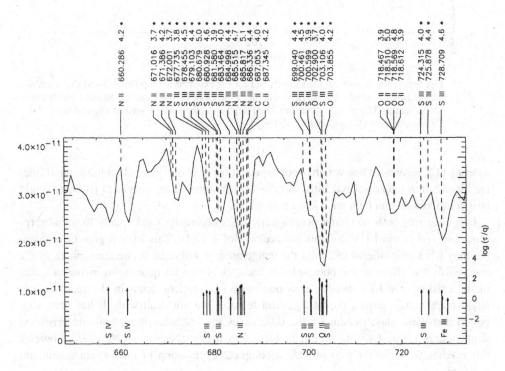

Fig. 6.4. *(cont.)*

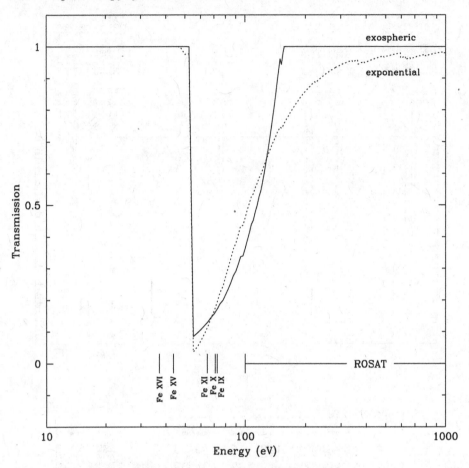

Fig. 6.5. Transmission of the stellar wind for the exospheric (solid curve) and exponential (dashed curve) wind models. The location of the EUV Fe lines and the ROSAT bandpass are indicated. The models are normalised to provide the same degree of attenuation at the Fe line complex (from Cohen *et al.* 1996).

a strong O V resonance line which is not expected in the photosphere. Candidate wind lines are identified in figure 6.5 and listed in table 6.2. In some cases, such as O III, there might be a contribution from both wind and photosphere.

It is clear that both wind and photospheric lines contribute significantly to the absorption observed in the EUV continuum spectrum of ε CMa. This high degree of blanketing may have a significant effect on the temperature distribution in the atmosphere of the star. While the effect of the photospheric lines should be adequately taken into account in the existing non-LTE models, provided there is sufficient detail in the input atomic physics and model atoms, the strong wind features are not dealt with. It has been suggested that these may account for the differences seen between the actual and predicted temperatures in the Lyman continuum formation region (Najarro *et al.* 1996); however this mechanism is unlikely to provide a complete explanation of the Lyman continuum problem.

Table 6.2. *Candidate wind absorption lines in ε CMa.*

Ion	λ (Å)	τ/q_{is}
C III	386	35.2
N III	452	5.0
	687	58.7
O III	508	122.1
	526	25.7
	600	42.3
	703	259.1
O IV	555	328.1
	609	93.2
O V	629	239.9
Ne III	489	35.4
Ne IV	543	9.2
Ne V	569	11.2
	572	4.4
S III	678	15.0
	700	27.7
	703	1.3
	725	13.1
	729	6.4
S IV	657	11.6
	661	11.1
Ar III	468	9.5
	476	15.6
Ar V	458	5.3
	460	2.9
	464	3.3
	512	2.5
	522	1.5
Fe IV	527	5.6

6.2.2 Emission lines

Many peaks can be seen in the Lyman continuum emission of ε CMa but, since the actual continuum flux level is so heavily blanketed by absorption lines, these might well be artificial. Certainly, the reality of any emission lines is difficult to establish in this region. However, at shorter wavelengths, where the photospheric flux is effectively zero, several emission features are clearly visible. The strongest of these arise from low ionisation stages of oxygen and helium, but highly ionised iron lines are also weakly detected (table 6.3). The O III resonance lines at 374 Å is very likely a result of the Bowen fluorescence mechanism, with the upper level of the transition pumped by the He II Lyman α line at 304 Å due to the overlap with a number of O III energy levels (Cassinelli *et al.* 1995).

The ionisation stages of the Fe lines are consistent with the range that might be expected based on ROSAT X-ray observations of the star (see Drew *et al.* 1994). However, the line strengths are weaker than might be predicted by a one or two temperature plasma model which matches the ROSAT data. This is probably the result of two effects: there is a distribution in

Table 6.3. *Emission lines in ϵ CMa.*

Ion	λ (Å)	$\log T_{\epsilon_{max}}$[a]	Flux ($\times 10^{-13}$ ergs cm^{-2} s^{-1})	Luminosity[b] ($\times 10^{-29}$ ergs s^{-1})
O III	374		3.39 ± 0.28	36.0
He II	304 (LW)		32.1 ± 1.1	240
He II	304 (MW)		25.3 ± 1.2	200
He II	256[c]		<1.5	<9.2
He II	243		<0.41	<2.4
			yr 1/yr 2	ave. of yr 1/yr 2
Fe IX	171.5	5.90	1.7/1.1	6.9
Fe X	174.5	6.02	0.8/1.2	4.3
Fe XI	180.6	6.09	1.0/0.6	4.4
Fe XV	284.2	6.32	1.0/1.5	10.3
Fe XVI	335.4	6.45	0.6/0.8	7.4

[a] Temperature of peak emissivity in a Raymond–Smith equilibrium model.
[b] Assuming a distance of 188 pc and corrected for interstellar attenuation with
  $N_H = 1 \times 10^{18}$ and $N_{He} = 1 \times 10^{17}$.
[c] Blended with Si X and S XIII.

the temperature of the gas producing the X-ray and EUV Fe line emission; and there must be some attenuation of the line emission by ionised material along the line-of-sight, perhaps from the stellar wind material.

The strongest emission line observed in the EUV spectrum, the He II Lyman α line at 304 Å, is difficult to explain. It cannot arise in the photosphere because, for a value of T_{eff} at 21 000 K, the non-LTE models predict an extremely low abundance of fully ionised He in the upper levels of the photosphere, where such a line could be formed. In addition, there would be too little He III in the wind, if the wind has a temperature in radiative equilibrium with the photosphere. However, accounting for the wind generated X-rays, as observed by ROSAT, there are sufficient photons for production of the 304 Å line, assuming that the X-rays arise from shocks distributed throughout the wind.

6.2.3   *The wind and the high energy flux from ϵ CMa*

Although Cassinell *et al.* (1995) achieved some measure of understanding concerning the origin of the high energy radiation from the wind, several problems remain. In particular, theoretical EUV spectra, based on two temperature model fits to the ROSAT data predict too much flux, especially in the iron line complex near 175 Å. A similar problem also appears to exist when comparing the predictions of stellar wind theory with observations. For example, while the general properties of these models (e.g. Owocki *et al.* 1988) agree reasonably well with the X-ray observations of O stars (Cooper 1994), the X-ray emission from B stars later than about B2 is greater than any wind shock model can explain (Cassinelli *et al.* 1994; Cooper 1994). Cohen *et al.* (1996) have used the combined data from EUVE spectroscopy and ROSAT PSPC observations of ϵ CMa to examine these problems. Although the EUVE and ROSAT detectors cover different photon energy ranges, for the most part they probe the same gas, when the EUVE range is confined to the shorter ($\approx$80–360 Å) wavelengths. Many

Table 6.4. *ROSAT and EUVE observation of ∈ CMa.*

Instrument	Observation dates	Exposure time (s)
ROSAT PSPC (unfiltered)	Mar 23–Apr 15 1992	2290
ROSAT PSPC (boron filter)	Mar 17–Apr 15 1992	9213
EUVE year 1	Jan 17–19 1993	62 323[a]
EUVE year 2	Mar 8–12 1994	143 337[a]

[a] Effective exposure for the MW spectrometer.

of the ion species which exist at temperatures of between 0.5 and 5 million K have several emission lines in both the ROSAT and EUVE energy ranges. Table 6.4 lists the observations included in the Cohen *et al.* study. All the high energy lines seen in the first EUVE observation, reported by Cassinelli *et al.*, were also detected in the second exposure. In addition, a further Fe line (Fe X 174.5 Å) not noted earlier was discovered. Interestingly, the Fe XII lines near 193 Å and 195 Å are not detected. These lines are among the strongest detected in the solar EUV spectrum (Thomas and Neupert 1994) and similar in intensity to the solar Fe IX 171 Å feature. The reason for this is that, while the Fe XII lines have similar emissivities to the 171 Å line, they are formed at slightly higher temperatures where there is less plasma. Furthermore, they are closer to the He II edge and, therefore, subject to more attenuation in the wind. The observed line strengths for both EUVE observations are included in table 6.3. It can be seen that the intensity of none of the five Fe lines varies by more than 50% between the two observations, indicating that the properties of the hot plasma from which lines originate are constant over time.

Cohen *et al.* (1996) have tried to reconcile the X-ray and EUV data with several empirical models. For example, the hot coronal plasmas observed in late-type stars are usually inter-preted by optically thin thermal plasma models such as those of Raymond and Smith (1977), Mewe *et al.* (1985) or Landini and Monsignori-Fossi (1990) and subsequent developments. Depending on the spectral resolution of the instrument one or more temperature components may be required to give a good match to the data. Typically with data from the ROSAT PSPC detector, a single temperature component fails to reproduce the observation and a minimum of two temperatures is required in the model. Alternatively, a power law distribution of tem-peratures can be assumed. In the case of ∈ CMa, a single temperature model certainly does not fit the PSPC data, while both a two temperature representation or power law temperature distribution are equally good. A more complex four temperature model is also a good, if not slightly better fit to the data, although there is no independent evidence that such a complex model is appropriate. However, none of these models are capable of matching the EUVE data. Cohen *et al.* find that the data in the combined spectral ranges can only be reconciled by invoking a plasma emission model with at least two temperatures, combined with attenuation by the wind.

Including wind absorption requires a calculation of the wind ionisation balance, obtaining the photoionisation absorption cross-sections and specifying the distribution of emitting and absorbing material. The wind is significantly ionised by the EUV/X-ray radiation and, therefore, has a rather different absorption cross-section per particle than neutral material. Cohen *et al.* estimate that about 20% of helium is in the form of He II and that heavier elements are predominantly in their second, third and fourth ionisation stages. This is illustrated in

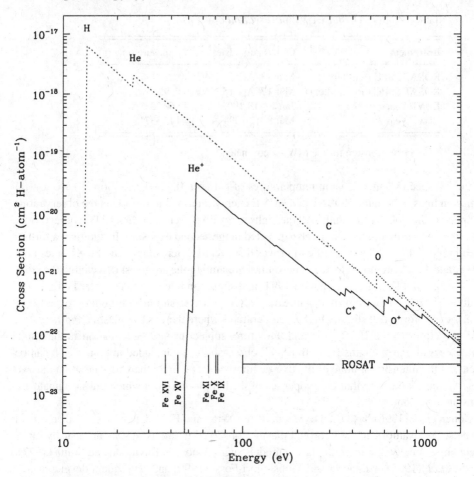

Fig. 6.6. The absorption cross-section per hydrogen atom for the ions in the stellar wind
(solid line) and for the neutral ISM (dotted line) as a function of photon energy (from
Cohen *et al.* 1996).

figure 6.6, which compares the absorption cross-section per hydrogen atom for ions in the
stellar wind with the neutral ISM as a function of photon energy. Although the neutral ISM has
a greater absorption cross-section than the wind, the total column density of wind material is
much larger. Hence, above the He II 54 eV edge (shortward of 228 Å), the wind is the primary
source of opacity.

The assumed spatial distribution of the absorbing plasma depends on the general model of
the wind under consideration. In a coronal model, all the emitting gas will be confined near
the surface of the star and all the absorbing material lies between this and the observer. In this
case a simple exponential absorption can be applied (see Cohen *et al.* 1996). Alternatively,
in a wind-shock model, the emitting and absorbing plasmas are interspersed and distributed
throughout the wind. In this case an exospheric approximation is used, where only the outer,
optically thin wind contributes to the observed luminosity. Inclusion of either of these models
makes the analysis more physically realistic, but only adds a single additional free parameter
dependent on the mass-loss rate and terminal velocity of the wind. Figure 6.5 shows how

the transmission of the stellar wind differs for the two models. Even with inclusion of one or other wind model, a single temperature emission component is unable to reproduce the data, while use of two temperatures can allow the ROSAT and EUVE data to be matched separately but not simultaneously. The power law differential emission measure model does work but only with the exospheric wind absorption.

A mass-loss rate of $3 - 8 \times 10^{-8} M_\odot \, y^{-1}$ is derived from the wind-shock model. This is consistent with theoretical predictions but about five times that estimated from the ROSAT data by Drew *et al.* (1994). This is because in the Cohen *et al.* model, beyond a certain radius, only a small fraction of the wind is emitting X-rays. Furthermore, Drew *et al.* assume that the mass-loss rate is governed by the X-ray emission whereas with the shock model it is primarily the attenuation by the wind that determines mass loss.

6.3 Observations of β CMa

β CMa (Mirzam = HD44743) is similar to ε CMa but slightly hotter (spectral type B1). Lying in the same low column density direction as ε CMa, it was also detected in all three channels of the EUVE spectrographs. The star is well studied and its physical properties are summarised, with those of ε CMa, in table 6.1. Both stars are X-ray sources and the properties of their winds are very similar, which is important because of the anomalously large Lyman continuum in ε CMa. A further interesting point for comparison is that β CMa lies in the β Cephei instability strip and pulsates while ε CMa lies just redward of this region and does not. Partly because β CMa is the brightest known member of the β Cephei class it was the first to be well studied, revealing evidence for more than one period (Meyer 1943). β Cephei stars (also called β Canis Majoris stars) are a class of early B star, showing photometric and radial velocity variations. Pulsation periods, which are often multiple, range from 3 to 6 h. The pulsations, which are driven by temperature sensitive ionisation and opacity changes in the mantle of the star, manifest themselves as expansions and contractions of the stellar atmosphere resulting in changes in the effective temperature and luminosity of the star. As the light curves of β Cephei stars lag behind the radial velocity curves by 0.25 cycles, that is the stars have maximum brightness when the radius is smallest, the variability is best explained as a periodic change in the effective temperature of the star. The fact that flux changes in the visible spectrum are small compared to those in the UV is consistent with this picture. Observations of the UV flux amplitudes in β CMa indicate a change in temperature of 180 ± 130 K (Beeckmans and Burger 1977). However, any changes in the EUV flux levels should reduce the large uncertainty in this figure.

Two EUVE observations of β CMa were made with exposure times of 52 600 s and 104 600 s during December 1993 and February/March 1994 respectively. Both observations were made using the 'dither' mode. As in ε CMa, the most striking feature is the strong Lyman continuum extending out to 750 Å (figure 6.7). In contrast to ε CMa, the spectrum of β CMa shows a strong He I photoionisation edge at 504 Å. A signal is detected shortward of this feature, but most, if not all, can be attributed to photons scattered by the instrument from above the edge. Hence, there is no evidence for any emission lines similar to those seen in ε CMa in the short and medium wave spectrograph channels.

An analysis of the β CMa EUV spectrum, using synthetic spectra, along the lines of that carried out for ε CMa reveals a similar problem in matching all the available data. Neither LTE nor non-LTE models are able to provide a consistent description of the flux at all wavelengths. If the EUV and UV spectral ranges are consistent there is a visible and IR flux deficit in the

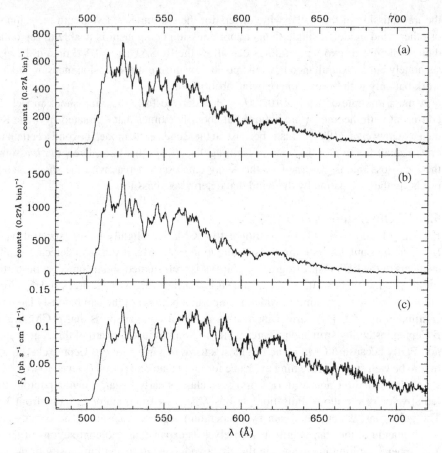

Fig. 6.7. The LW spectra of β CMa from (a) 1993 and (b) 1994. Panel (c) shows the combined data divded by the total effective exposure time and corrected for the effective area of the instrument (from Cassinelli *et al.* 1996).

models when compared with the data. If the models are adjusted to give agreement from the UV through to the near-IR there is a shortfall in the predicted Lyman continuum flux of about a factor of five.

In the range 500–700 Å the EUV spectrum is rich in absorption features which can be used to compare the temperature and ionisation structure of β CMa's atmosphere and wind with those of ϵ CMa. Although β CMa has an effective temperature that is several thousand degrees hotter than ϵ CMa, many of the same features are observed in both stars. However, in β CMa, the continuum level and extent of the absorption features are better determined because of the 'dithering' of the observations.

The absence of any emission lines shortward of the He I 504 Å edge in β CMa is a surprise. The most likely cause is a higher level of interstellar attenuation. The magnitude of the He I 504 Å edge can be entirely explained by photospheric absorption, providing no reliable evidence for He I in the ISM. However, this analysis is very sensitive to uncertainties in the atmosphere models. An estimate of the intrinsic luminosity in the strongest emission line expected (He II 304 Å, from its detection in ϵ CMa) allows a lower limit to the He I column to be determined. After taking into account the known H I column of 2.0×10^{18} cm^{-2}, Cassinelli *et al.* (1996) obtain a value of $N_{\mathrm{He\,I}} > 1.4 \times 10^{17}$ cm^{-2}, which is comparable to the amount of neutral hydrogen. One possible explanation is that the gas along the line-of-sight to β CMa is highly ionised. From the observed S II/S III interstellar line ratio and a standard Strömgren H II region model, Gry *et al.* (1985) estimated a total H II column density of 1.8×10^{19} cm^{-2}. Assuming a cosmic abundance ratio of 0.1 for He/H the implied He I column density is 2.0×10^{18} cm^{-2}, a value consistent with the Cassinelli *et al.* lower limit. Hence, the EUVE data agree with the findings of Gry *et al.* that there is an H II region associated with β CMa.

6.3.1 Pulsations

Three pulsation periods, all close to 6 h, have been identified through optical observations (Shobbrook 1973) but the amplitude of the variability is small (0.02 magnitudes). UV variability has also been measured by several satellites. Cassinelli *et al.* (1996) report the clear detection of pulsations in the EUV data. To investigate the potential variability, they created a light curve from the spectrum between 500 Å and 700 Å, filtering the data to remove times of bad attitude and sampling and subtracting the background count rate as for the spectral reduction. The resulting light curve (see figure 6.8) consists of a three-day long segment separated by 70 d from the later observation covering six days. Within these periods the light curve is interrupted by periods when the source is occulted by the Earth. Although the data are sampled rather unevenly, the power spectrum of the data reveals a strong signal near 6 h (figure 6.9). The peak near 90 mins is attributable to the spacecraft orbital frequency. The EUVE data do not span a long enough period for the 6 hr peak to be resolved into its multiple components. Hence, to determine the periods and amplitudes, Cassinelli *et al.* fit an empirical pulsation model comprising a superposed sum of sine and cosine functions for the three known optical periods. The best fit model is shown in figure 6.8 and its parameters listed in table 6.5. The amplitude of the strongest mode ($p = 0.251065$ d) is 0.084 mag, approximately four times that seen in the optical, which is consistent with the idea that the brightness variations are driven by temperature changes.

The amplitude of the EUV pulsations can be translated directly into temperature changes. For the primary period this corresponds to a variation of $108 + 31/- 32$ K with much smaller values for the other periods (see table 6.5). This is a smaller temperature change than found in the UV by Beeckmans and Burger (1977) but falls within their error bars. Therefore, it can be concluded that while there may be some damping of the pulsations between the deeper layers of the UV/optical continuum formation and those of the Lyman continuum formation, the pulsations do propagate to very high layers in the atmosphere. Indeed, the Lyman continuum is formed approximately six scale heights above the optical continuum in static models.

6.4   Coronal sources – the stellar zoo

The ultimate goal of a programme of EUV stellar spectroscopy is to cover equally all regions of the HR diagram, studying main sequence stars with a range of spectral types and examining evolved objects in the giant and supergiant luminosity classes. Complicating factors in this area may be the presence of a companion which might make any analysis ambiguous and, if the period is short enough, produce enhanced activity through tidal increases in the stellar rotation rates. Photometric observations of the kind discussed in chapters 3 and 4 have defined

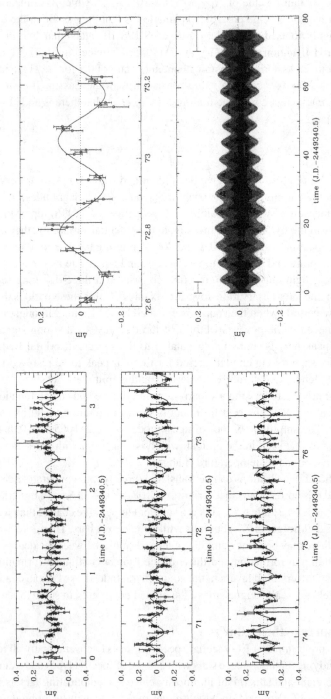

Fig. 6.8. (Left) The EUV light curve of β CMa constructed from the wavelength interval 500–700 Å with 500 s bins. The best fit three-period model is shown as the solid line. (Right) The top panel shows one small section of the fitted light curve and the bottom panel the full three-period model. The horizontal bars indicate the epochs for which data exist (from Cassinelli *et al.* 1996).

Table 6.5. *Best fit model light curve parameters.*

	Period (d)	t_0 (d)[a]	Δmag	ΔT_b(K) (range)
P1	0.251065	0.1043	0.084	108 (76–139)
P2	0.249951	0.2017	0.020	26 (14–59)
P3	0.239035	0.0517	0.022	28[b]

[a] The epoch of maximum light (JD–2 449 340.5).
[b] Fixed.

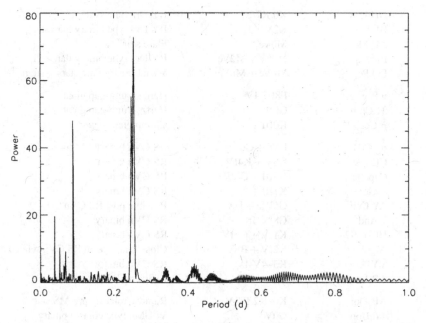

Fig. 6.9. The power spectrum of the 500–700 Å light curve of β CMa for the combined datasets (from Cassinelli *et al.* 1996).

the general relations between activity, rotation rate and stellar age across the HR diagram. However, the same data only provide a relatively crude indication of the physical state of the plasma in an individual star and it is detailed information concerning the temperature and density distribution together with the abundance found in coronal plasma that is needed to test the plausibility of any heating mechanisms proposed.

Haisch *et al.* (1994) provide a good summary of the outstanding problems that could be addressed by EUV spectroscopy.

- What is the scale, filling factor and density of material in stellar coronal loops?
- Is there a universal shape for the differential emission measure for all stars?
- What is the relationship between coronal heating and flares?
- Are all coronae of magnetic origin or are some wholly or partly acoustically heated?
- How do the coronae change as the stars evolve?
- Do coronal abundances vary from the photospheric?

Table 6.6. *A list of some important late-type stars observed spectroscopically with EUVE (MS = main sequence; WD = white dwarf).*

Star name	Spectral type/Class	Comments
Procyon	F5IV–V	Single star near MS, WD companion
ξ UMa B	G0V + ?	Single line spectroscopic binary
χ^1 Ori	G0V	Single MS star
αCen A + B	G2V + K1V	Wide binary, effectively single stars
ξ Boo A (+ B)	G8V (+ K4V)	Wide binary, A component dominates flux
ε Eri	K2 V	Chromospherically active isolated MS star
DH Leo	K0/K7/K5	BY Dra type system
BF Lyn	K2V + K	BY Dra type binary flare star
AU Mic	M0Ve	Flare star
FK Aqr	M2Ve + M2Ve	BY Dra type binary flare star
EQ Peg	M4Ve + M6Ve	Visual binary flare stars system
ν Peg	F8III–IV	Hertzsprung-gap giant
31 Com	G0III	Hertzsprung-gap giant
β Cet	K0III	Clump giant
σ^2 CrB	F6V + G0V	RS CVn binary
CF Tuc	G0V + K4IV	RS CVn binary
Capella	G5III + G0III	RS CVn binary
σ Gem	K1III + ?	RS CVn binary
AY Ceti	G5IIIe + DA	Possible post RS CVn?
λ And	G8IV–III	RS CVn binary
HR 1099	K1IV + G5IV	RS CVn binary
Algol	K2IV + B8V	Close binary with RS CVn related activity
VY Ari	K3–4V–IV	RS CVn binary
AR Lac		Eclipsing RS CVn binary
AB Dor	K0–K2 IV–V	Rapidly rotating pre-MS star
44i Boo	G0V	W UMa type contact binary

A strategic programme of studies at key points in the HR diagram would be the ideal, examining, for example, the region near the late A to mid-F stars which represents the upper temperature limit of coronal activity where convection zones become extremely shallow. Another important region is the coronal dividing line as evolving giants pass from G to early K spectral class and the coronae disappear from view. Although more than half the EUV sources are late-type stars, they are significantly less luminous than the white dwarfs. However, all the flux is concentrated in a few lines rather than spread across a continuum. In reality the strategy was determined mainly by which stars could be detected by the EUVE spectrographs. Accordingly, the actual observations have been rather heterogeneous. Even so, distinct aspects of stellar activity can be examined within this programme. Table 6.6 lists all the late-type stars observed spectroscopically by EUVE for which comprehensive analyses have been published, including notes about the spectral type, luminosity class and other relevant information about the system (e.g. binarity). It should be noted that this is not a complete list of all EUVE late-type star observations.

It can be seen in table 6.6 that all regions of the cool star portion of the HR diagram are represented apart from the very earliest (late A to early F) spectral types. The coverage is a little unbalanced in some senses, only six stars spanning the main sequence from F to K while there are five stars in the M dwarf group and a preponderance of RS CVn systems. This really arises from the relative EUV luminosities of the different types. RS CVn and M dwarf flare star groups have enhanced luminosities, arising from rapid rotation. Most lower activity stars fall below the sensitivity threshold of the EUVE spectrographs and, consequently, contribute a relatively small fraction of the total sample. Table 6.6 groups all the objects into several broad categories associated with the HR diagram – main sequence stars, M dwarf flare stars, giants and RS CVn systems plus some miscellaneous objects. Since distinct activity levels and behaviour can be associated with each group, they form a useful basis for a review of the results from the EUVE spectroscopic observations.

6.5 Main sequence dwarfs (F–K)

6.5.1 Procyon

The photospheric abundances in Procyon are well-known from a number of independent studies and are solar within experimental uncertainties ($<25\%$ e.g. most recently Edvardsson *et al.* 1993). Hence differences between photospheric and coronal abundances can be probed with higher precision than might be possible for other stars. Since it is relatively nearby (3.48 pc) and has a corona similar to that of the active Sun, it is the brightest effectively isolated star available in the EUV. It does have a DF white dwarf companion (Procyon B), but this is too far away from Procyon A to interact and also too cool to be a source of EUV emission itself. Drake *et al.* (1995a,b) have carried out a detailed analysis of the EUVE spectrum of Procyon to determine the pattern of low FIP and high FIP coronal abundances. These data can also be used to determine the emission measure distribution (see section 5.4.3). The star was observed for $\approx 100\,000$ s and, due to the low $\approx 1 \times 10^{18}$ cm^{-2} column density, was detected in all three spectrograph channels. Figure 6.10 shows the flux calibrated spectra, with identifications of the principal spectral features.

Superficially, the EUV spectrum of Procyon is very similar to the quiet Sun, with strong lines of Fe IX to Fe XIV dominating the SW and MW ranges. The assistance this afforded in improving the calibration of the EUVE wavelength scale allowed the identification of many weaker features. Drake *et al.* (1995b) measured the line intensities directly, by fitting Gaussian profiles to the data. This procedure allowed information to be extracted from partially blended as well as completely resolved features. The profile of any line is dominated by the instrumental resolution and is well-represented by a Gaussian, except at the short wavelength end of the MW band where there appears to be extra flux in the blue wings of the lines. As this is not present in the same lines observed in the long wavelength end of the SW spectrum, it must be an instrumental effect. In these cases the line intensities were estimated just by summing the pixels, ignoring any lines that appeared to be blended.

Emission measures were derived from the line intensities, after correction for ISM attenuation, following the prescription of C. Jordan *et al.* (1987; see also the discussion in section 5.4.3) and averaging the expression describing the line emissivity over the temperature interval $\Delta \log T = 0.3$. Electron densities derived from Fe XIV lines by Schmitt *et al.* (1996a) yield values in the range 3–5×10^9 cm^{-3}. Drake *et al.* (1995b) adopt a value of 10^{10} for their analyses, deriving two different emission measure distributions, one using the solar coronal abundances adopted by Feldman *et al.* (1992) and a second using the photospheric

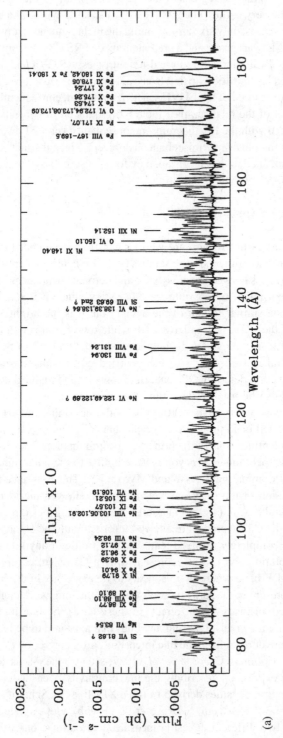

Fig. 6.10. Flux calibrated EUV spectra of Procyon. Individual lines and unresolved groups used in the analysis of Drake *et al.* (1995b) are labelled, together with a number of other features not used. The line identifier '2nd' refers to shorter wavelength lines seen in second order. (a) SW channel; (b) MW channel; (c) LW channel (from Drake *et al.* 1995b).

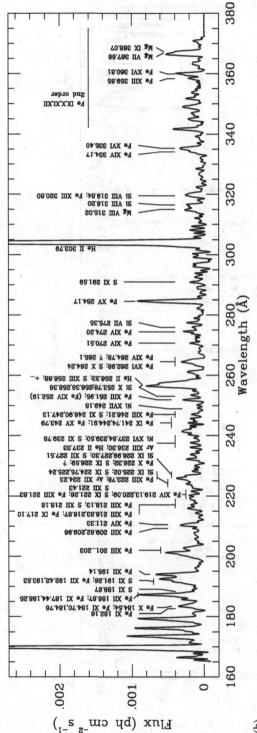

Fig. 6.10. (cont.)

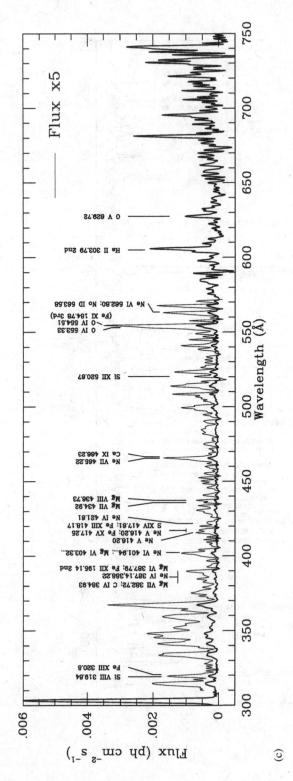

Fig. 6.10. (cont.)

(c)

Table 6.7. *Procyon photospheric and solar coronal abundances adopted by Drake et al. (1995b). Values are logarithmic relative to H = 12.00.*

Element	Procyon photosphere	Solar corona
H	12.00	12.00
He	10.99	10.9
O	8.93	8.89
Ne	8.07	8.08
Mg	7.58	8.15
Si	7.55	8.10
S	7.21	7.27
Ar	6.56	6.58
Fe	7.51	8.10
Ni	6.25	6.84

abundances for Procyon (see table 6.7). In such an analysis, if the assumed abundances are appropriate values for the star in question, the emission measures corresponding to low and high FIP species should be identical at any given temperature, within the experimental uncertainties. According to Laming *et al.* (1995; see also section 5.4.6), the FIP effect is only visible in the disc integrated solar spectrum at temperatures above $\approx 10^6$ K. If Procyon had the same pattern of FIP enhancement as the Sun, the emission measures derived using the photospheric abundances corresponding to low and high FIP species should overlap at temperatures below 10^6 K and bifurcate at higher temperatures. For emission measures derived using solar coronal abundances the opposite would occur.

Figures 6.11 and 6.12 show the emission measure distribution of Procyon, derived using solar coronal and Procyon photospheric abundances respectively. In both cases the open and filled symbols correspond to high and low FIP species respectively. When the solar coronal abundances are used, the distribution of points corresponding to the low FIP species lies significantly below that indicated by the high FIP species. This is particularly clear in the spline fits to the sets of points (lower panel of figure 6.11). The discrepancy amounts to about 0.3 to 0.6 in the log of the emission measure. In contrast, the same exercise with the photospheric abundances of Procyon (figure 6.12) shows good agreement between the high and low FIP emission measures. Hence, there is no evidence for the presence of a solar-like FIP effect in the corona of Procyon.

6.5.2 ε Eridani

Most of the coronal sources in the EUV catalogues are so-called active stars, with coronal X-ray and EUV luminosities several orders of magnitude greater than the Sun. However, of great importance in relating coronal physics to what is seen in detail on the Sun itself is the study of the few objects that can be considered to be solar-like in a strict sense. Such low luminosity stars can only be detected within the immediate vicinity of the Sun (within ≈ 5 pc). However, the stars in this volume are primarily M dwarfs with only a few of spectral type K or earlier. One of these is the K2V dwarf ε Eri ($d = 3.29$ pc). While its X-ray luminosity is greater than that of the Sun at solar maximum it is one to two orders of magnitude less than more active and more distant dKe stars. While very active stars have coronal temperatures in

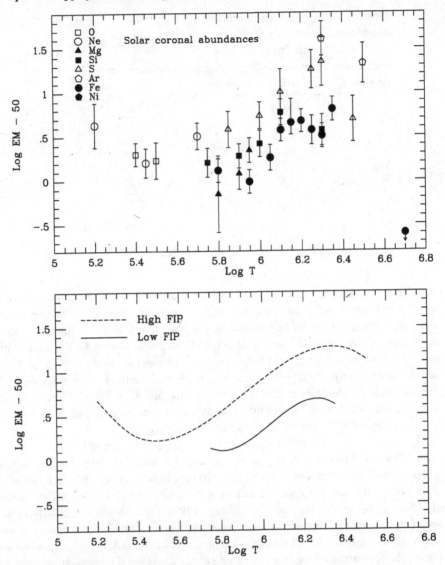

Fig. 6.11. The emission measures as a function of temperature derived using solar coronal abundances from individual ions observed in the EUVE spectra of Procyon (upper panel). Low FIP species are indicated by solid symbols while high FIP species are the open symbols. The lower panel shows the first order cubic spline fit to the points corresponding to low FIP (solid curve) and high FIP (dashed curve) species (from Drake *et al.* 1995b).

excess of 10^7 K, that of ϵ Eri is somewhat lower, although not as low as inactive stars such as Altair or Procyon. Therefore, ϵ Eri appears to be of intermediate activity.

The highest ionisation stage of Fe detected in the Procyon spectrum is Fe XVI. In ϵ Eri, Schmitt *et al.* (1996b) find emission lines from ionisation stages up to Fe XXI (the lowest is Fe IX as in Procyon). Hence, the corona of this star appears to be somewhat hotter than that of Procyon, covering a temperature range from 10^6 K to 10^7 K and confirming the existing view, expressed above, of their relative levels of activity. While the lines of the lowest and

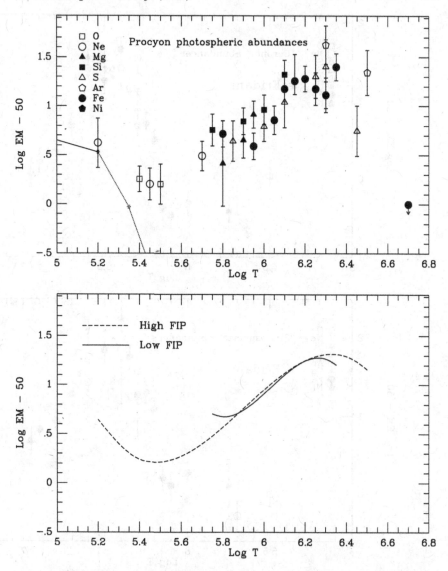

Fig. 6.12. The emission measures as a function of temperature derived using Procyon photospheric abundances from individual ions observed in the EUVE spectra of Procyon (upper panel). Low FIP species are indicated by solid symbols while high FIP species are the open symbols. The lower panel shows the first order cubic spline fit to the points corresponding to low FIP (solid curve) and high FIP (dashed curve) species (from Drake *et al.* 1995b).

highest ionisation stages of Fe are relatively weak, the strongest lines detected are of Fe XV and Fe XVI, which implies that there is a peak in the emission measure distribution near $\log T \approx 6.4$. Even so, the distribution does have significant emission at higher and lower temperatures (figure 6.13). Schmitt *et al.* (1996b) base their results purely on an analysis of the Fe lines, but Laming *et al.* (1996), who include emission lines from all detected species in their analysis, obtain very similar results.

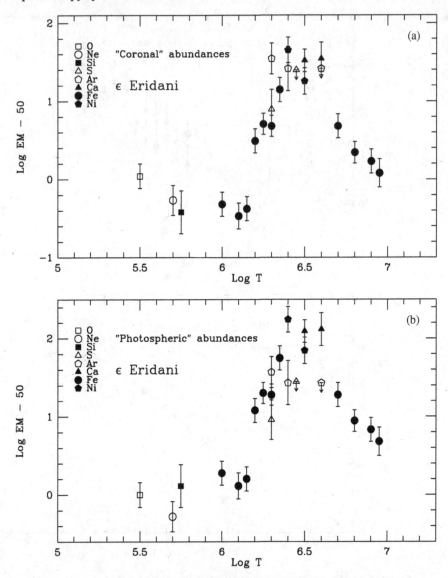

Fig. 6.13. Emission measures determined from the lines in the EUV spectrum of ϵ Eri for (a) solar coronal abundances scaled to the metallicity determined by Drake and Smith (1993) and (b) solar photospheric abundance scaled in the same way. The open symbols represent high FIP species and the filled symbols low FIP species (from Laming *et al.* 1996).

An important feature of the ϵ Eri spectrum (and Procyon, see above) is the presence of the density-sensitive Fe XIII and Fe XIV lines. Schmitt *et al.* (1996b) are able to use the line ratios to estimate the coronal density in the star, i.e. the ratio of Fe XII $\lambda202.04:\lambda203.79$ and Fe XIV $\lambda211.33:\lambda264.78$. Line fluxes are determined by fitting Gaussian profiles to the data, from which the ratios are then calculated (as in Drake *et al.* (1995b) and Schmitt *et al.* (1996a)). Comparing the observed ratios for each pair of lines with predicted values from N. S. Brickhouse *et al.* (1995) yields a nominal density of $2 \times 10^9 \text{ cm}^{-3}$, with an uncertainty spanning the range from $10^9 \, \text{cm}^{-3}$ to $10^{10} \, \text{cm}^{-3}$. This is similar to the typical density of $3 \times 10^9 \, \text{cm}^{-3}$ found in solar active regions.

Laming *et al.* (1996) have considered the issue of coronal abundance in ϵ Eri in the same manner first demonstrated for Procyon (Drake *et al.* 1995a,b). However, in this case, the results are less clear cut due to a lower signal-to-noise in the spectrum. Furthermore, the photospheric composition of ϵ Eri is not so well known as that of Procyon. However, the available data do not support significant deviations from a solar mixture, apart from a very small metal deficiency determined by Drake and Smith (1993) for Ca and Fe. Consequently, Laming *et al.* (1996) adopt $[X/H] = -0.1$ for all the elements. Figure 6.13a shows the emission measure determined for coronal abundances, scaled by this metallicity. A complete, smooth distribution is obtained from the large number of lines originating from low FIP species (Fe, Ca, Ni and Si). However, for temperatures above $\log T = 5.8$, there are just three high FIP ions measured (S XII, Ar XII and Ar XIII) and two upper limits (S XIV and Ar XVI). Apart from Ar XII, these points all lie close to the low FIP distribution. This suggests that there might be a similar FIP effect to that found in the Sun. When photospheric abundances are used in the exercise (figure 6.13b) a difference between low and high FIP data points becomes apparent, supporting this conclusion. The ϵ Eri result is far less conclusive than that obtained for Procyon. However, if a FIP effect is present, it cannot be substantially larger than that found in the Sun.

6.5.3   α Centauri

α Cen A is probably the star that is most similar to the Sun among the targets available to EUVE. However, its observation is complicated by the presence of its close companion, α Cen B, resulting in a composite spectrum, since the separation is only 21 arcsec. UV and X-ray observations indicate that both stars are likely to contribute substantially to the net EUV spectrum, because their levels of activity are comparable. Hence, the spectrum arises from a linear combination of the emission measure distributions from both components.

At a distance of only 1.33 pc, the amount of interstellar absorption is very small. Hence, the system is detected in all three EUVE spectrograph channels. Mewe *et al.* (1995) present a detailed analysis of the $\approx 110\,000 \, \text{s}$ exposure. Unlike, the analyses of Procyon and ϵ Eri discussed above, where the emission measure distribution was determined from the analysis of individual spectral lines, Mewe *et al.* (1995) adopt the alternative approach of using a spectral modelling code to fit the observation. In this case the Mewe *et al.* (1985) code was used with the modifications of Kaastra (1992). This was further extended with spectral lines between 300 Å and 2000 Å from Landini and Monsignori-Fossi (1990). Interstellar absorption was taken into account using the Rumph *et al.* (1994) absorption cross-sections. In performing their analysis, Mewe *et al.* (1995) identified a deficiency in the wavelength calibration for each spectrograph channel. Corrections derived from comparing apparent line wavelengths with predicted values taken from nine different stellar sources were applied to the data. Subsequently, these modifications were found to be in good agreement with the recalibration of the wavelength scale incorporated into subsequent versions of the processing software (egodata 1.8 and later).

In the analysis of broadband data, either a one/two temperature component emission model might have been considered or, alternatively, a differential emission measure specified by a simple function such as a power law. With limited information, more complex models would not have been justified. Now, with the enormous amount of information encapsulated in the lines found in an EUVE spectrum a more detailed emission measure distribution can be computed. However, in the absence of an *a priori* function which describes the distribution and which can be applied to the plasma code for comparison with the data, determining the actual emission measure distribution becomes a complex inversion problem. The adopted procedure is summarised by Mewe *et al.* (1995) and is described in detail by Press *et al.* (1992). It relies on assuming that the total emission from the star is specified by a linear combination of isothermal components. The differential emission measure is then the function that specifies the relative contributions in each temperature bin. Mathematically, the output spectrum can be expressed as a matrix–vector product, which must be inverted. The 2-D matrix specifies the predicted flux for each wavlength interval and temperature and the vector is the differential emission measure. The outcome of the inversion is a differential emission measure which can then be applied to the plasma code, folded through the instrument response and compared to the observed spectrum. A χ^2 statistic can then be calculated to give an *a posteriori* indication of the goodness of fit but, unlike the spectral fitting described earlier, and applied to the analysis of white dwarf spectra (see section 5.4.1), this is not an iterative method which searches for a minimum in the value of χ^2.

Figure 6.14 shows the differential emission measure distribution determined by Mewe *et al.* (1995). This shows that the observed flux arises mainly from plasma below $\approx 5 \times 10^6$ K, with a broad peak around $\approx 3 \times 10^6$ K. The width of this peak is only slightly larger than the expected resolution of about a factor of two in temperature. A weaker tail extends from 2×10^6 K down to 5×10^5 K, with very little emission from plasma between the latter value and 10^5 K. There is a significant contribution from plasma below 10^5 K and evidence for a very hot component exceeding several tens of millions of degrees.

The high temperature component in the differential emission measure, lying above $\approx 3 \times 10^7$ K, reflects the presence of a featureless continuum superimposed on the line spectrum. This is most pronounced in the short wavelength spectrograph channel (figure 6.15). When the high temperature component is removed from the best fit model (figure 6.15b), the SW continuum cannot be acounted for. However, the shape of the emission measure distribution is poorly constrained above $\approx 2 \times 10^7$ K, because few of the lines seen in the EUV arise from ions formed at these temperatures. Hence, while there is a need for a high temperature component to explain the observation, its temperature is not well constrained.

Mewe *et al.* (1995) consider carefully the reality of the high energy tail. For example, the assumption that the coronal abundances are solar may be erroneous. Lowering the abundances of those elements that contribute to the lines (e.g. Fe, Si, Mg and Ne) increases the relative level of the continuum, compared to the lines but, at most, any changes in the abundances only account for $\approx 10\%$ of the continuum. In addition, decreases in the abundances are at odds with the fact that the stellar photospheric abundances seem to be higher than solar. However, it is important to note that the FIP effect can lead to apparent differences between photospheric and coronal abundances, as seen in Procyon (see section 6.5.1).

If this hot component is real, the lower part of the high temperature tail would produce stronger Fe XXIV lines than are present in the spectrum (e.g. at 192.02 Å). An alternative explanation of the continuum is that it is caused by blends of many unresolved lines of lower ions or lines that are not incorporated in the spectral model. Mewe *et al.* (1995) suggest this is unlikely in the SW range, although they are unable to completely rule it out. Jordan (1996)

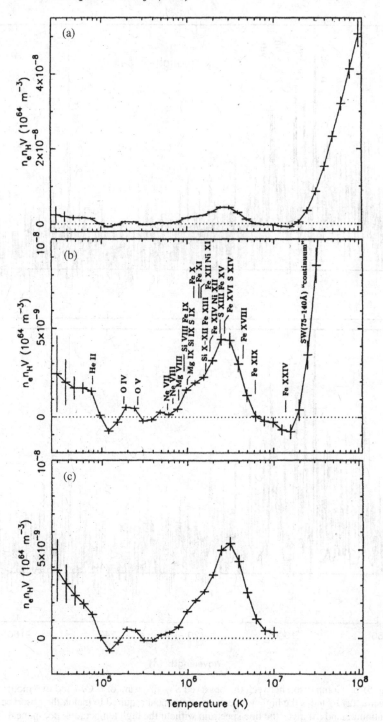

Fig. 6.14. Emission measure distribution of α Cen, determined by Mewe *et al.* (1995) from their analysis. Panels (a) and (b) show the result, after wavelength recalibration and relative effective area corrections, on different scales. Panel (c) is the distribution obtained by a fit with a reduced temperature range. The formation temperatures of ions contributing to the lines in the spectrum are noted in panel (b) (from Mewe *et al.* 1995).

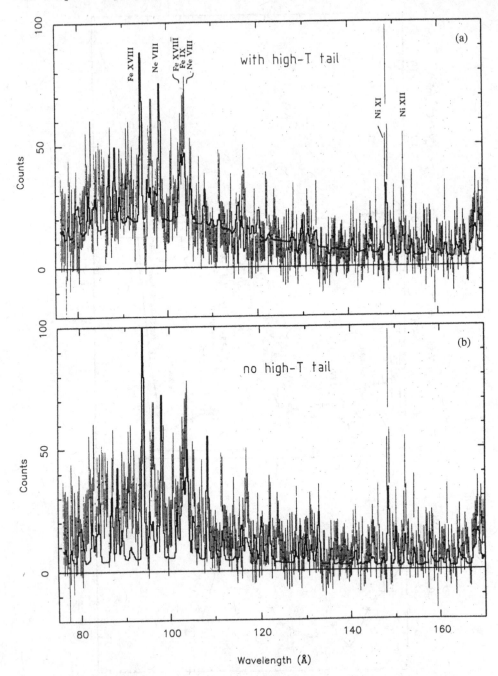

Fig. 6.15. Comparison between the observed SW spectrum of α Cen and two model spectra. (a) Includes the high temperature component required to match the observed continuum and (b) gives the line spectrum without the high temperature component (from Mewe *et al.* 1995).

also indicates, on similar grounds, that it would be premature to attribute any discrepancies to additional components or non-solar abundances.

Mewe *et al.* (1995) suggest that the most likely explanation of the continuum flux may be that it arises through resonant scattering of photons from the strongest lines of the spectrum in gas that is not completely optically thin. Line photons are expected to be scattered preferentially as compared to continuum photons. A fraction of the scattered flux will be intercepted by the stellar chromosphere and photosphere and destroyed by absorption. A similar effect is noted in the EUV spectrum of Procyon by Schrijver *et al.* (1995). The reality of the resonant scattering proposition is questioned by Schmitt *et al.* (1996c). They analyse both the EUV spectra and ROSAT PSPC pulse height distribution for Procyon. If the EUV and X-ray emission is assumed to arise in an optically thin plasma, the ROSAT PSPC spectrum is not consistent with any high temperature tail to the differential emission measure. If the EUV emission is treated as optically thick and the X-ray optically thin, the X-ray flux predicted above 0.5 keV is about an order of magnitude greater than observed in the PSPC spectrum. This cannot plausibly be explained by time variability. In addition, Schmitt *et al.* (1996c) demonstrate that the observed count fluctuations in the EUVE SW spectrum are not consistent with the hypothesis that the majority of the observed flux arises from a continuum. As a consequence, their conclusion argues that resonance scattering is unlikely to be relevant for the interpretation of soft X-ray and EUV spectra of cool stars and that the 'missing' line hypothesis is a much more reasonable explanation of the observations.

6.5.4 Other solar-like stars

ξ UMa B, χ^1 Ori and ξ Boo A are all main sequence G stars of similar activity and, consequently, would be expected to have similar EUV spectra. This is certainly so for the first two objects, which are analysed by Schrijver *et al.* (1995). Their respective spectra are shown in figures 6.16 and 6.17. Although some different lines are identified in each case, this appears mainly to arise from the fact that they lie close to the limit of the signal-to-noise in each spectrum. χ^1 Ori is a binary comprising a G0V star with a mid-M dwarf companion. The orbital period is long (5191 d), indicating that little in the way of enhanced activity could arise from tidal interactions and the binary nature of the system. Nevertheless, as the equatorial velocity of the G star ($9 \, \mathrm{km \, s^{-1}}$, Soderblom and Mayor 1993) is several times that of the Sun ($v \sin i = 2.0 \, \mathrm{km \, s^{-1}}$) it should exhibit some enhanced activity. ξ UMa is also a binary and each component is a single-lined spectroscopic binary. Both primaries are classed as G0V with the A component having a 669 d orbital period and low $v \sin i$ ($1 \, \mathrm{km \, s^{-1}}$) while ξ UMa B has a much shorter period of 3.98 d. The projected equatorial velocity of ξ UMa B is also low ($2.8 \, \mathrm{km \, s^{-1}}$) but with such a short orbital period, this is probably a result of a low inclination, i.e. the system is viewed nearly pole-on from Earth. The B component does appear to dominate the activity of the system as a whole, since it is responsible for the bulk of the Ca II H and K emission.

Schrijver *et al.* (1995) follow the differential emission measure analysis described earlier for α Cen and the resulting emission measure distributions for ξ UMa and χ^1 Ori are shown in figures 6.18 and 6.19 respectively. In studying ξ Boo A, Jordan (1996) adopts a different approach, computing the emission measure for individual line fluxes, as described in section 5.4.3 (figure 6.20). It should be noted that Jordan considers the emission measure distribution over height, while Schrijver *et al.* use the volume emission measure. Hence, the units of

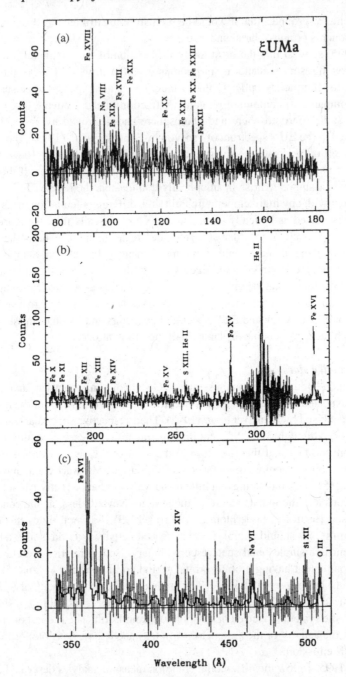

Fig. 6.16. EUVE background corrected count spectra of ξ UMa. (a) SW, (b) MW and (c) LW ranges. The observed spectrum is shown by the error bars, with the uncertainties on the observed number of counts, and the best fit spectrum from a differential emission measure analysis is represented by the solid histogram. Several prominent lines are indicated by the ions from which they originate (from Schrijver *et al.* 1995).

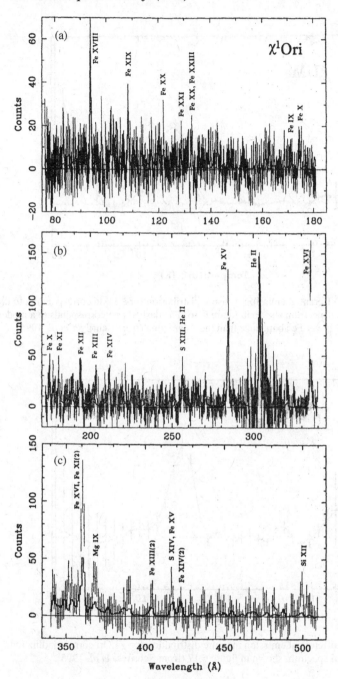

Fig. 6.17. EUVE background corrected count spectra of χ^1 Ori. (a) SW, (b) MW and (c) LW ranges. The observed spectrum is shown by the error bars, with the uncertainties on the observed number of counts, and the best fit spectrum from a differential emission measure analysis is represented by the solid histogram. Several prominent lines are indicated by the ions from which they originate (from Schrijver *et al.* 1995).

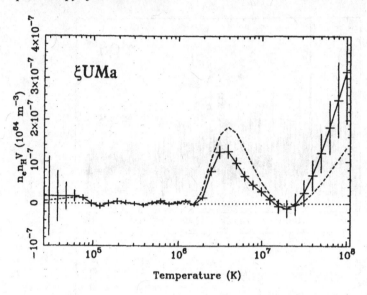

Fig. 6.18. Differential emission measure distribution for ξ UMa corresponding to the best fit theoretical spectrum shown in figure 6.16. The dashed line corresponds to a model with a factor of 3.3 lower Fe abundance than the solar value (from Schrijver *et al.* 1995).

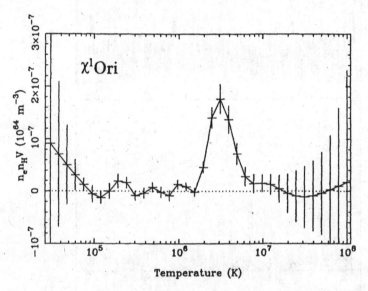

Fig. 6.19. Differential emission measure distribution for χ^1 Ori corresponding to the best fit theoretical spectrum shown in figure 6.17 (from Schrijver *et al.* 1995).

figure 6.20 and figures 6.18 and 6.19 are different. It is clear in all these cases that the emission measure distribution is dominated by plasma at temperatures in the range 10^6 to 10^7 K, as seen also in Procyon, α Cen and ϵ Eri. There are some differences in detail. For example, there is significant emission from α Cen down to $\approx 5 \times 10^5$ K while ξ UMa B and χ^1 Ori seem to have no component below $\approx 2 \times 10^6$ K. However, we must be careful

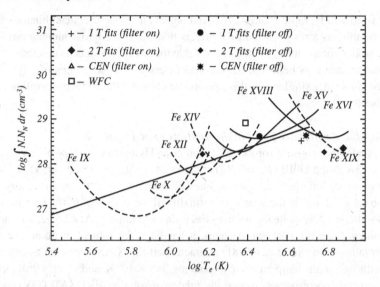

Fig. 6.20. The apparent emission measure distribution for ξ Boo A, from EUVE line fluxes. The dashed curves indicate upper limits and the thin line corresponds to the prediction of an energy balance model. Fits to the ROSAT PSPC and WFC data are shown for comparison, as indicated in the key (from Jordan 1996).

not to read too much into this. The α Cen spectrum has a signal-to-noise about four times that of the other stars. Hence, the lower temperature components, which have lower emission measure and, therefore, will typically yield lower flux may just be more easily detected in that case. More importantly, the emission measure distributions all peak at a similar temperature; $3–4 \times 10^6$ K, showing that the coronae of these 'solar-like' stars are rather similar. Their relationship to the Sun is illustrated by the Laming and Drake (1999) analysis of ξ Boo A. They study the emission measure of individual ions, as described earlier in this section, finding that the low FIP element emission measures are systematically higher than those derived from lines formed at similar temperatures but due to elements with high FIPs. The problem can be alleviated by adopting an abundance pattern similar to that which is seen in the solar corona. Laming and Drake interpret this as evidence that the same compositional fractionation mechanism at work in the solar corona is also present in this significantly more active star.

ξ UMa B shows a high temperature tail in the emission measure distribution similar to that seen in α Cen A. No such component is seen in χ^1 Ori. Bearing in mind the earlier discussion of the possible nature of such a component, it is not possible to comment definitively on its reality. However, there is no direct evidence that this conflicts with the possible interpretation that there is a real hot component.

6.6   Active systems

RS CVn systems and highly active stars such as M dwarf flare stars or rapidly rotating objects like AB Dor represent very different stages of stellar evolution, including young stars newly arrived on the main sequence and objects that have evolved into subgiants and giants. However, a common thread that can link all these is the relationship between rotation and activity. Young stars, which have yet to dissipate their angular momentum through mass loss in a stellar wind will be rapid rotators. Evolved objects in close or contact binaries will also be rapidly rotating as a result of the tidal effects that transfer angular momentum from the orbit to the stellar components. Hence, if the behaviour of the coronae is determined mainly by the stellar dynamo, as believed, the coronae of each of these groups of objects should appear to be somewhat similar. For this reason, we consider the EUV observations of these stars together.

6.6.1   Capella: a reference point in the analysis of active stars?

Capella (α Aur) is the brightest coronal EUV source and is an RS CVn type binary comprising G5III ($R_* = 13R_\odot$) and G0III ($R_* = 7R_\odot$) components. With an orbital period of 104 d, it is among the group of long period systems. The G0III secondary is the most rapidly rotating ($v \sin i = 36$ km s^{-1}), and is the source of virtually all the emission in the transition region and chromospheric lines, while the primary has $v \sin i = 5$ km s^{-1}. At 12.5 pc, it is the closest RS CVn binary. Over the past two decades, it has been the subject of numerous UV and X-ray observations. Swank *et al.* (1981) characterised the X-ray spectrum as arising from a corona with two main temperature components, $\approx 5 \times 10^6$ K and $\approx 3 \times 10^7$ K, using the Einstein solid state spectrograph. Later, the modest resolution EXOSAT TGS observation was able to detect some lines but could not to resolve the line blends. Hence, the models of Lemen *et al.* (1989), applied to these data, were simple parameterisations based on global spectral fitting. The EUVE spectra reveal strong, resolved emission lines from Fe IX, XV, XVI and XVII–XXIV, as reported by Dupree *et al.* (1993). They were able to construct an emission measure distribution finding that it is continuous in the range 10^5 K to $10^{7.8}$ K, with a minimum near 10^6 K. This is different from the solar atmosphere, where the minimum occurs near $10^{5.2}$ K.

The ratios of several Fe XXI lines are sensitive to the electron density (e.g. Doschek 1991). Dupree *et al.* (1993) consider the ratios of $\lambda 142.16/\lambda 128.73$ (=0.51), $\lambda 145.65/\lambda 128.73$ (=0.24) and $\lambda 102.35/\lambda 128.73$ (=0.21), which imply log electron densities of 13.2, 12.8 and 11.6 cm^{-3} respectively. Hence, it appears that the density in Capella is $\approx 4 \times 10^{11} - 10^{13}$ cm^{-3}, implying that the scale of the emitting volume is small, $\approx 10^{-3} R_*$. This implies that the electron pressures exceed those in the solar transition region and lie at or beyond the upper bound in solar flares (Doschek 1991).

Following the work of Dupree *et al.* (1993), Capella has been the subject of a more detailed emission measure distribution analysis by both Brickhouse (1996) and Schrijver *et al.* (1995). It is interesting to compare all these results (including the work of Dupree *et al.* 1993), since the distribution of Schrijver *et al.* (1995) is different to either of the others. Figures 6.21 and 6.22 show the distributions of Dupree *et al.* and Schrijver *et al.* respectively. Since that of Brickhouse is essentially identical to the Dupree *et al.* result it is not included here. A peak in the distribution near 3–7 $\times 10^6$ K is found by both authors. However, while Dupree *et al.* see significant flux at temperatures below a minimum in the distribution at 10^6 K, Schrijver *et al.* find no similar contribution. Furthermore, the dominant component above 2×10^6 K in Schrijver *et al.*'s distribution is not present in the Dupree/Brickhouse analysis. The latter do find an increasing slope in the distribution towards high temperatures, but at a lower level, more in keeping with the reduced Fe abundance model of Schrijver *et al.*

Schrijver *et al.* interpret the hot tail as result of a real high temperature (1.0–1.5 $\times 10^7$ K) component with the strongest lines weakened by resonant scattering in the corona. A similar effect was invoked by Mewe *et al.* (1995) for α Cen but dismissed by Schmitt *et al.* (1996c).

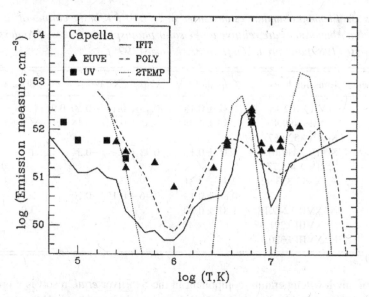

Fig. 6.21. The emission measure distribution of Capella determined by Dupree *et al.* (1993). The triangles correspond to measurements from the EUV line fluxes, the squares are results from UV line fluxes while the solid line is continuous emission measure distribution which best represents the observed fluxes. The dashed and dotted curves correspond to two temperature and polynomial fits to the EXOSAT TGS spectrum of Lemen *et al.* (1989).

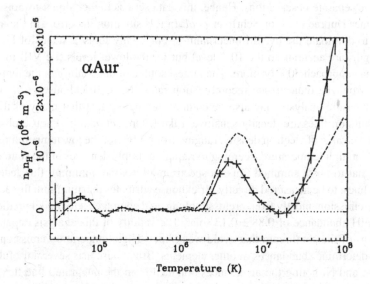

Fig. 6.22. Differential emission measure distribution computed for Capella by Schrijver *et al.* (1995). The solid curve correponds to their best fit solution assuming a solar abundance plasma while the dashed line is the case with an Fe abundance reduced by a factor of 3.3.

Table 6.8. *Element abundances measured at* $T_e \approx 10^{6.8}$ *K in the corona of Capella. Abundances are relative to the solar photospheric values of Anders and Grevesse (1989) and are taken from Brickhouse (1996).*

Element	Emission lines	Capella EUV	Capella X-ray	Solar corona
O	O VII λ102.45	0.42 ± 0.18	$0.13 + 0.14/-0.10$–0.9	
Si	Si XII λ499.40	1.99 ± 0.38	$0.85 + 0.22/-0.18$	3.6
	Si XII λ520.67			
S	S XIV λ417.61	1.15 ± 0.18	$0.73 + 0.41/-0.34$	1.1
	S XIV λ445.77			
Ar	Ar XVI λ353.92	2.41 ± 0.31		1.0
Fe	all	0.88 ± 0.13	$0.46 + 0.15/-0.10$	2.7
Ni	Ni XVII λ249.18	1.81 ± 0.20		3.9
	Ni XVIII λ291.97			
	Ni XVIII λ320.54			

The absence of any low temperature component in the Schrijver *et al.* result is a problem, since Dupree *et al.* clearly identify the spectral lines associated with this low temperature emission (see figure 6.21). One possible explanation for both effects may be differences in the plasma codes used for the respective analyses. Both Dupree *et al.* and Brickhouse make use of new Fe emissivities from N. S. Brickhouse *et al.* (1995), which makes use of the ionisation equilibria of Arnaud and Raymond (1992; see also section 5.5.5).

N. S. Brickhouse's (1996) analysis deals not with a single Capella spectrum but the result of coadding five separate observations. Hence, this data set has higher signal-to-noise than available to either Dupree *et al.* or Schrijver *et al.* In this situation the errors in the atomic models begin to dominate the overall uncertainties. For highly ionised stages of Fe some collision strengths are accurate to the 10% level but for the lower levels (Fe VIII to XIV), inaccuracies may approach 30% or more. The largest source of uncertainty in the ionisation balance arises from the dielectronic recombination rates (N. S. Brickhouse *et al.* 1995). The emission measure analysis may also be complicated by the fact that many of the low temperature emission lines are density sensitive. Like Dupree *et al.*, Brickhouse obtains a large spread in the value of electron density, ranging from $\approx 10^9$ for the low temperature lines to $10^{11.1}$–$10^{12.5}$ at high temperature, seeming to require multiple densities at the source.

The high signal-to-noise summed Capella spectrum allows the minimum flux level, between spectral lines, to be measured directly. Brickhouse attributes the minimum flux level to real continuum emission, mostly from bremsstrahlung, and compares it to model predictions to derive an Fe/H abundance of 0.88 ± 0.13 solar. The validity of this analysis depends on selecting regions of the continuum that avoid weak lines. High temperature emission lines can be used to determine abundances of other elements. Brickhouse lists several useful lines for O, Si, S, Ar and Ni noting the abundances obtained from the integrated line flux measurements (table 6.8). These are compared with abundances obtained from an X-ray study by S. A. Drake *et al.* (1994) and with values for the solar corona from Feldman *et al.* (1992). The X-ray values all lie below those derived from the EUV analysis, indicating that the abundance deficiencies reported from X-ray spectroscopy with ASCA may not be real.

Taking the EUV measurements alone, no consistent abundance trend can be seen. For example, Si, Fe and Ni are all low FIP species. While Si and Ni have enhanced abundances relative to solar photospheric values, Fe does not. Furthermore, the Si and Ni enhancements are not as extreme as in the solar corona. Hence, if a FIP effect is present, which is not clear, it is much weaker than observed on the Sun or in ϵ Eri (see section 6.5.2). Looking at the high FIP elements O, S and Ar, O and S both have roughly solar abundance but Ar, surprisingly, appears to be enhanced by a factor ≈ 2.5.

6.6.2   Coronal element abundances in RS CVn systems

The apparent deficiency in coronal element abundances first viewed by ASCA has been labelled 'metal abundance deficiency' (MAD, see e.g. Schmitt *et al.* 1996d). In a review of stellar spectroscopy with EUVE, Drake (1996) lists the stars for which coronal abundance measurements exist. All the evidence for FIP type abundance anomalies is drawn from EUV spectroscopy but most of the MAD results are from X-ray spectroscopy with ASCA. Since the EUVE results for Capella contradict the X-ray spectroscopy an important question is whether or not MAD effects are seen in any EUV spectra. Two EUVE observations, of CF Tucanae and Algol, are noted by Drake as lying in this category. Algol is particularly interesting as it has also been observed by ASCA.

Strictly speaking, Algol is not an RS CVn system, since only one component is a giant (K2IV), the primary being a B8 dwarf. Indeed, it is the prototype of an entirely separate group of objects. Nevertheless, the X-ray emission shows many of the characteristics of RS CVn systems, including the occurrence of large flares (e.g. Stern *et al.* 1992). ASCA observations of the system by Antunes *et al.* (1994) indicate that the quiescent coronal Fe abundance is $\approx 1/3$ that of the solar photosphere. However, this interpretation is based on an overly simple two temperature component plasma. Stern *et al.* (1995b) use the line to continuum ratio measured from the EUVE spectrum, obtained during the quiescent period (82 ks) of a total observation lasting ≈ 100 ks, to estimate the abundance of Fe. This analysis yields an Fe abundance ranging from 0.17 to 0.34 of the Anders and Grevesse (1989) solar photosphere values. While abundances less than 0.15 can be ruled out at the 99% confidence level, models containing a normal solar abundance cannot be statistically eliminated. However, if a solar abundance plasma is assumed, either most of the emitting plasma must be at an unrealistically high temperature ($T > 3 \times 10^7$ K), or the implied interstellar column density is a factor of five larger than the known upper limit.

In the case of Algol, the EUV result is thus consistent with the X-ray work. However, the abundance comparison is made with solar values, whereas the actual Algol photospheric abundances, particularly of Fe, are not well determined. While most analyses of optical and UV data assume that the abundances are solar, there is evidence that C may be depleted relative to the solar value in the B star (e.g. Tomkin *et al.* 1993). CF Tucanae is interesting in this respect because its photospheric abundance peculiarities have been determined by Randich *et al.* (1993). Relative to solar abundance, the G star has Fe/H = 0.32 and the K star Fe/H = 0.13, the latter being the most extreme case of underabundance in their sample. Randich *et al.* emphasise caution in interpreting the results, since the apparently weak Fe lines may be a result of filling in by emission from plage regions or, particularly in the case of the K components, erroneous assumptions concerning the spectral types.

Schmitt *et al.* (1996d) report an EUVE observation of CF Tuc, spread over several cycles of the 2.8 d orbital period. The system is partially eclipsing and the EUV flux is clearly modulated, with minimum light coinciding with phase zero, when the smaller G star is in front of the larger K star (figure 6.23a). Although large flares have been seen in this system in

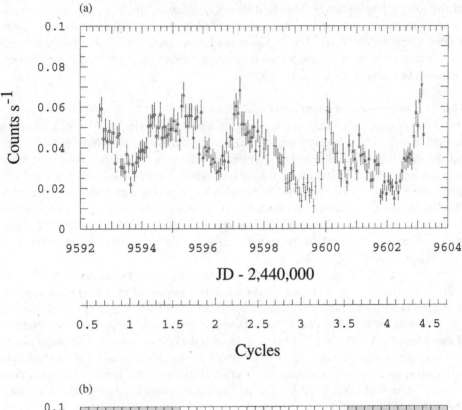

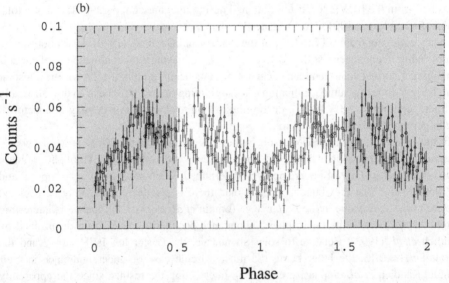

Fig. 6.23. (a) The deep survey Lexan count rate of CF Tuc as a function of time and orbital phase (from Schmitt *et al.* 1996d). (b) The folded EUV light curve of CF Tuc; the different plot symbols refer to different data taken from the different binary orbits (from Schmitt *et al.* 1996d).

the past, there were none in this observation. The folded light curve (figure 6.23b) illustrated the high degree of reproducibility of the modulation indicating that the coronal structure was stable over several orbits. As can be seen from the deep survey count rate, CF Tuc is not a particularly bright source and a long exposure of 330 ks, spread over ten days, was needed to provide a spectrum of reasonable signal-to-noise. Consequently, no phase resolved information could be extracted and the spectrum analysed by Schmitt *et al.* (1996d) is averaged over the binary cycle.

Since CF Tuc provides a spectrum that is only weakly detected, correct treatment of the data, and the background in particular, is important. Schmitt *et al.* (1996d) outline an approach to this problem that is generally applicable to dealing with weak sources. In this observation the background has features of comparable strength to the recorded source + background signal. Hence, the background must be estimated in the immediate vicinity of the spectral trace. To accomplish this Schmitt *et al.* examined the variation of counts perpendicular to the dispersion direction, fixing the overall background level from this. They then compared the total recorded SW count rate with that of the deep survey Lexan band, a ratio $R_{DS}/R_{SW} \sim 11.4$ with the value expected from thermal line spectra for a range of temperatures. The background in the disperson direction was constructed from an average of measurements made either side of the spectrum, approximated with a second order polynomial. The source spectrum was extracted using a ten pixel wide box and summed over eight pixels in the dispersion direction.

After subtraction of the background the net CF Tuc spectrum seems rather featureless and any fluctuations lie within the level of the photon noise. However, the spectrum is very similar to that of Algol (see figure 6.24) and the 'peaks' at 118 Å and 134 Å coincide with the Fe XXII $\lambda 117.17$ and Fe XXIII $\lambda 132.85$ lines. Schmitt *et al.* (1996d) establish the reality of these features with a maximum likelihood analysis which yields a significance greater than 0.999 in each case.

It is striking that only a small number of lines (two) are detected in the EUV spectrum of CF Tuc. The low line to continuum ratio can only be explained by a plasma temperature of 3×10^7 K or more, for a solar abundance composition. If the Fe abundance is less than solar, then the line and continuum fluxes can be reconciled at lower temperature. The EUVE data alone cannot be used to distinguish between these alternatives. However, the quiescent X-ray spectrum, recorded earlier by the ROSAT PSPC, provides a strong constraint, favouring the low temperature/low abundance model. Hence, CF Tuc represents another instance of the MAD syndrome. Importantly, in this case the low Fe abundance might be a direct consequence of an anomalously low photospheric abundance of this element rather than a differentiation mechanism in the corona.

UX Ari has been observed extensively by both ASCA and EUVE (Güdel *et al.* 1999). While the X-ray data covered both quiescent and flaring phases of the system, the EUV observation was acquired only during quiescence. The emission measure distributions obtained from the quiescent X-ray and EUV data are in remarkable agreement, with significantly sub-solar elemental abundance. In this case the Fe abundance is ≈17% of the solar photospheric value and the overall pattern is very similar to that seen in both Algol and CF Tuc. It is interesting to note that the coronal abundances of Mg, Fe, Si, S, Ni and Ne are observed to rise by a factor of two to six during the flare. The increase in low FIP abundances tends to be higher than in the high FIP elements.

Apart from the detailed spectroscopic information obtained by EUVE observations of RS CVns, the long exposures often required have produced an extensive by-product of

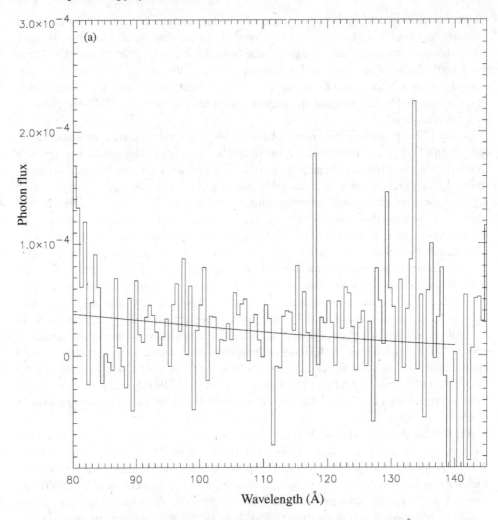

Fig. 6.24. (a) EUVE short wavelength spectrum of CF Tuc in the 80–140 Å region after background subtraction and effective area correction (from Schmitt *et al.* 1996d).

photometric monitoring data. For example, Osten and Brown (1999) have analysed 12.2 Ms of photometry of 16 systems, which included 4 Ms in which the flaring activity was recorded. Such an extensive data set allows systematic searches for phase dependent variations and study of flaring time scales and flux levels. Figure 6.25 shows the various light curves obtained by these authors, exhibiting several identifiable flares and other statistically significant variability.

A detailed analysis reveals a number of interesting properties. Although ER Vul, AR Lac and CF Tuc are partially or totally eclipsing systems, no EUV eclipses are evident in any of the light curves. However, CF Tuc does show clear evidence of a modulation of the signal on its orbital period of 2.8 d. It is interesting to note that no modulation is seen in the V711 Tau (HR1099) data that is coherent over the four year time scale of the observations, although

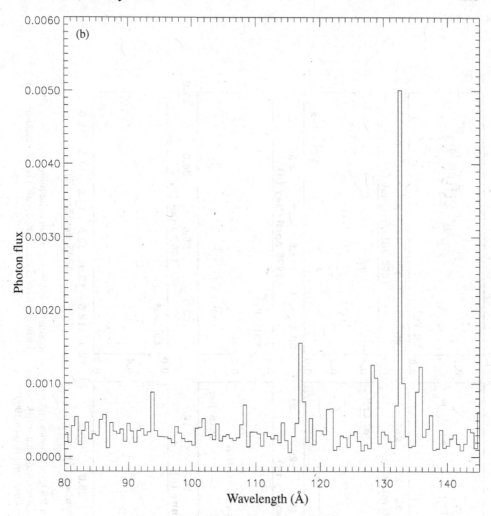

Fig. 6.24. (b) EUVE spectrum of Algol in the wavelength range 80–140 Å (from Schmitt *et al.* 1996d).

phase related variability was seen in both WFC (Barstow *et al.* 1992a) and EUVE (Drake *et al.* 1994b) sky surveys. This suggests that the reported variability arises from relatively short-lived active regions. Considerable statistically significant variability is seen in the light curve of the shortest period system, ER Vul. A comparison of the expected Poisson count rate distribution with that observed indicates that the variability cannot be due to Poisson processes alone, which Osten and Brown interpret as evidence of small-scale stochastic flaring. The peak ER Vul luminosity is on the low end of the range of the individual large flare peak luminosities.

A number of distinct flares were detected in the RS CVn light curves, of which Osten and Brown analysed 30 on nine individual systems. All the flares occurred over several satellite orbits (no small time scale microflaring behaviour was observed) with a positive

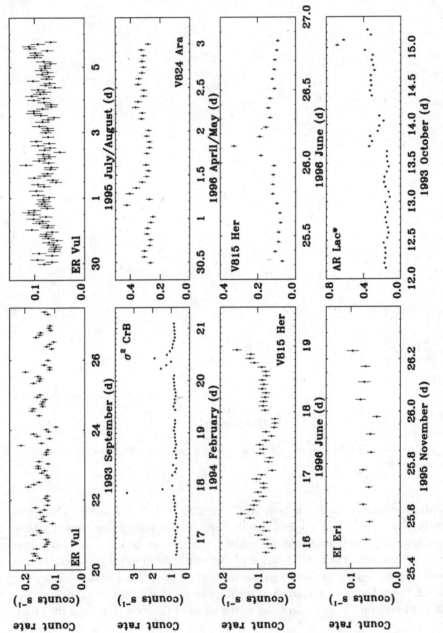

Fig. 6.25. EUV light curves for RS CVn binaries ordered by increasing orbital period. Each point represents one satellite orbit. Error bars are 1σ. Light curves with an asterisk next to the identifier were contaminated with a known dead spot on the Deep Survey detector (from Osten and Brown 1999).

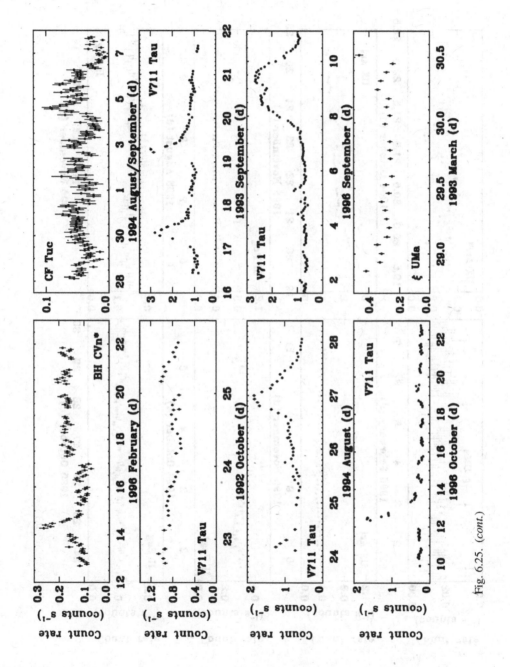

Fig. 6.25. (cont.)

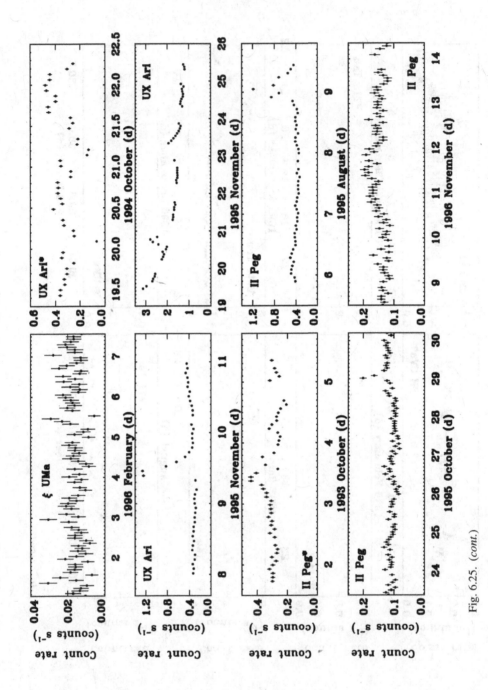

Fig. 6.25. (cont.)

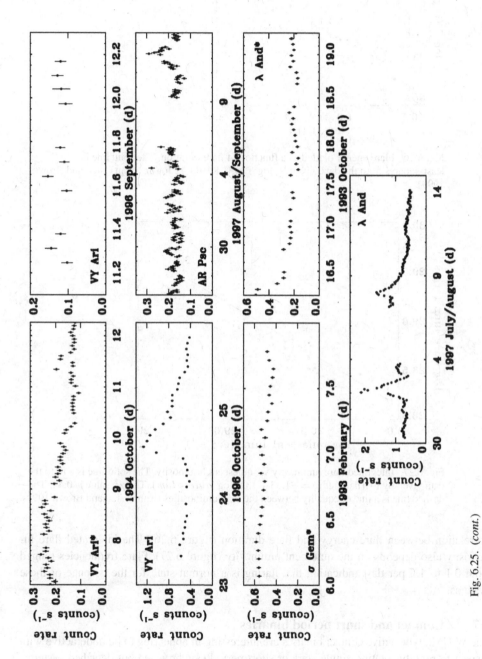

Fig. 6.25. (cont.)

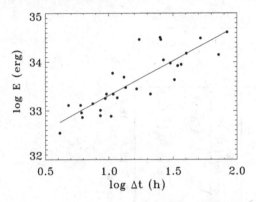

Fig. 6.26. Flare energy plotted as a function of flare duration. The solid line is a least-squares fit to the data, given by $\log E = 31.89 + 1\,42 \log \delta t$ (from Osten and Brown 1999).

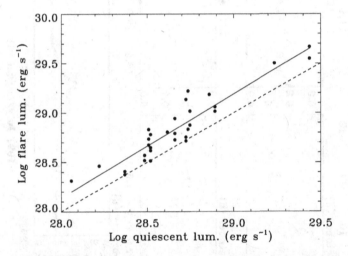

Fig. 6.27. Integrated flare luminosity vs. quiescent luminosity. The solid line is a fit to the data given by $\log(\text{flare lum.}) = -1.33 + 1.05 \log (quiesc\ lum)$. The χ^2 value is 0.44. The dashed line is a line of equality between the two luminosities (from Osten and Brown 1999).

correlation between flare energy and flare duration (figure 6.26). The integrated flare luminosity also depends on the quiescent luminosity (figure 6.27). Flare frequencies ranged from 0.1 to 1.5 per day, indicating that flaring is a normal state for the coronae of these systems.

6.7   Contact and short period binaries

The W UMa type active contact binaries are interesting in appearing to be underactive with respect to rapidly rotating single stars or short period synchronised but detached systems. Possible underlying causes include decreased differential rotation in the contact stars, shallower convection zones or mass transfer between the components. 44i Boo is the brightest such system detected in the EUV. Its DS light curve shows modulation on a period of 0.272652 ± 0.000141 d along with two brief episodes of enhanced flux (Brickhouse and Dupree 1998). The EUV period is slightly longer than the 0.26781662 d optical light variations. In addition, the EUV light curve does not follow the optical photometry, showing instead a well-defined sinusoidal variation with a single maximum compared to a double eclipse modulation. The maximum EUV flux occurs at phase 0.0, when the primary star eclipses the secondary, and the minimum EUV emission is at phase 0.5.

The phase averaged EUV spectrum reveals an emission measure distribution peaking near 10^7 K, indicating that the coronal plasma is typically hotter than in the RS CVn systems. Spectral line diagnostics also indicate extremely high electron densities (2×10^{13} cm^{-3}) and yield a lower limit on the coronal Fe abundance of 0.46 times the solar photospheric value. The high electron density and peak emission measure suggest that all the EUV flux arises from a small emitting volume with a scale length $\approx 0.004 R_*$. Brickhouse and Dupree suggest that there are two types of emitting region on 44i Boo: a diffuse corona with an emission measure distribution that depends on $\approx T_e^{3/2}$ and a localised active region of high density and field strength (≈ 1 kG).

Rucinski (1998) compares two W Uma systems, 44i Boo (discussed above) and VW Cep with the close but detached RS CVn ER Vul and the single rapid rotator AB Dor. All four systems have orbital periods less than 1 d and are expected to show saturated levels of chromospheric, transition region and coronal emission. A comparison of the line fluxes relative to the bolometric fluxes for selected strongest chromospheric, transition region and coronal emission lines shows consistency with period-independent (i.e. saturated) levels of activity for features formed at plasma temperatures below 10^5 K, while features formed at $\approx 10^7$ K are definitely weaker in AB Dor than in the three other binaries. However, it is not clear whether this is due to temporal variability or is a general distinction between the properties of binaries and isolated stars at extreme rotation rates.

6.8 The effect of stellar activity on EUV spectra

The sections above have dealt specifically with relatively normal main sequence stars, somewhat similar to the Sun and more active RS CVn type binaries. In broad terms, the main sequence stars have emission measure distributions which peak at temperatures of a few 10^6 K, while the RS CVns have peak temperatures approaching 10^7 K or more. Current evidence also suggests that the coronal abundances of the RS CVn systems are deficient in metals (Fe in particular) while some of the main sequence objects exhibit the FIP effect, although others do not. Simplistically, the temperature differences could be attributed to the enhanced activity in the binaries. Accordingly, other active systems might also be expected to show higher coronal temperatures. Transient phases of enhanced activity, such as stellar flares, may also have a different temperature structure to the quiescent corona.

Dupree (1996) compares the emission measure distributions of several binaries spanning a range of orbital periods (figure 6.28). All are RS CVn (or RS CVn-like) except 44i Boo, which is a W UMa contact binary. The enhanced emission measure feature is present in all these objects and occurs at very similar, if not identical temperatures in each case. Interestingly, the longest period system, λ And, does not show a particularly well-defined bump when compared to the others. Examination of the data from other active stars reveals similar results. For example, Rucinski *et al.* (1995) present a detailed study of quiescent emission from the rapidly rotating pre-main sequence star AB Dor. If the coronal composition is assumed to match solar photospheric abundances, the emission measure distribution looks very similar to the result Schrijver *et al.* (1995) obtained for Capella, with a high temperature tail. However, ASCA observations (Mewe *et al.* 1996) show no sign of a high temperature tail. Rucinski

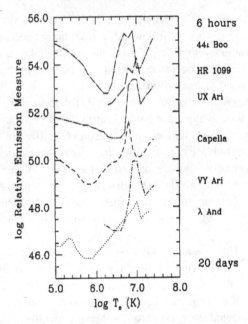

Fig. 6.28. The relative emission measure distribution for binary systems with a range of orbital periods. The curves have been shifted by arbitrary amounts in the vertical direction and are ordered by increasing period from top to bottom. The period for Capella is attributed to that of the rapidly rotating secondary rather than the orbital period, which is much longer (from Dupree 1996).

et al. show that this problem with the EUV analysis can be resolved by reducing the coronal abundances to match those obtained from the ASCA spectrum. The more detailed analysis by Mewe *et al.*, combining the EUVE and ASCA spectra, yields abundances of O, Mg, Si, S and Fe factors of two to three below solar photospheric values. Interestingly, the abundance of Ne is slightly enhanced. The emission measure distribution has two peaks (figure 6.29). The lower temperature one, at $\approx 7 \times 10^6$ K, is consistent with the distributions seen earlier in the RS CVns but the higher temperature peak ($\approx 3 \times 10^7$ K) is not observed in those systems (see figure 6.28; Dupree 1996). This may be due to the insensitivity of EUVE to the emission from higher temperature plasma. Typically, the EUV spectra only sample lines arising from temperatures up to $\approx 3 \times 10^7$ K. In the AB Dor analysis, it is the additional information on higher temperature components from ASCA that generates the second peak in the emission measure distribution. Note that all the distributions reported by Dupree (1996 and figure 6.28) show emission rising towards the plotted limit ($\approx 3 \times 10^7$ K). It seems likely that the RS CVns also have a high temperature emission measure peak similar to AB Dor. Importantly, it now seems clear that all these highly active systems have coronae that are depleted in heavy elements compared to the solar photosphere.

Apart from rotation related enhancement of general levels of activity on coronal sources, many stars also show significant flux enhancement in the form of stellar flares. Study of the energy balance in such events is of great importance in understanding the coronal heating mechanism (or mechanisms) and measurements of the plasma temperature and electron density distribution are an essential component of this work.

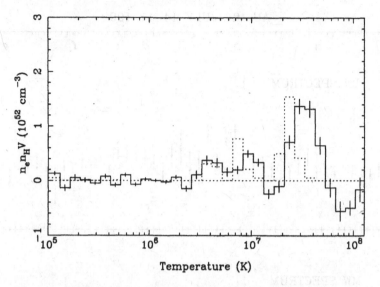

Fig. 6.29. The differential emission measure distribution of AB Dor reported by Mewe *et al.* (1996), combining simultaneous EUVE and ASCA data. The solid histogram was produced using a regularisation method and the dashed histogram by a clean algorithm.

Monsignori-Fossi *et al.* (1995, 1996) present observations of massive stellar flares in AU Mic and EQ Peg, recorded by the EUVE spectrographs. The light curve of AU Mic obtained from summing all the counts in each spectrograph channel reveals two flares separated by approximately 24 h, the second reaching roughly half the peak flux of the first (figure 6.30; Monsignori-Fossi *et al.* 1996). Seven time intervals were selected for analysis representing pre-flare 1 quiescence, flare 1 peak, flare 1 decay, pre-flare 2 quiescence, flare 2 peak, flare 2 decay and post-flare 2 quiescence. The physical conditions in the quiescent and flaring plasma can be traced through the evolution of the differential emission measure (DEM), calculated for each spectrum. During those phases that are clearly quiescent (figure 6.31a,g), the distribution peaks at $\approx 8 \times 10^6$ K. In the flares and the decay periods, Fe XXIV $\lambda 192.04$ emission is clearly detected, while it is not visible at other times. Formation of this line requires a temperature $\approx 5 \times 10^7$ K. Such a temperature is also needed to reproduce the observed continuum in the SW spectra. A secondary peak in the DEM associated with this higher temperature component is clearly seen in figure 6.31b–f. It is interesting to compare the individual DEMs with those obtained for other stars. For example, the DEM observed during flare is very similar to that seen in the quiescent corona of AB Dor (Mewe *et al.* 1996). In particular, the high temperature component associated with the flare in AU Mic is always present in AB Dor. The quiescent component of AU Mic is very similar to the quiescent emission observed in the similarly active RS CVn systems, but hotter than relatively inactive objects like Procyon or α Cen.

Some of the Fe lines seen in the EUVE spectrum of AU Mic are density sensitive, allowing Monsignori-Fossi *et al.* to estimate the time variation of the plasma density. Using the intensity ratio of Fe XXI (142.14 + 142.26/128.7) they find that during active phases the density exceeds 10^{12} cm^{-3}, increasing to $\approx 1.5 \times 10^{13}$ cm^{-3} during the flares. The latter value is

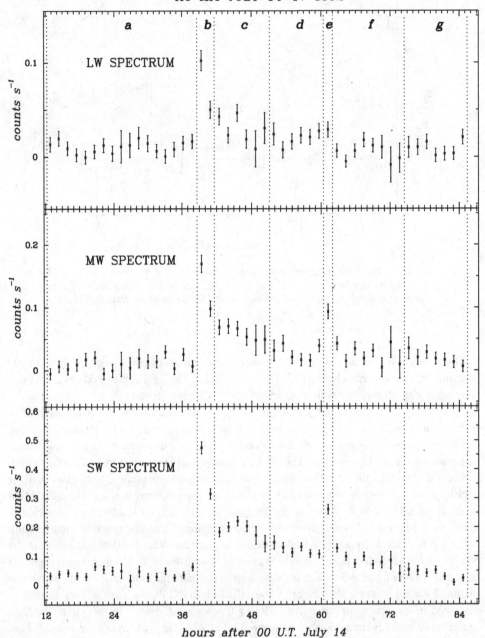

Fig. 6.30. The SW, MW and LW light curves of AU Mic. The vertical dotted lines show the individual time intervals selected by Monsignori-Fossi *et al.* (1996) for more detailed spectroscopic analysis (from Monsignori-Fossi *et al.* 1996).

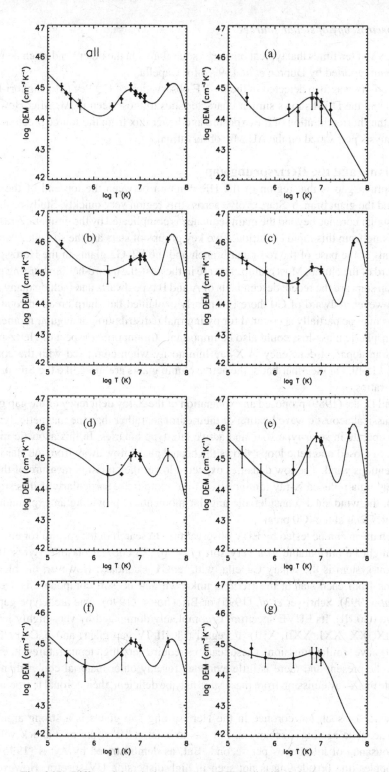

Fig. 6.31. Differential emission measure (DEM) distributions that reproduce the observed spectra of AU Mic during the July 1992 observation. Panel (a) shows the DEM for the entire observation, with successive panels following the time sequence listed in the text and shown in figure 6.30 (from Monsignori-Fossi *et al.* 1996).

around a factor of ten times that typical of a large solar flare. On the other hand, such densities have also been reported by Dupree *et al.* (1993) for Capella.

Flaring activity was also detected in the EUVE observation of EQ Peg (Monsignori-Fossi *et al.* 1995) and the DEM shows similar characteristics to those seen in AU Mic. However, in this case the shorter duration of the exposure and lower flux from the source precluded the detailed analysis performed on the AU Mic observation.

6.9   Giants and the Hertzsprung gap

The Hertzsprung gap is that region of the HR diagram between the top end of the main sequence and the giant branch. Stars evolve across this region very quickly. Study of the behaviour of stellar coronae beyond the main sequence is complicated by the dramatic structural changes arising from this rapid evolution. Two key groups of stars are the so-called 'clump' G8–K0 giants at the base of the red giant branch and the F2–G2 giants in the Hertzsprung gap, blueward of the clump. Most clump stars are in the post-flash core-helium burning phase while the gap stars are the recent descendants of A and B type dwarfs and include many rapid rotators. However, redward of G0 there is a poorly explained but sharp break in rotational speeds. This may be partially accounted for by internal redistribution of angular momentum but magnetospheric mass-loss could also be important. An important aspect of Hertzsprung gap stars is an apparent deficiency in X-ray luminosity when compared with the coronal proxy C IV $\lambda 1549$. On the other hand, the active clump giants are more like the Sun in their X-ray/C IV ratios.

Simon and Drake (1989) proposed an explanation of the X-ray deficiency of the gap giants based on classical acoustic wave heating of their coronae, rather than the magnetic dynamo believed to operate in solar-type stars and active late-type binaries. In this acoustic model, the absence of closed magnetic loops allows the heated gas to flow away from the star in the form of a tenuous wind. The low densities expected lower the emission measure of the hot plasma, yielding a reduced X-ray luminosity. At the same time, particularly if it is slightly magnetized, the wind sheds considerable angular momentum, providing an explanation of the spin-down seen at the 'G0 break'.

This mechanism can be tested by EUV observations, to search in the spectra for emission lines that might be attributable to a very soft ($T < 10^6$ K) coronal plasma (Ayres 1996). An important system is the binary Capella, with its G1 secondary. However, the blend of contributions from each star in the EUV is unknown and no soft component is seen by Dupree *et al.* (1993), Schrijver *et al.* (1995) or Brickhouse (1996). One archetype gap star is 31 Comae (G0 III). Its EUVE spectrum is completely dominated by very highly ionised iron (Fe XIX, XX, XXI, XXII, XIII). ν Pegasi (F8 III–IV gap giant) and β Ceti (K0 III clump giant) have similar emission. These features require very high temperatures, in excess of 10^7 K, to be present and there is little evidence for any other material cooler than this. Hence, while the X-ray emission from these stars may be deficient, their coronal temperatures are not.

The presence of such hot coronae in the Hertzsprung gap giants is a strong argument against the acoustic heating mechanism. For gas to exist at 10^7 K the post-shock velocity must be thousands of kilometres per second. But, as demonstrated by Ayres (1996), the implied Doppler line broadening is not seen in high dispersion UV spectra. However, in the absence of any signatures of activity it is also hard to invoke the magnetic heating scenario.

Ayres (1996) suggests that one possible explanation may be the nanoflare hypothesis (e.g. Parker 1988), where the magnetosphere of a typical Hertzsprung gap giant is continually flaring. These flares are numerous, relatively compact and widely dispersed events. The population would be equivalent to that proposed for heating the solar corona, but without the tail of larger events seen in the Sun.

The clump and gap stars are bridged by an intermediate evolutionary phase of rapid breaking where stars cross from the red edge of the gap to the base of the giant branch at K0 III. Here, the C IV activity drops substantially coinciding with a precipitous decline in rotational velocity redward of G0 III (see Ayres *et al.* (1999) for a summary and references therein). Since the lifetimes of these stars are short, such yellow giants are relatively rare. One of the few nearby examples, μ Velorum (HD93497; $d = 36$ pc) exhibited a remarkable flare when observed by EUVE (Ayres *et al.* 1999). The 1.5 d decay time is long compared to the few hours typical of active short period systems. Secondly, very few flares of any kind have been reported from single red or yellow giants. The physical characteristics of the flare point to a highly extended structure of moderate density ($n_e < 10^9$ cm^{-3}), perhaps similar to a solar coronal mass ejection. Such an event, magnetically coupled to the photosphere out to a large Alfvén radius, can carry away substantial angular momentum. Flare associated mass-loss might then be a major contributor to the as yet unexplained spin-down that occurs in this region of the HR diagram.

6.10 Physical models

Most of the work dealing with the EUV spectroscopy of stellar coronae has been phenomenological, associated with determining plasma temperatures, emisson measures, densities and abundances. Important discussions concerning departures from photospheric abundance in the coronae, dealing with the FIP effect or the MAD syndrome, have begun to establish patterns for stars with different levels of activity. However, little work has been done to link the observational parameters to physical mechanisms that might be relevant in explaining how stellar coronae are heated and how elemental abundance might be altered. Some initial work is now beginning to emerge that can be applied to the interpretation of EUV observations and we review this briefly here.

Stellar coronae of all levels of activity have often been modelled in terms of a complex of quasi-static magnetic loops confining the plasma to a particular geometry. Such models have been used to infer general properties of the plasma, such as how much of the stellar surface might be covered by such loops and their possible scale height compared to the stellar radius and, in binaries, the stellar separation. Van den Oord *et al.* (1996) have considered how the predictions of such a model match the observed differential emission measures derived from EUV spectroscopy. In a quasi-static coronal loop model, the emission measure distribution ($D(T)$) has the form

$$D(T) \approx A^2 p^2 T^{3/2} / |F_c|$$

where A is the cross-section of the loop, p is the pressure and F_c the conductive flux through the loop cross-section. In the transition region $|F_c|$ is large and consequently $D(T)$ is low. At higher temperatures $D(T)$ increases until the maximum temperature of the loop is reached, above which it equals zero. As the region of the loop where heating dominates over radiative losses is almost isothermal, this geometrically largest part of a loop

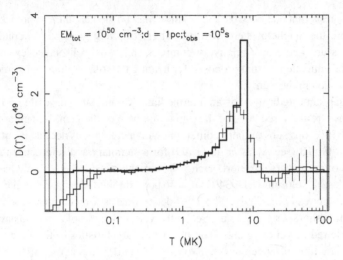

Fig. 6.32. DEM of a quasi-static coronal loop (thick solid line) and the recovered DEM from the simulated spectrum (line with error bars) (from van den Oord *et al.* 1996).

occupies only $\Delta \log T = 0.1 - 0.2$ in the DEM. Loop expansion tends to steepen the slope of $D(T)$.

An interesting comparison between the theoretical DEM and that obtained from an analysis of a simulated spectrum based on it has been carried out by van den Oord *et al.* (1996; figure 6.32). The extracted slope is in good agreement with the input model, but the DEM maximum is underestimated. This effect is taken into account in their subsequent analyses of the observed DEMs of α Cen, Capella, ζ UMa and χ^1 Ori (figure 6.33). For all but α Cen, the simple constant pressure/constant cross-section models do not match the tail of the observed DEM, indicating that the loops are expanding (i.e. do not have constant cross-section). Unfortunately, a detailed analysis is complicated, since different combinations of the loop parameters (such as the heat energy, pressure, length, expansion factor and gravity) can result in almost identical DEMs. Even so, the case for expanding loop geometries is strong.

The van den Oord *et al.* (1996) model was based around the presence of static loops in stellar coronae. However, an important component of stellar activity is found in the form of flares, for which such schemes are not appropriate. Livshits and Katsova (1996) have developed a stellar flare model in which the main energy release above the stellar chromosphere is through electron acceleration and impulsive heating of the plasma. The response of the chromosphere is expected to lead to the development of specific plasma motions that generate sources of EUV emission. Modern numerical methods can be used to solve the system of equations that describe the gas dynamics. Livshits and Katsova (1996) deal with two temperatures and a fluid plasma including the heating flux, radiative energy losses, thermal conduction flux and the energy exchange between the electron and ion plasma components.

In the initial phases of a flare, bulk motions are practically absent. Hence, the chromosphere is heated suddenly, without appreciable change in the total density, in the first 1/10 of a second. With a temperature in the range 10^4 to $<10^5$ K, this heated material emits predominantly in the UV 1000–2000 Å range and in He II 304 Å. The UV and EUV burst should be shorter than 1 s, although in the He II line it can be longer because of ionisation of He I. However, there is no direct observational evidence for such short UV and EUV spikes.

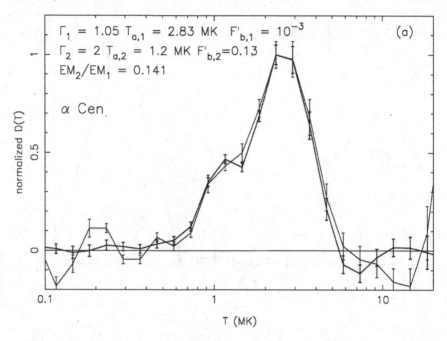

$$\Gamma_1 = 1.05 \quad T_{a,1} = 2.83 \text{ MK} \quad F'_{b,1} = 10^{-3}$$
$$\Gamma_2 = 2 \quad T_{a,2} = 1.2 \text{ MK} \quad F'_{b,2} = 0.13$$
$$EM_2/EM_1 = 0.141$$

α Cen

(a)

Fig. 6.33. Observed DEMs (error bars) for (a) α Cen, (b) Capella, and (c) ζ UMa compared to the expected DEMs (after accounting for the effects of the recovery algorithm) for constant pressure/cross-section loops (from van den Oord *et al.* 1996).

The general appearance of an AU Mic flare observed by EUVE is similar to large solar flares, as pointed out by Drake *et al.* (1994a). In particular, the long time scale for decay of the EUV flux requires substantial post-flare heating. Katsova *et al.* (1996) consider several alternative explanations for the event, including the presence of very high coronal loops and coronal mass ejection, as originally proposed by Cully *et al.* (1994).

The decay time of soft X-rays is not usually greater than two to three times the radiative cooling time (t_{rad}), which is given by

$$t_{\text{rad}} = 3kT/n^2 L(T)$$

where $n^2 L(T)$ is the radiative loss function for the range 1–2000 Å. Predicted values of t_{rad} times for loop models range from ≈ 100–2000 s (Katsova *et al.* 1996), well below the observed time of ≈ 12 h. Furthermore, the integrated 3×10^{35} ergs of power in the AU Mic flare exceeds that of typical red dwarf flares, besides the energy of solar flares.

The explanation that best matches the duration and magnitude of the post eruptive decay in AU Mic is that of a coronal mass ejection. Martens (1988) provides a theoretical framework where a coronal mass ejection event distends the magnetic loop field lines in a radial direction forming a large-scale vertical current sheet. Plasma instabilities and reconnection within this sheet are then able to accelerate particles up to energies of $\approx 10^{10}$ eV. However, solar observations indicate that the mass ejection alone is not sufficient to provide the observed EUV and soft X-ray fluxes over many hours. Cooling times depend strongly on the plasma density. Indeed, the fact that flaring loops cool down rapidly is due to their comparatively

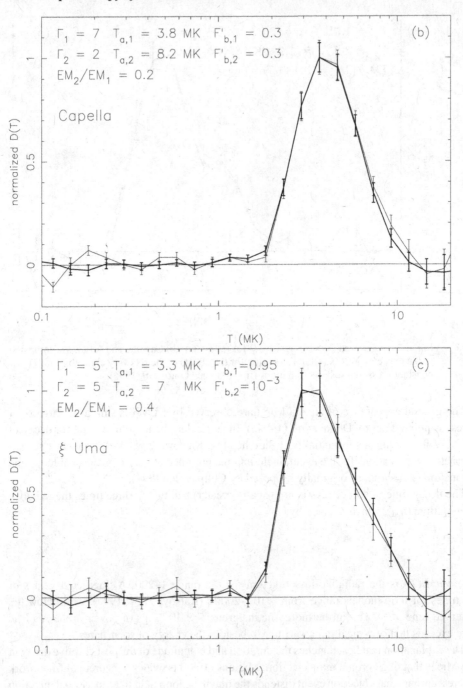

Fig. 6.33. *(cont.)*

high density ($n_e \approx 10^{13}$ cm^{-3}). Hence, for an event to last for a long time a mechanism is required that will compensate for the radiative losses.

Small-scale flaring processes cannot support additional heating with a total energy $\approx 10^{35}$ ergs. Therefore, Katsova *et al.* (1996) consider processes occurring on a larger scale, comparable to or greater than the stellar radius. The first pulse of the EUV light curve is considered to be caused by an impulsive flare. The gradual decay after this can be presented as the radiation of a system of loops which form after the coronal mass ejection. Extension of the magnetic field lines to form a current sheet makes available additional heating for the coronal loops, prolonging their lifetime. The large value of t_{rad} implies that the loop density is lower than that encountered during the impulsive phase. This model represents a new kind of surface activity on late-type stars, intermediate between the impulsive flares commonly seen on red dwarfs and the longer duration, more energetic events on RS CVn binaries.

7

Structure and ionisation of the local interstellar medium

7.1   A view of local interstellar space

It is generally accepted that the solar system resides inside a relatively dense local interstellar cloud a few parsecs across with a mean neutral hydrogen density of $\approx 0.1\,\mathrm{cm}^{-3}$ (e.g. Frisch 1994; Gry *et al.* 1995). This cloud, the so-called local interstellar cloud (LIC) or surrounding interstellar cloud (SIC), lies inside a region of much lower density, often referred to as the local bubble (figure 7.1). The general picture built up is one where this bubble has been created by the shock wave from a past supernova explosion, which would also have ionised the local cloud. The current ionisation state of the local cloud is then expected to depend on the recombination history of the ionised material, i.e. the length of time since the shock wave passed through. However, if the flux of ionising photons from hot stellar sources is significant, the net recombination rate may be reduced (Cheng and Bruhweiler 1990; Lyu and Bruhweiler 1996). The photometric data from the ROSAT WFC survey have been used to map out the general dimensions of the cavity by Warwick *et al.* (1993) and Diamond *et al.* (1995), as already discussed in section 3.9.2. However, this relatively crude interpretation of the observations probably hides greater complexity. For example, several studies of the lines-of-sight towards β and ϵ CMa had already demonstrated the existence of a low density tunnel some 200–300 pc in extent, even before any EUV observations were carried out (e.g. Welsh 1991). The predicted low column densities were dramatically confirmed by the EUVE observations of β and ϵ CMa described in section 6.1.

Prior to the launch of EUVE, few spectroscopic observations were available with which to directly study H and He in the local interstellar medium (LISM). Furthermore, certain models of the LISM predicted significant photoionisation from the decay of primordial neutrinos (Sciama 1990, 1993). Strong ionisation of hydrogen relative to helium was required to match the combined EXOSAT (Heise *et al.* 1988) and Voyager (Holberg *et al.* 1980) observations of HZ 43, which would support such models. However, the work of Kimble *et al.* (1993a) contradicted this result. In addition, sounding rocket and HUT spectra of G191–B2B, spanning the He I 504 Å photoionisation edge, concluded that hydrogen could not be highly ionised in the LISM (see Green *et al.* (1990) and Kimble *et al.* (1993b) respectively).

7.2   Spectral observations of the diffuse background

EUV astronomy offers two distinct ways of studying the LISM: indirectly from the effects of absorption on the EUV spectra of sources, or directly through emission from warm or hot components of the LISM. Although not specifically designed or optimised for observation of diffuse emission, the EUVE spectrometers were the most sensitive instrument yet flown for such work. Wire grid collimators were used to limit the diffuse flux from the major airglow

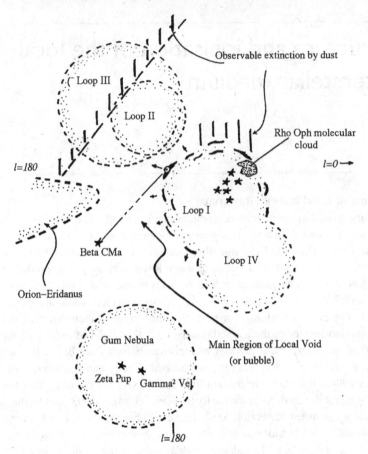

Fig. 7.1. Schematic diagram of the local interstellar medium (from Bruhweiler 1996).

lines of helium (at 304, 537 and 584 Å) to a small part of the spectrum. The resulting spectral resolutions of the medium and long wavelength spectrometers for diffuse emission were 17 and 34 Å respectively (Jelinsky *et al.* 1995). Diffuse emission is not spectrally resolved in the short wavelength channel, since the absence of strong background lines within its bandpass obviated the need for a collimator.

Jelinsky *et al.* (1995) have reported a deep observation of the EUV background made during the all-sky survey, while the spectrometer was pointing along the Earth shadow, reducing the contribution from the geocoronal background. A total effective exposure time of 575 232 s covered a $2.5° \times 20°$ field of view scanned from $(l^{II}, b^{II}) = (24°, -28°)$ to $(44°, -47°)$. After subtraction of the detector background component, the resulting spectrum reveals a continuum flux level on which are superimposed the broad emission lines from He II at 304 Å and He I at 584 Å (figure 7.2). The continuum flux, which rises towards shorter wavelengths, is possibly zero order scattering of the strong 584 Å line by the diffraction grating, although the shoulder feature shortward of the 584 Å line is a combination of the remaining He I Lyman series lines which converge at 504 Å. The intensities of the 584 and 537 Å lines are 1.04×10^5 and 3.16×10^3 photons cm^{-2} s^{-1} sr^{-1} respectively (or 1.30 and 0.04 R). The 584 Å flux is consistent with that expected from the scattered solar He I radiation (Chakrabarti *et al.* 1984).

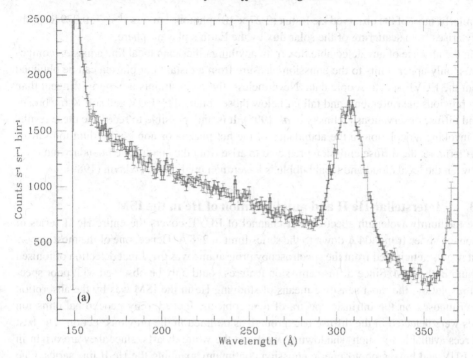

Fig. 7.2. (a) Background subtracted spectrum for the medium wavelength spectrometer (from Jelinsky *et al.* 1995).

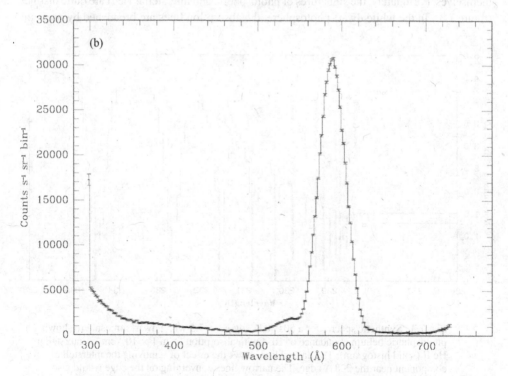

Fig. 7.2. (b) Background subtracted spectrum for the long wavelength spectrometer (from Jelinsky *et al.* 1995).

Similarly, most of the observed He II flux (2.27×10^3 photons cm^{-2} s^{-1} sr^{-1}, 0.029 R) probably arises from scattering of the solar flux by the Earth's plasmasphere.

In the absence of any detectable flux from anything other than local line emission components, only upper limits to the emission measure from a distant hot plasma can be obtained from the EUVE spectroscopic data. Nevertheless, these constraints are more stringent than any previous measurements and fall far below those obtained by EUV and soft X-ray broadband diffuse observations (Jelinsky *et al.* 1995). It is only possible to reconcile these results by invoking a depletion in the abundance of the hot plasma or non-equilibrium ionisation. Furthermore, the diffuse emission predicted to arise from the conductive boundary interface between the local cloud and local bubble is lower than predicted by Slavin (1989).

7.3   Interstellar He II and autoionisation of He in the ISM

The medium wavelength spectrometer channel of EUVE covers the entire He II series of resonance lines from 304 Å down to the series limit at 228 Å. Hence, one of the most important results anticipated from the spectroscopy programme was the direct detection of ionised helium in the ISM. Since diffuse emission features could only be observed with poor spectral resolution, the most sensitive means of studying He in the ISM was by the absorption superimposed on the intrinsic spectra of other objects. Indeed, only geocoronal emission lines were detected in the diffuse background, as outlined in the previous section. The best sources available for such 'shadowing' work are the white dwarfs, since they are bright in the EUV and have a photospheric emission continuum spanning the He II line series. One potential problem is the possible presence of He II features in the photosphere of the stars themselves. Fortunately, the signatures of photospheric and interstellar He II are quite distinct (figure 7.3). In the white dwarf photosphere, the absorption lines are broadened by pressure

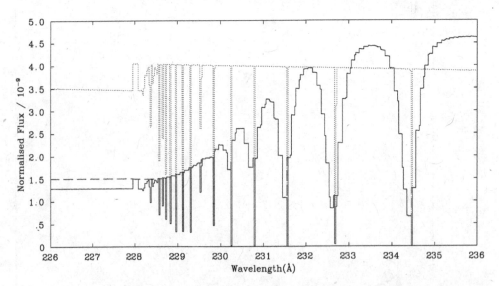

Fig. 7.3. Synthetic spectra for a 50 000 K, log g = 7.5 white dwarf atmosphere showing a photospheric helium abundance of 10^{-4} with absorption from 1×10^{17} cm^{-2} interstellar He II (solid histogram). The dashed line shows the effect of removing the interstellar component near the 228 Å edge. The narrow lines converging on the edge would also disappear. The dotted histogram shows a pure H model with an interstellar He II column at the above level.

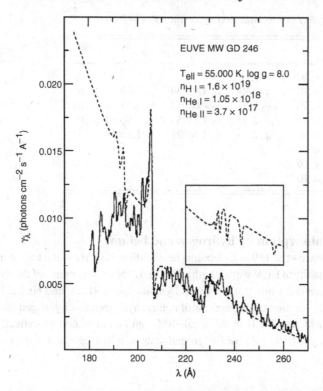

Fig. 7.4. Calibration observation of GD246 showing the region of the spectrum between 180 and 270 Å compared to a pure H model with interstellar absorption included using the model of Rumph *et al.* (1994).

and temperature. As they converge towards the series limit, they overlap so much that the absorption edge expected at the 228 Å series limit is smeared out and effectively eliminated. In contrast the interstellar lines are much narrower than their photospheric counterparts and the 228 Å edge is clearly seen.

The first direct detection of the interstellar He II edge was found in the spectrum of the hot DA white dwarf GD246, when it was observed during calibration of the EUVE spectrometers in August and September 1992 (Vennes *et al.* 1993). The absorption edge from He II can be seen in the data (figure 7.4) but there is an even stronger reverse edge at 206 Å, which is attributed to the autoionisation of He I. Autoionisation is a two stage process in which the absorption of an energetic photon by an inner shell electron results in a highly excited state, whose energy is larger than the ionisation potential, followed by the emission of an electron. Resonance features like this were known to occur in the photoionisation cross-sections of elements heavier than hydrogen before its detection in the spectrum of GD246, but such fine detail was not included in the EUV absorption cross-sections of Cruddace *et al.* (1974).

Since neutral helium is the primary source of interstellar opacity in the wavelength range between 30 and 300 Å, Rumph *et al.* (1994) carried out an extensive reevaluation of the EUV absorption in the interstellar medium. They have developed an effective cross-section for this absorption which takes into account both continuum and autoionisation contributions.

Table 7.1. *Helium I continuum and resonance fit coefficients.*

Continuum			Fano profile	
c_0 -2.953607×10^1				
c_1 $+7.083061 \times 10^0$	q_1 2.81	ϵ_1 4.42239	γ_1 2.64061×10^{-3}	
c_2 $+8.678646 \times 10^{-1}$	q_2 2.51	ϵ_2 4.68017	γ_2 6.20116×10^{-4}	
c_3 -1.221932×10^0	q_3 2.45	ϵ_3 4.73954	γ_3 2.56061×10^{-4}	
c_4 $+4.052997 \times 10^{-2}$	q_4 2.55	ϵ_4 4.76521	γ_4 1.32016×10^{-4}	
c_5 $+1.317109 \times 10^{-1}$				
c_6 -3.265795×10^{-2}				
c_7 $+2.500933 \times 10^{-3}$				

7.4   Interstellar absorption by hydrogen and helium

As pointed out by Cruddace *et al.* (1974), elements heavier than H or He make little contribution to the interstellar opacity at EUV wavelengths. Hence, in the neutral phase of the ISM, the dominant contributors are H I, from ≈ 300 Å to the Lyman edge at 912 Å, and He I at shorter wavelengths. The photoionisation cross-sections of neutral hydrogen and hydrogen-like ions, the most important of which is He II in the local ISM, can be calculated analytically. The photoionisation cross-section $\sigma(\lambda, Z)$ for the ground state of a hydrogenic ion with nuclear charge Z is

$$\sigma(\lambda, Z) = \frac{A_0}{Z^2} \left(\frac{\lambda}{\lambda_1} \right)^4 \frac{\exp[4 - (4\tan^{-1}\epsilon)/\epsilon]}{[1 - \exp(-2\pi/\epsilon)]} \quad \text{for } \lambda \le \lambda_1$$

with

$$A_0 = 6.3 \times 10^{-18}\,\text{cm}^2, \ \epsilon = \sqrt{(\lambda_1/\lambda) - 1}, \ hc/\lambda_1 = 13.606Z^2\,\text{eV}$$

Determination of the neutral helium cross-section is more complex and, while Cruddace *et al.* (1974) used calculations from Bell and Kingston (1967), more recent measurements and theoretical work have superseded these. For example, a critical compilation of cross-sections from several sources by Marr and West (1976) are in good agreement with subsequent theoretical and experimental determinations (see e.g. Fernley *et al.* (1987) and Chan *et al.* (1991), repectively). Marr and West give the neutral helium cross-section as a polynomial fit of the data to $\log \lambda$:

$$\log_{10} \sigma_{\text{He I,cont}}(\lambda) = \sum_{i=0}^{7} c_i (\log_{10} \lambda)^i$$

The polynomial coefficients (c_i) are shown in table 7.1.

The autoionizing doubly-excited transitions of helium between 172 Å and 210 Å show complex resonance structures (Oza 1986) with a characteristic shape following a Fano profile (Fano 1961). When these transitions are included, the total continuum cross-section,

Table 7.2. *Helium I resonance characteristics.*

i	λ_{min} (Å)	$fwhm_{min}$ (mÅ)	λ_{max} (Å)	$fwhm_{max}$ (mÅ)
1	205.885	503.0	206.080	159.0
2	194.675	95.1	194.713	35.5
3	192.257	37.5	192.271	14.5
4	191.227	19.0	191.235	7.44

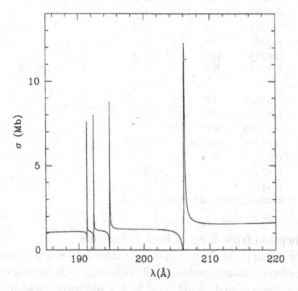

Fig. 7.5. The photoionisation cross-section of He I showing the first four autoionizing $(sp, 2n+)^1P^0$ resonances. $(1\ Mb = 10^{-18} cm^2.)$

according to Rumph *et al.* (1994), is given by

$$\sigma_{He\,I} = \sigma_{He\,I,cont} \prod_{i=1}^{4} \frac{(\Delta\epsilon_i - \frac{1}{2}\gamma_i q_i)^2}{(\Delta\epsilon_i)^2 + \frac{1}{4}\gamma^2}$$

where the energy ϵ, and coefficient γ, drawn from Oza (1986), are expressed in rydbergs and $\Delta\epsilon_i = \epsilon - \epsilon_i$. The values of q are taken from Fernley *et al.* (1987). All the coefficients are listed in table 7.1 and the He I photoionisation cross-section is shown in figure 7.5.

Only the first four resonances contribute significantly to the photoionisation cross-section since the higher ones in the series have characteristic widths <0.01 Å (Fernley *et al.* 1987). Hence, in their model, Rumph *et al.* only incorporate these in the effective cross-section. The wavelengths of the maximum and minimum values of the cross-section (λ_{max} and λ_{min}) of the four strongest resonances, together with the fwhm of the features are listed in table 7.2.

Table 7.3. *Selected EUV lines for He I and He II with an ionisation energy less than 100 eV.*

Ion	λ(Å)	f_{ij}	E_i(cm^{-1})	g_i, g_j
He I	506.200	0.00119	0.0	1, 3
	506.570	0.00150	0.0	1, 3
	507.058	0.00209	0.0	1, 3
	507.718	0.00275	0.0	1, 3
	508.643	0.00399	0.0	1, 3
	509.998	0.00593	0.0	1, 3
	512.098	0.00848	0.0	1, 3
	515.617	0.0153	0.0	1, 3
	522.213	0.0302	0.0	1, 3
	537.030	0.0734	0.0	1, 3
	584.334	0.2762	0.0	1, 3
He II	231.454	0.00318	0.0	2, 128
	232.584	0.00481	0.0	2, 98
	234.347	0.00780	0.0	2, 72
	237.331	0.01394	0.0	2, 50
	243.027	0.02899	0.0	2, 32
	256.317	0.07910	0.0	2, 18
	303.780	0.4162	0.0	2, 8

7.5   Interstellar absorption from lines of heavy elements

While elements heavier than H and He make only a modest contribution to the continuum opacity, they can in principle produce absorption lines in EUV spectra which might be detected spectroscopically. Similar lines have been detected in the far-UV spectra of hot white dwarfs (e.g. Bruhweiler and Kondo 1981, 1982). Rumph *et al.* (1994) have published a useful table of the strongest lines for the most cosmically abundant elements, giving wavelength, oscillator strength (f_{ij}), energy of the lower state (E_i) and statistical weights of both lower and upper levels (g_i and g_j). We reproduce here the data for the He I and He II lines, which are the only interstellar species detected in EUV spectra (table 7.3). The transition probability (A_{ij}) for any line can be calculated from the tabulated parameters

$$A_{ij} = f_{ij}(6.67 \times 10^{15}/\lambda^2)(g_i/g_j)$$

and interstellar column densities estimated from standard interstellar techniques, such as a curve of growth analysis. To date, however, no discrete interstellar absorption lines have been detected at EUV wavelengths.

7.5.1   *The helium ionisation fraction along the line-of-sight to GD246*

The interstellar opacity model developed by Rumph *et al.*(1994) is an essential tool for use in the analysis of EUV spectra, particularly when there is significant flux at wavelengths above $\approx$180 Å where the autoionisation and photoelectric absorption edges can be seen. The initial application to the analysis of an astronomical spectrum was carried out by Vennes *et al.*

(1993) in their study of GD246, the object in which the interstellar autoionisation features and photoionisation He II edge were first seen. Interpretation of the GD246 data is complicated by the presence of heavy element opacity in the photosphere of the star, but by considering just the spectral region between 200 and 270 Å, the underlying white dwarf spectrum can be represented by a pure hydrogen model at an effective temperature of 55 000 K and log $g = 8.0$. Vennes *et al.* obtain a value for the line-of-sight interstellar He II column density in the range $3.5\text{–}4.0 \times 10^{17}\,\text{cm}^{-2}$. The corresponding values of the He I and H I column densities are $1.05\text{–}1.25 \times 10^{18}\,\text{cm}^{-2}$ and $1.25\text{–}1.60 \times 10^{19}\,\text{cm}^{-2}$ respectively. The ionisation fraction of helium estimated from these data lies between 23% and 26%.

7.6 Measuring interstellar opacity with white dwarf spectra

7.6.1 Computation of ionisation fractions from column densities

It is important to note that the ionisation fractions are not direct measurements but must be calculated from measurements of H I, He I and He II column densities. This is straightforward, for helium but requires the assumption that the amount of He III is negligible, and is given by

$$f_{He} = N_{He\,II}\,(N_{He\,I} + N_{He\,II})$$

where $N_{He\,I}$ and $N_{He\,II}$ are the He I and He II column densities respectively.

To estimate the ionisation fraction of hydrogen it is necessary to make the further assumption that the *total* He/H ratio has its cosmic value of 0.1. The total H column density N_H is then

$$N_H = 10 \times (N_{He\,I} + N_{He\,II})$$

and the hydrogen ionisation fraction (f_H) is

$$f_H = (N_H - N_{H\,I})/N_H$$

7.6.2 Spectroscopic studies of interstellar absorption

As demonstrated so convincingly by Vennes *et al.* (1993), spectroscopic observations of white dwarfs in the EUV potentially provide sensitive measurements of the interstellar column densities of H I, He I and He II. Exactly what information can be obtained will depend on the wavelength range covered by the data and their signal-to-noise ratios, which are dependent upon the amount of interstellar opacity present. For example, a further detection of ionised He was reported by Holberg *et al.* (1995) along the line-of-sight to GD659. Since this object has a pure H atmosphere, they were able to determine ionisation fractions for both H and He of 0.36 ± 0.17 and 0.36 ± 0.06 respectively. Furthermore, a study of the long wavelength EUV spectra of stars by Dupuis *et al.* (1995) suggested that the ionisation state in the LISM is inhomogeneous and that He is more highly ionised than H.

The most detailed study to date is that of Barstow *et al.* (1997) and covers a sample of 13 DA white dwarfs drawn from the EUVE Guest Observer programme and the data archive. In contrast to the work of Dupuis *et al.* (1995) this work utilised the entire EUV spectrum from 60 to 550 Å, where possible, rather than just the long wavelength 400–550 Å data. Using the entire spectral range covered by EUVE gives independent information on both He I and He II columns from the size of their photoionisation edges. As noted by Vennes *et al.* (1993),

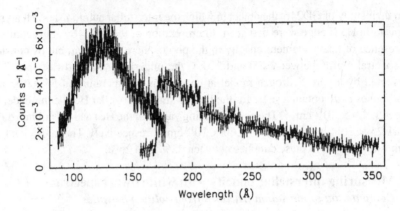

Fig. 7.6. EUV count spectrum of GD659 compared to the best fit ISM/homogeneous H+He model.

He I can generally be studied directly for a wide range of column densities since it is either visible through the 504 Å edge at low columns or the strongest autoionisation feature at 206 Å if the longer wavelengths are cut off. Many of the stars included in the study were observed with the dither mode but several spectra were obtained before its implementation. All but two of the 13 objects had more or less pure H envelopes and could be represented by simple H+He models. However, the hottest stars (PG1123+183 and GD246) contained significant quantities of heavy elements requiring fully blanketed non-LTE models containing C, N, O, Si, Fe and Ni to account for the known sources of photospheric opacity.

In comparing the theoretical models with the observed data, Barstow *et al.* used the XSPEC program and techniques described in section 5.4. To take into account absorption by the ISM they adopted the model of Rumph *et al.* (1994; see section 7.4). This was modified to take into account the converging series of lines near the He I (504 Å) and He II (228 Å) edges in a manner similar to that described by Dupuis *et al.* (1995). For each star the allowed values of T_{eff} and $\log g$ were restricted to the 1σ uncertainties in the optical measurement (see Marsh *et al.* 1997a). Normalisation was provided by a visual magnitude data point, using a central wavelength of 5500 Å and value of 3.64×10^{-9} (from Zombeck 1990) for $f_\lambda(0)$ to calculate the conversion to flux. Figure 7.6 shows, as an example, the spectrum of GD659 with the best fit model, clearly showing interstellar absorption from He I at 206 Å and He II at 228 Å.

Since both PG1123+189 and GD246 contain significant abundances of heavy elements in their photospheres, a restricted region of the spectrum from 200–260 Å was used in the analysis of these stars, coupled with new models containing C, N, O, Si, Fe and Ni computed for the analysis of the heavy element rich DA G191–B2B (Lanz *et al.* 1996). The measured column densities are summarised in table 7.4 for homogeneous H+He models, and table 7.5 for PG1123 and GD246, listing the best fit column densities and uncertainties. Stratified H+He models yield similar results.

He II is clearly detected in five of the white dwarfs – GD659, REJ1032, PG1123, REJ2156 and GD246 – although prior analyses of GD659 and GD246 had already been reported by Holberg *et al.* (1995) and Vennes *et al.* (1993) respectively. While the photospheric He II opacity is allowed to vary freely, the inferred levels are very low for almost all stars, indicating that the observed features are truly interstellar and have a shape that cannot be reproduced by He within the star itself. In fact, a 228 Å edge of the magnitude observed would yield

Table 7.4. *H I, He I, He II column densities, limits and uncertainties determined from analysis of the EUV spectra with homogeneous H+He models.*

Star	$N_{\mathrm{HI}}(\times 10^{18}\,\mathrm{cm}^{-2})$			$N_{\mathrm{HeI}}(\times 10^{17}\,\mathrm{cm}^{-2})$			$N_{\mathrm{HeII}}(\times 10^{17}\,\mathrm{cm}^{-2})$		
	Value	-1σ	$+1\sigma$	Value	-1σ	$+1\sigma$	Value	-1σ	$+1\sigma$
GD659	3.10	2.36	3.91	2.14	1.37	2.86	1.56	1.18	1.90
GD71	0.60	0.53	0.69	0.45	0.30	0.60	0.02	0.00	0.14
REJ0715−705	59.37	9.86	65.92	0.00	0.00	8.63	2.43	0.00	37.76
REJ1032+535	4.16	3.81	4.66	5.69	5.19	6.06	2.07	1.50	2.68
GD153	0.78	0.71	0.85	0.61	0.48	0.72	0.00	0.00	0.27
HZ 43	0.88	0.86	0.90	0.65	0.62	0.69	0.44	0.28	0.62
CoD−38 10980	16.72	0.00	18.86	0.00	0.00	3.82	0.00	0.00	12.36
BPM93487	1.97	0.00	3.06	0.00	0.00	2.71	2.38	0.01	3.78
REJ2009−605	20.86	9.03	31.49	7.99	0.00	14.10	0.35	0.00	10.06
REJ2156−546	5.89	4.48	7.28	5.45	4.07	6.95	3.47	2.71	3.49
REJ2324−547	4.33	3.40	8.10	8.84	5.39	10.18	1.29	0.00	2.86

Table 7.5. *H I, He I, He II column densities, limits and uncertainties determined from analysis of the 200–260 Å region of the spectra of PG1123+189 and GD246 using the predicted continuum fluxes from (top) homogeneous H+He models and (bottom) models incorporating the heavy elements.*

Star	$N_{\mathrm{HI}}(\times 10^{18}\,\mathrm{cm}^{-2})$			$N_{\mathrm{HeI}}(\times 10^{17}\,\mathrm{cm}^{-2})$			$N_{\mathrm{HeII}}(\times 10^{17}\,\mathrm{cm}^{-2})$		
	Value	-1σ	$+1\sigma$	Value	-1σ	$+1\sigma$	Value	-1σ	$+1\sigma$
PG1123+189	15.80	13.70	18.00	9.80	8.10	11.40	3.89	2.79	4.92
GD246	16.50	14.40	19.50	11.00	9.30	12.60	4.41	3.15	5.17
PG1123+189	10.20	7.30	13.10	14.30	12.20	16.30	3.64	2.82	4.25
GD246	13.10	10.50	13.70	12.60	10.30	14.50	3.68	2.93	4.41

strong He II Lyman series lines, principally at 256 and 304 Å, which are not observed and would depress the continuum flux below the edge, making the model inconsistent with the observation. In the remaining white dwarfs, the He II measurements could only be considered as limits in the absence of any direct detections.

The ionisation fractions of hydrogen (f_{H}) and helium (f_{He}) can be calculated from the measured column densities, as described in section 7.6.1 (see tables 7.6 and 7.7). In general, the stars observed can be divided into three groups. First, the 228 Å He II photoionisation edge is clearly detected in five objects (GD659, REJ1032, PG1123, REJ2156 and GD246), providing a direct measurement of the He ionisation fraction. There appears to be no particular relationship between direction and the degree of ionisation measured. The weighted mean ionisation fractions of He and H in these five directions are 0.27 ± 0.04 and 0.35 ± 0.1 respectively. Second, although the region of the He II edge is included in the spectra of GD153 and GD71, the edge is not detected and only an upper limit to the He II column density and, consequently, the He ionisation fraction, is available. Interestingly, although the edge is not seen in HZ 43 either, the best fit models do require the presence of a finite He II column, with a

Table 7.6. *The ionisation fractions of helium and hydrogen in the interstellar medium, calculated from the line-of-sight column density measurements using homogeneous H+He photospheric models.*

Star	f_H			f_{He}		
	Value	-1σ	$+1\sigma$	Value	-1σ	$+1\sigma$
GD659	0.16	0.00	0.50	0.42	0.29	0.58
GD71	0.00	0.00	0.28	0.04	0.00	0.32
REJ0715−705	0.00	0.00	0.79	1.00	0.00	1.00
REJ1032+535	0.46	0.30	0.56	0.27	0.20	0.34
GD153	0.00	0.00	0.28	0.00	0.00	0.36
HZ 43	0.19	0.00	0.34	0.40	0.29	0.50
CoD−38 10980	0.00	0.00	1.00	0.00	0.00	1.00
BPM93487	0.17	0.00	1.00	1.00	0.00	1.00
REJ2009−605	0.00	0.00	0.63	0.04	0.00	1.00
REJ2156−546	0.34	0.00	0.57	0.39	0.28	0.46
REJ2324−547	0.57	0.00	0.74	0.13	0.00	0.35

Table 7.7. *The ionisation fractions of helium and hydrogen in the interstellar medium, calculated from the line-of-sight column density measurements towards PG1123+189 and GD246 using (top) homogeneous H+He photospheric models and (bottom) models incorporating heavy elements.*

Star	f_H			f_{He}		
	Value	-1σ	$+1\sigma$	Value	-1σ	$+1\sigma$
PG1123+189	0.00	0.00	0.16	0.28	0.20	0.38
PG1123+189	0.43	0.13	0.65	0.20	0.15	0.26
GD246	0.00	0.00	0.19	0.29	0.20	0.36
GD246	0.20	0.04	0.45	0.23	0.17	0.30

lower error bound that does not extend down to zero, but in the absence of a positive detection this measurement is also treated as an upper limit. Third, in the third group of stars (REJ0715, BPM93487, CoD−38, REJ2009, REJ2324) the He II spectral region is not sufficiently well exposed to obtain either direct measurements of any feature or well-determined upper limits. However, the data *imply* that ionised material may be present since the models which provide the best fit to the spectra often include a finite quantity of He II. It should be noted that in some cases the uncertainties in the measurement of the He II column are very large, encompassing zero, as are the corresponding estimates of the ionisation fractions.

Coupling these measurements of He II column density with the upper limits and implied values for other stars now allows a limited systematic study of photoionisation in the LISM to be made. A number of simple questions can be raised. Does the He II ionisation fraction depend on viewing direction (i.e. location)? Is the amount of ionised helium correlated with either interstellar volume density or line-of-sight column density? Figure 7.7 shows the distribution of stars in the study of Barstow *et al.* (1997) on the sky in the galactic coordinate system. Superimposed are the contours of constant distance to the edge of the 'local bubble',

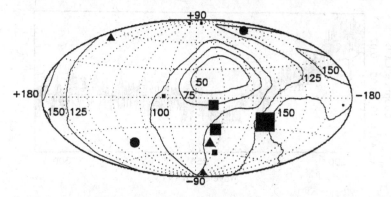

Fig. 7.7. Hammer–Aitoff plot in galactic coordinates, showing the distribution of white dwarf positions. The sizes of the symbols correspond to the H I column density along the line-of-sight (proportional to $H I^{0.5}$). Squares are observations where a 228 Å edge is not detected; triangles are 228 Å edge detections analysed with H+He models; circles are 228 Å edge detections analysed with heavy element models (i.e. GD246 and PG1123+189). The contours included indicate the distance (pc) to the edge of the local cavity, as determined by Warwick *et al.* (1993).

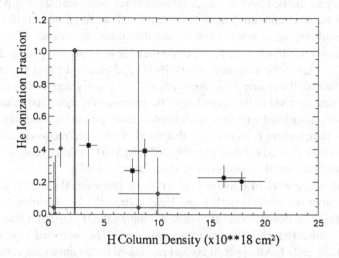

Fig. 7.8. The observed fractional ionisation of He for each star in the sample as a function of the total column density of hydrogen along the line-of-sight. The shaded shapes indicate those stars where He II is detected directly, squares corresponding to objects analysed with H+He models (GD659, REJ1032, REJ2156) and circles objects analysed with heavy element models (PG1123 and GD246). The open diamonds represent those stars where He II was not detected and only broad limits on the ionisation fraction obtained.

determined by Warwick *et al.* (1993), and the five stars with detected ionisation are marked. Three of these lie towards the south galactic pole but they are separated by ≈20–30° on the sky indicating that the ionised region could be very extensive. The dependence of the He ionisation fraction on the average column density of hydrogen along the line-of-sight is illustrated in figure 7.8. The total quantity of H is estimated from the calculated total He (He I+He II) according to the cosmic H/He ratio. There is no evidence for any correlation between the He ionisation fraction and H column or volume density. For example, if this material were circumstellar, a correlation of $N_{He II}$ with stellar temperature might be expected.

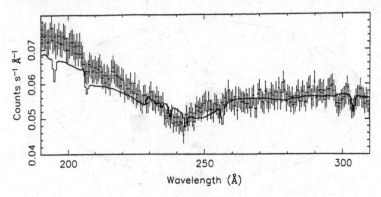

Fig. 7.9. Section of the EUV spectrum of HZ 43 near the He II Lyman series at 228 Å, illustrating the magnitude of He II edge expected for an ionisation fraction of 40%, which would be present if the weak 304 Å absorption 'feature' were completely accounted for by an interstellar component.

The data points appear to be clustered into the three groups identified earlier (ionisation upper limits, measurements and implied best fit values without direct detection). This is perhaps not too surprising as the total column density has a strong influence on what features are potentially visible. Furthermore, in low column density directions, the absence of detected He II features could easily be a threshold effect because the edge is simply too weak to detect. For example, figure 7.9 shows the magnitude of the He II edge expected along the line-of-sight to HZ 43 if the interstellar material had the ≈40% He ionisation fraction seen in other directions. It is clear that the level of the upper limits, the errors on the measured values and the range of uncertainty associated with the values implied from the model comparisons all overlap substantially. Hence, there is no evidence that the He ionisation fraction is not close to the weighted mean (0.27 ± 0.04) of the GD246, GD659, REJ1032, PG1123 and REJ2156 measurements in any of the directions probed with these observations.

An important result of the work of Barstow *et al.* (1997) is that when the total H column density is predicted from the combined He I and He II values, the H ionisation fraction inferred from the H I column density is always positive. Plotting the H ionisation fraction as functions of total H column (figure 7.10) reveals no correlation. The measured values have a weighted mean of 0.35 ± 0.1. All upper limits and associated uncertainties are consistent with this level of ionisation, which is also similar to that measured for He. This contradicts earlier statements (e.g. Vennes *et al.* 1993; Dupuis *et al.* 1995; Bowyer 1996) that the hydrogen ionisation fraction in the LISM is far lower than that of helium. To a large extent these statements are based on the study of GD246. The most plausible explanation may be the use of an inappropriate pure H model atmosphere to set the level of the EUV continuum flux in that work.

Concluding that observations are consistent with constant H and He ionisation fractions in the LISM is not the same as asserting that H and He ionisation fractions are constant within the LISM or that these quantities do not differ for different lines-of-sight. The weighted mean values, however, do provide very useful approximations representing the average ionisation state of that portion of the LISM which contributes the bulk of the EUV interstellar opacity. Another quantity which is widely used in discussions of the LISM is the ratio N_{HI}/N_{HeI} or alternately, its reciprocal. Dupuis *et al.* (1995) measured N_{HI} and N_{HeI} for six white dwarfs observed with EUVE, including four from the Barstow *et al.* sample. They find a

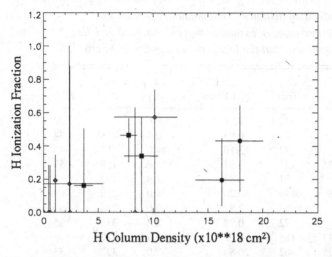

Fig. 7.10. The inferred fractional ionisation of H for each star in the sample as a function of the total column density of hydrogen along the line-of-sight. The shaded shapes indicate those stars where He II is detected directly: squares corresponding to objects analysed with H+He models (GD659, REJ1032, REJ2156) and circles to objects analysed with heavy element models (PG1123 and GD246). The open diamonds represent those stars where He II was not detected and only broad limits on the ionisation fraction obtained.

value of $N_{\mathrm{HI}}/N_{\mathrm{HeI}} = 14$ from a visual fit to the plot of N_{HeI} vs. N_{HI} for their sample. Kimble *et al.* (1993a) also used HUT 1 observations of G191−B2B to determine $N_{\mathrm{HI}}/N_{\mathrm{HeI}} = 11.6 \pm 1.0$. Barstow *et al.* use their data to estimate $N_{\mathrm{HI}}/N_{\mathrm{HeI}}$ by two somewhat independent methods. If the weighted estimates of the H and He ionisation fractions are used directly, then $N_{\mathrm{HI}}/N_{\mathrm{HeI}} = 8.9 + 1.7/-1.3$ from the relation between the H and He ionisation fractions and the cosmic H to He ratio. Alternatively, following Dupuis *et al.* the individual N_{HI} and N_{HeI} measurements can be used to determine this quantity, but taking only the nine stars where measured values of the H I and He I densities were obtained, i.e. excluding the three stars having only limits on N_{HeI} and REJ2009−605 where the formal limits on N_{HI} and N_{HeI} both include 0.0. A maximum likelihood estimation of the slope of a N_{HeI} vs. N_{HI} plot, including the uncertainties on both axes, gives a value of $0.084 + 0.014/-0.12$. This corresponds to $N_{\mathrm{HI}}/N_{\mathrm{HeI}} = 11.9 + 2.0/-1.7$. This result is marginally consistent with the estimate obtained from the H and He ionisation fractions as well as those of Dupuis *et al.* and Kimble *et al.* It is, however, unsatisfactory from a statistical standpoint. The formal chi-squared per degree of freedom for this fit is 3.5, thus the assumed model of a completely uniform LISM can be rejected, as might well be expected. An even more serious objection is the dominance of the result by just three stars having the lowest column densities and lowest relative uncertainties, HZ 43, GD153 and GD71 (in addition, HZ 43 and GD153 share virtually the same line-of-sight). These stars largely explain the higher ratio found by Dupuis *et al.* A more acceptable way of determining the average conditions in the LISM would be to take a simple unweighted geometric average of the $N_{\mathrm{HI}}:N_{\mathrm{HeI}}$ ratios for the nine stars, which leads to $N_{\mathrm{HI}}/N_{\mathrm{HeI}} = 10$.

The distribution of the absorbing material along the lines-of-sight is important in considering the physical mechanisms that might give rise to the levels of ionisation observed. For example, Frisch (1994) estimates the electron density (0.22 to 0.44 cm^{-3}) in the so-called local cloud (or surrounding interstellar cloud, SIC), in which the solar system is embedded

Table 7.8. *Estimated distance to the boundary of the local interstellar cloud (d_{cloud}), assuming that all observed H I lies within the cloud and that the local volume density is 0.1 cm^{-3}.*

Star	d_{stars} (pc)	H I ($\times 10^{18}$ cm^{-2})	d_{cloud} (pc)	l^{II}	b^{II}
BPM93487	42	2.88	9.6	34	2
HZ 43	71	0.83	2.8	54	84
GD246	72	13.10	44.0	87	−45
REJ1032+535	135	3.97	13.0	158	53
GD71	51	0.84	2.8	192	−4
PG1123+189	139	10.20	34.0	232	69
GD659	57	2.87	9.6	294	−84
GD153	72	0.78	2.6	317	85
REJ2324−547	185	4.02	13.4	327	−58
REJ2009−605	62	20.86	69.0	337	−33
REJ2156−546	129	5.87	20.0	339	−48

and predicts that the hydrogen ionisation fraction should lie in the range 69% to 81%. A more definitive measurement is that of Wood and Linsky (1997), who use the C II absorption lines to obtain a value of $n_e = 0.11 + 0.12/-0.06$ along the line-of-sight to Capella. In addition, recent measurements, made in the direction of ϵ CMa, indicate that the electron density in the local cloud is rather lower, ≈ 0.09 cm^{-3}, giving a H ionisation fraction no greater than 50% and more in keeping with the EUVE result (Gry *et al.* 1995). In addition, the nominal 'constant' He and H ionisation fractions seen in the EUVE data are consistent with the predictions of Lyu and Bruhweiler's (1996) time dependent ionisation calculations, considering the recombination of the LISM following shock heating from a nearby supernova. On the basis of the earlier studies of GD246 and GD659 they estimate an elapsed time of 2.1 to 3.4 million years since the onset of the recombination phase. The more recent results also fit within this time envelope, the increased number of data points serving to strengthen the case for this theory.

In the light of this apparent agreement, it is necessary to consider that the observations might be sampling material both in the local cloud and in the surrounding medium. The latter is more tenuous but the effects are integrated over greater distances. The local interstellar cloud is considered to be only a few parsecs across with a mean neutral hydrogen volume density of ≈ 0.1 cm^{-3} (e.g. see Frisch 1994; Gry *et al.* 1995). It is possible to consider whether or not the interstellar material observed really does lie in the local cloud by estimating the distance to the cloud boundary on the assumption that all the H I is contained within, at the density noted above. Considering just those stars where there are relatively unambiguous estimates of N_{HI} (i.e. ignoring high column directions where He I and He II are not directly observable), it is clear that while some distances are small enough to be consistent with our assumption (e.g. GD153, HZ 43, GD71) most imply a boundary at significantly greater distance than believed (table 7.8). Consequently, either the local interstellar cloud is rather larger than reported, which is unlikely, or a significant fraction of the interstellar absorption observed lies outside the cloud in most of the lines-of-sight examined. It is important to note that this is so for all those directions in which ionised He is detected. Hence, the ionisation fractions reported may not represent the conditions in the local cloud and the agreement with the measurements of Gry *et al.* (1995) and predictions of Lyu and Bruhweiler (1996) could be purely fortuitous.

A number of broad conclusions can be drawn from an EUV spectroscopic survey of the LISM. A range of total column densities are seen, which are consistent with the previous understanding of the local distribution of material and ionised He II is found along the lines-of-sight to five stars. However, the non-detection of He II in low column directions is possibly a threshold effect as the 228 Å edge is predicted to be too weak to detect for equivalent levels of ionisation. In very high column density directions, He II cannot be seen because the interstellar opacity cuts off the spectra at wavelengths shorter than 228 Å. Where He II is directly detected, the observed ionisation fractions are not correlated with direction or with the volume/column density of material along the line-of-sight. Furthermore, the limits on the amount of He II established in all other directions completely encompass the range of observed values. Indeed, all the data can be consistent with constant He and H ionisation fractions corresponding to the weighted means of the directly measured values of 0.27 ± 0.04 and 0.35 ± 0.1 respectively, throughout the local ISM.

8

Spectroscopy of white dwarfs

8.1   The importance of EUV spectra of white dwarfs

It is clear, from chapters 3 and 4 (sections 3.6, 3.7, 4.3.2), that the ROSAT and EUVE sky surveys have made significant contributions to our understanding of the physical structure and evolution of white dwarfs. Among the most important discoveries are the ubiquitous presence of heavy elements in the very hottest DA stars (above 40 000–50 000 K), the existence of many unsuspected binary systems containing a white dwarf component and a population of white dwarfs with masses too high to be the product of single star evolution. In each case, however, the detailed information that could be extracted from the broadband photometric data was often rather limited. For example, although simple photospheric models (e.g. H+He) could often be ruled out, it was not possible to distinguish between more complex compositions with varying fractions of He and heavier elements. Furthermore, rather simplistic assumptions needed to be made about the relative fractions of the H and He in the interstellar medium besides the degree of ionisation of each element, to restrict the number of free parameters to a tractable level in any analysis. Direct spectroscopic observations of gas in the LISM (see sections 7.3 and 7.6) indicate that the convenient assumption of a cosmic He/H ratio (0.1) and minimal ionisation is unlikely to be reasonable.

Spectroscopic observations of white dwarfs in the EUV can address a number of important questions. The presence, or not, of trace helium has not yet been definitively studied. While the cooler ($<40\,000$ K) DAs mostly appear to have more or less pure H envelopes, there could still be astrophysically important quantities of He, with an abundance $\approx 10^{-5}$ or below. Study of the He II Lyman series between 228 and 304 Å provides an opportunity of detecting the presence of He at lower levels than is possible with either UV/optical spectroscopy ($\approx 10^{-4}$) or EUV/soft X-ray photometry, or at least providing upper limits to its presence that can be compared with the predictions of evolution theory. EUV spectra should also reveal the signatures of elements heavier than He so that the species which block the emergent EUV flux can be identified. Although many ions have been identified from UV spectra some do not have strong transitions in that wavelength range. Furthermore, depth dependent abundance profiles might be revealed by comparing measurements made in different wavebands. Apart from the photospheric composition, the emergent EUV flux is a strong function of the effective temperature. While T_{eff} (and also log g) can often be determined from the Balmer line profiles, these data are not always routinely available and may be compromised by the presence of a companion or an unknown stellar composition. Hence, EUV spectrophotometry can be an important tool in measuring T_{eff} for white dwarfs in binaries. Consistency between the values determined from the EUV continuum and Balmer lines in isolated stars is an essential test in establishing the white dwarf temperature scale.

8.2   Measuring effective temperature from EUV continua

Binary systems are astrophysically important for several reasons. Some may be useful as laboratories for binary interactions, particularly from stellar winds, and some represent the results of common-envelope evolution. The results of such evolution may be apparent in the physical remnants. In binaries with primary stars in the spectral range A to K the hydrogen Balmer lines of the white dwarf are not accessible to us, because in general, the optical data are dominated by the primary flux (see section 3.7). Although Lyman α profiles can, in part, solve this problem, it is not generally possible to obtain unique, independent values for T_{eff} and log g, since different combinations of these parameters can yield similarly shaped profiles (see Napiwotzki *et al.* 1993). The additional constraint of an estimated distance can narrow the allowed range of T_{eff} and log g but its usefulness depends on how accurately the primary spectral type is known (and hence its luminosity and radius).

Section 7.6, in discussing the use of hot DA white dwarf observations to probe the state of the LISM, illustrates in detail the basic principles of applying the stellar models to fit the spectrum of the star in combination with the detailed ISM model, which incorporates the effects of He I and He II. In that example, T_{eff} and log g were confined within narrow bounds determined from optical data. Here, these must be treated as completely free. It is still helpful to apply an independent normalisation to the data, if possible, and, in the absence of an accurate V magnitude measurement, a far-UV flux can be used. An example of the usefulness of EUV data in the study of a binary system is the observation of HR8210 (IK Peg), in which the white dwarf was identified through its detection in the ROSAT survey (Landsman *et al.* 1993; Wonnacott *et al.* 1993; Barstow *et al.* 1994a). HR8210 consists of an A8m primary and DA white dwarf secondary with an orbital period of 21.7 d. Lyman α measurements give T_{eff} in the range 26 700 to 35 600 K, depending on the surface gravity of the white dwarf. Assuming that the white dwarf really is physically associated with the A star, the estimated distance to the system favours the higher temperature and log $g \approx 9.0$. The EUV/soft X-ray photometry is consistent with a more or less pure H envelope.

Figure 8.1 shows the complete EUVE spectrum of HR8210, together with a stratified H+He model which best represents the data. The parameters of the model and associated 1σ uncertainties are summarised in table 8.1. There is no evidence for any absorption in the spectrum, other than those attributable to interstellar He I at 206 and 504 Å, confirming that the white dwarf photosphere is best represented by a pure H composition. The results also confirm the suspected high gravity (log $g = 8.95$) for this object and confine the value of T_{eff} to a very narrow range (34 450–34 620 K), independently of any information concerning the A star. The total error range of 200 K for T_{eff} is more than an order of magnitude lower than the 3 000 K obtained from the Lyman α line.

From these results it is possible to obtain an independent measurement of the mass and distance of the white dwarf, based on the value of $(R_*/d)^2$ and an estimated R_* from the evolutionary models of Wood (1995), for the appropriate temperature and gravity. The estimated mass of the white dwarf is $1.12M_\odot$ (Barstow *et al.* 1994c) and the corresponding radius $0.006R_\odot$, yielding a distance of 44 pc from the best fit solid angle. The uncertainty in the calculation is dominated by the error in the radius (10%), arising from log g, which is much larger than that of the solid angle (2%). Hence, the resulting error in the distance determination is ±4.5 pc, yielding a possible range of 39.5–48.5 pc, in very good agreement with the 43–54 pc inferred from the spectral type and luminosity class of the primary.

Table 8.1. *Best fit stratified models for HR8210 and HD15638. A single value in the error columns corresponds to upper limit of grid.*

Parameter	HR8210		HD15638	
	Best fit value	1σ uncertainty	Best fit value	1σ uncertainty
χ^2_{red}	0.72		0.49	
T (K)	34540	34450–34620	46000	45150–47790
$\log g$	8.95	8.89–9.0[a]	8.50	8.30–8.60
H mass ($\times 10^{-14} M_\odot$)	1.45	1.29–1.45[a]	6.2	4.1–9.8
$(R_*/d)^2$ ($\times 10^{-23}$)	1.03	1.02–1.07	0.60	0.57–0.62
H I column ($\times 10^{18}$ cm^{-2})	2.69	2.28–3.43	12.0	2.60–22.0
He I column ($\times 10^{18}$ cm^{-2})	0.224	0.147–0.269	2.00	1.30–2.80
He II column ($\times 10^{18}$ cm^{-2})	0.083	0.048–0.121	0.73	0–1.10

[a] Limit of model grid.

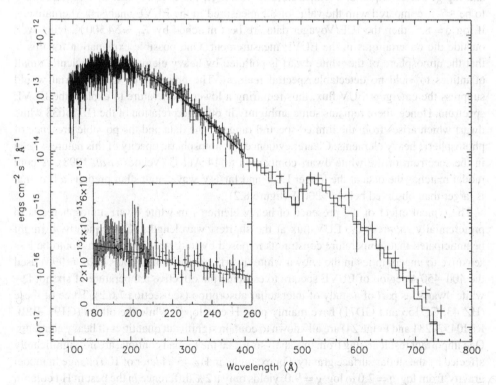

Fig. 8.1. EUV spectrum of HR8210 after deconvolution with the EUVE instrument response (error bars). The solid histogram shows the best fitting stratified model corresponding to the parameters listed in table 8.1. The inset shows the wavelength range from 180 to 260 Å, covering the possible He II Lyman series and He I autoionisation resonance features (from Barstow *et al.* 1994c).

To a certain extent, such a good result is dictated by individual circumstances. The HR8210 spectrum has good signal-to-noise over a wide wavelength range, lies along a low column density line-of-sight and has no source of photospheric opacity apart from hydrogen. Hence, the predicted shape and flux level of the EUV spectrum is much more sensitive to the effective temperature and surface gravity than any other free parameter. An interesting contrast is the binary HD15638, which, through greater interstellar absorption, is not detectable beyond 400 Å. In addition, the signal-to-noise of this spectrum is too poor to reveal any of the He features in the 180–310 Å range. As a consequence, the overall uncertainties in the model parameters that best match the data are much larger (table 8.1). In particular, the value of T_{eff} spans a range of $\approx 2\,600$ K (from 45 150–47 790 K), a factor of ten greater than that for the HR8210 temperature measurement.

Interestingly, the EUV spectroscopic temperature determination for HD15638 suggests that the white dwarf is significantly cooler than estimated by either Landsman *et al.* (1993) or Barstow *et al.* (1994a), although their errors are substantial and do overlap with the EUVE result. More important is a discrepancy between the value of log g determined from the EUVE data and that from the IUE and Voyager Lyman line observations. In an analysis of Lyman α data alone, values of T_{eff} and log g are often highly correlated in a positive sense, i.e. models with high T_{eff} and log g are seen to fit equally well with models having lower T_{eff} and log g. For example, at a temperature of 46 000 K, the IUE/Voyager data require log g to be ≈ 7.7, compared with the value of 8.5 measured in the EUVE analysis. Alternatively, if log $g = 8.5$, then the IUE/Voyager data are best matched by $T_{eff} = 54\,500$ K, lying well outside the uncertainties in the EUVE measurement. One possible explanation for this is that the atmosphere of the white dwarf is polluted by heavy elements in sufficiently small quantities to yield no detectable spectral features. The presence of such material would suppress the emergent EUV flux, thus requiring a lower temperature to explain the EUVE spectrum. Hence, there remains some ambiguity in our interpretation of the HD15638 white dwarf which arises from the limited spectral range of the data and the possible presence of photospheric heavy elements. Clearer evidence of photospheric opacity of this nature is seen in the spectrum of the white dwarf companion to 14 Aur C (Vennes *et al.* 1998). A pure model matches the data at the longer EUV and far-UV wavelengths but predicts a flux that is larger than observed below ≈ 200 Å (figure 8.2).

The typical effect of the presence of heavy elements in white dwarf atmospheres is to preferentially suppress the EUV flux at the shortest wavelengths. Consequently, it might be anticipated that temperature constraints imposed by long wavelength data would be less sensitive to uncertainties in the known white dwarf composition. Dupuis *et al.* (1995) used the 400–450 Å region of EUVE spectra to estimate the effective temperature of six hot DA white dwarfs, as part of a study of interstellar absorption (see section 7.6.2). Three of these (HZ 43, GD153 and GD71) have mainly pure H envelopes, while the others (G191–B2B, REJ0457–281 and Feige 24) are all known to contain significant quantities of heavy elements. One important result arising from this analysis is that the Lyman continua are not significantly affected by the stellar surface gravity. For example, in HZ 43 a factor of 100 change in model gravity, from log $g = 7.0$ to log $g = 9.0$, yields only a 2% difference in the best fit H I column density. Values of T_{eff} obtained for HZ 43, GD153 and GD71 are in good agreement with optical and UV measurements (table 8.2).

In contrast, for those stars containing significant heavy element opacity, the temperatures measured from the Lyman continuum differ significantly from the optical and UV results

Table 8.2. *Lyman continuum measurements of* T_{eff} *for hot DA stars (from Dupuis et al. (1995).*

Name	Model	T_{eff} (optical, UV) ($\times 10^3$ K)	T_{eff} (EUV) ($\times 10^3$ K)
HZ 43	pure H	50.0, 49.0	51.1 ± 0.5
GD153	pure H	41.2, 40.0	39.7 ± 0.3
GD71	pure H	32.8, 33.6	32.4 ± 0.5
REJ0457−281	pure H	60.7	68.7 ± 1.3
	H+CNOFe		63.4 ± 0.9
Feige 24	pure H	54.0	63.1 ± 1.0
	H+CNOFe		58.7 ± 0.9
G191−B2B	pure H	55.7, 64.1	57.9 ± 1.5
	H+CNOFe		54.0 ± 0.8

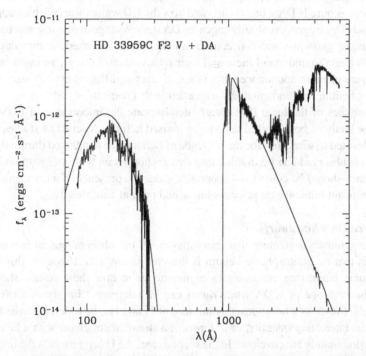

Fig. 8.2. EUV spectrum of 14 Aur C spanning EUV and far-UV wavelength ranges. The smooth curve represents a pure H model atmosphere that gives the best match to the data above ≈ 200 Å (from Vennes *et al.* 1998).

(table 8.2). Dupuis *et al.* attribute this to a backwarming effect due to short wavelength opacity, which gives rise to an increase in the absolute flux level above that predicted by a pure H model at wavelengths longward of ≈ 304 Å. Re-evaluation of the expected Lyman continuum flux using LTE model spectra, incorporating the blanketing effects of C, N, O and

Fe, yields substantially different values for T_{eff} which are much closer to those determined from the optical and UV data (table 8.2).

It is clear that EUV spectroscopy is an important method for determining the effective temperature and, to a lesser extent, log g for hot white dwarfs. However, the results are very sensitive to knowledge of the photospheric composition. Hence, the technique works best for stars known to have pure H atmospheres. However, if heavy element abundances are known from observations made in other wavebands, important constraints on T_{eff} can still be obtained.

8.3   Photospheric helium in hot white dwarfs

The division of white dwarfs into H-rich and He-rich groups continues to pose a problem for evolution theory. It is generally believed that the quantity of residual hydrogen (and even helium) is determined by the number of times the white dwarf progenitor ascends the red giant branch. However, the details are not completely understood. If we accept that the majority of white dwarfs originate from H-rich or He-rich post asymptotic giant branch (AGB) stars, then the main difficulties lie in explaining why planetary nebulae with a mixture of H and He eventually appear as pure H DA white dwarfs and how the DO white dwarfs, which appear to be largely devoid of hydrogen, eventually appear as DA stars. A range of possible mechanisms, including diffusion, mass-loss and convection, can alter the photospheric composition of a white dwarf. To begin to undertand these and their relative contributions, an important first step is to measure element abundances in a range of stars at different evolutionary stages. Observations of helium are fundamentally important in this respect.

The EUVE studies of the local ISM clearly demonstrate the importance of EUV spectroscopy for the study of both neutral and singly ionised helium. A series of studies of DA white dwarfs devoted to searching for the presence of helium was conducted during the mission. The objects observed can be divided into two groups – stars with and without known companions. Each should be considered separately, since the presence of a companion may have had a significant influence on prior evolution and present behaviour.

8.3.1   Isolated DA white dwarfs

The immediate summary statement that encompasses all the observations of isolated DA white dwarfs is that no photospheric helium is detected in any star. However, this simple report hides some interesting and complex arguments as to how these results should be interpreted. The envelope of a DA white dwarf can be interpreted in terms of two quite different models. Either as a homogeneous mixture of H and He, in which case the helium abundance is the interesting quantity, or in terms of a stratified structure, with a layer of H overlying a predominantly He envelope. In the second case, the H layer mass is the important variable and any He observed is what diffuses across the H/He boundary. This model might be considered to be the most physically realistic, since, in the high gravitational field, He might be expected to sink out of the atmosphere. Even so, a measurement of He 'abundance' remains an important parameter when comparing different objects.

The potential presence of interstellar absorption from helium in the spectrum of any star might, at first sight, appear to pose a significant problem for interpretation of any data. Fortunately, a happy accident of physics makes the signatures of photospheric and interstellar He quite different. Interstellar material is of low density, with most ions populating the ground state. Hence, interstellar absorption from He II is dominated by narrow weak absorption features converging at the photoionisation edge at 228 Å (see figure 7.3). In contrast, in a white dwarf atmosphere, the excited states have substantial populations giving rise to strong absorption lines associated with all members of the He II Lyman series. Furthermore, the high gravity and, consequently, atmospheric pressure, makes the lines broader than typical interstellar features. Overlapping of the pressure broadened higher order members of the Lyman series generally smooths out the He II absorption 'edge'.

As one of the brightest white dwarfs throughout most of the EUV wavelength range, HZ 43 is probably the best studied of all hot DAs and has been observed by sounding rockets (e.g. Malina *et al.* 1982), EXOSAT spectrometers (Paerels *et al.* 1986b) and the Einstein objective grating spectrometer (OGS), prior to the ROSAT or EUVE missions. While Malina *et al.* reported the detection of an He II absorption edge at 228 Å, this was not confirmed in the EXOSAT data which placed a 90% confidence upper limit on the abundance of He of 3×10^{-6}. A more recent analysis of the same data, using more up-to-date LTE model atmosphere calculations gives a slightly lower limit of 2×10^{-6} (Barstow *et al.* 1995b). The improved spectral resolution and signal-to-noise available from a $\approx 150\,000$ s exposure recorded with the EUVE spectrometers, in May and June 1994, might be expected to provide a direct detection of He or, at least, a substantially improved limit. With data recorded using the dither mode to minimise the systematic errors arising from the fixed pattern efficiency variation, the observation represents the limit of what was achievable with EUVE.

Figure 8.3 shows the complete flux corrected spectrum of HZ 43 compared with the prediction of a homogeneous LTE model atmosphere. The effective temperature and surface gravity

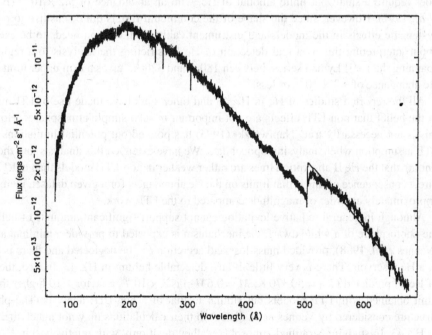

Fig. 8.3. EUVE spectrum of the DA white dwarf HZ 43 after deconvolution with the instrument response (error bars). The model histogram corresponds to the best fitting homogeneous H+He synthetic spectrum with He/H = 3.2×10^{-7}.

were constrained to lie between the optically determined limits ($T_{\text{eff}} = 47\,000-51\,000$ K, $\log g = 7.5-7.9$) and the model which gives the best match to the data has a He/H abundance of 3.2×10^{-7}. A closer view of the He II Lyman series region (see figure 7.9) shows a weak, but distinct, absorption dip at the position of He II 304 Å with an equivalent width of 200 ± 100 mÅ. However, as there are other 'features' in the spectrum that are only a little weaker, the 304 Å feature may not be real. Even so, it gives a good indication of the sensitivity of the observation to the presence of photospheric He. No other lines in the He II Lyman series are visible in the spectrum but these are predicted to be weaker than 304 Å and might plausibly remain hidden in the noise. If attributed to an interstellar component the observed 304 Å equivalent width would correspond to an He II column $\approx 5 \times 10^{17}$ cm^{-2}. Such a large He II column, in addition to producing a prominent absorption edge is also nearly equivalent to the measured H I column of HZ 43. Hence, any He present should be predominantly photospheric.

Both homogeneous and stratified H+He models give good agreement with the EUVE data – matching the overall shape of the continuum, the 584 Å He I interstellar edge and the absence of an He II 228 Å interstellar edge simultaneously (figure 8.3; see Barstow *et al.* 1995b, 1997). In order to match the continuum well, the stratified model requires that the H layer mass is greater than $6.8 \times 10^{-13} M_{\odot}$ but the predicted strength of the 304 Å line is much weaker than can be allowed by the data. A limit to the H layer mass determined from the region of the He 304 Å line alone is less constraining than that imposed by the continuum. For example, a model with an H layer mass of $5.0 \times 10^{-14} M_{\odot}$, matching the features, would yield a significant shortfall (more than 30%) in the predicted flux below ≈ 120 Å, which cannot be removed by varying the remaining parameters. The best fit homogeneous model does require a small but finite amount of He, with an abundance of 3.2×10^{-7} (Barstow *et al.* 1997). However, since the shape of the continuum flux is quite sensitive to possible systematic effects in the models and instrument calibration, caution needs to be exercised about interpreting this as a real detection of He. Restricting the analysis to a region just spanning the He II Lyman series, between 180 Å and 320 Å, gives a firm upper limit on the He abundance of $\approx 3 \times 10^{-7}$ or less.

All the reported studies of He in HZ 43 and other stars have made use of LTE models, in the belief that non-LTE effects are not important in such simple atmospheres. However, this is not necessarily true. Napiwotzki (1995) has pointed out potential limitations of the LTE assumption when analysing optical data. We have extended this analysis into the EUV finding that the He II absorption lines are rather weaker in non-LTE models than in LTE. The direct consequence of this is that limits on the He abundance, for a given dataset, increase by approximately an order of magnitude compared to the LTE work.

Although in general radiative levitation cannot support significant amounts of helium in the photosphere of a white dwarf, the mechanism is expected to provide a minimal amount (Vennes *et al.* 1988), provided mass-loss and accretion can be neglected and there is access to a He reservoir. There is very little, if any, detectable helium in HZ 43. Indeed, the value of He/H predicted ($T_{\text{eff}} = 50\,370$ K; $M = 0.6 M_{\odot}$) is 3×10^{-6}, a factor ≈ 10 higher than the limit obtained with LTE models but similar to that of a non-LTE analysis. The physical structure considered by Vennes *et al.* (1988) in their calculations may not match the reality in HZ 43. In studying stratified atmospheres, they dealt only with relatively thin H layers, with masses less than $10^{-13} M_{\odot}$. Stellar evolution calculations predict that DA white dwarfs should be formed with massive $\approx 10^{-4} M_{\odot}$ H layers (e.g. Iben and Tutukov 1984; Iben and

Table 8.3. *Limits on the photospheric helium abundance and H layer mass, determined using LTE homogeneous and stratified H+He models respectively.*

Star	He/H[a] Value	-1σ	$+1\sigma$	M_H ($M_\odot$)[b] Value	-1σ	$+1\sigma$
GD659	1.2×10^{-7}	1.0×10^{-8}	7.2×10^{-7}	1.2×10^{-12}	5.4×10^{-13}	1.2×10^{-12}
GD71	1.0×10^{-8}	1.0×10^{-8}	3.6×10^{-8}	9.0×10^{-13}	3.8×10^{-13}	1.6×10^{-12}
REJ0715–705	7.8×10^{-8}	1.0×10^{-8}	1.3×10^{-6}	3.9×10^{-13}	3.1×10^{-13}	4.7×10^{-13}
REJ1032+535	1.5×10^{-6}	5.4×10^{-7}	2.4×10^{-6}	6.1×10^{-13}	4.3×10^{-13}	9.6×10^{-13}
GD153	1.0×10^{-7}	1.0×10^{-8}	5.0×10^{-7}	1.2×10^{-12}	8.0×10^{-13}	2.7×10^{-12}
HZ 43	3.2×10^{-7}	6.5×10^{-8}	1.0×10^{-6}	6.8×10^{-13}	6.4×10^{-13}	1.8×10^{-12}
CoD–38 10980	1.0×10^{-8}	1.0×10^{-8}	1.7×10^{-6}	6.6×10^{-13}	2.2×10^{-13}	6.6×10^{-13}
BPM93487	1.0×10^{-8}	1.0×10^{-8}	1.7×10^{-6}	9.1×10^{-13}	4.7×10^{-13}	1.6×10^{-12}
REJ2009–605	1.3×10^{-5}	1.0×10^{-8}	2.4×10^{-5}	1.7×10^{-13}	1.1×10^{-13}	2.8×10^{-13}
REJ2156–546	6.5×10^{-7}	1.0×10^{-8}	1.9×10^{-6}	3.2×10^{-13}	2.8×10^{-13}	8.5×10^{-13}
REJ2324–547	2.0×10^{-5}	1.7×10^{-5}	2.5×10^{-5}	1.6×10^{-13}	1.3×10^{-13}	1.8×10^{-13}

[a] $\times 10^{-8}$ is the lower limit of the homogeneous grid.
[b] When value and upper limit are equal, the value lies at the limit of the grid.

MacDonald 1985; Koester and Schönberner 1986). With such a thick layer, it is possible to envisage that the amount of helium which reaches the regions where the EUV lines and continuum are formed might be lower than expected because the boundary between H and He layers is deeper than in the Vennes *et al.* (1988) calculations. Further theoretical work is needed to test this proposition directly. A wider survey of white dwarfs with more or less pure H envelopes has been carried out by Barstow *et al.* (1997). Table 8.3 lists both best fit He abundances and corresponding H layer masses. None of these measurements represent real detections of He and can only be interpreted as upper limits. It should be noted that non-LTE model calculations will increase the limiting He abundance or H layer mass by an order of magnitude.

8.3.2 *White dwarfs in binaries*

While the intense search for photospheric helium in isolated white dwarfs has been rather fruitless, almost unlooked for, helium was detected in the atmosphere of the ultra-massive DA white dwarf GD50 (Vennes *et al.* 1996a). GD50 is one of the most massive ($1.2M_\odot$; Bergeron *et al.* 1991) of its class. Previous analyses of soft X-ray observations showed that for reasonable upper limits to the ISM column density ($n_H < 10^{19}$ cm^{-2}) the photosphere must contain trace heavy elements (Kahn *et al.* 1984; Vennes and Fontaine 1992). This was confirmed using the ROSAT soft X-ray/EUV survey data by Barstow *et al.* (1993b), who obtained a helium abundance in the range $(1–3) \times 10^{-4}$, if the photospheric opacity was entirely attributed to this element. However, the ROSAT data could not rule out contributions from other species. Furthermore, diffusion theory in high gravity atmospheres predicted no detectable traces of helium in such massive stars. For example, Vennes *et al.* (1988) concluded that at $T_{\text{eff}} < 40\,000$ K and $M > 0.8M_\odot$, the He abundance should not exceed 10^{-8}. Chayer *et al.* (1995a) predict similarly low abundances for heavier elements.

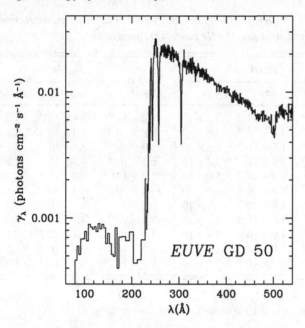

Fig. 8.4. EUV spectrum of GD50 showing the He II ground state Lyman line series and the He I 504 Å photoionisation edge (from Vennes *et al.* 1996a).

The detection of GD50 in the longest wavelength bandpasses of the EUVE sky survey (Bowyer *et al.* 1994) clearly indicated that the interstellar column density must be very low and that most of the observed EUV opacity must reside in the star. To investigate the nature of this opacity, GD50 was observed using the EUVE spectrometers by Vennes *et al.* (1996a). Their spectrum (figure 8.4) shows the prominent ground state He II Lyman line series together with the He I photoionisation edge from the ISM. Fixing the surface gravity of the star at $\log g = 9.0$, Vennes *et al.* obtain $T_{\text{eff}} = 40\,300 \pm 100$ K and He/H $= 2.4 \pm 0.1 \times 10^{-4}$, which is consistent with the ROSAT photometric results. However, although the continuum shape of the spectrum is correctly predicted, some of the detailed characteristics are not reproduced. Most importantly, the predicted line cores of all the He II lines, except the first (shortest wavelength) members of the series, are sharper than observed. Vennes *et al.* make the interesting point that rotational broadening of the model with a velocity of $v \sin i = 1\,000$ km s^{-1} gives much better agreement (see figure 8.5).

The presence of a large abundance of helium in GD50 is a paradox. Vennes *et al.* rule out the possibility of a faint red companion, which might be a source of accreted helium down to a spectral type later than dM7 to dM8. Accretion from the ISM is also ruled out, given the very low average H I density (≈ 0.01 cm^{-3}) along the GD50 line-of-sight. Massive white dwarfs like GD50 are often considered to be the products of a stellar merger. The most plausible explanation of the helium is that fast rotation, arising from such an event, may induce a large meridional current, dredging up He from the envelope, if the H layer is sufficiently thin.

An interesting group of objects that could make a useful comparison with GD50 are the three DAO+dM binaries detected in the EUV surveys – REJ0720–318, REJ1016–053 and REJ2013+400 – where He has already been spectroscopically detected in the visible waveband. In principle, the signature of this He should also be visible in the EUV but REJ2013

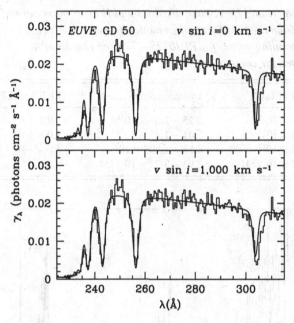

Fig. 8.5. EUV spectrum of GD50 in the region of the He II Lyman line series compared with a model having no rotational broadening (upper panel) and a model including broadening corresponding to $v \sin i = 1\,000\,\mathrm{km\,s^{-1}}$ (lower panel) (from Vennes *et al.* 1996a).

is too faint to be observed by EUVE at all and the spectrum of REJ1016 is cut off by interstellar absorption in the region of the He II Lyman series (see Vennes *et al.* 1997b). Fortunately, the He II region is accessible in the EUVE spectrum of REJ0720 and does reveal the signature of photospheric material (figures 8.6 and 8.7). In addition, a distinct sharp He II 228 Å edge can be seen, which must arise from either interstellar or circumstellar material. However, the most striking feature is a saturated He I absorption edge at 504 Å. Nothing like this has been seen in any other EUVE white dwarf spectra (e.g. Dupuis *et al.* 1995; Barstow *et al.* 1997). Comparison of the data with a homogeneous model spectrum can produce a good match to the data, including the He I edge, down to the He II series limit. However, the predicted short wavelength continuum does not match the observed flux (figure 8.6; Burleigh, Barstow and Dobbie 1996). On the other hand, a stratified configuration gives extremely good agreement (figure 8.7; table 8.4.).

Two important results emerge from this analysis of REJ0720–318. First, while the optical He II line can only be matched by a homogeneous H+He model spectrum the EUV spectrum cannot and is best reproduced with a stratified configuration. The implied mass of the H layer is $3 \times 10^{-14} M_\odot$. Secondly, compared to the H I column, the line-of-sight interstellar He I density is unusually high. The implied hydrogen ionisation fraction is 90%, if all the material is attributed to the ISM and a cosmic He/H ratio of 0.1 is assumed. This peculiarity is emphasised by Dupuis *et al.* (1997), who compare the object with the apparently normal DA white dwarf REJ0723–277 which lies along a nearby line-of-sight. Despite the small angular separation (4°) the column densities are strikingly dissimilar, although the reported H and He ionisation fractions (65% and 62% respectively) are both different from typical values for the

Table 8.4. *Interstellar H and He column densities and H layer mass derived for REJ0720–318 from the best fit stratified model (from Burleigh et al. 1996). The corresponding values for REJ0723–277 from Dupuis et al. (1997) are also listed for comparison.*

Parameter	Best fit value	1σ uncertainty	REJ0723–277
$n_{\rm H}$	2.30×10^{18}	$2.25 - 2.36 \times 10^{18}$	9×10^{17}
$N_{\rm He\,I}$	1.44×10^{18}	$1.41 - 1.47 \times 10^{18}$	1.0×10^{17}
$N_{\rm He\,II}$	9.78×10^{17}	$8.58 - 10.03 \times 10^{17}$	1.6×10^{17}
$M_{\rm H}$	$3.07 \times 10^{-14} M_\odot$	$2.93 - 3.21 \times 10^{-14} M_\odot$	–

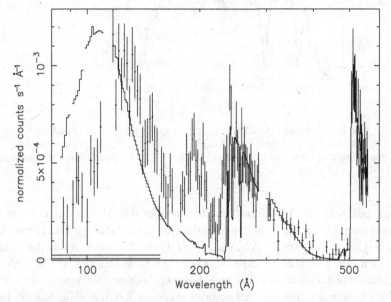

Fig. 8.6. EUV spectrum of REJ0720–318 with the best fitting homogeneous model (every second data point has been removed for clarity; from Burleigh *et al.* 1996).

ISM (see Barstow *et al.* 1997). Interestingly, Dupuis *et al.* are able to match the observed EUV spectrum of REJ0720 with a homogeneous H+He structure (He/H = 3.2×10^{-5}), provided heavier elements (e.g. C, N, O, Si, S and Fe) are included in the model at a level of $\approx 1/100$ of the solar mixture. This indicates that the stratified interpretation proposed by Burleigh *et al.* is not necessarily unique. A detailed comparison of the quality of agreement between both models and the data might establish whether or not one is better than the other. However, visual inspection of the results presented by both groups of authors suggests that the stratified model gives rather better agreement with the detailed shape of the short wavelength flux distribution than the homogeneous H+He/heavy element model.

The He I/H I ratio towards REJ0720 (≈ 1) is the highest measured in any direction in the sky from EUVE data. Taking this together with the large implied H ionisation fraction forces us to question whether the absorbing gas resides in the local ISM or in the close vicinity of the binary system. It is possible that most of the helium is associated with the

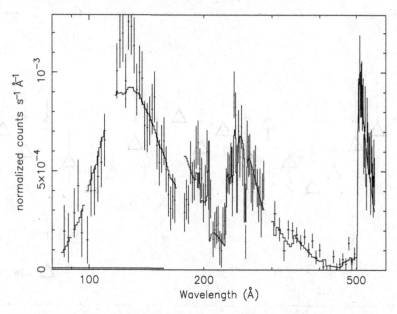

Fig. 8.7. EUV spectrum of REJ0720–318 with the best fitting stratified model (every second data point has been removed for clarity; from Burleigh *et al.* 1997).

REJ0720 system in the form of circumbinary gas. Terman and Taam (1996) suggest that circumbinary discs are likely to form in post common-envelope systems. The spiral-in process decelerates so rapidly in the final stages of common-envelope evolution that material in the immediate vicinity of the binary cores is not expected to be ejected from the system. Instead, this forms a disc in the orbital plane of the binary. Since the radial velocity and mass function data suggest that the orbital inclination of this object may be as high as 85° (Barstow *et al.* 1995c), it is possible that the white dwarf is viewed through an optically thin disc.

The stratified He/H interpretation of REJ0720–318 was called into question by Dobbie *et al.* (1999), who considered a more complex structure including heavy elements, since photospheric C IV is clearly detected in the GHRS spectra from the HST. It was also apparent from EUV photometry that the flux is variable on a period of 0.463 ± 0.004 d (figure 8.8), interpreted as being due to the rotation of the white dwarf. In the absence of this information, the work of Burleigh *et al.* was carried out on a time averaged spectrum and did not include the effect of carbon opacity.

Dobbie *et al.* find that a simple H+He+C non-LTE model is unable to explain the observed EUV spectrum, irrespective of whether or not the He is treated in a homogeneous or stratified configuration (note that Burleigh *et al.* used LTE models). Instead, the best match to the average spectrum is achieved with a homogeneous model containing He and C at the levels determined by the UV data together with a mixture of heavier elements (figure 8.9). As these mostly predict lines outside the wavelength ranges covered by the GHRS data or have expected strengths below the sensitivity limits, the relative abundances are somewhat arbitrary. However, it is clear that the Fe abundance is at least a factor of ten lower than that predicted by radiative levitation theory (e.g. Chayer *et al.* 1995b).

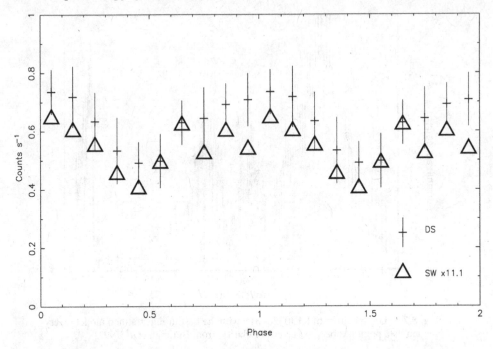

Fig. 8.8. Deep survey and short wavelength spectrometer light curves for REJ0720–318 (taken from Dobbie *et al.* 1999).

Since the EUV flux is variable, Dobbie *et al.* investigated the phase dependence of the spectrum by binning the data by rotation phase. They found that the change in the EUV flux is confined to wavelengths below 240 Å, which is consistent with an apparent change in photospheric opacity levels. The non-uniform distribution of helium and heavier elements suggests that the material is most likely being accreted from the wind of the companion star. The accretion rate predicted by the red dwarf mass-loss rate and relative wind velocity ($\approx 1.1 \times 10^{6} M_{\odot}/y^{-1}$) is somewhat larger than implied by the observed abundances, which Dobbie *et al.* interpret as due to selective mass-loss by a weak white dwarf wind. They also suggest that the anomalous $N_{H\,I}/N_{He\,I}$ ratio is due to the location of the system in an extended region of ionised gas lying in the direction of the CMa ISM tunnel, rather than being caused by a circumbinary disc.

8.3.3   *Comments on the role of He in hot DA white dwarfs*
As outlined in the previous sections, a theme of many EUV spectroscopic studies of hot DA stars has been the search for evidence of photospheric helium as a way of studying the division of white dwarfs into H-rich and He-rich groups. The presence of trace He in otherwise H-rich objects would be an important indication that at least some DA white dwarfs evolve from the DOs by upward diffusion of residual H. However, in most cases there has been no compelling evidence for the presence of photospheric helium and it has only been possible to measure upper limits for the He abundance, if homogeneous H+He models were used, or place lower limits on the H layer mass with stratified models (Barstow *et al.* 1994b, 1994c, 1995b). In only two H-rich objects observed by EUVE has photospheric He

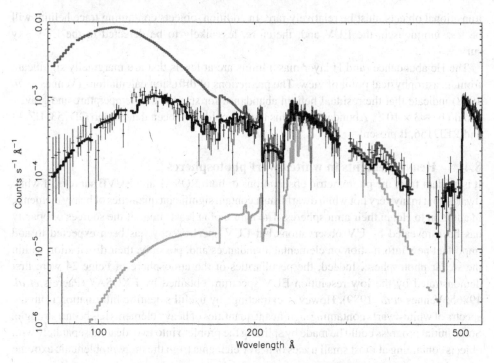

Fig. 8.9. The non-LTE model fits to the EUV spectrum of REJ0720–318: (a) the homogeneous H+He+C (dark grey), log (He/H) = −3.8, log (C/H) = −5.5; (b) the stratified H+He+C (light grey), log (H-layer) = −14.15, log (C/H) = −5.6; (c) homogeneous model incorporating additional heavy elements (black).

been detected, but both stars are unusual. The EUV spectrum of GD50 (Vennes *et al.* 1996a) shows strong He II absorption, which can be accounted for by including a homogeneous He/H abundance of 10^{-4} in the model atmosphere. GD50 is one of the most massive white dwarfs known and possibly the result of a binary merger. The presence of He may be related to its earlier evolution. Similarly, the DAO white dwarf REJ0720–318, in which He had already been identified spectroscopically with optical observations, shows absorption from photospheric He II (Burleigh *et al.* 1996; Dobbie *et al.* 1999). This star is part of a pre-cataclysmic binary system, having a dM companion. As in GD50, the presence of the He is probably a result of past evolution, perhaps from the common-envelope phase, or may be due to more recent accretion episodes from the M dwarf wind. The picture emerging from this work is one where all the isolated DA white dwarfs appear to be devoid of photospheric helium. The published results confirm this and enlarge the sample of observations by a factor ≈3–4.

Of particular importance are the group of stars with temperatures in the range 40 000–50 000 K, which include REJ0715, REJ1032, HZ 43, REJ2009, REJ2156 and REJ2324. These objects lie near the 45 000 K upper temperature limit of the DO–DB gap. If they had recently evolved from the DO population then traces of He might still be visible. Clearly, no such evidence is present and this work poses yet another difficulty when trying to explain the gap in the He-rich cooling sequence. One reason may be a statistical one, since there are approximately seven times as many DA white dwarfs as DOs and transitional objects must be relatively rare. In addition, objects containing trace helium will be less luminous in the EUV and, therefore, less likely to be detected in the EUV sky surveys.

The He abundance (and H layer mass) limits are at levels that are marginally significant from an astrophysical point of view. The predictions of diffusion calculations (Vennes *et al.* 1988) indicate that the residual helium abundance for stars of this temperature and gravity should be $\approx 3 \times 10^{-6}$. Abundances at this level should have been detected in GD153, HZ 43 and REJ2156, if present.

8.4   Heavy elements in white dwarf photospheres

It is evident from the photometric observations of both ROSAT and EUVE surveys of white dwarfs, that many very hot white dwarfs must contain significant quantities of heavy elements, in addition to He, in their atmospheres. The nature of at least some of the sources of opacity has been revealed by UV observations but EUV spectroscopy has been expected to add important new information on elemental abundances and, possibly, their distribution within the stellar photosphere. Indeed, the peculiarities of the atmosphere of Feige 24 were first demonstrated by the low resolution EUV spectrum obtained by EXOSAT (Paerels *et al.* 1986b; Vennes *et al.* 1989). However, extracting any useful scientific information from the spectra of white dwarfs containing significant quantities of heavy elements has been a struggle. Some initial progress could be made by splitting the problem into two, dealing separately with objects containing at most small traces of heavy elements from the more problematic extreme cases.

8.4.1   *Limits on heavy element abundances in 'pure H' white dwarfs*

Where only small traces of heavy element opacity must be considered, theoretical model atmosphere calculations are perturbed only slightly from the pure H structure. As a consequence, the synthetic spectra can be calculated using a pure H model with a sprinkling of heavy elements added only when computing the radiation transfer, to place limits on the possible quantities of heavy elements in objects such as HZ 43. For example, Barstow *et al.* (1996b) compare abundance limits for C, N, O, Si and Fe, determined from EUVE and IUE high dispersion spectra of HZ 43 ($T_{eff} = 49\,000$ K) and GD659 ($T_{eff} = 35\,300$ K; table 8.5). The strengths of EUV absorption lines were calculated for a range of abundances, using the TLUSTY and SYNSPEC non-LTE codes (Hubeny 1988; Hubeny and Lanz 1992, 1995), and compared with the data. Any lines with a depth approximately twice the amplitude of the signal-to-noise of the spectra should have been readily detected and the upper limits were defined as the abundance at which the strongest features just reach this threshold. While IUE or HST data provide the best constraints on C and Si, the EUV data are much more sensitive to N and O. At the effective temperatures in question, N IV and O IV are the most populated ionisation states and most of their bound–bound transitions are found in the EUV rather than the far-UV.

Theories considering the effects of radiative levitation and diffusion make specific predictions about the expected heavy element content in hot DA white dwarfs as a function of both T_{eff} and surface gravity (e.g. Chayer *et al.* 1994, 1995a, 1995b). As the predicted abundances increase towards higher temperature it becomes increasingly important to carry out the spectral modelling with completely self-consistent calculations of the atmospheric structure.

Table 8.5. *Limits on heavy element abundances in 'pure H' DA white dwarfs.*

	HZ 43		GD659	
Element abundance	IUE high res.	EUVE	IUE high res.	EUVE
log C/H	−8.5	−5.5	−7.7	−4.6
log N/H	−5.0	−6.0	−4.2	−5.4
log O/H	−6.0	−6.5	–	−6.0
log Si/H	−8.0	−6.0	−9.0	−7.0
log Fe/H	−4.5	−4.5	–	–

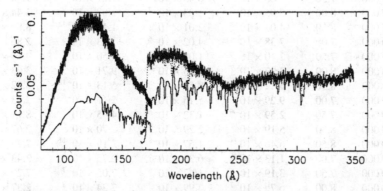

Fig. 8.10. A comparison of the EUV spectrum of HZ 43 with model spectra including heavy element abundances of $1\times$ (solid histogram) and $0.1\times$ (dotted histogram) the values predicted by the radiative levitation theory.

Barstow *et al.* (1997) applied the TLUSTY non-LTE code to this problem, considering a temperature range from 35 000 K, above which the predicted heavy element abundances are expected to begin to have a significant effect on the photospheric opacity, extending up to 55 000 K, the point at which Fe absorption begins to dominate. The heavy element abundances incorporated in the calculations were taken from the predictions of Chayer *et al.* (1995a; see table 8.6). In addition, calculations were carried out for 0.5, 0.1, 0.05 and 0.01 times these abundances, maintaining the relative fractions of each element constant and equal to the predicted ratios.

In their analysis Barstow *et al.* (1997) fixed T_{eff} and log g within the range of their optically determined values, allowing the heavy element abundance parameter (the fractional scaling of the predictions of Chayer *et al.* (1995a)) and the ISM column densities to vary freely. The heavy element abundance in the best fit models is constrained by both the strength of any lines present and the combined effect of all the opacity sources on the shape of the continuum flux. Figure 8.10 illustrates this, comparing two models with abundances of 0.1 (dotted histogram) to 1.0 (solid histogram) times the predicted values with the pure H spectrum of HZ 43. In fact the best fit to the HZ 43 data has an abundance 1/14 of the level expected by the radiative levitation theory. Results for HZ 43 and six other stars are summarised in

Table 8.6. *C, N, O and Si abundances of the non-LTE model grid.*

T_{eff}	log g	log C:H	log N:H	log O:H	log Si:H
35 000	7.00	4.24×10^{-6}	2.05×10^{-6}	6.46×10^{-7}	3.52×10^{-6}
35 000	7.50	1.07×10^{-6}	4.89×10^{-7}	4.45×10^{-8}	1.11×10^{-6}
35 000	8.00	1.90×10^{-7}	7.41×10^{-8}	6.17×10^{-11}	3.54×10^{-7}
35 000	8.50	2.34×10^{-8}	9.33×10^{-11}	6.17×10^{-14}	8.93×10^{-8}
36 000	7.00	4.98×10^{-6}	2.47×10^{-6}	8.99×10^{-7}	3.37×10^{-6}
36 000	7.50	1.28×10^{-6}	6.47×10^{-7}	1.53×10^{-7}	1.07×10^{-6}
36 000	8.00	2.41×10^{-7}	9.93×10^{-8}	4.40×10^{-10}	3.54×10^{-7}
36 000	8.50	3.17×10^{-8}	4.72×10^{-10}	2.69×10^{-13}	8.93×10^{-8}
37 000	7.00	5.86×10^{-6}	3.14×10^{-6}	1.22×10^{-6}	3.10×10^{-6}
37 000	7.50	1.63×10^{-6}	8.45×10^{-7}	2.21×10^{-7}	9.98×10^{-7}
37 000	8.00	3.05×10^{-7}	1.33×10^{-7}	1.89×10^{-9}	3.44×10^{-7}
37 000	8.50	4.04×10^{-8}	2.01×10^{-9}	1.17×10^{-12}	8.89×10^{-8}
38 000	7.00	7.38×10^{-6}	4.02×10^{-6}	1.63×10^{-6}	2.86×10^{-6}
38 000	7.50	2.10×10^{-6}	1.08×10^{-6}	3.50×10^{-7}	9.14×10^{-7}
38 000	8.00	3.97×10^{-7}	1.75×10^{-7}	9.71×10^{-9}	3.03×10^{-7}
38 000	8.50	5.12×10^{-8}	8.57×10^{-9}	5.13×10^{-12}	8.02×10^{-8}
39 000	7.00	9.29×10^{-6}	5.15×10^{-6}	2.11×10^{-6}	2.63×10^{-6}
39 000	7.50	2.59×10^{-6}	1.32×10^{-6}	4.95×10^{-7}	8.11×10^{-7}
39 000	8.00	5.19×10^{-7}	2.29×10^{-7}	2.70×10^{-8}	2.67×10^{-7}
39 000	8.50	6.46×10^{-8}	1.57×10^{-8}	2.19×10^{-11}	7.29×10^{-8}
40 000	7.00	1.13×10^{-5}	6.38×10^{-6}	2.75×10^{-6}	2.43×10^{-6}
40 000	7.50	3.19×10^{-6}	1.67×10^{-6}	7.05×10^{-7}	7.29×10^{-7}
40 000	8.00	6.76×10^{-7}	2.99×10^{-7}	7.24×10^{-8}	2.37×10^{-7}
40 000	8.50	8.00×10^{-8}	2.50×10^{-8}	9.55×10^{-11}	6.78×10^{-8}
41 000	7.00	1.33×10^{-5}	7.83×10^{-6}	3.58×10^{-6}	2.08×10^{-6}
41 000	7.50	3.78×10^{-6}	2.10×10^{-6}	9.91×10^{-7}	6.62×10^{-7}
41 000	8.00	8.45×10^{-7}	4.09×10^{-7}	1.24×10^{-7}	2.13×10^{-7}
41 000	8.50	1.02×10^{-7}	3.82×10^{-8}	6.03×10^{-10}	5.83×10^{-8}
42 000	7.00	1.53×10^{-5}	9.59×10^{-6}	4.57×10^{-6}	1.75×10^{-6}
42 000	7.50	4.40×10^{-6}	2.58×10^{-6}	1.36×10^{-6}	6.03×10^{-7}
42 000	8.00	1.04×10^{-6}	5.61×10^{-7}	1.93×10^{-7}	1.93×10^{-7}
42 000	8.50	1.41×10^{-7}	5.73×10^{-8}	3.01×10^{-9}	4.69×10^{-8}
43 000	7.00	1.67×10^{-5}	1.17×10^{-5}	5.82×10^{-6}	1.46×10^{-6}
43 000	7.50	5.12×10^{-6}	3.16×10^{-6}	1.66×10^{-6}	5.31×10^{-7}
43 000	8.00	1.26×10^{-6}	7.76×10^{-7}	2.77×10^{-7}	1.63×10^{-7}
43 000	8.50	1.80×10^{-7}	8.49×10^{-8}	1.34×10^{-8}	3.53×10^{-8}
44 000	7.00	1.82×10^{-5}	1.41×10^{-5}	7.40×10^{-6}	1.23×10^{-6}
44 000	7.50	5.61×10^{-6}	3.96×10^{-6}	2.07×10^{-6}	4.51×10^{-7}
44 000	8.00	1.44×10^{-6}	1.01×10^{-6}	3.91×10^{-7}	1.33×10^{-7}
44 000	8.50	2.32×10^{-7}	1.21×10^{-7}	3.35×10^{-8}	2.77×10^{-8}
45 000	7.00	1.98×10^{-5}	1.60×10^{-5}	8.89×10^{-6}	1.04×10^{-6}
45 000	7.50	6.14×10^{-6}	4.79×10^{-6}	2.53×10^{-6}	3.84×10^{-7}
45 000	8.00	1.64×10^{-6}	1.24×10^{-6}	5.18×10^{-7}	1.08×10^{-7}
45 000	8.50	2.84×10^{-7}	1.67×10^{-7}	7.03×10^{-8}	2.22×10^{-8}
46 000	7.00	2.14×10^{-5}	1.82×10^{-5}	1.07×10^{-5}	8.77×10^{-7}
46 000	7.50	6.70×10^{-6}	5.32×10^{-6}	3.11×10^{-6}	3.26×10^{-7}
46 000	8.00	1.88×10^{-6}	1.45×10^{-6}	6.71×10^{-7}	8.75×10^{-8}
46 000	8.50	3.32×10^{-7}	2.23×10^{-7}	1.02×10^{-7}	1.79×10^{-8}

Table 8.6. (*cont.*)

T_{eff}	log g	log C:H	log N:H	log O:H	log Si:H
47 000	7.00	2.18×10^{-5}	2.06×10^{-5}	1.28×10^{-5}	7.41×10^{-7}
47 000	7.50	7.08×10^{-6}	6.01×10^{-6}	3.74×10^{-6}	2.77×10^{-7}
47 000	8.00	2.05×10^{-6}	1.69×10^{-6}	8.71×10^{-7}	7.00×10^{-8}
47 000	8.50	3.71×10^{-7}	2.96×10^{-7}	1.38×10^{-7}	1.40×10^{-8}
48 000	7.00	2.23×10^{-5}	2.28×10^{-5}	1.52×10^{-5}	6.28×10^{-7}
48 000	7.50	7.48×10^{-6}	6.87×10^{-6}	4.37×10^{-6}	2.36×10^{-7}
48 000	8.00	2.17×10^{-6}	1.99×10^{-6}	1.06×10^{-6}	5.50×10^{-8}
48 000	8.50	4.14×10^{-7}	3.86×10^{-7}	1.82×10^{-7}	1.06×10^{-8}
49 000	7.00	2.27×10^{-5}	2.40×10^{-5}	1.75×10^{-5}	5.31×10^{-7}
49 000	7.50	7.89×10^{-6}	7.74×10^{-6}	5.09×10^{-6}	1.79×10^{-7}
49 000	8.00	2.31×10^{-6}	2.34×10^{-6}	1.27×10^{-6}	4.32×10^{-8}
49 000	8.50	4.61×10^{-7}	4.54×10^{-7}	2.25×10^{-7}	7.91×10^{-9}
50 000	7.00	2.32×10^{-5}	2.53×10^{-5}	2.00×10^{-5}	4.38×10^{-7}
50 000	7.50	8.30×10^{-6}	8.69×10^{-6}	5.98×10^{-6}	1.35×10^{-7}
50 000	8.00	2.44×10^{-6}	2.65×10^{-6}	1.52×10^{-6}	3.38×10^{-8}
50 000	8.50	5.11×10^{-7}	5.35×10^{-7}	2.79×10^{-7}	3.08×10^{-9}
51 000	7.00	2.39×10^{-5}	2.70×10^{-5}	2.29×10^{-5}	3.57×10^{-7}
51 000	7.50	8.59×10^{-6}	9.40×10^{-6}	7.33×10^{-6}	9.84×10^{-8}
51 000	8.00	2.58×10^{-6}	3.00×10^{-6}	1.82×10^{-6}	2.63×10^{-8}
51 000	8.50	5.51×10^{-7}	6.31×10^{-7}	3.66×10^{-7}	3.35×10^{-10}
52 000	7.00	2.46×10^{-5}	3.00×10^{-5}	2.62×10^{-5}	2.92×10^{-7}
52 000	7.50	8.87×10^{-6}	1.04×10^{-5}	8.20×10^{-6}	7.33×10^{-8}
52 000	8.00	2.69×10^{-6}	3.40×10^{-6}	2.17×10^{-6}	2.05×10^{-8}
52 000	8.50	5.86×10^{-7}	7.59×10^{-7}	4.97×10^{-7}	1.00×10^{-10}
53 000	7.00	2.54×10^{-5}	3.32×10^{-5}	2.91×10^{-5}	2.29×10^{-7}
53 000	7.50	9.18×10^{-6}	1.16×10^{-5}	9.23×10^{-6}	5.79×10^{-8}
53 000	8.00	2.80×10^{-6}	3.67×10^{-6}	2.58×10^{-6}	1.36×10^{-8}
53 000	8.50	6.25×10^{-7}	9.04×10^{-7}	5.79×10^{-7}	6.61×10^{-11}
54 000	7.00	2.62×10^{-5}	3.68×10^{-5}	3.18×10^{-5}	1.73×10^{-7}
54 000	7.50	9.42×10^{-6}	1.29×10^{-5}	1.05×10^{-5}	4.78×10^{-8}
54 000	8.00	2.91×10^{-6}	3.97×10^{-6}	3.05×10^{-6}	1.02×10^{-8}
54 000	8.50	6.64×10^{-7}	1.05×10^{-6}	6.87×10^{-7}	4.37×10^{-11}
55 000	7.00	2.70×10^{-5}	3.98×10^{-5}	3.47×10^{-5}	1.34×10^{-7}
55 000	7.50	9.53×10^{-6}	1.40×10^{-5}	1.14×10^{-5}	3.82×10^{-8}
55 000	8.00	3.03×10^{-6}	4.29×10^{-6}	3.48×10^{-6}	7.80×10^{-9}
55 000	8.50	7.05×10^{-7}	1.18×10^{-6}	8.49×10^{-7}	2.88×10^{-11}

table 8.7, listing the best fit values and uncertainties for the abundance parameter with the preferred values of T_{eff} and log g.

As discussed earlier, soft X-ray and EUV surveys of white dwarfs have revealed that most DA stars hotter than 40 000 K have significant quantities of heavy elements in their atmospheres. In particular, Marsh *et al.* (1997b), with a large sample of 89 stars, were able to study in detail the temperature range between 40 000 and 50 000 K, a region which was sparsely populated in earlier work. It is clear that, when compared with the predictions of pure H atmospheres, many stars in this temperature range show evidence of significant opacity in the EUV, but at levels lower than found in stars at higher temperatures (>55 000 K). The

Table 8.7. *Limits on heavy element abundances determined from the non-LTE models. The abundance parameter (A) tabulated is a scale factor multiplying the abundances of C, N, O and Si predicted by Chayer et al. (1995a), listed here in table 8.6.*

Star	A Value	-1σ	$+1\sigma$
REJ0715–705	0.11	0.09	0.15
REJ1032+535	0.10	0.06	0.11
GD153	0.10	0.02	0.13
HZ 43	0.07	0.06	0.09
REJ2009–605	0.22	0.01	1.00
REJ2156–546	0.06	0.01	0.07
REJ2324–547	0.20	0.17	0.28

photometric data are unable to yield any further information on the nature of the opacity sources and, in particular, decide whether photospheric or interstellar absorption (or both) is responsible. Furthermore, the abundances of heavy elements, if really present, do not yield UV features with sufficient strength to be detected by IUE. Hence, the acquisition of EUV spectra for several of these stars was an important opportunity to determine the nature of the EUV opacity.

Most of the stars listed in table 8.7 fall in the 40 000–50 000 K range, with GD153 (37 900 K) lying just below. Apart from the ISM components, there are no detectable absorption features which could be attributed to the presence of heavy elements. The upper limits reported then arise from the need to simultaneously keep the predicted line strengths within the noise level of each observation and match the general shape of the EUV continua. The results indicate that significant, although small, quantities of heavy elements could be present in the atmospheres of these stars hidden by the signal-to-noise of these data. For example, applying the scale factor estimated for GD153 in table 8.7 to abundances interpolated from table 8.6 gives approximate individual limits of 3.0×10^{-8}, 2.5×10^{-8}, 7.2×10^{-9} and 1.8×10^{-8} for C/H, N/H, O/H and Si/H respectively. However, it is important to remember that the ratios of C:N:O:Si are fixed at their predicted values, so these numbers should not be taken too literally. There is no correlation between the abundance limits and any physical parameter of the stars, except that the higher limits are associated with those spectra having the worst signal-to-noise. The lowest limit achieved is for HZ 43, which, as the brightest EUV source in this sample, has the best signal-to-noise spectrum.

In the initial analysis of the ROSAT soft X-ray and EUV survey data by Barstow *et al.* (1993b), HZ 43 was a unique object in the 40 000–60 000 K temperature range through having a pure H envelope. However, in the enlarged sample of Marsh *et al.* (1997b) it was clear that several other objects probably had similar composition. Several crucial objects have subsequently been examined spectroscopically (REJ0715, REJ1032, REJ2009, REJ2156 and REJ2324), which appeared to have significant EUV opacity and, therefore, were supposed not to have pure H envelopes. In fact it is clear from the column densities measured from the ISM features (table 7.4, section 7.6.2), the limits on the possible photospheric helium content (section 8.3) and the upper limits to the heavy element abundances, that the dominant absorption is in the ISM and not the photosphere of these stars. Hence, in the EUV, it appears that, like HZ 43, these stars probably have more or less pure H atmospheres. Interestingly, there is growing evidence that some heavy elements may exist in a thin layer near the surface of some of these stars, which does not contribute to any opacity in the EUV. This is discussed further in section 8.4.4.

8.4.2 White dwarfs with intermediate heavy element opacity

While it has been possible to place limits on the heavy element content in the photospheres of stars where the abundances are very small, calculating reliable models for objects where significant quantities of heavy elements are known to exist, from UV observations, has proved to be particularly problematic (see e.g. Barstow *et al.* 1996b). In general, the predicted synthetic EUV spectra generated from models incorporating the abundances measured using UV data failed to reproduce either the detailed shape of the observed spectrum or the absolute flux level. The predicted flux was almost always greater than that observed. A complicating factor is an ambiguity in the exact nature of some heavy element absorption features found in UV spectra; for example, whether or not high ionisation lines are associated with photospheric absorption or circumstellar material. Unless the photospheric velocity is known from independent measurements, these possible sources of absorption cannot be distinguished.

A good illustration of this problem is the DA white dwarf GD659. High dispersion IUE spectra show lines of N V (1239/1243 Å doublet), C IV (1548/1551 Å doublet) and Si IV (1394/1403 Å doublet). These species are too highly ionised to be attributed to the local ISM and, furthermore, are shifted in velocity with respect to the known ISM lines by ≈ 25 km s^{-1} (Vennes *et al.* 1991; Holberg *et al.* 1995). The EUV spectrum of GD659 is entirely consistent with the star having an atmosphere composed mainly of hydrogen. A photospheric velocity obtained by Wegner (1974) is -37 km s^{-1}, implying that the highly ionised features are *redshifted*. From the combined evidence, Holberg *et al.* (1995) concluded that the observed features cannot be photospheric and suggest a circumstellar origin. However, a revision of the stellar radial velocity based on new optical data (Holberg, private communication) shows that the highly ionised features are indeed in the photosphere. To avoid affecting the emergent EUV flux, the material must lie in a thin layer near the surface of the star (see also section 8.4.4).

Vennes *et al.* (1991) drew similar conclusions about the nature of Si lines in the IUE spectrum of GD394. However, in contrast to GD659, the ROSAT survey data gave a clear indication that significant photospheric opacity was present in this object, the coolest DA white dwarf to exhibit such an effect (Barstow *et al.* 1993b). This is confirmed by the EUVE spectrum of the star, since neither stratified nor homogeneous H+He models can match the flux level and shape of the continuum. These data also show that the opacity detected by ROSAT must be attributed to elements heavier than He, as no He II lines are observed (figure 8.11; Barstow *et al.* 1996a).

The anomaly between the supposed origin of the heavy element lines and the EUV observations prompted Barstow *et al.* (1996a) to re-examine the hypothesis that the material was in fact photospheric. Using non-LTE line blanketed models computed with the TLUSTY code (see section 5.5.1), the IUE echelle data were reanalysed together with more recent HST observations, of higher signal-to-noise, covering the regions of the Si III (1293.5–1303.5 Å)

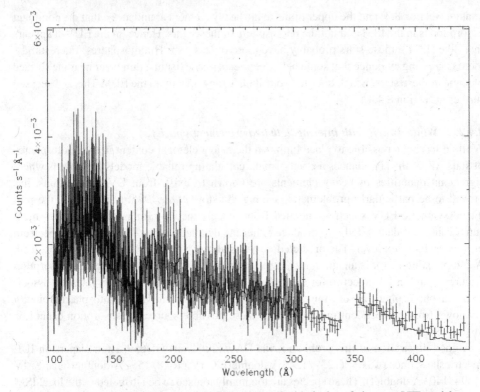

Fig. 8.11. EUVE spectrum of GD394 compared with the best fitting non-LTE model calculation H I $= 4.39 \times 10^{18}$, He I $= 4.24 \times 10^{17}$, He II $= 2.42 \times 10^{17}$, log He/H $= -8$, log C/H $= -7.5$, log N/H $= -5.6$, log O/H $= -5.6$, log Si/H $= -5.6$, $T = 37\,900$ K, log $g = 7.9$.

and Si IV (1390–1406 Å) lines (e.g. figure 8.12). Complete lists of the lines identified in the IUE and HST spectra are found in tables 8.8 and 8.9 respectively.

The results of this work demonstrate that the observed Si III and Si IV lines can be adequately explained by a photospheric model with a Si abundance (relative to H) of 3×10^{-6}. In their earlier analysis, Vennes *et al.* obtained a similar result but went on to exclude the photospheric explanation on the basis of the predicted strength of the Si II, which is not detected, requiring an abundance below 10^{-7} to be consistent with the observations. Examining the predicted strengths of the Si II lines at 1260.4, 1264.7 and 1264.99 Å within the temperature range allowed by the optical data is a sensitive test of the viability of the photospheric model. A Si abundance of 10^{-5}, a factor ≈ 3 higher than measured, would yield Si II line strengths within the noise of the IUE spectrum, supporting the conclusion that the Si III and Si IV features are indeed photospheric. The contradiction between this result and that of Vennes *et al.* probably arises from the use of non-LTE models, which predict weaker Si II lines than do LTE models containing the same Si abundance. The strong evidence that the highly ionised silicon features are photospheric was still contradicted by old measurements of a $+94$ km s^{-1} radial velocity for the photosphere. This problem was solved by new Hα observations which yield a lower velocity of $+17.5$ km s^{-1}, which is marginally consistent with the UV lines (see Barstow *et al.* 1996a).

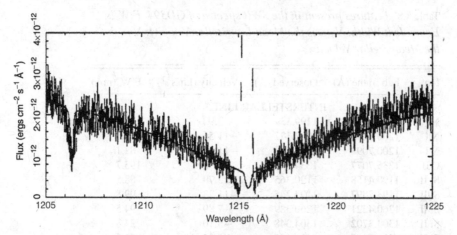

Fig. 8.12. (a) High dispersion HST spectra of GD394, recorded with the GHRS and compared with the best fit photospheric model with log Si/H $= -5.6$, $T = 39\,700$ K and log $g = 7.9$. 1205–1225 Å. Line identifications are listed in table 8.9.

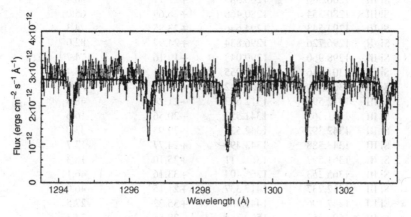

Fig. 8.12. (b) As (a) but for 1293.5–1303.5 Å.

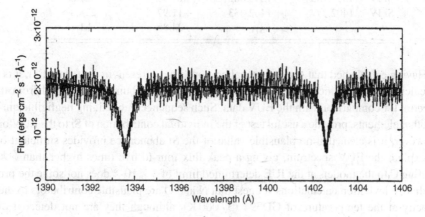

Fig. 8.12. (c) As (a) but for 1390–1406 Å.

Table 8.8. *Features present in the SWP spectra of GD394. E.W. is Equivalent Width; V_{ISM} is the Mean Interstellar Velocity; V_{STL} is the Mean Stellar Velocity.*

Ion	Lab frame (Å)	Observed (Å)	Velocity (km s^{-1})	E.W. (mÅ)
		INTERSTELLAR LINES		
N I	1199.5496	1199.534	−3.90	98.3
N I	1200.2233	1200.177	−11.56	56.7
N I	1200.7098	1200.663	−11.68	82.1
C II	1335.7077	1334.530	−0.52	193.7
Si II	1190.4158	1190.365	−12.79	88.5
Si II	1193.2897	1193.245	−11.23	99.8
Si II	1260.4221	1260.389	−7.89	94.5
Si II	1304.3702	1304.348	−5.10	24.8
Si II	1526.7066	1526.687	−3.85	52.4
		V_{ISM}	−7.61	±1.40
		PHOTOSPHERIC LINES		
Si III	1206.500	1206.589	+22.11	238.4
Si III	1280.354	1280.468	+26.69	65.0
Si III	1294.543	1294.646	+23.85	74.3
Si III	1296.726	1296.834	+24.97	82.0
Si III	1298.960	1299.043	+19.16	174.7
Si III	1301.146	1301.255	+25.11	55.4
Si III	1303.320	1303.442	+28.06	57.0
Si III	1312.590	1312.683	+21.24	48.4
Si III	1341.465	1341.557	+20.56	16.6
Si III	1342.392	1342.526	+29.93	33.0
Si III	1343.388	1343.499	+24.77	17.7
Si III	1361.597	1361.711	+25.10	16.3
Si III	1365.253	1365.404	+33.16	46.1
Si III	1417.237	1417.337	+21.15	46.0
Si III	1447.196	1447.366	+35.22	27.5
Si III	1500.253	1500.384	+28.58	34.3
Si III	1501.150	1501.266	+23.17	26.1
Si IV	1393.755	1393.882	+27.32	427.6
Si IV	1402.777	1402.833	+11.97	228.3
		V_{STL}	+24.85	±1.15

Having established that some photospheric opacity is present in the form of Si, it is useful to calculate models including this element spanning the abundance inferred from the UV spectra, for comparison with the EUV data. Such sequences, containing negligible quantities of other elements, provide a useful test of the individual contribution of Si to the EUV opacity. However, it is clear that no plausible value of the Si abundance provides sufficient opacity to explain the EUV spectrum, giving a peak flux four to five times higher than observed. Adding C to the models at the IUE determined limit of 3×10^{-8} does not solve the problem. Radiative levitation calculations suggest that N and O are plausible contributors to the EUV opacity at the temperature of GD394 (37 000 K), although they are not detected directly in the UV data. In the absence of any observational information on the relative abundance

Table 8.9. *Features present in the GHRS spectra of GD394.*

Ion	Lab frame (Å)	Observed (Å)	Velocity (km s^{-1})	E.W. (mÅ)
		INTERSTELLAR LINES		
N I	1199.5496	1199.540	-2.40	57.7
N I	1200.2233	1200.226	$+0.67$	54.7
N I	1200.7098	1200.684	-6.44	26.4
O I	1302.1685	1302.160	-1.96	90.3
Si II	1304.3702	1304.348	-2.57	27.6
		PHOTOSPHERIC LINES		
Si III	1206.500	1206.633	$+33.05$	253.9
Si III	1207.517	1207.598	$+20.11$	8.8
Si III	1210.496	1210.588	$+22.78$	16.7
Si III	1294.543	1294.674	$+30.34$	66.4
Si III	1296.726	1296.884	$+36.53$	64.5
Si III	1298.960	1299.007	$+10.85$	131.6
Si III	1301.146	1301.290	$+33.18$	63.2
Si III	1303.320	1303.461	$+32.43$	77.9
Si III	1312.590	1312.732	$+32.43$	32.3
Si III	1417.237	1417.396	$+33.63$	38.6
Si IV	1393.755	1393.884	$+27.75$	411.1
Si IV	1402.777	1402.927	$+32.06$	262.9
		V_{STL}	28.76	±7.35

Table 8.10. *Best fit heavy element model to the EUVE data of GD394.*

Parameter	Value	1σ range
H I column	4.39×10^{18}	$(2.05 - 6.55) \times 10^{18}$
He I column	4.24×10^{17}	$(3.54 - 5.86) \times 10^{17}$
He II column	2.42×10^{17}	$(0.95 - 2.81) \times 10^{17}$
log He:H	-8	fixed
log C:H	-7.5	fixed
log N:H	-5.60	-5.59 to -5.61
log O:H	-5.60	-45.59 to -5.61
log Si:H	-5.5	fixed
T_{eff}	37900	$37500 - 38200$

of N and O, Barstow *et al.* set their abundances to be equal and allowed to vary in step, providing a single free compositional parameter to consider in the analysis. Fixing C at the upper limit and Si at the observed abundance, the additional opacity from N and O allows the synthetic spectra to reproduce the EUV data (figure 8.11). The best fit parameters are listed in table 8.10. The nominal abundances of N, O (both equal to 2.5×10^{-6}), C and Si do not give rise to any spectral features strong enough to be detected in the EUVE data. Indeed, limits that can be placed on the abundance of N and O by the spectral features predicted to be most prominent (N IV 225.1/225.2 Å; O IV 212.4 and 222.8 Å) are of a similar magnitude to the values incorporated in the best fit model. The line strengths predicted in the far-UV

(47 mÅ – N V 1240; 50 mÅ – N V 1243; <1 mÅ – O V 1371) are also below detectable limits, although the N V might be seen with a factor of two to three improvement in signal-to-noise. It is interesting to note that the small formal uncertainty quoted for the best fit value of the abundance of N and O is a true reflection of just how sensitive the predicted EUV flux is to the opacity included in the model. However, this is not a true constraint on the abundances, since it depends upon the assumption that N and O are present and that their abundances are equal.

It cannot be claimed that a unique solution exists for the composition of GD394 because, although models containing C, N, and O are consistent with the EUV and UV observations, there is no direct evidence that these elements are actually present in the photosphere, in the absence of any detectable absorption lines or edges which can be attributed to them. The best that can be said is that the abundances included represent upper limits. This does, however, demonstrate that the EUV spectrum of GD394 can be explained entirely by C, N, O and Si at levels consistent with the detection or non-detection of absorption lines. While no other heavy elements that might be present were considered, they are not needed to account for the observations. However, if other elements do contribute, the abundances of C, N, and O required would be reduced. Indeed, Chayer *et al.* (2000) report the detection of P V and Fe III in the FUSE spectrum of the star, which are likely to make a contribution to the EUV opacity.

The most important result from the study of GD394 is the demonstration that the highly ionised silicon features seen in the IUE echelle and HST GHRS spectra arise from photospheric rather than circumstellar material. Furthermore, other elements heavier than H or He must be present in the star to account for the magnitude of the EUV continuum flux but at levels below those which would give rise to detectable absorption features in either far-UV or EUV wavebands. Consequently, GD394 was the first white dwarf in which an opacity dominated EUV spectrum could be successfully matched by a model atmosphere, containing a mix of heavy elements, for which there is established a quantitative link between abundances measured in the UV and the EUV opacity. In contrast, in other instances, such as the hot metal-rich DA G191-B2B, it has proved difficult even with reasonably constrained heavy element abundances to model the EUV region of a DA with strong opacity. At the time of the EUVE study GD394 was by far the coolest DA white dwarf detected in the EUV to contain photospheric heavy elements; all other stars found to have such material were at temperatures well in excess of 50 000 K. However, there is a growing list of exceptions to this statement, e.g. the 20 000 K DAs Wolf 1346 and EG102, which show Si II features near the stellar velocity (Bruhweiler and Kondo 1983; Holberg *et al.* 1997a). However, both these stars are too cool to possess detectable EUV spectra. Recent HST spectra have revealed that the EUV sources REJ1614–085 and REJ1032+532, with $T_{eff} \approx 45\,000$ K, both contain heavy elements in their photospheres (Holberg *et al.* 1997a, 1999a). This is particularly intriguing in the case of REJ1023+532 as, like GD659, its EUV spectrum is well-matched by a pure H model, without any hint of photospheric opacity. The EUV and far-UV observations can only be reconciled if the observed photospheric nitrogen is stratified, residing only in the outermost layers of the envelope (Holberg *et al.* 1999a).

8.4.3   *White dwarfs with extreme heavy element opacities*
It was already clear before the launch of EUVE, from both EUV photometry with the ROSAT WFC and UV spectroscopy with IUE and HST, that the group of hot DA white dwarfs with effective temperatures in excess of 55 000 K all contained significant quantities of heavy

elements in their atmospheres. Species revealed by UV spectroscopy include C, N, O, Si, Al, Fe and Ni (see section 3.6.2 and references therein). More recently, ORFEUS observations of G191–B2B and REJ0457–281 also found evidence of photospheric P and S (Vennes *et al.* 1996b). While heavy elements were readily detected in the most extreme, visually brightest stars (G191–B2B, Feige 24, REJ2214–492 and REJ0623–371), with IUE, other similar but fainter and less heavy element rich objects (e.g. RE0457–281 and REJ2334–471) yielded limited information. Consequently, access to EUV spectra of all these stars was eagerly anticipated as an alternative and complementary means of studying the heavy element content.

The first attempts to match heavy element rich spectra using the same models already used to analyse and intepret the UV data proved to be rather disappointing, giving only poor agreement between synthetic spectrum and data (e.g. Barstow *et al.* 1996b). Some progress was made for objects with less extreme abundances. For example, using LTE models, Jordan *et al.* (1996) were able to match the shape of the EUVE spectrum of PG1234+482, although not the observed flux level. Nevertheless, these results demonstrated the sensitivity of EUV data to Fe abundances ($\approx 2.5 \times 10^{-7}$) significantly below the limits achievable with UV spectroscopy (a few $\times 10^{-6}$). However, with considerable blanketing from Fe group lines contributing to the overall opacity, non-LTE effects were expected to be important. Unfortunately, at first, non-LTE models of stars like G191–B2B were completely unable to match the observations whereas LTE models did yield a reasonable representation (see Koester 1996). The problem seemed to stem from the difficulties of dealing with a large enough number of heavy element lines in non-LTE. First non-LTE calculations (Lanz and Hubeny 1995) typically used the Kurucz line list, considering only the 36 000 lines between observed levels with some 6 000 contributing to the EUV opacity in the range 25 to 600 Å, but could not match the steep fall in the spectrum below 250 Å (dashed line in figure 8.13). Taking all the lines predicted by Kurucz for the higher levels of Fe IV, Fe V and Fe VI increases the total number of lines to about 730 000. About 300 000 of these contribute significantly to the opacity and a new calculation yields a large improvement in the level of agreement between model (thin solid line in figure 8.13) and data (thick solid line).

Lanz *et al.* (1996) provided the first successful non-LTE model of the G191–B2B spectrum by extending the Fe and Ni model atoms substantially. All the levels predicted by Kurucz (1988) were included in the models, totalling over 70 000 individual energy levels grouped into 235 non-LTE superlevels. The effect of over 9.4 million iron and nickel lines was then accounted for, representing a significant improvement over the previous non-LTE line blanketed models for hot white dwarfs (Lanz and Hubeny 1995; Barstow *et al.* 1996b). In the latest work, each transition between superlevels is represented by one or several ODFs (see Hubeny and Lanz 1995), with the line data to generate the necessary Opacity Distribution Functions (ODFs) taken from Kurucz (1988). To construct the ODFs, the lines included were computed with Voigt profiles. An early version of the iron photoionisation cross-sections, calculated by Pradhan and Nahar in the framework of the Iron Project, was used, while approximate hydrogenic cross-sections were employed for nickel photoionisation. Collisional rates were calculated with the general formulae of Seaton (1962) and van Regemorter (1962).

Having established that an estimate of effective temperature, from the Balmer line profiles of G191–B2B, was only weakly sensitive to the exact heavy element abundances incorporated in the non-LTE models, Lanz *et al.* explored the range of model compositions which

Table 8.11. *Abundances (relative to H) contained in the non-LTE model grid.*

Element	Model A	Model B	Model C	Model D (Best fit)
Helium	1.0×10^{-5}	1.0×10^{-5}	1.0×10^{-5}	1.0×10^{-5}
Carbon	5.0×10^{-6}	1.1×10^{-6}	2.0×10^{-6}	2.0×10^{-6}
Nitrogen	8.0×10^{-6}	1.7×10^{-7}	1.6×10^{-6}	1.6×10^{-7}
Oxygen	4.0×10^{-6}	6.8×10^{-7}	1.6×10^{-6}	9.6×10^{-7}
Silicon	3.0×10^{-7}	4.2×10^{-7}	3.0×10^{-7}	3.0×10^{-7}
Iron	1.0×10^{-5}	3.0×10^{-6}	1.0×10^{-5}	1.0×10^{-5}
Nickel	1.0×10^{-6}	1.5×10^{-6}	2.0×10^{-6}	2.0×10^{-6}

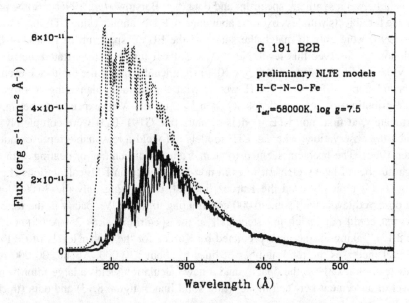

Fig. 8.13. EUVE spectrum of G191–B2B (thick solid line) compared with two non-LTE line blanketed models for $T = 58\,000$ K, log $g = 7.5$, C/H $= 2 \times 10^{-6}$, N/H $= 1.75 \times 10^{-7}$, O/H $= 1.0 \times 10^{-6}$ and Fe/H $= 6.5 \times 10^{-5}$. Dashed line 6 000 lines; thin solid line 300 000 lines. Interstellar absorption is included with H I $= 2.5 \times 10^{18}$ cm^{-2}, He I/H I $= 0.09$ and He II/H I $= 0.01$.

gave the best match to both the far-UV and EUV spectra holding temperature constant. Both spectral ranges were dealt with in an iterative manner. First, an initial model composition was constructed from sources in the literature (e.g. Vennes *et al.* 1992; Holberg *et al.* 1994; Vidal-Madjar *et al.* 1994) and compared with the IUE data. Apart from He, all the elements included (He, C, N, O, Si, Fe and Ni) were detected spectroscopically by IUE. However, contributions from the P or S discovered in ORFEUS observations (Vennes *et al.* 1996b) were not incorporated into the models, because their measured abundances (P/H $= 1 \times 10^{-8}$, S/H $= 1 \times 10^{-7}$) are so low that they are expected to make little contribution to the EUV continuum opacity. Improved estimates of individual element abundances were then obtained by matching the observed line profiles to the predictions of a synthetic spectrum calculated from the temperature and gas density structure of the original model. Table 8.11 summarises

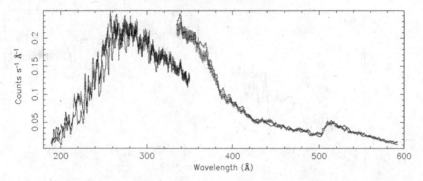

Fig. 8.14. EUV count spectrum of G191–B2B covering the wavelength range 180–600 Å. The data points (error bars) are compared with the predictions of non-LTE model D (histogram, see table 8.11 for abundances) including the effects of interstellar absorption as detailed in the text (H I , He I and He II column densities are 2.1×10^{18}, 1.8×10^{17} and 7.9×10^{17} cm^{-2} respectively). The discontinuity near 320 Å, where the MW and LW ranges overlap arises from differing spectrometer effective area, for which these data are not corrected.

the composition for each of the models used in the work (models A–C). Once agreement was obtained with the far-UV data, a final model (D) was calculated incorporating the abundances derived from the best fit to the IUE spectrum (see table 8.11). The He abundance adopted in the models was purely nominal and would not yield a detectable 1640 Å feature at the signal-to-noise of the IUE data. The upper limit to the amount of He that can be present is $\approx 2 \times 10^{-5}$, corresponding to an equivalent width of 140 mÅ. Uncertainties in the abundance estimates for individual species are approximately a factor of two.

When combined with the interstellar absorption cross-sections of Rumph *et al.* (1994) (modified to take account of the converging series of He lines near He I and He II edges, as described in section 7.6), all the models listed in table 8.11 yield equally good agreement with the data. However, this is only true when a significant quantity of He II is included to suppress the predicted flux below 228 Å, with model D giving the best match (figure 8.14). The quality of the fit to models A and C, which have the same Fe abundance, is not significantly different to that for model D. However, model B (lower Fe abundance) is significantly poorer at the 3σ confidence level indicating that the shape of the predicted spectrum is very sensitive to the amount of Fe included. Indeed, a degradation in the quality of the agreement is seen at the 90% level if the ratio of Fe/H is increased or lowered by only 20%.

In the EUV analysis it was initially assumed that the necessary He II opacity is completely interstellar; the best fit values for the H I , He I and He II column densities were then 2.1×10^{18}, 1.8×10^{17} and 7.9×10^{17} cm^{-2} respectively. Figure 8.15a compares the predictions of model D with the data in the 190–330 Å region, incorporating these interstellar components. If the He II lies along the line-of-sight in the interstellar medium and the amount of He III is assumed to be negligible, the implied ionisation fraction of He is 80% ($\pm 20\%$) and that of H equally large, far greater than anything reliably reported elsewhere (e.g. Holberg *et al.* 1995; Barstow *et al.* 1997; see section 7.6.2). In the absence of any measurement against which to compare these figures, it is difficult to assess their plausibility. Such large ionisation fractions could be avoided if the He II were assumed instead to be circumstellar. At the time of this study there was no evidence from the far-UV of a circumstellar component to the

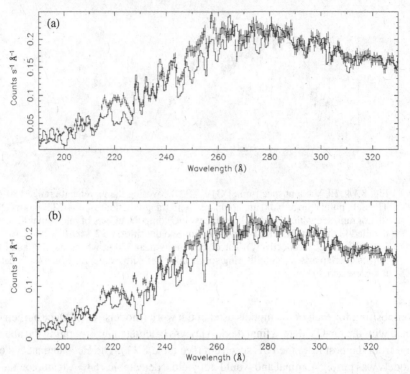

Fig. 8.15. Region of the MW spectrum of G191–B2B including the He II Lyman series. The data points (error bars) are compared with the predictions of non-LTE model D (histogram, see table 8.11 for abundances) including the effects of interstellar absorption as detailed in the text (H I and He I column densities are 2.1×10^{18} and 1.8×10^{17} respectively). (a) shows the best fit non-LTE model D assuming that the necessary He II opacity is completely interstellar, with column a density of 7.9×10^{17} cm^2. (b) The best fit non-LTE model D with no interstellar component and a photospheric He/H abundance of 5.5×10^{-5}.

observed absorption features in G191–B2B. However, observations with the STIS instrument on HST show blueshifted components to the C IV resonance doublet at 1550 Å, which appear to be circumstellar (Bannister *et al.* 2001).

The He II contribution need not be entirely in the line-of-sight material and could reside in the photosphere of the star. Indeed, the many lines of heavy elements present in the EUV spectrum might easily be blended with members of the He II Lyman series, masking the signature of photospheric He. Fixing heavy element abundances at the values determined from the IUE analysis (model D), the He/H ratio was allowed to vary from a nominal 10^{-5} up to 10^{-3}, to test this proposition. The best match in this case (figure 8.15b) requires no interstellar He II and yields a He abundance of 5.5×10^{-5}. In terms of the overall comparison with the EUVE data, it gives a significantly better fit than a model including interstellar He II. However, the predicted strength of the He II 1640 Å (267 mÅ) is above the limit of detection in the IUE spectrum (140 mÅ, He/H $\approx 2 \times 10^{-5}$). If the photospheric abundance of He is fixed at the UV limit, 5.5×10^{17} cm^{-2} of interstellar He II is still required for a good agreement in the EUV.

Placing all the necessary He II opacity in the photosphere we avoid the problem of the high implied interstellar He ionisation, but, with the photospheric He/H content then required, the predicted He II 1640 Å line should be detected. All the models used in this study have made the assumption that the He and heavier elements are distributed homogeneously. Although radiative levitation can account for the presence of the C, N, O, Si, Fe and Ni that are in the atmosphere (but not necessarily the observed abundances), the mechanism cannot prevent He from sinking out under the influence of gravity (Vennes *et al.* 1988). Hence, the photosphere should become stratified with a thin layer of H (containing heavy elements in this case) overlying a predominantly He envelope and the absence of a detectable 1640 Å line could be explained by such an atmosphere.

The most important advance made in the work of Lanz *et al.* (1996) is that, by treating all major absorbers simultaneously and incorporating the blanketing effect of more than nine million Fe and Ni lines in non-LTE, the overall shape and absolute flux level of the EUV spectrum of a heavy element rich white dwarf, like G191–B2B, could be closely matched by the models for the first time. Furthermore, the majority of the observed features are reproduced by the synthetic spectrum. In some cases the predicted depths are not perfect but this might equally be a problem with the detailed shape of the underlying continuum. Indeed, for example, the line shapes and strength of the features seen at 255 and 260 Å mirror the observed lines very accurately apart from a modest depression in the absolute flux level compared with the rest of the data. Almost all the apparently distinct lines are really unresolved blends, with transitions of Fe V the dominant contribution. Interestingly, the line at 246 Å (whose shape is also accurately reproduced by the model) is expected to be a blend of C IV, N IV and O III. Several predicted strong absorption lines are not detected, including features near 200, 220, 250 and 275 Å. The few discrepancies between data and synthetic spectra that are seen are probably associated with deficiencies in the atomic data, particularly for Fe and Ni, incorporated in the models which cannot be addressed until improved calculations are carried out by the Iron Project (see Pradhan 1996). However, to maintain the good match between model and data below the He II 228 Å Lyman limit, it has been necessary to incorporate additional He II opacity, either in the photosphere or as absorption along the line-of-sight. Since there is no independent evidence that this He really exists in any form, we must carefully question this hypothesis. Nevertheless, the possible implications of the existence of absorbing He are important.

Extending the analysis applied to G191–B2B to other stars gives similar results. For example, an explanation of the EUV spectrum of its spectroscopic 'twin', REJ0457–281, also requires a comparatively high He II column density (Barstow *et al.* 1997). Interestingly, although Fe and Ni are not directly detected in the UV spectrum of the star, finite abundances ≈50% of those seen in G191–B2B are required to match the EUV continuum flux. These levels are consistent with the upper limits imposed by the IUE data.

A comprehensive study of a large sample of DA white dwarf EUV spectra has been carried out by Wolff *et al.* (1998), including white dwarf groups having a range of photospheric opacities. They obtained similar results to Lanz *et al.* in a reanalysis of the G191–B2B spectrum and then used G191–B2B as a template, scaling its abundances by a 'metallicity' factor for modelling the other stars. From a visual comparison of the spectra, Wolff *et al.* define four groups of stars, based on their overall shape and implied metallicity (figure 8.16). The first group consists of six stars similar to G191–B2B, while they take GD246 as the

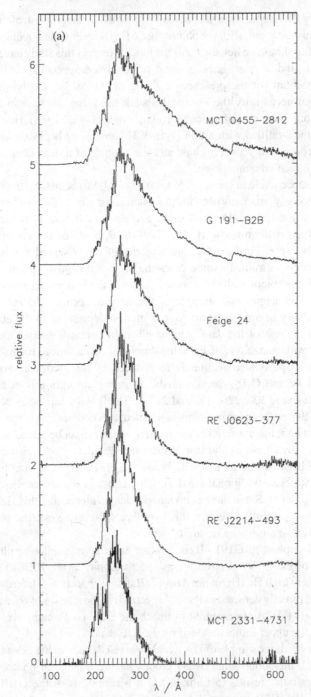

Fig. 8.16. (a) EUVE spectra of the G191–B2B group (from Wolff *et al.* 1998).

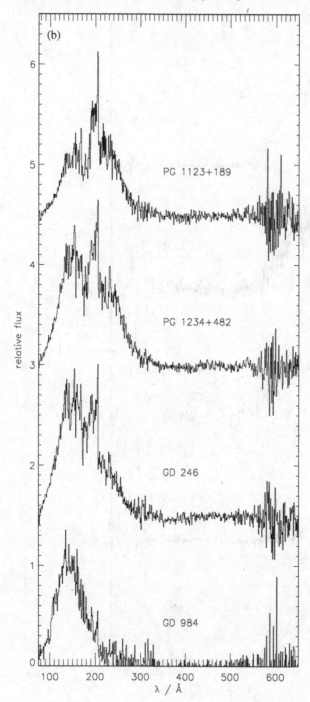

Fig. 8.16. (b) EUVE spectra of the GD246 group (from Wolff *et al.* 1998).

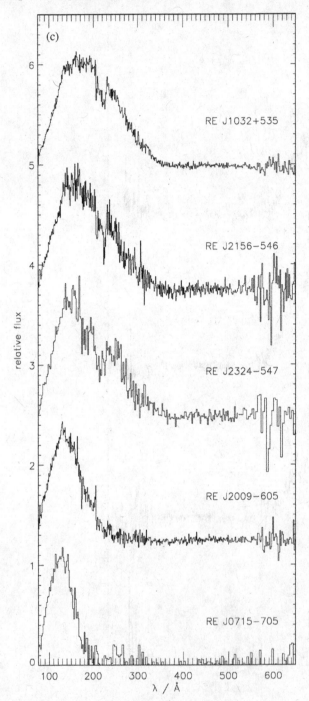

Fig. 8.16. (c) EUVE spectra of pure H atmosphere group (from Wolff *et al.* 1998).

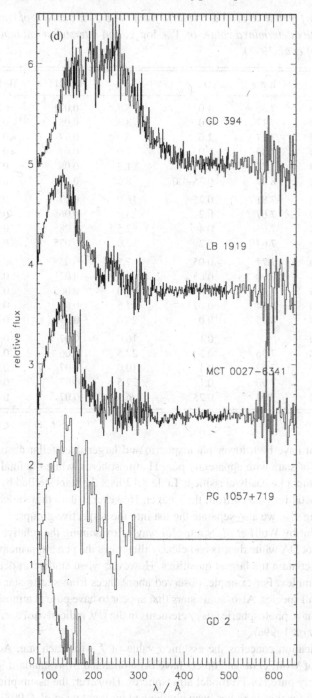

Fig. 8.16. (d) EUVE spectra of miscellaneous objects (from Wolff *et al.* 1998).

Table 8.12. *DA white dwarf metallicity (m) determined from the EUVE analysis of each star. Also listed are the optically determined values of* T_{eff}, log g, *and interstellar column densities (data from Wolff et al. 1998).*

Name	T_{eff} (K)	log g	m	N_{HI}	He I /H I	He II /H I
REJ2214−493	66 000	7.38	4.0	5.8	0.07	<0.05
Feige 24	62 000	7.17	1.0	3.3	0.05	<0.2
REJ0623−377	58 000	7.27	2.0	5.0	0.07	<0.1
G191−B2B	56 000	7.49	1.0	2.0	0.07	<0.2
REJ0455−281	56 000	7.77	1.0	1.3	0.08	<0.15
REJ2331−473	56 000	8.07	0.75−1.0	8.5	<0.07	0.05−0.1
GD246	59 000	7.81	0.25	16.0	0.07	0.025
PG1234+482	56 000	7.67	0.2	13.0	0.065	0.03
PG1123+189	54 000	7.63	0.4	12.5	0.085	0.025
GD984	50 000	7.67	0.2	22.0	0.075	0.03
REJ1032+535	47 000	7.77	<0.05	7.5	0.055	0.045
REJ2324−547	45 000	7.94	0.05	9.5	0.07	0.08
REJ2156−546	44 000	7.91	0.0	7.0	0.065	0.05
REJ2009−605	44 000	8.14	<0.025	17.5	0.07	0.025
REJ0715−703	44 000	8.05	0.0	26.0	0.07	0.02
LB1919	69 000		0.1	16.0	0.07	0.06
REJ0027−634	64 000	7.96	0.2	27.5	0.06	0.02
GD2	49 000	7.63	<0.1	110.0	0.07	0.02
PG1057+719	41 500	7.90	0.1	27.5	0.07	0.0
GD394	39 600	7.94	0.25	6.5	0.07	0.07

prototype of a group that have both lower photospheric and larger interstellar absorption. The third group consists of stars with apparently pure H atmospheres, while the final group comprises objects that cannot be easily classified. Table 8.12 lists the stars studied by Wolff *et al.* and includes the metallicity taken from their paper. However, where they ordered the information by decreasing T_{eff}, we also separate the list into the respective groups.

The analysis carried out by Wolff *et al.* is a useful way of examining the relative heavy element abundances in hot DA white dwarfs and clearly illustrates the general principle that the hottest white dwarfs contain the largest quantities. However, when studied in detail the picture is much more complex. For example, observed abundances from star to star do not scale in a linear way for all species. Also, some stars that appear to have pure H atmospheres in the EUV are shown to have photospheric heavy elements in the UV, indicative of a stratified envelope (e.g. Holberg *et al.* 1999a).

One particular complication concerns the assumed value of T_{eff} for each star. Adopted values used in the work of Wolff *et al.* and others have been obtained from standard Balmer line analyses, using mainly pure H LTE model atmospheres. However, the assumption that such models are adequate for this work has been challenged by Barstow *et al.* (1998). They found that the Balmer line profiles predicted by the synthetic spectra are sensitive to both the assumption of LTE and inclusion of heavy elements in the model calculations. Consequently, LTE Balmer line measurements overestimate T_{eff} by a few thousand degrees. Inclusion of heavy elements at the abundances observed in G191−B2B reduces the measured value further,

Table 8.13. *Values of* T*eff* *and log g for all stars in the hot DA sample, determined from analysis of the Balmer lines using Koester LTE pure H,* TLUSTY *pure H LTE and* TLUSTY *pure H non-LTE models.*

Star	Koester pure H LTE $T_{\text{eff}} \pm 1\sigma(\text{K})$ $\log g \pm 1\sigma$	TLUSTY pure H LTE $T_{\text{eff}} \pm 1\sigma(\text{K})$ $\log g \pm 1\sigma$	TLUSTY pure H non-LTE $T_{\text{eff}} \pm 1\sigma(\text{K})$ $\log g \pm 1\sigma$
Feige 24	62000(+1500/ − 1860) 7.46(+0.11/ − 0.10)	68000[a](+0[a]/ − 1790) 7.35(+0.10/ − 0.12)	64000(+2000/ − 1250) 7.35(0.12/ − 0.13)
REJ0457−281	57450(+2460/ − 1150) 7.92(+0.12/ − 0.10)	62680(+2500/ − 1840) 7.82(+0.12/ − 0.09)	58270(+1670/ − 1890) 7.78(+0.11/ − 0.08)
G191−B2B	58200(+1190/ − 1070) 7.58(+0.07/ − 0.08)	63180(+1460/ − 1110) 7.44(+0.06/ − 0.08)	59160(+1270/ − 1070) 7.36(+0.08/ − 0.07)
REJ0623−371	63190(+2270/ − 2080) 7.29(+0.11/ − 0.13)	68000[a](+0[a]/ − 1360) 7.12(+0.13/ − 0.11)	66880(+1120[a]/ − 2580) 7.00[a](+0.11/ − 0.00[a])
PG1123+189	57400(+1300/ − 1150) 7.71(+0.08/ − 0.07)	62770(+1260/ − 1230) 7.56(+0.06/ − 0.10)	58450(+1120/ − 1050) 7.53(+0.03/ − 0.10)
PG1234+482	57860(+1110/ − 1200) 7.58(+0.07/ − 0.07)	61990(+1270/ − 1170) 7.42(+0.07/ − 0.07)	58930(+970/ − 1220) 7.41(+0.07/ − 0.07)
REJ2214−492	69500(+2520/ − 2540) 7.44(+0.20/ − 0.13)	68000[a](+0[a]/ − 420) 7.40(+0.18/ − 0.14)	68000[a](+0[a]/ − 1060) 7.28(+0.11/ − 0.12)
GD246	58200(+1160/ − 1120) 7.94(+0.08/ − 0.07)	64420(+1150/ − 1700) 7.75(+0.07/ − 0.06)	59280(+980/ − 1340) 7.73(+0.07/ − 0.06)
REJ2334−471	60250(+1660/ − 2160) 7.70(+0.10/ − 0.11)	66020(+1880[a]/ − 2070) 7.57(+0.10/ − 0.10)	60900(+2400/ − 1340) 7.54(+0.10/ − 0.14)

[a] Value or error bar at the limit of the model grid.

by a similar amount. This indicates that earlier measurements of T_{eff} were overestimated by 10–15%. Tables 8.13 and 8.14 compare the values obtained by Holberg *et al.* (1999a) using various models, including LTE pure H calculations, with those from heavy element-rich non-LTE models. Ultimately, it will be necessary to perform a self-consistent analysis to completely characterise each individual star, dealing with the measurement of T_{eff} in conjunction with abundance determinations from EUV or UV spectra.

8.4.4 Heavy element stratification
The presence of photospheric heavy elements in stars such as REJ1032+535 (Holberg *et al.* 1999a), which have apparently pure H atmospheres at EUV wavelengths demonstrates that opacity sources are not necessarily uniformly distributed at all depths within the atmosphere. In the case of REJ1032+535, nitrogen observed in the UV resides in a thin layer overlying the region in which the EUV flux is formed and, therefore, does not contribute to the EUV opacity. Furthermore, the radiation levitation calculations predict that each individual element present should have a different distribution, depending on the cross-sections of the various species.

While the comprehensive analyses of G191–B2B and similar stars made significant progress in understanding hot white dwarfs, some problems remained to be answered. The most obvious was the high He II opacity needed to suppress the flux below 228 Å. Barstow

Table 8.14. *Values of* T_{eff} *and log g for all stars in the hot DA sample, determined from analysis of the Balmer lines using* TLUSTY *non-LTE H+He, weakly blanketed non-LTE and fully blanketed heavy element-rich non-LTE models.*

Star	H+He non-LTE $T_{eff} \pm 1\sigma$ (K) log $g \pm 1\sigma$	Low metallicity non-LTE $T_{eff} \pm 1\sigma$ (K) log $g \pm 1\sigma$	High metallicity non-LTE $T_{eff} \pm 1\sigma$ (K) log $g \pm 1\sigma$
Feige 24	64310(+1710/ − 1440)	64140(+1690/ − 1620)	56370(+1530/ − 760)
	7.34(0.12/ − 0.12)	7.35(+0.12/ − 0.13)	7.36(0.11/ − 0.12)
REJ0457–281	58750(+1700/ − 1590)	57850(+2150/ − 1400)	53640(+630/ − 690)
	7.79(+0.09/ − 0.10)	7.80(+0.10/ − 0.09)	7.80(+0.10/ − 0.11)
G191–B2B	59190(+1410/ − 820)	59060(+1130/ − 1090)	53840(+400/ − 160)
	7.36(+0.07/ − 0.07)	7.36(+0.08/ − 0.07)	7.38(+0.07/ − 0.08)
REJ0623–371	66740(+1260[a]/ − 2370)	66570(+1430[a]/ − 2400)	59740(+1690/ − 1700)
	7.00[a](+0.11/ − 0.00[a])	7.01(+0.11/ − 0.01[a])	7.00[a](+0.11/ − 0.00[a])
PG1123+189	58800(+1010/ − 1070)	58210(+1280/ − 980)	52740(+150/ − 180)
	7.52(+0.07/ − 0.10)	7.53(+0.07/ − 0.09)	7.52(+0.07/ − 0.10)
PG1234+482	59030(+1090/ − 1020)	58640(+1040/ − 1100)	53860(+200/ − 170)
	7.41(+0.07/ − 0.07)	7.42(+0.07/ − 0,07)	7.42(+0.08/ − 0.07)
REJ2214–492	68000[a](+0[a]/ − 1030)	68000[a](+0[a]/ − 1020)	62050(+2290/ − 2150)
	7.28(+0.11/ − 0.13)	7.28(+0.12/ − 0.10)	7.23(+0.12/ − 0.11)
GD246	59300(+1350/ − 830)	59070(+940/ − 1330)	53740(+220/ − 310)
	7.73(+0.06/ − 0.07)	7.74(+0.06/ − 0.07)	7.74(+0.06/ − 0.06)
REJ2334–471	60930(+2490/ − 1200)	61160(+1840/ − 1860)	54630(+860/ − 760)
	7.57(+0.09/ − 0.14)	7.54(+0.10/ − 0.13)	7.58(+0.10/ − 0.10)

[a] Value or error bar at the limit of the model grid.

and Hubeny (1998) investigated a possible solution to this by treating the photospheric helium in a more realistic gravitationally stratified model. This was partially successful, yielding a predicted 1640 Å line strength which was consistent with the UV upper limit but still requiring a high interstellar ionisation fraction.

Another major difficulty which received considerably less attention and was almost ignored is the spectral region below ≈ 180 Å. Significant flux is detected shortward of this but was not dealt with in the Lanz *et al.* (1996) or Wolff *et al.* (1998) studies. However, as Barstow *et al.* (1999) point out, in general the predicted short wavelength flux is between five and ten times that observed. This is illustrated in figure 8.17 which shows the non-LTE model which gives the best match to the data at wavelengths above 180 Å. Any attempt to optimise the fit across the complete wavelength range yields poor overall agreement between model and data.

However, Barstow *et al.* noted that better agreement could be achieved by treating the short wavelength and longer wavelength sections of the spectrum with separate models containing different Fe abundances, prompting them to consider using models with stratified Fe abundances. Several different models were tested, using two or three 'slabs' of differing Fe abundance as well as a crude attempt to test radiative levitation by applying an arbitrary acceleration term to counter downward gravitational diffusion. All other elements heavier than He were treated as homogeneous and their abundances set at the values previously obtained from the far-UV spectra. The best match to the full EUV spectrum and the UV data

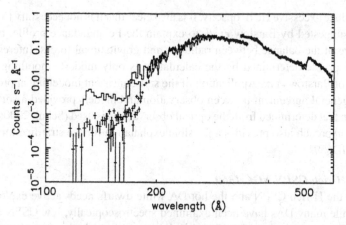

Fig. 8.17. Homogeneous non-LTE atmosphere model which gives the best agreement with the spectrum of G191–B2B at wavelengths above 180 Å (from Barstow *et al.* 1999).

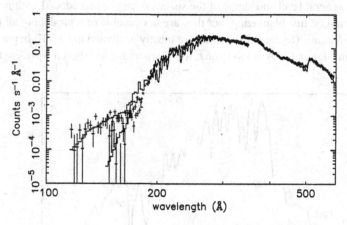

Fig. 8.18. Stratified non-LTE atmosphere model which gives the best agreement with the spectrum of G191–B2B at all wavelengths (from Barstow *et al.* 1999).

was achieved with a model comprising two discrete layers. The Fe abundance (Fe/H) in the upper layer was 10^{-6}, extending to a depth of $6.2 \times 10^{-16} M_\odot$, while that in the deeper layer (down to $8.3 \times 10^{-13} M_\odot$) was 4×10^{-5} (figure 8.18).

While this empirical model of the Fe abundance profile is relatively crude, it does give a good match to the data for the entire spectral range from the short wavelength EUV through to the optical band. This implies that the true Fe abundance profile probably has a sharp step. An important consequence of the stratified approach is that the He II interstellar opacity and, as a result, the implied He ionisation fraction are much more in keeping with the measurements made on other stars (see chapter 7).

Dreizler and Wolff (1999) have improved on the Barstow *et al.* analysis by considering model atmospheres where gravitational settling and radiative levitation are dealt with self-consistently within the calculations. They find that their synthetic spectra can reproduce the complete EUVE spectrum and the Fe lines visible in the UV. They also find that it is not necessary to introduce excessive He II opacity. It is also clear that it is not necessary to invoke weak mass loss, suggested by Barstow *et al.* to explain the Fe abundance profile, but that it is a consequence of the balance between radiative and gravitational forces. Interestingly, the Fe abundance profile determined by the calculation is only modest smoothing of the sharp slab profile of Barstow *et al.* Application of the self-consistent modelling approach to other stars gives general agreement between observation and models provided a somewhat higher gravity than that determined from the optical observations is used (Schuh 2000; Schuh *et al.* 2001). This approach also provides a possible explanation for the stratification found in REJ1032 and GD659.

8.4.5   *The hot H-rich CSPN NGC 1360*

The link between the H-rich CSPN and the hot DA white dwarfs needs active exploration. Unfortunately, while many DAs have been examined spectroscopically, few CSPN are detectable in the EUV and only one, NGC 1360, is bright enough to obtain a spectrum. Hoare *et al.* (1996) report an analysis of this, in conjunction with new optical Balmer line data. Strong continuum absorption edges due to OV are visible, as well as a number of absorption line features. The general level and shape of the spectrum can be reproduced with non-LTE models incorporating Fe line blanketing, but they are not capable of reproducing all the observed absorption features (figure 8.19), which are mostly attributed to Fe VII. In particular, a model absorption edge at 120 Å is too strong, too sharp and at too short a wavelength. The

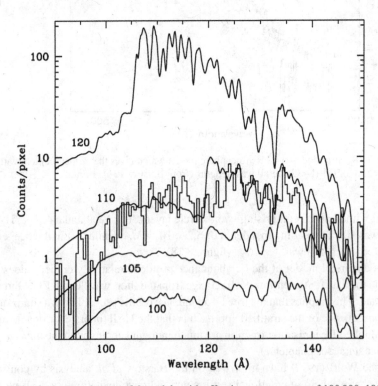

Fig. 8.19. Four log $g = 6.0$ models with effective temperatures of 100 000, 105 000, 110 000 and 120 000 K, compared to the observed spectrum, including attenuation by a neutral hydrogen column of 8.5×10^{19} cm^{-2} (from Hoare *et al.* 1996).

authors propose that this is due to inclusion of an insufficient number of levels in the O V model atom.

As can be seen in figure 8.19, the EUV spectrum does constrain the stellar effective temperature. It also places limits on the allowed surface gravity and line-of-sight column density. The best match to data, with $T_{eff} = 11\,000$ K, $\log g = 6.0$ and $N_{HI} = 8.5 \times 10^{19}$ cm^{-2} is consistent with the range of parameters determined from a Balmer line analysis, assuming a solar value for the He/H abundance.

8.5   Hydrogen-deficient white dwarfs

The majority of hot white dwarfs detected in the EUV waveband are hydrogen-rich DA stars and most of the discussion on white dwarfs in this book has concentrated on these objects. Nevertheless, a handful of hydrogen-deficient objects have been detected in the surveys, including mainly PG1159 type stars (which are also C/O rich) and a single DO white dwarf, REJ0503–289 (Barstow *et al.* 1994d). The ability to interpret the DA white dwarf EUV photometry, from which so much information has been gleaned, has depended on the comparatively simple photospheric composition of their atmospheres. Even where significant quantities of heavy elements are present, the comparison between the fluxes expected for a pure H atmosphere and those observed has been informative, because of the large statistical sample studied. In the case of the DOs and PG1159 stars, such simple comparisons have not been possible. First, the number of objects detected in the EUV is very small, since the DAs outnumber the DO/PG1159s by a factor of about 7:1 and the additional opacity from He (and other elements) suppresses the EUV flux, requiring higher stellar temperatures for significant detections. Second, the physical understanding of their atmospheres is much less advanced. For example, Werner *et al.* (1991) demonstrated emphatically that the composition of the PG1159 star atmospheres is dominated by C and O, not He, and that detailed non-LTE calculations are essential for reliable measurements of composition and effective temperature. Subsequent studies have revealed a far from coherent pattern of abundances in both PG1159 and DO groups of stars (see Werner 1992; Dreizler and Werner 1996).

EUV spectroscopy on the other hand, as has been demonstrated here from the studies of DA white dwarfs, provides much more information than photometry when the targets are bright enough to obtain good signal-to-noise data. A handful of the H-deficient white dwarfs were expected to make suitable targets and some important results have emerged from their study.

8.5.1   The temperature range of the GW Vir instability strip

A significant subset of the PG1159 stars are non-radial mode pulsators, including the proto-type PG1159–035 itself and which also has the variable star designation GW Vir. An important question in all studies of pulsating white dwarfs is the range of temperatures in any group of objects within which pulsationally unstable stars are found, the edges of the 'instability strip'. In addition, there is the problem of why some apparently similar objects pulsate, but others do not. A particular example of this is the apparent 'twin' nature of PG1159–035 and PG1520+525. Werner and Heber (1993) were able to derive a precise value for the effective temperature of PG1159–035 of $140\,000 \pm 5\,000$ K, from the HST FOS spectrum. However, no useable spectrum was recorded for PG1520+525. Instead, Werner *et al.* (1996) used a 155 ks EUVE exposure of the latter star to constrain its effective temperature. Only the short wavelength section of the resulting spectrum (figure 8.20) is useful, due to the high interstellar column density. If the star had a similar temperature to PG1159–035 a prominent O V edge

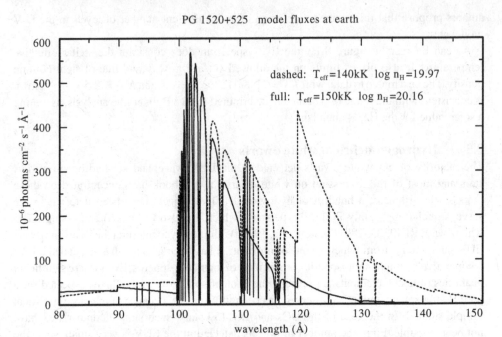

PG 1520+525   model fluxes at earth

dashed: T_{eff}=140kK log n_H=19.97

full: T_{eff}=150kK log n_H=20.16

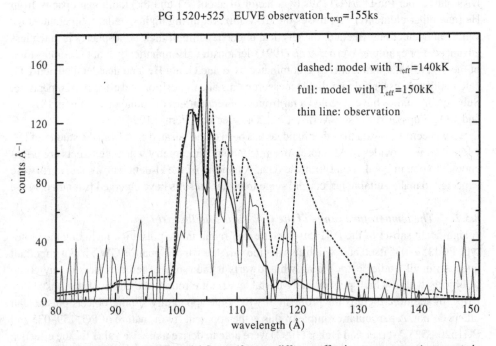

PG 1520+525   EUVE observation t_{exp}=155ks

dashed: model with T_{eff}=140kK

full: model with T_{eff}=150kK

thin line: observation

Fig. 8.20. Top: non-LTE model fluxes (for two different effective temperatures) attenuated by the ISM. Note the strong O V absorption edge at 120 Å in the cooler model, which almost disappears when T_{eff} is increased by only 10 000 K. Bottom: both models folded through the EUVE SW instrument response and compared with the observation of PG1520+525 (from Werner *et al.* 1996).

should be visible at 120 Å, but is not detected (see figure 8.20). To depopulate the O V and weaken the edge requires a higher temperature. A value of 150 000 K gives good agreement with the observed spectrum. Hence the blue edge of the GW Vir instability strip lies between this and the effective temperature of PG1159–035.

8.5.2 The composition and temperature of REJ0503–289

The hot helium-rich DO white dwarf REJ0503–289 lies in a region of extremely low interstellar hydrogen density (Barstow *et al.* 1994d). As the only DO white dwarf detected in both ROSAT WFC and EUVE sky surveys, it was seen as a very important target for further study with the EUVE spectrometer. The star is also bright enough to be observed at high dispersion with IUE and detection of weak P Cygni profiles in one spectrum is evidence for the presence of an episodic weak wind in the object (Barstow and Sion 1994).

The IUE and EUVE data show that, in addition to the carbon already noted on its discovery, the atmosphere of the star contains significant quantities of nitrogen, oxygen and silicon. There is no sign of any Fe, found in DA white dwarfs of similar temperature, but many of the weaker features in the spectrum coincide with Ni V transitions. More detailed analysis of the photospheric composition of REJ0503–289 proved difficult. It is possible to estimate photospheric abundances by comparing non-LTE model spectra with the measured UV line strengths. However, the EUV fluxes predicted by the same models which match the UV data are up to an order of magnitude greater than observed. A consistent solution can be achieved if the assumed temperature of the star is lowered to 63 000 K (cf. 70 000 from earlier work) and a C abundance of 0.1%, a factor of ten below that originally estimated. Abundances of N, O and Si are all ≈0.01% and that of Ni 0.001%, with respect to helium. With these parameters, the EUV model spectrum is now a very good match to the data (figure 8.21),

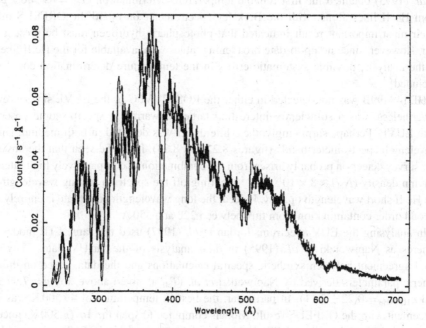

Fig. 8.21. EUVE spectrum of REJ0503–289 compared to the best fit model discussed in the text.

although the predicted He II line profiles are rather broader than observed. This probably arises from the absence of a satisfactory broadening theory for the He II Lyman series and the problem will need to be addressed by theorists working on the atomic data.

Unfortunately, a serious anomaly remains when comparing this result with the detailed optical analysis of Dreizler and Werner (1996). At 63 000 K, the predicted strength of the He I 4471 Å line is much greater than observed. Consistency between observation and theory can only be achieved by adopting the higher value of 70 000 K for $T_{\rm eff}$. Analysis of the HST GHRS spectrum by Barstow *et al.* (2000) confirms the detection of photospheric Ni and places tight limits on the possible presence of Fe. However, new EUV models based on the photospheric abundances determined from the HST data are still unable to reproduce the EUVE spectrum. This is a problem that remains to be resolved.

8.5.3   The cool DO white dwarf HD149499B

The well-known DO white dwarf is also one of the coolest ($T_{\rm eff} \approx 50\,000$ K), lying close to the DO–DB temperature gap. Therefore, establishing an accurate temperature and composition is important in considering the general relationship between H- and He-rich white dwarfs and seeking an explanation of the gap itself. Unfortunately, optical spectroscopic observations, from which this information might be obtained (see Dreizler and Werner 1996), are not possible because the nearby red companion (2 arcsec separation) is three magnitudes brighter and swamps the signature of the white dwarf in the optical spectrum. Furthermore, although the UV continuum could be used to determine a value for $T_{\rm eff}$ the slope is very insensitive to temperature changes. Sion *et al.* (1982) obtained $T_{\rm eff} = 55\,000 + 5\,000/ - 15\,000$ K from a fit to the He II lines and continuum flux together, but the uncertainty in the measurement is too large to provide much insight into the relation of the star to the DO–DB gap. Napiwotzki *et al.* (1995) obtained the first reliable temperature determination ($T_{\rm eff} = 49\,500 \pm 500$ K), from He II lines alone, with the Berkeley spectrometer on board the ORFEUS mission. Their most important result indicated that photospheric hydrogen must be present in the star. However, since no up-to-date broadening tables were available for the He II lines used in the analysis, possible systematic errors in the temperature determination could not be excluded.

HD149499B was not detected in either the ROSAT WFC or the EUVE sky surveys but, nevertheless, was a sufficiently interesting target to warrant a spectroscopic observation with EUVE. Perhaps surprisingly, the white dwarf was detected in both medium and long wavelength spectrometer bands (figures 8.22 and 8.23). It can be seen that the absence of any survey detection probably arises from the combination of comparatively high interstellar column density ($N_{\rm HI} \approx 8 \times 10^{18}$ cm^{-2}), cutting off the spectrum at long wavelengths, and the He II short wavelength opacity. Indeed, the long wavelength count rate is entirely due to second order contamination from flux between 228 and 350 Å.

In analysing the EUV spectrum, Jordan *et al.* (1997) used the same LTE model atmospheres as Napiwotzki *et al.* (1995) in their analysis of the far-UV data. They obtain good agreement between synthetic spectral calculations and the data, based on the atmospheric parameters derived by Napiwotkzi *et al.* ($T_{\rm eff}$ as noted above, log $g = 7.97 \pm 0.08$ and $n_{\rm H}/n_{\rm He} = 0.22 \pm 0.11$). In particular, the best fit temperature of 49 500 K was in full agreement with the ORFEUS result. Models computed to span the 1σ (± 500 K) uncertainties in $T_{\rm eff}$ derived from the ORFEUS show large deviations from the EUV spectrum. Hence,

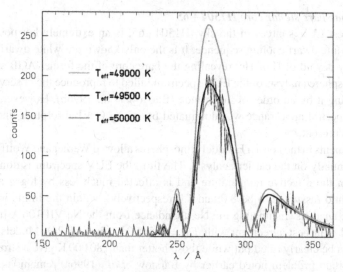

Fig. 8.22. Comparison of the EUVE medium wavelength spectrum of HD149499B with theoretical models for $T_{eff} = 49\,000$, $49\,500$ and $50\,000$ K (from Jordan *et al.* 1997).

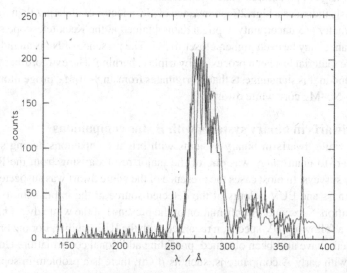

Fig. 8.23. Comparison of the EUVE long wavelength (second order) spectrum of HD149499B with theoretical models for $T_{eff} = 49\,000$, $49\,500$ and $50\,000$ K (from Jordan *et al.* 1997). Note that the flux longward of 456 Å is due to second order contamination from the flux between 228 and 350 Å.

the EUVE observation improves the accuracy of the temperature determination to ± 200 K. Unfortunately, the EUV data do not provide any additional information about the H/He ratio within the uncertainties in the other photospheric parameters. On the other hand, Jordan *et al.* are able to show that elements heavier than H or He are much less abundant than predicted by the radiative levitation calculations of Chayer *et al.* (1995a).

8.5.4   The unique bare stellar core H1504+65

One of the brightest EUV sources in the sky, H1504+65, is an extremely hot post-AGB star entering the white dwarf cooling sequence. It is the only known pre-white dwarf with a surface completely devoid of H or He, revealing the bare core of the former AGB star. An initial model atmosphere analysis of the EUV spectrum failed to reproduce the observed flux level, overestimating it by an order of magnitude (Barstow *et al.* 1996b). However, it was obvious that the spectral appearance was dominated by strong O VI absorption lines with some Ne VII lines present.

Recent developments in the non-LTE model atmospheres allowed Werner and Wolff (1999) to improve substantially on the earlier analysis. The fit to the EUV spectrum is dominated by uncertainties in the effective temperature and is affected much less by log g and the C/O abundance ratio (assumed to be 8.0 and 1.0 respectively), within the limits imposed by the optical line analysis. Estimating the Ne abundance from the Ne VII 106 Å line, the absolute flux level is in best agreement with $T_{eff} = 175\,000$ K (figure 8.24). Models cooler than 170 000 K can be clearly ruled out while those hotter than 180 000 K give a strong flux excess. One important problem noted earlier by Barstow *et al.* (1996a) remains unsolved by Werner and Wolff (1999): the predicted flux shortward of 100 Å is distinctly lower than observed. Werner and Wolff suggest that this may be caused by an underestimate of the O VI line broadening or as yet unidentified opacity sources.

The EUV analysis indicates a high 2% abundance of Ne, a factor of 20 larger than the solar value, but with a factor ≈ 3 uncertainty. Optical data obtained at the Keck telescope suggest that the Ne abundance may be even higher, closer to 5%. The presence of Ne is an indication that the stellar core material has been processed by triple α burning. The extraordinarily high Ne abundance found in this star suggests that it originates from an $8–10 M_\odot$ progenitor which would form an O–Ne–Mg core white dwarf.

8.6   White dwarfs in binary systems with B star companions

The existence of white dwarfs in binary systems with bright companions having spectral classes in the dwarf to giant range was one of the major results arising from the ROSAT and EUVE all-sky surveys. In most cases the presence of the white dwarf was suspected from the relative brightness and EUV colours of the detected source or the implausibility of the apparent identification. Subsequent confirmation of the presence of the white dwarf has usually resulted from a follow-up UV spectrum recorded with the IUE satellite. For the brightest systems EUV spectra have also been obtained, providing additional confirmation. However, in those binaries with early A companions, such as β Crt, there is a problem in separating the contribution of the white dwarf from its companion in the UV range covered by IUE.

Any suspected binary system with an O or B companion would pose even more difficulty. In the IUE waveband, the white dwarf would be completely overwhelmed by the flux of its partner. To find a spectral region where the white dwarf can be discriminated from the companion it is necessary to move to even shorter wavelengths, into the EUV and X-ray bands. O and B stars are also X-ray/EUV sources in their own right. Fortunately, the task is simplified as O and B star emission is expected to arise from hot plasma in a stellar wind, producing an emission line spectrum in contrast to the strong continuum expected of a white dwarf.

Three potential white dwarf/B star binary systems were found in the EUV source catalogues but, interestingly, no O star candidates, as noted by Burleigh (1999). These systems are particularly important from an astrophysical point of view, having the earliest, most massive

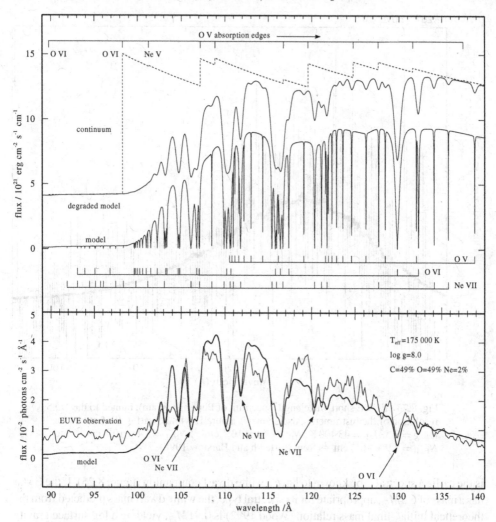

Fig. 8.24. EUVE observation of H1504+65. Top panel: emergent flux of a $T_{eff} = 175\,000$ K, log $g = 8$ model. Shifted upward is the same model spectrum degraded by EUVE instrument response. Overplotted (dashed line) is the continuum flux computed without occupation probabilities in order to display the position of the absorption edges. They are smeared out by atomic level dissolution and broad absorption lines. The spectrum is dominated by strong and broad O VI lines and less prominent O V and Ne VII lines. Bottom panel: comparison of the observed EUVE spectrum with this model, attenuated by an interstellar column density $N_{HI} = 5.1 \times 10^{19}$ cm^{-2} with He II/H I $= 0.068$ and He I/H I $= 0.052$ (from Werner and Wolff 1999).

primary companions, which are necessarily less evolved than the white dwarf progenitor. Such a progenitor star must have been more massive than the B star shedding light on the upper mass limit for white dwarf production and the initial final mass relation.

Spectroscopy with EUVE has confirmed the existence of the suspected white dwarf companions to the bright B stars y Pup (HR2875, B5V, $V = 5.4$) and θ Hya (HR3665, B9.5V, $V = 3.9$). HR2875, discussed by both Burleigh and Barstow (1998) and Vennes *et al.* (1997a),

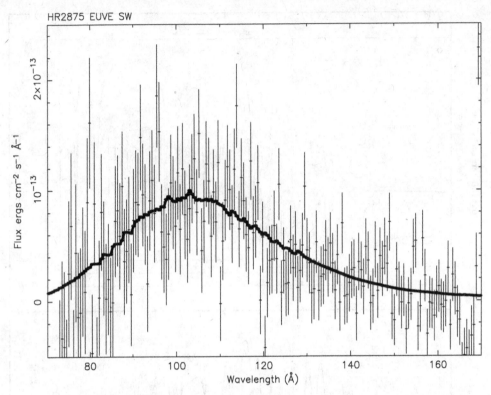

Fig. 8.25. EUVE short wavelength spectrum of Pup (histogram), binned to the ≈ 0.5 Å resolution of the instrument. Also shown is a pure H white dwarf + ISM model for $\log g = 8.5$, $T_{\mathrm{eff}} = 43\,400$ K, $N_{\mathrm{HI}} = 2.1 \times 10^{19}$ cm^{-2}, $N_{\mathrm{HeI}} = 2.4 \times 10^{18}$ cm^{-2} and $N_{\mathrm{HeII}} = 8.9 \times 10^{17}$ cm^{-2} (from Burleigh and Barstow 1998).

is the earliest spectral type known to have a white dwarf companion (figure 8.25). If the B star has a mass of $6.5 M_{\odot}$, appropriate for its spectral type, the white dwarf mass predicted from the theoretical initial–final mass relation (Wood 1992) is $0.91 M_{\odot}$, yielding a log surface gravity of above 8.5. Coupling this with the constraint of the *Hipparcos* distance $(170 + 17/ - 13\,\mathrm{pc})$ the EUV spectrum can be fit to a pure H model with an effective temperature of 43 400 K. At this temperature it is unlikely that the white dwarf has any heavy elements in its photosphere, although it is possible, if the system is a close binary, that earlier interaction of the components may have altered the composition of the white dwarf envelope. It is certainly clear, from the detection of the spectrum in the SW wavelength range only, that the white dwarf's atmosphere cannot be as extreme as the hot G191–B2B like DAs.

Vennes (2000) has subsequently discovered that the B star is a double lined spectroscopic binary, making the system triple. Echelle spectroscopy reveals that the primary and secondary spectral types are B3.5 and B6 respectively. With an earlier primary spectral type than originally assumed, the limit on the mass of the white dwarf progenitor is increased to $\approx 5.5 M_{\odot}$. The orbit is highly eccentric ($e = 0.68$) and has a period of 15.0811 d. The location of the white dwarf relative to the pair of B stars remains unknown.

The second system, θ Hya, is a high proper motion star, lying at a distance of 39.5 ± 1.5 pc, according to *Hipparcos* (see Burleigh and Barstow 1999; figure 8.26). The white dwarf is

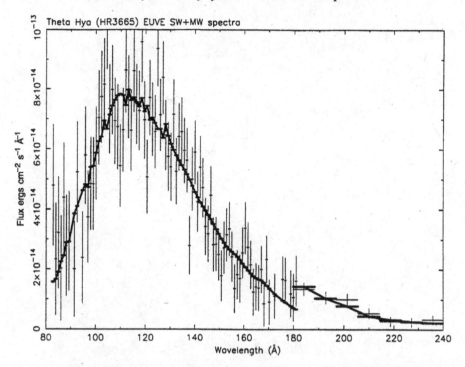

Fig. 8.26. EUVE short wavelength spectrum of θ Hya (histogram), binned to the ≈ 0.5 Å resolution of the instrument. Also shown is a pure H white dwarf + ISM model for $\log g = 8.5$, $T_{\text{eff}} = 28\,500$ K, $N_{\text{H I}} = 6.6 \times 10^{18}$ cm^{-2}, $N_{\text{He I}} = 7.4 \times 10^{17}$ cm^{-2} and $N_{\text{He II}} = 2.8 \times 10^{17}$ cm^{-2} (from Burleigh and Barstow 1999).

cooler (25 500–30 000 K) than that in the y Pup system. With a less massive primary, the constraint on the progenitor mass is not so severe. The limit on the white dwarf mass is also lower at $0.68 M_\odot$.

Burleigh and Barstow (1999) also discuss a third possible B star plus white dwarf binary system. 16 Dra is one member of a bright resolved triple system (16 Dra = HD150100, B9.5V; HD150118, AV; HD150117, B9V). The *Hipparcos* parallaxes confirm that all three stars lie at the same distance of ≈ 120 pc. 16 Dra is also a WFC and EUVE source. Given its similarity to y Pup and θ Hya, it is likely that 16 Dra also has a white dwarf companion. However, 16 Dra is considerably fainter than the other B star systems in the EUV and would require a very long (400–500 ks) exposure time to be detected by the EUVE spectrometer. If it is assumed that the EUV flux arises entirely from a white dwarf component, the EUV and soft X-ray photometric observations indicate that its temperature lies in the range 25 000–30 000 K.

Cataclysmic variables
and related objects

Compared to the large numbers of coronal sources and white dwarf stars found in the EUV all-sky survey catalogues, the number of cataclysmic variable detections is small, numbering only ≈25 objects. Nevertheless, many of these systems were bright enough to have been suitable targets for spectroscopic observations with EUVE, but these studies were complicated by the fact that the sources are highly variable and the exposure times needed to produce good signal-to-noise were at least 50 ks for the brightest sources, much longer than the binary and rotation periods. Hence, with EUVE it was only possible to obtain observations of the time averaged spectrum, except in periods of outburst, which lasted for several days.

Cataclysmic variables can be conveniently divided into two groups: (i) those where the white dwarf does not have a significant magnetic field (<0.1−1 MG) and accretion onto its surface occurs via an accretion disc (e.g. figure 9.1); (ii) the polars–systems in which the white dwarf does have a strong magnetic field (≈10−100 MG) which prevents the formation of an accretion disc and channels accrete material onto the magnetic poles of the white dwarf along the field lines (figure 9.2). A small number of magnetic systems, the intermediate polars, have weaker fields (B ≈ 1−10 MG) allowing a partial disc to accumulate which is disrupted by the field in its inner regions. As in the polars, accretion onto the white dwarf follows the field lines onto the magnetic poles (figure 9.2).

9.1 Emission mechanisms in CVs

As in other sources we have discussed, the basic EUV emission is thermal, arising from heating of gas falling into the gravitational potential well of the white dwarf. However, while we can identify a single mechanism in white dwarfs (the photosphere) and coronal sources (optically thin plasma) that can be modelled very well, CVs are more complex systems. There are several distinct regions that are possible sources of EUV radiation or may at least modify the EUV emission in some way. The hot gas primarily responsible for the high energy radiation is heated material from the donor star, which gains kinetic energy as it falls towards the white dwarf. This energy must be released in some way and most models predict this to be in the form of X-ray radiation.

In the case of the non-magnetic white dwarfs, the accreted material flows through a disc and a boundary layer onto the equatorial regions of the white dwarf. The inner part of the disc and the boundary layer are the main EUV sources. During outburst, in standard accretion disc theory the disc is optically thick and has a temperature in the range 20 000–40 000 K, depending on the accretion rate. Most of the luminosity of the disc emerges in the UV. In quiescence, the disc should, in contrast, be optically thin and cool (2 000–7 000 K). In a steady-state disc, half the accretion energy ($GM\dot{M}/r$) is radiated as material traverses the

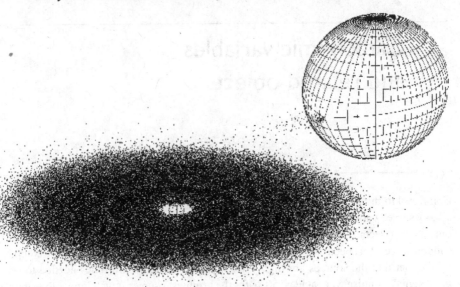

Fig. 9.1. Computer simulation of a cataclysmic variable system in which a non-magnetic white dwarf accretes matter from a disc (courtesy G. Wynn).

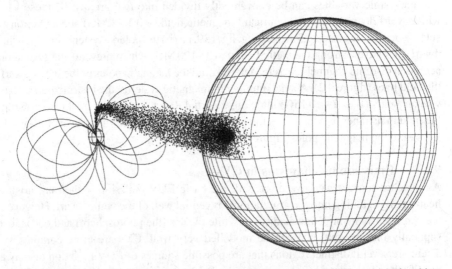

Fig. 9.2. Computer simulation of a cataclysmic variable system in which a magnetic white dwarf accretes matter from the donor star along the magnetic field lines onto a spot on the magnetic pole (courtesy G. Wynn).

disc, but the remaining half reaches the inner edge of the disc and is radiated away in the boundary layer. Since the boundary layer is smaller than the disc, its effective temperature is higher, in the range 100 000 to 300 000 K, radiating primarily in the EUV (Pringle 1977). However, in quiescence the boundary layer should be optically thin and relatively hot (10^7 K).

When the white dwarf has a significant magnetic field (polars and intermediate polars), the accretion stream is channelled onto one or both magnetic poles via the field lines, forming a column of material. If the accretion rate is sufficiently high ($\dot{M} > 0.1\,\mathrm{g\,cm^{-2}\,s^{-1}}$), the flow passes through a shock near the white dwarf surface (e.g. King and Lasota 1979). The shock temperature is given by

$$kT_s = 3\mu m_\mathrm{H}\,GM_\mathrm{wd}/8R_\mathrm{wd}$$

but in practice will be less than this due to the finite height of the shock above the white dwarf surface. Post-shock material is heated to several times 10 keV and cools by radiating infrared and optical cyclotron radiation and hard X-ray thermal bremmstrahlung. The surface of the white dwarf is irradiated by the bremmstrahlung. Some will be reflected but part will be reprocessed in the white dwarf atmosphere. In addition, self-absorption in the column will strongly modify the intrinsic spectrum.

9.2 Spectral modelling

Despite the potential complexities in the soft X-ray and EUV spectrum of the various kinds of CVs, arising from the effects outlined above, only limited progress was made in producing detailed emission models. Until the launch of EUVE only relatively coarse spectral resolution had been available from X-ray instrumentation, coupled with broadband photometry in the EUV. Hence, simple models have mostly been employed to interpret observations. For example, the high energy emission has often been treated as a blackbody source with interstellar absorption. In some cases a thermal bremmstrahlung or optically thin plasma emission model has been applied to modelling the boundary layer. However, in reality these are only parameterisations of the X-ray and EUV emission. They can be useful in comparing the properties of individual objects and groups of objects but do not say much about the underlying physical processes.

The EUV spectra of the CVs appear to divide fairly neatly into two groups according to their physical classification. The non-magnetic dwarf novae have spectra dominated by strong, broad emission lines (e.g. U Geminorum, SS Cygni; see figure 9.3 and figure 9.11), while the magnetic systems have smooth continua with much weaker features (e.g. AM Herculis, figure 9.4). If detectable, these features appear in absorption rather than emission. However, although great progress has been made on models for stellar coronae and white dwarfs, for interpretation of EUV spectra, only limited progress has been made in the CV area. In fact no suitable models currently exist for the non-magnetic systems. In contrast, some progress has been made in understanding the physical mechanisms underlying the EUV emission from polars.

As the prototype of the subclass of magnetic cataclysmic variables, AM Herculis has been studied extensively at a wide range of wavelengths. It was a key target for EUVE but was also the only CV observed with the EXOSAT TGS. Consequently, this system has been the main target for models developed by van Teeseling and co-workers (e.g. van Teeseling *et al.* 1994, 1996). As described earlier, in a magnetic CV, the accretion stream is diverted along the magnetic field lines, preventing the formation of an accretion disc, and onto the magnetic poles of the white dwarf. The infalling material is believed to form a shock-front just above the surface of the white dwarf. Hence, the X-ray and EUV emission is expected to predominantly arise from the shock-heated material, the settling flow beneath and a hot spot surrounding the region where the accretion column meets the white dwarf surface (figure 9.5). However, the spectrum viewed by the observer is modified by several interactions, including

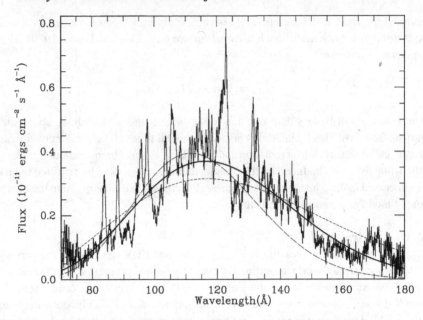

Fig. 9.3. EUVE spectrum of the dwarf nova U Geminorum. The smooth curve is a best fit blackbody with a temperature of 125 000 K, assuming a column density of 3×10^{19} cm^{-2} (from Long 1996).

irradiation of the white dwarf photosphere by the X-ray emission (which is reprocessed) and self-absorption in the accretion column.

Van Teeseling *et al.* (1994) have carried out a number of model atmosphere calculations for white dwarf atmospheres irradiated at the surface by hard X-rays. These models take into account Compton scattering within the stellar atmosphere and deal with the continuum absorption cross-sections of H, He, C, N, O, Ne, Na, Ca, Si, P, S, Ar, Fe and Ni at cosmic abundances. They find that the atmosphere divides into three distinct regions. At large optical depths the temperature of the atmosphere is not affected by the irradiation of the surface. At lower optical depths the temperature gradient is reduced when the surface is irradiated. Above these layers is a region where the plasma is optically thin with a temperature $\approx 10^6$ K. This arises from the large optical depth at short wavelengths due to K and L shell photoionization of abundant elements. Incoming radiation is absorbed leading to a backwarming of the outer layers of the atmosphere.

The temperature inversion and change in gradient produce interesting effects on the emerging spectrum and its angular dependency. For example, limb darkening is smaller (or even becomes limb brightening) at certain frequencies. At small angles the observed spectrum is harder than an un-illuminated atmosphere. The most prominent features are absorption edges from O VI (90 and 98 Å), Ne VII (60, 64 and 69 Å) and Ne VIII (52 and 56 Å), along with some smaller edges of Si and Fe (figure 9.6).

Van Teeseling *et al.* (1994) applied their irradiated model to an analysis of the EXOSAT TGS spectrum of AM Her, comparing it with a simple blackbody model and with a non-irradiated atmosphere calculation. The best fit blackbody model gives a temperature of

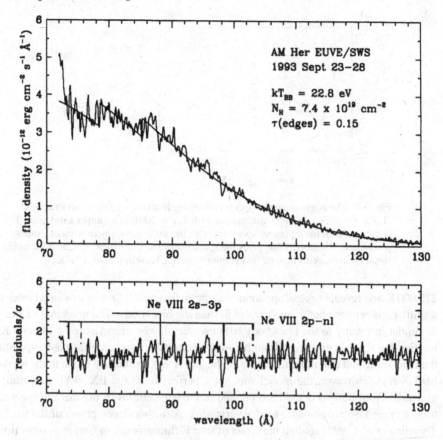

Fig. 9.4. EUVE spectrum of the prototype magnetic CV AM Herculis. The smooth curve is a best fit 250 000 K blackbody on which Ne VI absorption edges have been superimposed (from Paerels *et al.* 1996b).

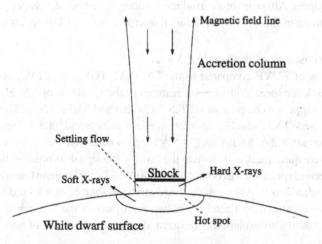

Fig. 9.5. Simplified accretion geometry at the magnetic pole of the white dwarf. The shock height is small compared to the hot spot, which in turn is small compared to the total white dwarf surface area (from van Teeseling *et al.* 1994).

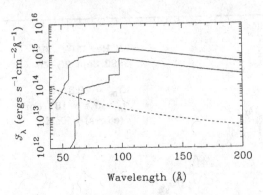

Fig. 9.6. The angle-averaged spectra emerging from the surface of an irradiated $(1.3 \times 10^{17} \, \mathrm{ergs\,s^{-1}\,cm^{-2}})$ atmosphere with $T_{\mathrm{eff}} = 200\,000$ K (upper solid line). This is compared with the emergent spectrum from the same atmosphere without irradiation (lower solid line). The dashed line gives the spectrum of the irradiation flux, which continues to shorter wavelengths beyond this graph (from van Teeseling *et al.* 1994).

270 000 K and reveals several apparent absorption features. The non-irradiated model yields a similar temperature but a much worse fit than the blackbody. The irradiated model gives a fit similar in quality to the blackbody but at a much lower temperature ($<100\,000$ K). This implies that the flux emerging from below the atmosphere has an effective temperature less than this value and that more than 96% of the soft X-ray luminosity is due to the reprocessed hard X-rays. However, the model was not a perfect fit to the EXOSAT spectrum, which may arise from limited detail incorporated in the calculations or the fact the spectrum was time-averaged over significant source variability. A further development of the work of van Teeseling *et al.* (1994) studied the effect of iron K fluorescence in Compton reflection when added to the original models (van Teeseling *et al.* 1996). However, the contribution from these is most significant at high X-ray energies, well beyond EUV wavelengths, and for the purposes of analysing EXOSAT or EUVE spectra does not significantly perturb the models discussed in the 1994 papers. Although basic irradiated models have been developed to study magnetic CVs, they remain of limited use for detailed interpretation of EUV spectra.

9.3   EUVE spectroscopy of magnetic CVs

The improved resolution of EUVE compared to the EXOSAT TGS allowed Paerels *et al.* (1996b) to directly search for specific absorption features in the spectrum of AM Her. Two Ne VI absorption edges appear to be present at 78.5 Å ($2s^2 2p$) and 85.3 Å ($2s2p^2$) but there is no evidence for expected O VI 2s and 2p absorption edges at 89.8 and 98.3 Å. Detections of possible Ne VIII lines at 98.2 Å, 88.1 Å and 74.5 Å are also noted. However, the features observed (figure 9.7) were quite weak and close to the limit of the signal-to-noise of the non-dithered spectrum. A second spectrum was obtained later (1998, cf. 1993) but no features were reliably detected (Christian 2000). Although the exposure was shorter, this would suggest that the absorption lines identified by Paerels *et al.* were artifacts of the detector quantum efficiency variation. It would also explain the apparent anomaly of detection of both Ne VI and Ne VIII features.

Mauche (1999) has carried out the most comprehensive study of magnetic CVs, as a group, in the EUV. In this work, the observed spectra are compared with a range of models, from

Table 9.1. *EUVE observations of magnetic CVs.*

Source	Exposure (ks)	Dithered?	s/n	Reference
AM Her	123.3	Yes	46	Paerels *et al.* 1996b
AR Uma	93.7	Yes	22	
BL Hyi	39.8	No	8	Szkody *et al.* 1997
EF Eri	95.7	No	7	Paerels *et al.* 1996a
QS Tel	69.5	No	13	Rosen *et al.* 1996
REJ1844−741	134.6	No	8	
UZ For	78.5	Yes	7	Warren 1998
VV Pup	43.6	No	5	Vennes *et al.* 1995b
V834 Cen	41.3	No	8	Warren 1998

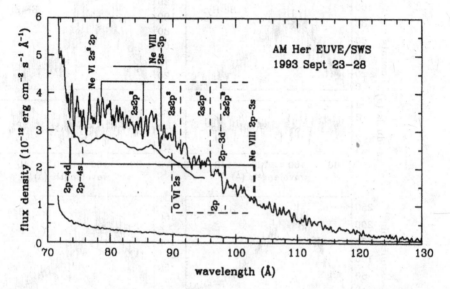

Fig. 9.7. EUVE spectrum of AM Her summed over all binary phases. The lower curve is an estimate of the uncertainty in the flux, based on photon counting statistics. The positions of the Ne VI edges and Ne VII absorption lines discussed in the text are shown, as are the position of expected, but undetected, O VI edges. Also plotted is part of the spectrum smoothed over a 2.2 Å sample size to emphasise the Ne VI edges (scaled by 0.8 for clarity; from Paerels *et al.* 1996b).

simple blackbodies to more sophisticated solar abundance white dwarf model atmosphere calculations. Of 17 separate systems observed only 11 had useful spectra and Mauche rejected two other objects as having too low signal-to-noise for fruitful analysis. Table 9.1 summarises the EUVE spectra of the nine CVs studied in detail by Mauche. The references are to previously published EUVE observations of these sources. Figure 9.8 shows the count spectra for all nine sources, along with the best fit blackbody model, incorporating an interstellar absorption component from the work of Rumph *et al.* (1994), taking H:He I:He II abundance ratios of 1:0.1:0.01 as typical for the diffuse local ISM. However, if significant absorption arises in material nearby the stars, these values may not be completely appropriate.

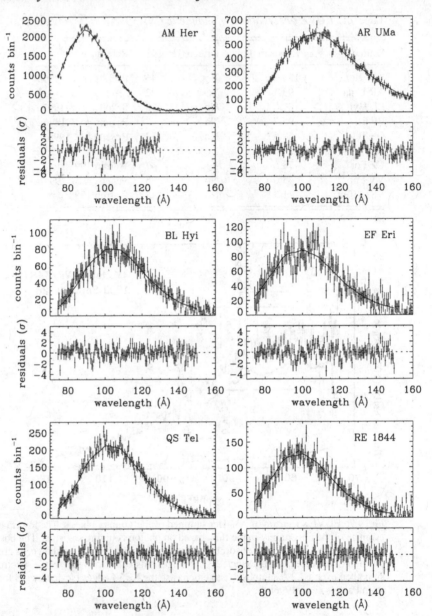

Fig. 9.8. EUVE count spectra of nine magnetic CVs and the residuals relative to the best fit blackbody model (from Mauche 1999). The spectral bin size is 0.54 Å.

The sources clearly vary in hardness, with AR UMa having the softest spectrum, peaking at 110 Å. In contrast AM Her has the hardest spectrum, which is largely cut off at 140 Å, the longer wavelength 'tail' arising from second order contamination by shorter wavelength flux.

Mauche uses the simple blackbody model to parameterise the systems studied. In addition, since the function is smooth the residuals in the fit can be used to detect spectral features such

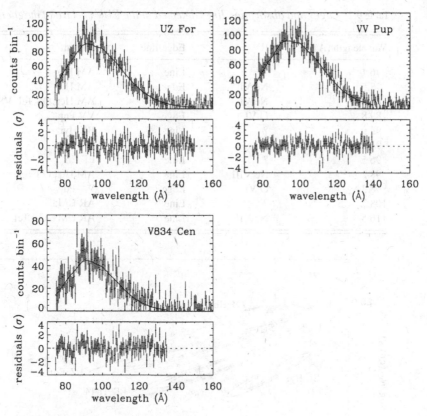

Fig. 9.8. (*cont.*)

as lines or edges. Evidence for OVI edges at 89.8 Å and 98.3 Å is seen in VV Pup (reported originally by Vennes *et al.* 1995b), while Ne VI is seen in AM Her. The Ne VI 85.2 Å edge is also seen in QS Tel (Rosen *et al.* 1996) and V834 Cen. A number of absorption lines are also detected, particularly in the highest signal-to-noise exposures. Both AM Her and QS Tel show Ne VIII 2p–3d at 98.2 Å, while QS Tel and AR UMa both have a feature at 116.5 Å, which has been identified with the Ne VII 2s2p–2s3d transition. A number of other apparently real features are found in AR UMa (108.7 Å), BL Hyi (92.9 Å), EF Eri (96.5 Å) and VV Pup (94.1 Å) but these are not convincingly identified. Table 9.2 summarises all features detected in the EUV in magnetic CVs.

Figure 9.9 shows the confidence contours for the blackbody fits to the EUVE spectra in the kT/N_H plane. Overall the temperatures range from 13.4 to 20.3 eV (156 000 to 236 000 K) with column densities lying between 8×10^{18} cm^{-2} (AR UMa) and 9×10^{18} cm^{-2} (AM Her). However, as pointed out by Mauche, the blackbody models give inconsistent results for different observations of AM Her in overlapping, but not identical, spectral ranges. Hence, the values of the parameters obtained should not be taken too literally. In a formal sense the fits to the highest signal-to-noise (s/n) observations (AM Her, AR UMa, QS Tel and EF Eri) have an unacceptably high value for the reduced χ^2, indicating that the models need to include some of the detected features.

Table 9.2. *Edges and absorption lines detected in the EUV spectra of magnetic CVs.*

Wavelength (Å)	ID	Edge/line	System
76.1	?	Line	AM Her
78.5	Ne VI	Edge	AM Her
85.2	Ne VI	Edge	AM Her, QS Tel, V834 Cen
89.8	O VI	Edge	VV Pup
92.9	?	Line	BL Hyi
94.1	?	Line	VV Pup
96.5	?	Line	EF Eri
98.2	Ne VIII	Line	AM Her, QS Tel
98.3	O VI	Edge	VV Pup
108.7	?	Line	AR UMa
116.5	Ne VII	Line	AR UMa, QS Tel

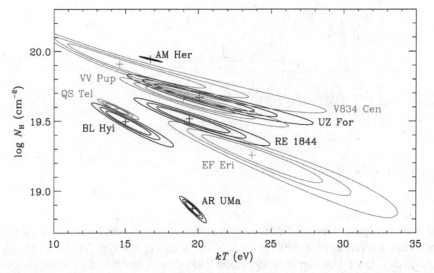

Fig. 9.9. 68, 90 and 99% confidence contours for blackbody fits to the EUVE spectra of magnetic CVs (from Mauche 1999).

The pure H stellar atmosphere models, which are also a smooth continuum in this spectral range, give a similar quality of agreement with the data. However, the temperatures obtained are systematically lower (2.4–7.5 eV; 28 000–87 000 K) and the column densities are correspondingly higher. The lower temperature requires much larger emission regions to account for the observed flux. In the cases of AM Her and QS Tel, this would be larger than the entire star, excluding the pure H model atmosphere for these objects. In other objects, such as BL Hyi (Szkody *et al.* 1997), the pure H model predicts a far-UV flux much lower than observed, providing a lower limit to the temperature that is significantly greater than the best fit pure H model.

In comparison with the pure H models, most of the fits to a solar abundance atmosphere model give an unacceptably high reduced χ^2. Consequently, the parameters derived from these fits can only be taken as a rough guide. Nevertheless, the effective temperatures are higher (22.8–27.4 eV; 265 000–318 000 K) and the column densities lower than for either blackbody or pure H atmosphere fits. These lower columns are more likely to be consistent with interstellar values and the higher temperatures more compatible with the observed soft X-ray fluxes.

The main reason the solar abundance atmospheres fail is the predicted strength of the O VI absorption edges, which are much larger than observed. It may be that inclusion of irradiation effects would solve this problem, as discussed by van Teeseling *et al.* (1994).

9.4 Non-magnetic CVs

In the non-magnetic systems, the material drawn off the companion star by the white dwarf forms an accretion disc in the orbital plane of the binary, arriving on the surface of the white dwarf at its equator via a boundary layer. It is in this region that the rotational kinetic energy of the disc is converted into radiation. These dwarf novae undergo outbursts which are understood to be the result of instability in the rate of mass transfer through the disc. The instability can be triggered at large or small disc radii, resulting in normal, fast rise outbursts or anomalous, slow rise outbursts respectively. The start of an outburst is announced by a rise in the optical flux followed by a rise in the UV flux as material passes through the disc. An EUV flux rise occurs when the material then passes through the boundary layer. Overall, the spectra of dwarf novae are complex, with contributions from the red dwarf, accretion disc, boundary layer and white dwarf. In general, the EUV flux is expected to be dominated by the emission from the boundary layer.

In contrast to the magnetic systems, which can be described by blackbody or stellar atmosphere models, the non-magnetic CVs do not show smooth spectra. A good example is the archetypal dwarf nova SS Cygni, which has been observed extensively in the EUV (Mauche *et al.* 1996; Mauche 1997). Typically, SS Cyg has been observed in outburst with the best published observation being of a slow rise time 'anomalous' outburst. In this, the exposure covered both the rise and plateau phases with a 100-fold increase in the flux in the short wavelength spectrometer over ≈6 d. Mauche *et al.* (1996) note that despite this large change in flux, the observed spectral features (figure 9.10) remain largely unaltered. The net spectrum, accumulated over 4.2 d, shows large departures from a smooth curve (figure 9.10). Although the observation was not performed in the 'dither' mode, the departures are much larger than the typical high frequency pixel to pixel variations associated with the fixed pattern QE fluctuations.

It is difficult to interpret the spectrum in figure 9.10 without a feasible model against which to compare it. It could be treated as either a spectrum with highly broadened emission lines or as a continuum with strong, broad absorption features. In fact, the spectrum is very reminiscent of that of the hot H- and He-deficient white dwarf H1504+65, which has an EUV spectrum dominated by strong oxygen and neon lines, or NGC1360, where oxygen absorption is prominent. However, at this stage no model calculations have been published with which to examine the SS Cyg spectrum. Furthermore, few spectra have been published for any similar systems which might allow us to examine whether or not the SS Cyg spectrum is representative of dwarf novae in general.

Long (1996) has discussed an EUVE observation of the dwarf nova U Geminorum (figure 9.11), comparing it directly with the spectrum of SS Cyg. In broad terms, the spectra seem to be similar, with a number of apparent 'emission' features appearing in both. The

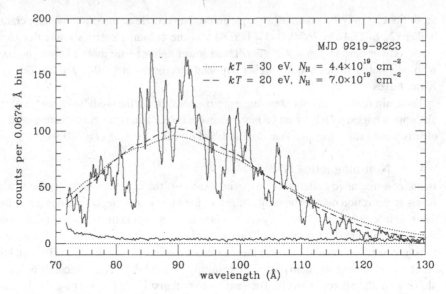

Fig. 9.10. SW spectrum of SS Cygni, accumulated between MJD 9218.9 and MJD 9223.1. The lower trace plots the 1σ statistical error associated with the spectrum. Flux density blackbody curves are shown by the dotted ($kT = 30$ eV and $N_H = 4.4 \times 10^{19}$ cm^{-2}) and dashed ($kT = 20$ eV and $N_H = 7.0 \times 10^{19}$ cm^{-2}) (from Mauche *et al.* 1996).

colour temperature of SS Cyg appears to be somewhat higher than U Gem but, as the SS Cyg spectrum is cut off at wavelengths longer than 130 Å, this difference may be due to different levels of interstellar opacity.

The level of activity in a dwarf nova may have a strong influence on the observed EUV spectrum. For example, Mauche and Raymond (2000) report an observation of OY Carinae during a superoutburst. In this case the spectrum is clearly an assembly of emission lines (arising from N V, O V–O VI, Ne V–Ne VII, Mg IV–Mg VI, Fe VI–Fe VIII) and can be represented by a collisional plasma model, although not all the emission features are accurately reproduced. Good fits to the spectrum are obtained with a model where radiation from the boundary layer and accretion disc is scattered in the line-of-sight by the system's photoionized accretion disc wind. Such a model is similar to that used for Seyfert 2 galaxies and yields a boundary layer temperature in the range 90 000–130 000 K.

9.5   Intermediate polars

The intermediate polar, or DQ Her, cataclysmic variables have a weaker magnetic field than the AM Her types. Hence, while the field is sufficient to channel the accretion flow onto the primary, the accretion disc is not completely disrupted and the spin and orbital periods of the primary not completely synchronised. EX Hydrae is one of only two eclipsing members of this group. However, its 98 min spin and 67 min orbital period are closer to synchronisation than most members of the class. This system was observed in quiescence by EUVE for 180 ks on May 9 1994, fully covering both orbital and spin cycles. Unlike the AM Her systems and SS Cyg, the EUV spectrum of this system is not dominated by a strong continuum, but consists of narrow emission lines, similar to a stellar coronal plasma (Hurwitz *et al.* 1997; figure 9.12).

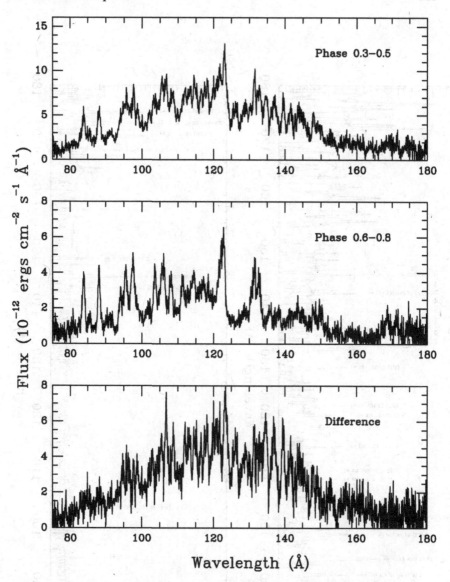

Fig. 9.11. Flux calibrated SW spectrum of U Gem, as a function of orbital phase. The emission is partially eclipsed near phase 0.7. The upper, middle and lower panels show the uneclipsed spectrum, the eclipsed spectrum and the difference between the two (from Long 1996).

Figure 9.12 shows two spectra divided by the phase of the white dwarf light curve. In the faint phase ($0.07 < \phi < 0.43$), only a few lines are detected associated with plasma temperatures in the $10^5 - 10^6$ K range. In the bright phase many high ionization lines from Fe and Ne are detected, most being characteristic of a 10^7 K plasma temperature. Interestingly, the lines identified at wavelengths shorter than 140 Å are all seen in the active coronal source HR1099 (Brown 1994). Spectral model fitting gives consistent strengths for the strongest Fe lines, with a 10^7 K solar abundance plasma. However, the interstellar column density to this

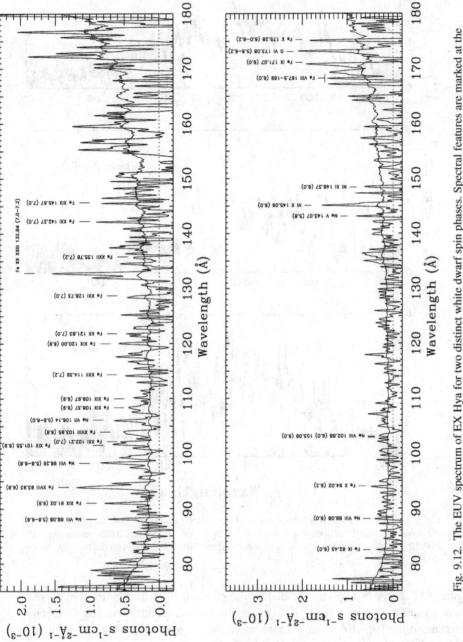

Fig. 9.12. The EUV spectrum of EX Hya for two distinct white dwarf spin phases. Spectral features are marked at the expected wavelength of the ionic species. Numbers in parentheses indicate the log temperatures where each line is expected to be bright (according to Mewe *et al.* 1986). Also plotted (smoother curves) are the 1σ uncertainties. Top: bright phase $0.6 < \phi < 0.9$. Bottom: faint phase, $0.07 < \phi < 0.43$. Figure taken from Hurwitz *et al.* (1997).

source is very uncertain, complicating the analysis. A distance of 125 pc and column density of 2×10^{19} cm^{-2} were assumed by Hurwitz *et al.* (1997). There are several lines detected at wavelengths below 110 Å that do not have a counterpart in the 10^7 K model. Addition of a cooler 10^6 K component creates some plausible candidates, without perturbing significantly the other spectral features, indicating that the plasma may have more than a single temperature component.

9.6 Summary

Compared with other classes of objects observed in the EUV, such as white dwarfs and coronally active stars, fewer high quality spectra of CVs are available to form a clear view of the general properties of these systems in this spectral range. Nevertheless, between the three main types – the polars, intermediate polars and dwarf novae – there seem to be very distinct differences in the type of emission observed. The polars have more or less smooth continua with relatively weak absorption features that can be reasonably described by blackbody or stellar atmosphere models, although accurate representations of the spectra have not yet been developed. In contrast, the emission from the intermediate polars appears to arise from optically thin plasma, similar in nature to that found in coronal sources. There are hints that the plasma may have multiple temperature components but the quality of the available spectra is not good enough to provide a definitive analysis. The spectra of the dwarf novae seem to arise from strong continua with emission lines superimposed or may, alternatively, be continua with strong absorption lines. However, the observation of OY Carina in outburst is a clear example of an emission line spectrum, favouring this model for other systems.

10

Extragalactic photometry and spectroscopy

10.1 Active galaxies

While extragalactic objects such as normal and active galaxies are certainly the most EUV luminous sources discovered, they are also among the most difficult to observe. Their great distance, coupled with the absorbing effect of the intergalactic medium and the interstellar gas in our own galaxy, yields fluxes fainter than most of the more local EUV sources. Hence, the acquisition of EUV spectra requires comparatively long exposure times. With the capability of EUVE, typical minimum exposure times were a few hundred thousand seconds, approaching the practical limit imposed by the instrument background, beyond which no further improvement in signal-to-noise could be achieved. Consequently, the number of extragalactic objects for which spectroscopic observations have been feasible is small. Furthermore, these sources are only visible in the short wavelength region of the EUVE short wavelength spectrometer. Table 10.1 lists those objects which have published EUV spectra, noting their exposure times and classification. Two BL Lac objects and 5 Seyfert galaxies, probably all type I, are listed.

The now commonly accepted explanation of the various different types of AGN is the so-called 'Unified Model', which adopts a common physical mechanism for the source, the AGN types representing different viewing angles. In this model, the central energy source is a massive black hole accreting matter from its host galaxy via a disc. This disc is surrounded by a thick torus of material as shown in figure 10.1. A relativistic jet of material has an outflow axis perpendicular to the plane of the disc and this is surrounded by a cone of beamed radiation from the jet. A disc-driven hydromagnetic wind may propel photoionised clouds outwards from the plane of the disk. Within this model, BL Lac objects are those viewed by the observer along the beaming cone, while in Seyfert 1 galaxies the line-of-sight is between this cone and the torus. In Seyfert 2 galaxies the torus obscures the direct radiation from the central source. EUV spectroscopy might be expected to probe the combined effects of emission and absorption by material along the line-of-sight. In particular, detection of absorption lines would provide evidence for the presence of absorbing clouds above the accretion disc and measurements of Doppler broadening would provide dynamical information concerning their location, if present. Observation of line emission would be a direct indication of the presence of ionised plasma associated with the 'warm absorber' that gives rise to absorption features in the X-ray spectra of these objects (e.g. Nandra *et al.* 1991, 1993).

The form of the observed EUV spectrum is then of crucial importance. Is it just an extension of the power law form that is ubiquitous at higher energies or are more complex features visible? A definitive answer to this question remains elusive, since the AGN spectra themselves are subject to poor signal-to-noise, despite the long exposures. Hence, the results are sensitive to the reliability of the background subtraction process and the detector fixed

Table 10.1. *Extragalactic sources with published EUV spectra.*

Name	Type	Exposure (ks)	≈Max λ (Å)	Reference
Mrk 421	BL Lac	300	100	Fruscione *et al.* 1996
		242	100	Kartje *et al.* 1997
PKS 2155–304	BL Lac	133	100	Königl *et al.* 1995
		157		Kartje *et al.* 1997
NGC5548	Seyfert 1	292	90	Kaastra *et al.* 1995
		559	90	Marshall *et al.* 1997
Mrk 478	Seyfert 1	250	100	Marshall *et al.* 1996
				Leidahl *et al.* 1996
				Hwang and Bowyer 1997
Mrk 279	Seyfert 1	200	100	Hwang and Bowyer 1997
Ton S180	Seyfert ?	80	90	Hwang and Bowyer 1997
RXJ0437.4–4711	Seyfert 1	496	70	Halpern and Marshall 1996
3C273	QSO	130		Marshall 1996

pattern efficiency variation (these observations were not conducted with the 'dither mode'). A good example of the current debate can be found in different interpretations of the observation of NGC5548 (figure 10.2). Kaastra *et al.* (1995) claim to have detected lines near 72 Å and 90 Å. They associate the 90 Å feature with a (cosmologically redshifted) blend of Ne VII 88.13 Å and Ne VIII 88.09 Å transitions, while the 72 Å line is associated with Si VII 70.02 Å. However, if real, the lines have to be significantly broadened to match the observed spectrum, by a fwhm corresponding to 3800 ± 1700 km s^{-1}. Assuming that this arises from Keplerian motion around the central black hole places the material at a distance similar to that of the optical broad-line region. In contrast, Marshall (1996) and Marshall *et al.* (1997) are sceptical about the evidence for the presence of emission features in the NGC5548 spectrum. Marshall *et al.* (1997) carried out a number of tests to quantify their conclusion that emission lines were not present. They found a low level pattern in the background that could account for irregularities in the spectrum that might be interpreted as real features. The weak source signal and the dominance of systematic errors make any formal analysis of the spectrum difficult and Marshall *et al.* (1997) restrict themselves to comparing the EUV data with the existing spectral models derived from X-ray analyses.

An interesting approach to solving this problem has been taken by Hwang and Bowyer (1997). First, they developed a new method for evaluating the background, performing a least-squares polynomial fit to the background along the imaging axis (perpendicular to the dispersion direction). Then they coadded the spectra of the Seyfert 1 galaxies MrK 478, Mrk 279 and Ton S180, after correcting for their relative red shifts. The result is a composite spectrum of higher signal-to-noise than the individual components which shows a significant emission feature at a rest-frame wavelength of 79 Å together with a weaker one at 83 Å (figure 10.3). However, the origin of these 'lines' is difficult to explain. An optically thin plasma of the kind invoked by Kaastra *et al.* (1995) in their analysis of NGC5548 cannot generate lines at these wavelengths. Hwang and Bowyer (1997) propose iron fluorescence as a possible mechanism but more supporting evidence needs to be gathered.

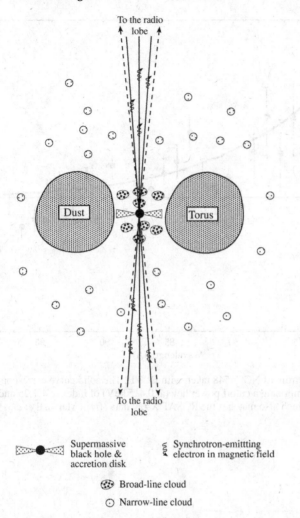

Fig. 10.1. Schematic representation of the unified AGN model, comprising central black hole, accretion disc, obscuring torus and beamed relativistic jet. Different classes of AGN correspond to different lines-of-sight.

The pair of BL Lac objects observed by EUVE, Mrk 421 and PKS2155–304 are strikingly similar, both showing an apparent deficit in the flux at the very shortest wavelengths, when a power law model is constrained by the X-ray flux and the galactic column density (figures 10.4 and 10.5). Consistent fits to the EUV data alone can be achieved in each case (see work by Kartje *et al.* (1997) and Königl (1996) respectively) but the spectral indices lie in the range 3–4. Such steep power laws are inconsistent with the X-ray results. A straightforward explanation of this discrepancy is the presence of a strong absorbing component in the 65–75 Å range. Both Königl and Kartje *et al.* suggest that this may arise from outflowing clouds of photoionised gas, where the absorption line profiles are smeared by Doppler effects.

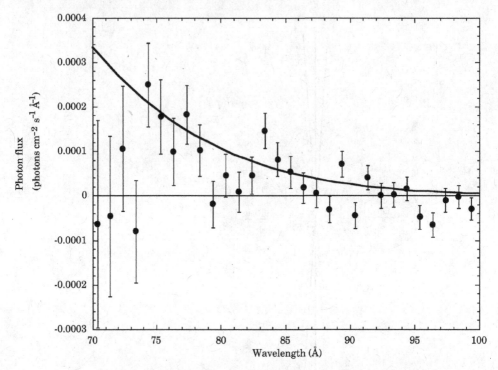

Fig. 10.2. EUV spectrum of NGC5548 taken with EUVE. The solid curve corresponds to a model prediction comprising a cutoff power law (at 0.415 keV) of index $\alpha = 1.25$ and $n_H = 1.93 \times 10^{20}$ cm^{-2}, which also matches the ROSAT X-ray data (from Marshall *et al.* 1997).

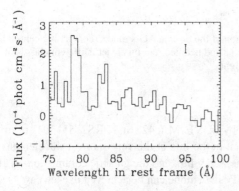

Fig. 10.3. Composite spectrum from the EUVE spectra of the Seyfert galaxies MrK 478, Mrk 279 and Ton S180 (from Hwang and Bowyer 1997).

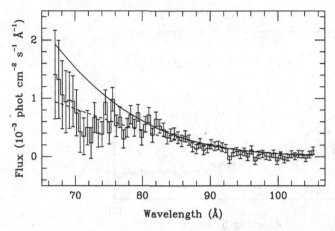

Fig. 10.4. EUV spectrum of Mrk 421, averaged over four days. The dashed curve is the best fit power law ($\alpha = 3.5$) with galactic column density while the solid curve corresponds to a power law of index $\alpha = 1.32$, extending between the simultaneously observed fluxes at 85 Å and 1.5 keV, assuming Galactic absorption (from Kartje *et al.* 1997).

The observed spectrum of Mrk 421 can be explained by total line-of-sight cloud column densities of $\approx 5 \times 10^{21}$ cm^{-2} and cloud ionisation parameter in the range 1.5 to 3.5 (figure 10.6). For PKS2155–304, the required ionisation is similar but the inferred cloud column density about an order of magnitude larger. The EUV line widths constrain the cloud velocities along the line-of-sight to a narrow range between 0.05 and 0.1 times the speed of light, mainly determined from the complex of Mg VIII and Mg IX lines near 71 Å. An important conclusion from this work is that the physical characteristics of the ionised absorbing gas can be attributed to broad absorption line region clouds in the broad absorption line QSOs (e.g. Hammann *et al.* 1993).

10.2 Extragalactic source variability

A by-product of the long observation times needed to obtain EUV spectra of extragalactic sources is that of long term monitoring of the integrated EUV flux over a period of one to two weeks. From light travel time arguments, studies of source variability can yield information on the dimensions of an emitting region and comparison with data from other wavebands can shed light on the physical relationships between different regions in a source, particularly if an event observed in two different spectral ranges experiences a delay. Indeed, much of the work reporting EUV spectra outlined in the previous section also deals with analyses of the concurrent light curves.

In general, all the sources show significant variations in flux on timescales of less than a day, in keeping with observations at other wavelengths. The most exciting result arises from a multiwavelength study of PKS2155–304, conducted simultaneously with ASCA, ROSAT, EUVE and IUE (Urry *et al.* 1997). The normalised fluxes obtained with each instrument are shown in figure 10.7. The most striking feature is the large amplitude X-ray flare seen with ASCA. However, as the ASCA coverage is rather limited it is not clear whether this corresponds to the UV flare of May 21. Given the additional presence of an

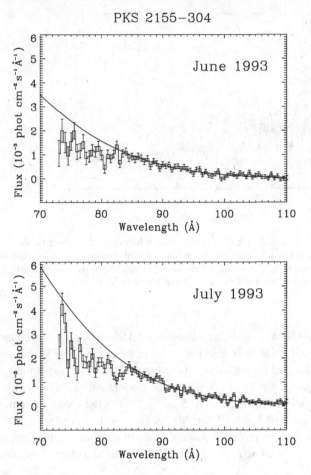

Fig. 10.5. EUV spectrum of PKS2155–304 for the June and July 1993 observations. The solid curve corresponds to a power law of index $\alpha = 1.65$, modified by galactic absorption (from Königl *et al.* 1995).

EUV flare feature, Urry *et al.* propose that the correlations and delays represent the real propagation of a flare from short to long wavelengths. The UV then lags ≈ 2 days behind the X-ray and ≈ 1 day behind the EUV. Apart from the obvious broad flare, the flux rise at the onset of the UV coverage lines up with an increase in the EUV, if a similar delay is assumed.

The amplitudes and shape of the flares differ in each waveband. It is sharpest in the X-ray, with a duration of no more than a day with an approximately equal rise and fall time of 0.5 d. In the UV the rise time is slower (≈ 4 d), with a fall time of about 2.5 d. The shape of the EUV flare is not so well determined but the rise time is around 1 d. The EUV flare ends during the decay, making its decay time and total duration difficult to determine.

According to earlier multiwavelength studies (see Edelson *et al.* 1995 and references therein), the emission mechanism from the UV through to the medium energy X-ray range is synchrotron radiation. The relative X-ray to EUV/UV lags (1–2 d) and the ≈ 1–2 h lag between soft and hard X-ray photons approximately match a $\nu^{-1/2}$ relation, suggesting

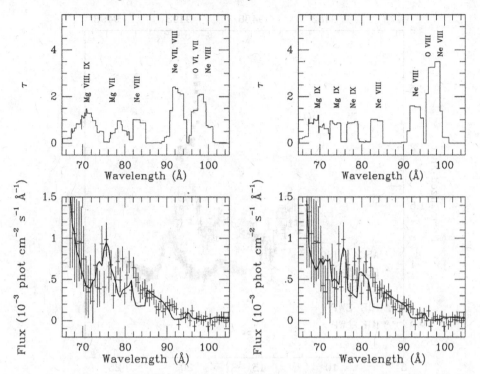

Fig. 10.6. Model fits to the EUV spectrum of Mrk 421. The left panels correspond to the smeared absorption profile optical depths and resulting spectrum for an outflow with cloud ionisation parameter of 1.58. The right panels correspond to outflow with an ionisation parameter of 3.36. In the lower panels the crosses represent the data points while the solid curve is the combination of the outflow models with a power law of index 1.32 and galactic absorption (from Kartje *et al.* 1997).

they are related to radiative losses. Urry *et al.* use the fact that the strong X-ray flare progresses to longer wavelengths to rule out a stochastic acceleration process in a homogeneous volume. They suggest that the feature could be explained by an instantaneous injection of high energy particles near the X-ray emitting energies with subsequent energy degradation in a homogeneous radiating synchrotron source. However, the magnetic field is constrained to a very low value, requiring a very high Doppler beaming factor for the model not to exceed the highest observed gamma ray flux. Alternatively, an inhomogeneous jet model, incorporating acceleration by shock can match both the broadband spectrum and observed spectral evolution of the flare with a higher magnetic field. In this case, the flare is caused by a propagating disturbance affecting different regions at different times.

PKS2155–304 is a BL Lac object. Subsequently, similar observations have been reported for the Seyfert 1 galaxies NGC4051 (Uttley *et al.* 2000) and NGC5548 (Chiang *et al.* 2000). In the former paper, a long duration EUVE time series is compared with observations made by the Rossi X-ray Timing Explorer (RXTE). The NGC5548 study utilises the ASCA observatory in addition to these two instruments. In both cases the EUV and X-ray variations

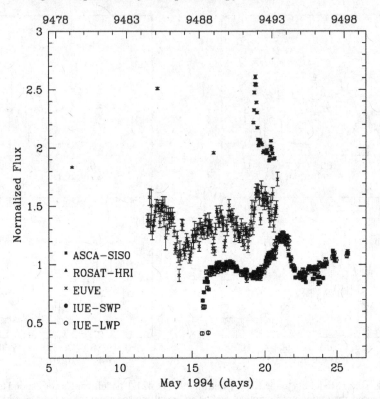

Fig. 10.7. Multiwavelength light curves of PKS2155–304 obtained during May 1994. The ASCA, ROSAT and EUVE fluxes are averages over one satellite orbit. Each light curve has been normalised to its mean intensity (for the IUE data this was done after discarding the first six flux points from SWP and LWP light curves). For clarity, the EUV and X-ray light curves have been shifted vertically by 0.4 and 1.1 respectively. The upper axis shows MJD (MJD9487.5 = May 15 1994 0000UT) (from Urry *et al.* 1997).

are highly correlated. In NGC5548 the EUV light curve leads the X-ray data by ≈10–30 ks, in the same sense but shorter than the lag reported in PKS2155–304. In contrast, the lag between EUV and X-ray in NGC4051 is less than 20 ks and may be as low as 1 ks. Such a short lag would imply that the radius of the Comptonizing region is less than 20 Schwarzschild radii for a black hole mass greater than $10^6 M_\odot$. The fact that, in general, the EUV variability precedes that seen in the X-ray is a strong indication that the EUV photons are not produced by reprocessing of the higher energy X-ray emission as suggested in some AGN models.

11

EUV astronomy in the 21st century

11.1 Looking back

The astrophysical research discipline we now know as Extreme Ultraviolet astronomy is approximately 30 years old. An observational technique once dismissed as impossible has become established as a significant branch of space astronomy and a major contributor to our knowledge of the Universe. In several areas, the science obtained from EUV observations is unique. For example, the presence of the He II Lyman series in this spectral range provides a diagnostic tool for the study of the second most abundant element in the Universe in the atmospheres of hot stars and in interstellar space. The determination of the ionisation fraction of helium in the local ISM could not have been carried out in any other spectral range.

EUV astronomy has passed through the development phases that might be deemed typical of a discipline depending on access to space. Beginning with the sounding rocket borne experiments of the early 1970s, the longer duration Apollo–Soyuz Test Project highlighted the potential of the field, with the first reported source detections in 1975. However, it was a further 15 years before the next major advance with the first EUV all-sky survey of the ROSAT WFC (in 1990) followed by the wider spectral coverage of the EUVE survey in 1992. The underlying reasons for this hiatus had more to do with national and international politics, together with the economics of funding opportunities and even the launch delays following the Challenger disaster, than technological limitations. However, the ROSAT and EUVE missions were certainly able to take advantage of the development in mirrors, detectors and filters that were made during the late 1970s and early 1980s, providing instrumentation of greater sensitivity than might have been possible had the surveys been carried out earlier. It is notable that the first EUV sky surveys were performed with imaging telescopes, in contrast to the first X-ray surveys. Indeed, the first imaging X-ray survey was that carried out by the ROSAT mission, with the WFC on board.

The level of development in the WFC and EUVE instrumentation arising from all the delays may, in the end, have placed the future of EUV astronomy on a firmer footing than would have been possible earlier. In particular, had the ROSAT WFC been launched in 1986, as originally planned, the telescope would have carried thicker thin film filters than were ultimately flown. As a consequence, the achieved sensitivity would have been significantly lower and the number of sources detected much smaller – probably less than one quarter of the published survey catalogues. At this level, with $\approx$25 white dwarfs, and $\approx$75 late-type stars, the statistical basis of the studies of these populations would have been seriously undermined. Furthermore, of the less populous source categories such as CVs and AGN, only a very few examples would have been detected. In the event, the $\approx$1000 now known EUV sources have provided a rich scientific output, forming a substantial basis for future EUV astronomy.

A foresighted enhancement of the EUVE mission, made possible by the long delays to the project, with the addition of a spectroscopic capability, has played a major role in defining what the future of EUV astronomy should be. This particular instrument has enabled ≈ 0.5–1.0 Å spectroscopy to be carried out on almost all the brightest EUV sources, allowing the study of the composition of white dwarf photospheres, stellar coronae, density and ionisation of the local interstellar medium and emission mechanisms in CVs and AGN. Without this, the next step in EUV astronomy would have been the first spectroscopic instrument and a consequent delay to the rapid progress that has been made during the past ten years.

11.2   Limitations

Although the ROSAT WFC and EUVE missions can be claimed as outstanding successes, their telescopes were of comparatively limited aperture. Typical effective areas for the survey instruments ranged from ≈ 10–20 cm^2, while the peak effective area of the EUVE spectrometer was ≈ 1 cm^2. These figures can be contrasted with the >200 cm^2 effective area of the ROSAT XRT with its PSPC detector. Clearly, it would be desirable to construct new EUV instrumentation with much higher effective areas than already flown.

The practicality of enhanced EUV astronomy missions depends on a combination of science, politics, funding and technology. X-ray astronomy has an impressive track record of continuing development of larger and larger telescope collecting areas accommodated on satellites of increasing size and complexity. Continuity of these missions seems to be assured well into the future. It would be nice to enjoy a similar situation in the field of EUV astronomy.

Unfortunately, the nature of the subject of EUV astronomy encompasses a few problems that make it more difficult to justify the magnitude of expenditure required by the large X-ray observatories. The principal limitation is the absorption by the interstellar medium that defines the horizon within which sources can be observed. While, as seen from the all-sky surveys, there are a few 'windows' in the galactic ISM through which extragalactic sources can be observed, these are small in number. Therefore, although in principle EUV observations are important for many astrophysical environments, they are mostly restricted to a local region of our galaxy. The source environment may play an additional role, since absorption by surrounding material may prevent the escape of EUV photons in detectable quantities. Since EUV astronomy is going to be of limited interest to non-stellar astronomers, except for studies of the ISM, it may never be possible to justify an expenditure on future missions similar to the large X-ray or UV observatories.

Nevertheless, it is clear from the content of this book that future EUV observations will be tremendously important for the study of white dwarfs, stellar coronal plasma, CVs and the local ISM. Hence, following the end of the ROSAT WFC and EUVE missions it is important to consider what new instrumentation should be developed and flown. However, such instrumentation will probably need to fit within the cost envelopes of the smaller missions rather than observatory class satellites, implying that the payloads will not be much more massive than those already flown.

11.3   New EUV science

Two general aims can probably be stated for EUV astronomy, although these must be justified in detail for individual science goals. The first is a requirement for greater effective areas to allow the observation of fainter sources. The second is the need for higher resolution spectroscopy than available from EUVE. While greater source sensitivity is a goal that can exist in isolation, increased collecting area is an essential adjunct to improving spectral resolution, to maintain the signal-to-noise.

11.3.1 A deeper sky survey

Approximately 1100 sources of EUV radiation are now catalogued, residing mainly in the region of the galaxy defined by the local bubble (see chapter 7). The bubble defines a volume beyond which the interstellar column density increases so rapidly that very few sources are detectable. Therefore, a new, more sensitive survey would probably be limited to searching for less luminous sources than those currently known out to distances of 100–200 pc. There are certainly many nearby stars that were not detected by the WFC that might be revealed in such a survey (see section 3.8). Only 25% of all stars within 10 pc were detected by the WFC and fewer than 5% of all those in the CNS3 nearby star catalogue (distance limit 25 pc).

One of the main surprises from both main sky surveys was the low number of white dwarf detections ($\approx$100) compared with the 1000–2000 expected from earlier predictions. This was eventually interpreted as arising from the presence of heavy elements in the photospheres of those stars with effective temperatures above $\approx$40 000 K (see section 3.6.2). However, those hotter DA white dwarfs that were detected do not necessarily represent the most extreme cases of heavy element opacity. Therefore, the existing survey data have only probed the tip of what must be a large iceberg in studying the pattern of photospheric opacity across the population of hot white dwarfs.

11.3.2 High resolution EUV spectroscopy

The spectral resolution of EUVE ranged from 0.4 Å at the shortest wavelengths to 2.5 Å at the longest. While enormously superior to the very limited spectral information available from broadband photometry, this resolution places restrictions on the scientific information that can be extracted. Resolving powers ($R = \lambda/\Delta\lambda$) in the range 10 000–30 000 have been available routinely in the far-UV with IUE and HST. With these instruments it has been possible, for example, to resolve the individual heavy element absorption lines in stellar atmospheres, resolve multiple components of interstellar absorption lines and study the dynamics of stellar systems through radial velocity variations. As EUV observations provide some unique information about a number of astronomical objects, it will be important to develop a similar spectroscopic capability in this band.

An important diagnostic measurement to be made on any spectrum is the radial velocity of an individual absorption line, with respect to the observer. Differences in line velocities can allow the discrimination of stellar and interstellar components or chart dynamical motions of the components of a stellar system or motions of gas within it. Hence, the important point is how a particular spectral resolution translates into velocity space. A resolving power of 10 000 is equivalent to a velocity of 30 km s^{-1}, equal to 0.15 Å at a far-UV wavelength of 1500 Å. In the EUV, this velocity discrimination corresponds to a wavelength resolution of 0.015 Å at 150 Å.

To define what performance may be required by a future EUV spectrometer, it is necessary to look at what current astrophysical problems may be addressed particularly by EUV observations and what resolution is needed to solve them. A good guide is to consider the questions left unanswered by the EUVE observations discussed in this text.

11.3.3 Photospheric opacity in white dwarf atmospheres

EUV astronomy has clearly played an important role in understanding the presence of opacity sources in the atmospheres of hot white dwarfs, in the H-rich DA stars in particular (see section 3.6, chapter 8). In conjunction with far-UV observations, EUV spectra have proved to be important diagnostic tools for determining general abundance levels (e.g. Lanz *et al.* 1996) and, uniquely, for studying the depth dependence of the absorbing material (e.g. Holberg *et al.* 1999a; Barstow *et al.* 1999). However, the limited spectral resolution of EUVE was unable to resolve the individual absorption lines in the EUV; those features that could be seen were in fact blends of large numbers of individual lines. This presents a particular problem in understanding the role of helium, whether in the photosphere or ISM, in those hot DAs which have significant quantities of heavy elements in their atmospheres. While, in principle, the He Lyman series lines should be detectable in the EUV spectra, the large number of other absorption features can easily mask their presence. Furthermore, although a component of He II is needed to completely explain the EUV spectral shape of G191–B2B, it is not known whether this is photospheric, circumstellar or interstellar.

11.3.4 Stellar coronae and coronal dynamics

A large number of coronal spectra were obtained by EUVE. The identification of many individual emission lines associated with high atomic ionisation stages provided important diagnoses of the structure and composition of the hot plasma (see chapter 6). Since the coronal plasma features are in the form of emission lines, set against a low level or even negligible continuum flux, the EUVE spectral resolution is less problematic, with all the most prominent emission lines being well separated in most cases. Only a few lines are blended (see table 5.3). However, these data are limited in other ways. The coronal sources are much less luminous than the white dwarfs. Even though the fluxes are confined to a number of narrow emission lines, long exposures have been required to obtain spectra of sufficient signal-to-noise for detailed analysis. However, coronal sources are highly variable, either through the natural cycles similar to the variability of activity on the Sun (including stellar flares), by rotational modulation of emitting regions or by interaction and eclipses in active binary systems. Hence, in most spectroscopic observations only the global time averaged coronae have been examined, compared to the shorter time scale variability.

Many of the most coronally active stars are in binary systems. In principle, the spectra of the individual stars could be separated by their relative Doppler shifts but this was not possible with EUVE. Furthermore, the long exposures often required (compared with the binary orbital periods) will have tended to smear out the line emission components in any case. In single stars, where rotational velocities are high, changing Doppler shifts of the emission from localised active regions would be lost within the EUVE spectral resolving power. Even if stellar motions are too slow for dynamical Doppler effects to be important, there should be information in the Doppler width of the emission lines, if this could be resolved.

11.3.5 Cataclysmic variables and related objects

The issues here are similar in a number of ways to the limitation of coronal source observations. Like these objects, CV-type objects are intrinsically faint in the EUV but are also highly variable objects, exhibiting outbursts, spin/rotation-modulated emission and flickering effects. Apart from the long duration flux changes associated with outbursts, the exposures required for good s/n spectra average out all the other effects. Some CV spectra contain what appear to be broad emission lines. However, it is not clear whether these are true emission features or artifacts produced by large numbers of blended absorption features. Higher resolution observations are essential to solve this problem. Acquiring exposures over shorter time-scales is also essential to limit any smearing affects associated with the dynamics of these systems, which otherwise limit the resolution that can be achieved.

11.3.6 Interstellar medium

EUV studies of the ISM have concentrated on the detection of absorption edges of He I and He II, with analysis of the continuum absorption by H (see chapter 7). The absorption edges are actually composed of a series of converging absorption lines, which give a rounded shape to the edge at the resolution of EUVE. However, in principle, the width of the lines has information about the temperature of the absorbing gas. The analysis of the EUVE data is not very sensitive to assumptions about the value of the Doppler broadening parameter, but high spectral resolution would make measurement of this feasible.

Interpretation of all the ISM results recorded with EUVE has to assume that all the absorbing gas is associated with a single component. However, this is not necessarily true. Far-UV observations often show ISM absorption at several different velocities that might be associated with discrete interstellar clouds. The work of Barstow *et al.* (1997) indicates that the total column density measured along the lines-of-sight to several white dwarfs is larger than might be expected if the absorbing material was primarily associated with the local interstellar cloud. Therefore, a higher resolution may well reveal multiple absorbing clouds of helium along these lines-of-sight.

It had been anticipated that, apart from the detection of the He resonance absorption edges, other interstellar absorption lines would be detected in the EUVE spectra. However, apart from the discovery of the He I autoionisation transitions (Rumph *et al.* 1994), no such lines have been detected. Interstellar lines are intrinsically very narrow, and convolution with the EUVE spectral response would weaken them with respect to the photospheric continuum against which they are viewed. Many of the interstellar absorption lines that might be observed in the EUV are associated with hot gas and could provide important information about the ionisation balance of the local cloud and bubble. If viewed against more distance objects, such as AGN, it should be possible to study the hot halo believed to surround our own galaxy. A particularly exciting scientific programme would be to search for absorption associated with the He^3 isotope in the local ISM, to study the He^3:He^4 ratio, which is an important cosmological parameter.

The above discussion places its emphasis on probing the ISM by studying its absorption on the spectra from distant objects. However, it is also possible to detect the emission from hot gas directly. The EUVE spectrometers were optimised to study point sources rather than diffuse emission and were of limited use in studying the hot interstellar gas. The Cosmic Hot Interstellar Plasma Spectrometer (CHIPS, figure 11.1) is a University-Class Explorer (UNEX) mission funded by NASA. It will carry out all-sky spectroscopy of the diffuse background at wavelengths from 90 to 260 Å with a peak spectral resolution of $\lambda/150$. CHIPS data will help scientists determine the electron temperature, ionisation conditions, and cooling mechanisms of the million-degree plasma believed to fill the local interstellar bubble. The majority of the luminosity from diffuse million-degree plasma is expected to emerge in the poorly explored CHIPS band, making CHIPS data of relevance in a wide variety of galactic and extragalactic

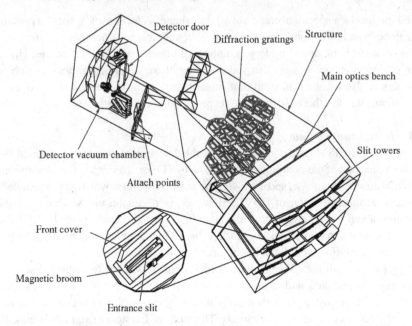

Fig. 11.1. Three-dimensional layout of the CHIPS spectrograph (courtesy M. Hurwitz).

astrophysical environments. The CHIPS instrument will be carried into space on board a dedicated spacecraft and is expected to be launched in 2002.

11.3.7 Requirements for new EUV instrumentation

Astronomers always require bigger and better telescopes for the next phase of their research. A corollary of this statement is that what astronomers actually get is a compromise between the available resources and the ideal, reduced to the minimum required to make a large enough scientific advance to justify the expenditure. The scientific questions/problems outlined in the preceding sections allow us to define such minimum requirements of the next generation of instrumentation. From the point of view of spectral resolution, a factor of 20 improvement (to $R \approx 4000$ on the resolution available with EUVE is needed to resolve the heavy element complexes in white dwarf atmospheres. With a corresponding velocity discrimination of $75 \, \mathrm{km \, s^{-1}}$, this would not be adequate to separate ISM and photospheric absorption features on velocity grounds, where differences are typically a few to a few tens of $\mathrm{km \, s^{-1}}$. It would probably be possible to determine the nature of any He II absorption lines found, as those in the photosphere will be broadened compared to the ISM. However, to truly resolve ISM and photospheric components will require a further factor of ten improvement in spectral resolution to $R \approx 40\,000$, similar to that available with the current generation of far-UV instrumentation.

Similar levels of performance to those discussed above will also be needed to obtain dynamical information. For example, a velocity resolution of $75 \, \mathrm{km \, s^{-1}}$ will allow the study of relatively close binary systems such as CVs or pre-CVs, which have radial velocity amplitudes of a few hundred $\mathrm{km \, s^{-1}}$. However, in those systems without a compact object the radial velocities will usually be much lower, requiring a correspondingly higher resolution.

While the resolution requirements are straightforward, depending directly on the scientific goals, it is more difficult to specify the effective area demands without carrying out detailed simulations for specific categories of source. In simplistic terms, if the spectral resolution is enhanced by a particular factor, the effective area must increase by the same amount to maintain the signal-to-noise in each pixel. However, when the spectral resolution is improved for a system of otherwise unresolved lines, the contrast of the line cores to continuum level is enhanced, so scientific goals may still be achieved without an automatic enlargement of the effective area. An increased ability to carry out time resolved studies, where the exposure times are determined by the time scale of the phenomenon being studied, is much more dependent on improvements in telescope effective area. Unlike the exposures of non-variable sources, any decrease in signal-to-noise commensurate with increased dispersion cannot be counteracted by increasing the exposure time.

11.4 Advanced instrumentation for EUV astronomy

The future of EUV astronomy depends significantly on advances in technology. Development of grating technology is needed to improve the achievable spectral resolution, coupled with improvements in efficiency as a contribution to enhancing the overall effective area. Similarly detector spatial resolution needs to evolve to accommodate the better spectral resolution. However, there does not currently seem to be much prospect of increasing detector quantum efficiencies. A primary limitation on the achievable effective area is the mirror technology. As in X-ray astronomy, the use of grazing incidence telescopes, like those employed on the ROSAT WFC and EUVE is the standard technique. Increasing the effective area significantly implies larger, more massive telescope systems, which must then be carried on board larger, more expensive satellites. As discussed earlier, the X-ray astronomy route of increasingly large observatory class missions is probably not practical for EUV astronomy. Therefore, more innovative approaches to mirror design are required.

It is well understood that grazing incidence systems are relatively inefficient light collectors, a single mirror only intercepting light in a thin annulus subtended by the leading element, when compared with the use of the full aperture by normal incidence telescopes. The ALEXIS mission has pointed the way forward by being the first experiment to use multilayer normal incidence optics for non-solar EUV astronomy. As discussed in section 4.4.1, the penalty for improved effective area, produced by the multilayers, is a decrease in the practical wavelength coverage. However, it is possible to envisage the production of low cost, highly efficient, EUV instruments that have wavelength range tuned to addressing particular astrophysical problems. We discuss here recent developments that have adopted this approach in producing the next generation of instruments.

11.4.1 The Joint Plasmadynamic Experiment

The Joint Plasmadynamic Experiment (J-PEX) was a collaborative project, led by the US Naval Research Laboratory (NRL) with contributions from the US Lawrence Livermore National Laboratory (LLNL), University of Leicester UK and the Mullard Space Science Laboratory (MSSL) of University College London. Its main aim was to answer the question regarding the presence of He II in the spectrum of G191–B2B, by resolving the individual He lines from the other heavier element species. As discussed in section 11.3.7, this requires an instrument with a resolving power $\approx \lambda / \Delta \lambda = 5000$, higher than previous instrumentation has delivered.

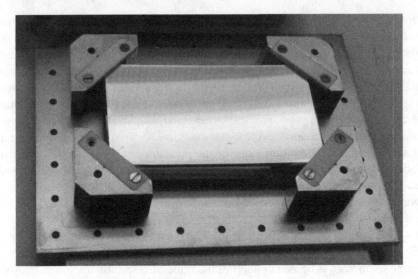

Fig. 11.2. Single J-PEX diffraction grating manufactured by Carl Zeiss.

The J-PEX experiment was a sounding rocket borne telescope, one of the first in a new generation of instruments employing optics operating at near-normal incidence. The conventional grazing incidence mirrors found in earlier instruments were replaced by MoSi multilayer coated reflection gratings which acted as both dispersion and light collection elements (Kowalski *et al.* 1999). Manufactured by Carl Zeiss, using multilayer coatings developed at NRL and LLNL, four gratings (figure 11.2) had a total geometric area of 512 cm^2 and had a spherical figure with a 2 m focal length. This arrangement enabled J-PEX to achieve an effective area in excess of 5 cm^2 and a resolving power of $\approx$4000 at 235 Å, more than ten times that of EUVE. The optical layout of the telescope is shown in figure 11.3.

The diffracted radiation was focused onto a microchannel plate detector, mounted along the grating optical axis to minimise optical aberrations. Developed and built at the University of Leicester, the detector design consisted of two 37 mm square Photonis MCPs with 6 μm pores, mounted in a chevron configuration. The front MCP was coated with a CsI photocathode to enhance detection efficiency. Imaging was by means of a progressive geometry vernier conductive anode readout which, along with the electronics and HV supply, has been developed and supplied by MSSL. The entire detector assembly was mounted in its own vacuum chamber with an integral ion pump, to allow operation of the detector on the ground and protect the photocathode. In the flight configuration, the front MPC was replaced by a circular 33 mm diameter IKI plate, having 10 μm pores, due to problems with the quantum efficiency of the Photonis MCPs (see Bannister 2001). The spatial resolution achieved by the detector was better than 20 μm (Bannister *et al.* 2000b; Lapington *et al.* 2000).

The J-PEX instrument was a slitless spectrometer. Hence, the absolute pointing requirements were not very critical. However, to achieve the high spectral resolution, the attitude drift needed to be very stable to avoid smearing of the spectral features in the dispersion direction. A new digital attitude control system (ACS) was utilised, coupled with an optical telescope to track residual drifts. The optical telescope consisted of a 12.5 cm diameter, 2 m focal length spherical mirror and a CCD detector. By imaging the target throughout the entire

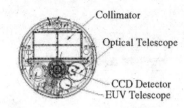

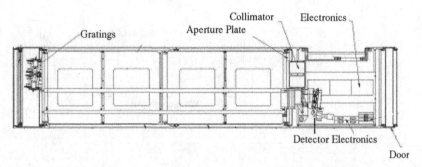

Fig. 11.3. Schematic view of the J-PEX high resolution spectrometer payload, designed and constructed for launch by a Terrier boosted Black Brant IX sounding rocket.

period of observation, this optical telescope ensures that post-flight aspect reconstruction can be performed, thus recovering resolution. A normal incidence multilayer EUV imaging telescope, coaligned with the spectrometer and focused on the spectrometer MCP detector, provided a backup system for attitude reconstruction by monitoring the position of the EUV target image.

The primary goal of J-PEX was to acquire an EUV spectrum of sufficient resolution to separate lines of He II, if present, from the large numbers of features from heavier elements such as C, N, O, Fe and Ni. The measured resolution of the instrument matched that required for this task. Furthermore, by analysing the width and depth of any He II features, the location of the material (photospheric or circumstellar/interstellar) can be determined.

J-PEX was first flown from White Sands Missile Range (WSMR) on a two-stage Black Brandt IX sounding rocket on February 25 2000, with an expected exposure time $\approx$300 s. Unfortunately, the rocket veered off course and had to be destroyed shortly before burn-out. However, although the instrument pointing could not be controlled and no science data were acquired, the telescope operated correctly and was recovered intact. The telescope was flown for a second time on February 21 2001 (NASA 36.195), completing its mission flawlessly.

Analysis of the J-PEX high resolution EUV spectrum of G191–B2B reveals some exciting results when comparing the observed spectrum (error bars in figure 11.4) with predictions based on a homogeneous composition stellar atmosphere and including interstellar H I, He I and He II absorption. The H I and He I column densities were fixed at values obtained from analysis of the broader band, lower resolution EUVE spectrum and the temperature and gravity taken from the latest optical analysis ($T = 54\,000$ K, $\log g = 7.5$; see Barstow *et al.* 1999 and 1998 respectively). The interstellar He II column density and photospheric He abundance were allowed to vary freely and the best match to the model obtained by a χ^2 minimisation technique. The best fit model folded through the J-PEX instrument response, assuming a nominal 0.05 Å resolution (fwhm), is shown in figure 11.4 (solid line histogram).

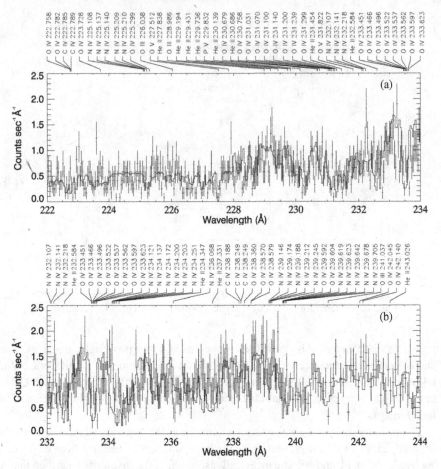

Fig. 11.4. (a) The spectrum of G191–B2B obtained by the J-PEX spectrometer, spanning the wavelength range 222–234 Å (error bars). The solid histogram is the best fit theoretical model of the star and ISM, as described in the text. The strongest predicted lines of He, C, N, O, and P are labelled with their ionisation state and wavelength. Lines of Fe and Ni are too numerous to include and account for unlabelled individual features and broader absorption structures. (b) As (a) but spanning the wavelength range 232–244 Å.

The good agreement between model and data is striking, with the most prominent resolved feature a strong O IV absorption line at 233.5 Å. Many other features are present, which are mainly blends of large numbers of Fe V and Ni V lines. Between ≈227 and 232 Å is a broad 'bump' which is a characteristic of the overlapping series of interstellar He II absorption lines superposed on a continuum. Taken with the strong depression of the flux below 227 Å, this provides conclusive proof that interstellar or circumstellar He II is present along the line-of-sight. With a total column density of 3.8×10^{17} atoms cm^{-2}, the implied He ionisation fraction of ≈0.7 is substantially higher than typical of the LISM (0.25–0.5). However, a circumstellar C IV component in the far-UV has recently been identified (Bannister *et al.* 2001a) and suggesting that part of the ionised He II may be circumstellar. The best fit model spectrum implies that there is also photospheric helium present. However, the strongest

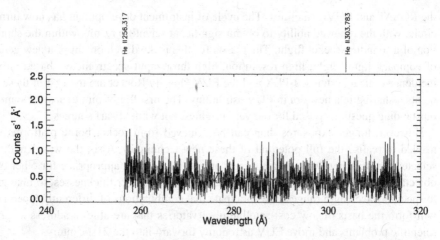

Fig. 11.5. The spectrum of G191–B2B obtained by the Colorado EUV opacity rocket spanning the wavelength range 255 to 310 Å (error bars). The solid histogram shows the bet fit theoretical model. The predicted He II lines are labelled along with their wavelength.

predicted photospheric line at 243.3 Å is only marginally detected with the current signal-to-noise. The inferred He abundance is 8.0×10^{-5}, about four times the limit imposed by the absence of detectable He II at 1640 Å in the STIS spectrum. In this case the photospheric He must be in a stratified configuration as suggested by Barstow and Hubeny (1998).

11.4.2 *The Colorado EUV opacity rocket*
In parallel with the development of the J-PEX telescope, a research team from the University of Colorado built a similar instrument, also based on normal incidence reflection grating optics, coupled with a MCP imaging detector. The 'EUV opacity rocket' (Gunderson *et al.* 1999, 2001) was aimed at similar scientific goals, but had a smaller peak effective area (2.5 cm², imposed by a smaller available payload diameter) and was tuned to a slightly different spectral range (254–317 Å, compared with 225–250 Å for J-PEX). The spectral resolving power of ≈2500–3000 was about half that of the J-PEX design. This telescope was flown from WSMR on September 27 1999. The spectrograph consisted of two adjacent, spherically curved, holographically ruled, ion etched multilayer coated diffraction gratings in near normal incidence Wadsworth f/11 mounts. The instrument obtained an exposure of 209 s at altitudes above 200 km, producing a detectable spectrum for G191–B2B, but with low signal-to-noise (figure 11.5), containing approximately ten counts per 0.1 Å spectral bin. Consequently, no individual absorption features could be positively identified, although it was possible to correlate the expected distribution of opacities with the overall shape of the spectrum. A detailed statistical analysis of the data identified likely absorption from OI V, Fe IV and Fe V. Only upper limits on the presence of He II 256.32 Å and 304.78 Å features were obtained.

11.5 Concluding remarks
Remarkable progress has been made in the field of EUV astronomy in its first 30 years, from the days of the first sounding rocket flights through to the major contributions from

the ROSAT and EUVE satellites. The cycle of instrument development has now turned full circle, with the renewed ability to obtain significant scientific results within the short duration of a sounding rocket flight. The reason for this is the development of a new generation of compact, lightweight, high resolution, high throughput spectrometers based on normal incidence optical systems. J-PEX and the EUV Opacity Rocket are the first of these instruments, ushering in a new era in EUV astronomy. The first flights aim to answer some of the outstanding questions posed for the very brightest hot white dwarf sources.

However, longer exposures than can be achieved from rocket borne platforms will be needed to realise the full potential of these new techniques. Also, the wavelength ranges selected for studying white dwarf targets are not necessarily appropriate for other types of object. Fortuitously, it is possible, by adjusting the multilayer thicknesses, to tune gratings to specific wavelength ranges. We anticipate that combinations of differently tuned gratings will form the basis of low cost satellite observatories that are able to address a variety of scientific problems and move EUV astronomy forward into the 21st century.

Appendix. A merged catalogue of Extreme Ultraviolet sources

We have merged all the ROSAT WFC and EUVE source lists to produce a composite catalogue of EUV sources. The data are presented in three tables:

- Table A.1. All-sky survey source list with position information and optical identification, together with spectral type and visual magnitude where appropriate.
- Table A.2. All-sky survey source list with count rates in the WFC S1, WFC S2, EUVE 100 Å, EUVE 200 Å, EUVE 400 Å and EUVE 600 Å filters.
- Table A.3. Composite EUVE deep survey source list.

Table A.1. *All-sky survey source list with position information and optical identification, together with spectral type and visual magnitude where appropriate.*

Name	RA$_{2000}$	DEC$_{2000}$	ID1	ID2	Spec. Type	M_V
EUVE J0001+515	00:01:15	+51:33.6	NO ID			
EUVE J0001+276	00:01:27	+27:38.5	NO ID			
EUVE J0002−495	00:02:50	−49:34.6	NO ID			
EUVE J0003+242	00:03:51	+24:17.4	NO ID			
2RE J0003+433	00:03:57	+43:35.4	WD 0001+433		DA1	16.8
EUVE J0005−500	00:05:49	−50:5.7	JL 156	HE 0003−5023	Sy	14
EUVE J0006+201	00:06:35	+20:10.4	[SPB96] 26		UV	10.9
EUVE J0006+290	00:06:37	+29:0.3	HD 166	BD+28° 4704	K0 V	6.13
2RE J0007+331	00:07:32	+33:17.6	WD 0004+330	GD 2	DA1	13.82
EUVE J0008+208	00:08:56	+20:48.8	G131−26	LTT 10045	M4	13.54
2RE J0012+143	00:12:31	+14:34.3	SAO 91772		K0	8.6
2RE J0014+691	00:14:35	+69:13.6	AG+68° 14		F8	10.3
EUVE J0016+198	00:16:14	+19:51.6	G32−7	LTT 10088	M4.	13.22
2RE J0018+305	00:18:21	+30:57.2	BD+30° 34		G5	8.6
EUVE J0019+064	00:19:43	+06:24.4	WD 0017+061	PHL 790	DA2	15.9
2RE J0024−741	00:24:45	−74:12.9	CPD−74° 35		F7V	9.7
2RE J0029−632	00:29:57	−63:25.1	WD 0027−636		DA1	15
2RE J0030−481	00:30:27	−48:13	TD1 264	HD 2726	F2V	5.69
2RE J0035−033	00:35:14	−03:35.6	13 Cet A		F7V+G4V	5.2
2RE J0039+103	00:39:40	+10:39.4	BD+09° 73		K5	10.6
EUVE J0040+307	00:40:17	+30:46.5	NO ID			
2RE J0041+342	00:41:17	+34:25.8	EUVE J0041+344		Ge	9.5
2RE J0042+353	00:42:48	+35:32.9	FF And		M0V:e	10.38
2RE J0043−175	00:43:34	−17:59.3	β Cet	HD 4128	K0III	2.04
2RE J0044+093	00:44:02	+09:33.3	BD+08° 102		K	10
EUVE J0045−287	00:45:53	−28:42.4	NO ID			
EUVE J0046−170	00:46:22	−17:1.5	NO ID			
2RE J0047−115	00:47:03	−11:52.4	NGC 246		Plan Neb	11.78
2RE J0047+241	00:47:20	+24:15.9	ζ And	HD 4502	K1IIe	4.06
EUVE J0051−410	00:51:24	−41:04.5	NO ID			
2RE J0053−743	00:53:05	−74:39.5	CF Tuc		G3:V	7.6
2RE J0053−325	00:53:17	−33:0.2	WD 0050−332	GD 659	DA1.5	13.38
2RE J0053+360	00:53:39	+36:1.8	WD 0053+360J		DA2	15.1
2RE J0054−142	00:54:56	−14:26	WD 0052−147	GD 662	DA2	15.12
EUVE J0057−223	00:57:18	−22:22.9	Ton S 180	KUV 00549−2239	Seyfert	15
EUVE J0102−027	01:02:21	−02:45.7	NO ID			
EUVE J0103+623	01:03:24	+62:22.2	1RXS J010318.0+622146	G243−55	M5	13.66
2RE J0103−725	01:03:49	−72:54.8	NO ID			
EUVE J0106−637	01:06:15	−63:47.1			dM4e	13.8
2RE J0106−225	01:06:47	−22:51.5	HD 6628		G5V	7.68
2RE J0108−353	01:08:20	−35:34.9	WD 0106−358	GD 683	DA2	15.8
EUVE J0113+624	01:13:05	+62:28.8	NO ID			
EUVE J0113−738	01:13:19	−73:49	NO ID			
EUVE J0115+269	01:15:56	+26:54.2	NO ID			
2RE J0116−022	01:16:35	−02:30.2	AY Cet		G5IIIe	5.41
EUVE J0122−603	01:22:01	−60:22.5	NO ID			
EUVE J0122−567	01:22:16	−56:46.6	HD 8435	CPD−57° 296	G6/G8III	8.8
2RE J0122+004	01:22:49	+00:00	BD−0° 210		G5V:+G5V:	8.6
2RE J0122+072	01:22:57	+07:25.4	AR Psc	1H 0123+075	G5	7.2
EUVE J0123+178	01:23:52	+17:52.5	NO ID			
EUVE J0129−185	01:29:26	−18:31.4	NO ID			
EUVE J0134+536	01:34:10	+53:37.3	NO ID			
2RE J0134−160	01:34:24	−16:7.2	WD 0131−163	GD 984	DA1+dM	13.96
2RE J0135−295	01:35:01	−29:54.9	GJ 60A+B		K3V+K4V	7.9
EUVE J0136+414	01:36:47	+41:25.2	HD 9826	BD+40° 332	F8	4.1
2RE J0137+183	01:37:39	+18:36.1	BD+17° 232		dK2e	9.9
2RE J0138+252	01:38:52	+25:23.1	WD 0136+251	PG 0136+251	DA1	15.83
2RE J0138+675	01:39:00	+67:54.1	NO ID			
2RE J0138−175	01:39:00	−17:57.3	UV Cet		M5.5V:e	12.52

Table A.1. (*cont.*)

Name	RA$_{2000}$	DEC$_{2000}$	ID1	ID2	Spec. Type	M_V
2RE J0140−675	01:40:58	−67:53.1	BL Hyi	1H 0136−681	CV	18
EUVE J0142−241	01:42:51	−24:11.4	LTT 925	G 274−85	K	12.6
2RE J0148−253	01:48:07	−25:32.2	WD 0148−255J	GD 1401	DA2	14.69
2RE J0151+673	01:51:06	+67:39.8	WD 0147+674	GD 421	DA1.5	14.42
EUVE J0153+295	01:53:06	+29:35.2	α Tri	TD1 1077	F6IV	3.41
2RE J0153+603	01:53:49	+60:36.2	GSC4032.01402			10.7
EUVE J0154+208	01:54:36	+20:48.7	HD 11636	TD1 1090	A5 V	2.64
2RE J0155−513	01:55:54	−51:36.7	χ Eri	TD1 1117	G8IIIbCNv	3.7
EUVE J0202+105	02:02:30	+10:35.0	1RXS J020228.9+103455		X	13
2RE J0205+771	02:05:15	+77:16.6	47 Cas		F0Vn	5.27
EUVE J0205+093	02:05:22	+09:20.3	NO ID			
EUVE J0208−100	02:08:43	−10:05.5	HD 13191	BD−10° 438	G5	9.9
EUVE J0210−472	02:10:12	−47:16.2	NO ID			
EUVE J0211+043	02:11:52	+04:24.0	BD+03° 301	SAO 110399	K5	9.1 :
EUVE J0213+057	02:13:17	+05:46.0	BD+05° 300	SAO 110413	G0	9.4 :
EUVE J0213+368	02:13:26	+36:49.6	1RXS J021320.6+364837		X	
2RE J0214+513	02:14:20	+51:34.9	GSC3293.00848			14.4
EUVE J0215−095	02:15:50	−09:30.8	1RXS J021559.9-092913		X	
2RE J0218+143	02:18:48	+14:37	WD 0216+143	PG 0216+143	DA2	14.53
2RE J0222+503	02:22:33	+50:34.3	BD+49° 646		star	9.4 ·
EUVE J0223+650	02:23:01	+65:2.3	NO ID			
EUVE J0223+371	02:23:15	+37:12	NO ID			
EUVE J0223−264	02:23:28	−26:28.4	NO ID			
2RE J0228−611	02:28:21	−61:18.3	HD 15638		F3IV/V	8.8
2RE J0230−475	02:30:53	−47:55.7	WD 0229−481	LB 1628	DA1	14.53
2RE J0232−025	02:32:47	−02:55.5	GSC4701.00192			14.8
2RE J0234−434	02:34:23	−43:47.9	CC Eri		M0Vp	8.7
2RE J0235+034	02:35:08	+03:44.6	WD 0232+035	Feige 24	DA1+dM1.5	12.4
2RE J0237−122	02:37:26	−12:20.9	WD 0235−125	PHL 1400	DA1.5	15.3
2RE J0238−525	02:38:43	−52:57.1	HD 16699B		K	7.5
2RE J0238−581	02:39:00	−58:11.4	GJ 1049	CD−58° 538	M0Ve	9.65
EUVE J0242−196	02:39:03	−19:36.6			dM0e	12.8
EUVE J0239−196	02:39:15	−19:38.1			dM0e	12.8
2RE J0239+500	02:39:50	+50:04.1	WD 0239+500J		DA	16
EUVE J0241−006	02:41:16	−00:41.4	NGC1068	3C 71	Seyfert 2	32
2RE J0241−525	02:41:44	−52:59.9			dMe+dMe	10.7
2RE J0243−375	02:43:25	−37:55.4	UX For		G6V	8.1
EUVE J0245−185	02:45:00	−18:35.9	τ^1 Eri	HD 17206	F5/F6V	4.50
2RE J0248+310	02:48:45	+31:7.1	VY Ari	HD 17433	K0	6.76
EUVE J0249+312	02:49:44	+31:13.7	NO ID			
EUVE J0249+099	02:49:53	+09:56	NO ID			
EUVE J0250+312	02:50:58	+31:14.7	NO ID			
EUVE J0251+312	02:51:02	+31:14.5	NO ID			
EUVE J0251−373	02:51:22	−37:19.1	NO ID			
EUVE J0252−127	02:52:33	−12:45.6	HD 17925	TD1 1733	K1V	6
EUVE J0254−511	02:54:32	−51:9.2			dM3e	
2RE J0254−051	02:54:39	−05:19.1	HD 18131 B		K0+DA	7.2
2RE J0255+474	02:55:43	+47:46.5			K+K	10.3
EUVE J0256+080	02:56:05	+08:3.7	NO ID			
EUVE J0304+029	03:04:35	+02:59.3	WD 0302+027	Feige 31	DA1.5	14.88
2RE J0308+405	03:08:10	+40:57.7	β Per	HD 19356	B8V	2.12
EUVE J0309+496	03:09:06	+49:37.3	ι Per	HD 19373	G0	4.05
EUVE J0309−009	03:09:51	+00:00	NO ID			
EUVE J0310−443	03:10:48	−44:22.3	NO ID			
EUVE J0311−229	03:11:21	−22:54.8	NO ID			
EUVE J0311−318	03:11:33	−31:53.1	NO ID			
2RE J0312−285	03:12:04	−28:58.9	α For	TD1 1986	F8IV	3.85
2RE J0312−442	03:12:26	−44:24.8	TD1 2005		F7III	5.93
2RE J0314−223	03:14:14	−22:34.7	EF Eri	1H 0311−227	CV	13.7
2RE J0317−853	03:17:02	−85:32.7	WD 0317−855J	LB 09802	DAp	13.9

Table A.1. (*cont.*)

Name	RA$_{2000}$	DEC$_{2000}$	ID1	ID2	Spec. Type	M_V
EUVE J0318−336	03:18:56	−36:24.6			dM4e	17.2
2RE J0319+032	03:19:17	+03:22.9	κ Cet	HD 20630	G5Ve	4.8
EUVE J0320−122	03:20:06	−12:12.3	PHL 8628			18.5
2RE J0320−764	03:20:13	−76:49.2	GSC9363.00867			14.7
2RE J0322−534	03:22:14	−53:45.1	WD 0320−539	LB 1663	DA	14.9
2RE J0324+234	03:24:05	+23:47.1	AC+23° 368−59		M0V	10.3
EUVE J0325+012	03:25:17	+01:17.5	NO ID			
2RE J0326+284	03:26:36	+28:43.3	UX Ari		G5IV	6.47
2RE J0327+094	03:27:08	+09:43.6	ξ Tau		B9Vn	3.7
EUVE J0328+041	03:28:12	+04:08.8	HD 21497	AG+03° 397	F8	8.7
2RE J0328+040	03:28:13	+04:09.6	SAO 111210		K0	8.8
EUVE J0328−296	03:28:34	−29:41	NO ID			
2RE J0332−092	03:32:56	−09:26.7	ε Eri		K2V	3.73
2RE J0333+461	03:33:16	+46:15.6	HD 21845		K2	8.5
EUVE J0333−258	03:33:18	−25:51.9	NGC 1360	TD1 31132	Plan Neb	11.33
EUVE J0335−257	03:35:30	−25:43.6	UZ For	EXO 0333.3−2554	CV	18.2
2RE J0335+311	03:35:31	+31:13.9	HD 22179		G0	9.1
EUVE J0335−348	03:35:34	−34:48.5	WD0335−348		DA1.5	16.4
2RE J0336+003	03:36:48	+00:00	V711 Tau	1H 0327+000	G9V	5.71
EUVE J0336+004	03:36:53	+00:24.3	HR 1101	HD 22484	F9IV−V	4.28
EUVE J0336+005	03:36:48	+00:34.9	V711Tau	HD 22468	G5	6.2
2RE J0337+255	03:37:11	+25:59.5	BD+25° 580	V837Tau	G2V+KV	8.1
EUVE J0337−419	03:37:20	−41:55.8	WD 0337−419J		DA1	17.0
EUVE J0346−575	03:46:36	−57:32.7	NO ID			
2RE J0348−009	03:48:37	−00:59.5	WD 0346−01	GD 50	DA1	14.06
EUVE J0349−477	03:49:18	−47:43	NO ID			
2RE J0350+171	03:50:24	+17:14.8	WD 0347+171	V471 Tau	DA1.5	9.2
EUVE J0352−492	03:52:40	−49:13.8			dK5e	
EUVE J0356−366	03:56:34	−36:40.4	WD 0354−368	1E 0354.6−3650	DA1+G2V	18.5
EUVE J0357+286	03:57:08	+28:37.8	RE J0357+283	V1092 Tau	K2 V	13.0
2RE J0357−010	03:57:30	−01:9.2	GJ 157A/B		K5V+M3Ve	8
EUVE J0359−398	03:59:30	−39:53.7	HD 25300	CF 1154	K0	9.9
2RE J0402−001	04:02:36	+00:00	TD1 2766	HD 25457	F5V	5.38
EUVE J0404+310	04:04:51	+31:1.7	NO ID			
EUVE J0405+220	04:05:26	+22:02.4	HR 1262	HD 25680	G5V	5.90V?
EUVE J0407−525	04:07:33	−52:34	AG Dor	HD 26354	K1Vp	8.6
2RE J0407+380	04:07:32	+38:4.4	GJ 160.1AB		K2+dK2e	7.3
2RE J0409−711	04:09:11	−71:17.8	VW Hyi		Dwarf Nov	
2RE J0409−075	04:09:40	−07:53.1	EI Eri		G5IV	7.08
2RE J0410+592	04:10:55	+59:25	WD 0410+594J		DA	14.4
EUVE J0412+236	04:12:00	+23:38.4	BD+23° 635	HD 284163	K0	9.35
EUVE J0414−624	04:14:25	−62:28.4	α Ret	TD1 3005	G8II−III	3.35
EUVE J0414+224	04:14:35	+22:25.0	HD 26737	BD+22° 657	F5V	7.04
2RE J0415−073	04:15:22	−07:38.5	o² Eri C	1E 041258−0745.9	M5e	11.17
2RE J0415−402	04:15:38	−40:22.7	WD 0415−402	1RXS J0415.7−4022	DA1	16.8
EUVE J0415+658	04:15:46	+65:52.5	NO ID			
2RE J0418+231	04:18:11	+23:16.4	HD 284303	BD+22° 669	K0V	9.5
EUVE J0419+217	04:19:41	+21:45.6			dM2e	11.7
EUVE J0419+156	04:19:47	+15:37.5	γ Tau	HD 27371	K0 III	3.65
EUVE J0422+150	04:22:45	+15:03.6	HD 27691	BD+14° 690	G0	6.99
EUVE J0423+149	04:23:54	+14:55.7	1RXS J042350.4+145514		M2.5	13.32
EUVE J0424+147	04:24:15	+14:45.3	HD 27836	BD+14° 693	G1 V	7.62
EUVE J0425−068	04:25:09	−06:52.8	NO ID			
2RE J0425−571	04:25:37	−57:13.8	LB 01727	1ES 0425−57.3	AGN Seyf.1	15.1
EUVE J0426+155	04:26:09	+15:33.8	HD 28034	BD+15° 624	G0	7.49
EUVE J0426−379	04:26:22	−37:56.7			dK4e	11.5
2RE J0426+153	04:26:24	+15:36.5	V777 Tau		A8Vn/F0V	4.5
2RE J0427+740	04:27:39	+74:7.3	WD 0427+741J		DA1	15
2RE J0428+165	04:28:38	+16:58.2	WD 0425+168		DA2	14.2
EUVE J0429+194	04:29:23	+19:26.5	NO ID			

Table A.1. (*cont.*)

Name	RA$_{2000}$	DEC$_{2000}$	ID1	ID2	Spec. Type	M_V
EUVE J0432+054	04:32:09	+05:24.9	TD1 3277	1E 0429.4+0518	F5V	6.4
2RE J0436+270	04:36:48	+27:7.9	V833 Tau		dK5e	8.4
2RE J0437−022	04:37:37	−02:28.7	HR 1474	HD 29391	F0V	5.2
EUVE J0439−342	04:39:28	−34:17.6	NO ID			
EUVE J0438+231	04:38:50	+23:08.4	HD 29419	BD+22° 721	F5	7.53
EUVE J0439−716	04:39:39	−71:40.6	NO ID			
EUVE J0440−418	04:40:37	−41:51.6	α Cae	TD1 3476	F2V	4.45
EUVE J0441+209	04:41:20	+20:54.0	V834Tau	HD 29697	K3 V	7.98
EUVE J0441+483	04:41:31	+48:18.5	HR 1482	HD 29526	A0V	5.67
2RE J0443−034	04:43:08	−03:46.9	WD 0443−037		DA1	16
EUVE J0444−278	04:44:26	−27:52.5	NO ID			
EUVE J0444−262	04:44:44	−26:17.8	NO ID			
2RE J0446+763	04:46:08	+76:36.4	HR 1468	HD 29329	F7V	6.5
EUVE J0446+090	04:46:49	+09:00.4	HD 30311	BD+08° 759	F5	7.26
2RE J0447−275	04:47:22	−27:50.3			dMe	13
EUVE J0447−169	04:47:41	−16:57.3	58 Eri	TD1 3566	G3V	5.5
EUVE J0449+057	04:49:35	+05:42.0	NO ID			
EUVE J0449+236	04:49:47	+23:40.3	EM LkCa 18			?
EUVE J0449+069	04:49:51	+06:58.1	HD 30652	TD1 3592	F6 V	3.19
EUVE J0450−086	04:50:30	−08:38.4			dK4e	13.2
EUVE J0451−391	04:51:28	−39:6.3	NO ID			
2RE J0453−421	04:53:27	−42:13.1	RS Cae		CV	−1
2RE J0453−555	04:53:30	−55:51.4	GJ 2036A/B		M2Ve+MVe	11.1
EUVE J0456+326	04:56:13	+32:37.8	NO ID			
2RE J0457−280	04:57:14	−28:7.7	WD 0455−282		DA1	13.95
EUVE J0457−063	04:57:19	−06:22.2	1RXSJ045717.3-062138		X	
2RE J0458+002	04:58:17	+00:00	V1198 Ori		G5	7.1
EUVE J0459−030	04:59:14	−03:04.3	NO ID			
2RE J0459+375	04:59:15	+37:53.2	BD+37° 1005A/B		A1V+F9	5
2RE J0459−101	04:59:52	−10:15.7	63 Eri	HD 32008	G4V	5.38
2RE J0500−362	05:00:04	−36:24.1	GSC7053.00682			13.7
2RE J0500−345	05:00:47	−34:50.5	GSC7050.00934			15.2
EUVE J0500−057	05:00:50	−05:45.8	HD 32147	BD−05° 1123	K3 V	6.2
EUVE J0501−064	05:01:56	−06:26.7	HD 32307	BD−06° 1064	A2	8.9
2RE J0501+095	05:02:00	+09:58.3	RBTS 285		M3Ve	11.5
2RE J0503−285	05:03:56	−28:54.4	WD 0501−289		DO1	13.95
EUVE J0504−039	05:04:47	−03:54.9	HD 32704	BD−04° 1032	G8 V	9.0
2RE J0505−572	05:05:31	−57:28.2	ζ Dor	HD 33262	F7V	4.7
2RE J0505+524	05:05:33	+52:49.7	WD 0501+527	G191−B2B	DA1	11.78
EUVE J0505−181	05:05:33	−18:8.2	NO ID			
2RE J0506−213	05:06:50	−21:35	BD−21° 1074		M2e	10.1
EUVE J0510−578	05:10:59	−57:51.2	NO ID			
EUVE J0511+225	05:11:13	+22:35.7	NO ID			
2RE J0512−004	05:12:08	+00:00	WD 0509−007		DA2	13.8
2RE J0512−115	05:12:17	−11:52.3	HD 33802 B	BD−12° 1095B	G5Ve	10.8
2RE J0512−414	05:12:24	−41:45.1	WD 0512−417		DA1	17.26
2RE J0514−151	05:14:27	−15:14.7	GSC5902.00227			15.3
2RE J0515+324	05:15:25	+32:41	KW Aur C	HD 33959 C	A2+DA1	7.95
2RE J0516+455	05:16:43	+45:59.9	α Aur	HD 34029	G5IIIe	0.08
2RE J0517−352	05:17:24	−35:22.4			dMe	13
EUVE J0520+363	05:20:22	+36:18.6	NO ID			
2RE J0520−394	05:20:35	−39:45.2	HD 35114		F6V	7.3
2RE J0521−102	05:21:19	−10:29.1	WD 0518−105		DA1.5	15.9
EUVE J0521+395	05:21:50	+39:31.8	HD 278078		A5	10
2RE J0524+172	05:24:26	+17:22.8	111 Tau	HD 35296	F8V	5
2RE J0527−115	05:27:06	−11:53.8	HD 35850	TD1 4647	F7V	6.35
EUVE J0527+654	05:27:41	+65:26.3	NO ID			
2RE J0528−652	05:28:44	−65:26.9	AB Dor		K1IIIp	6.83
2RE J0530−191	05:30:18	−19:16.7	SAO 150508		G5	9
EUVE J0531−036	05:31:25	−03:40.1	GJ 205	HD 36395	M1.5V	7.92V?

Table A.1. (*cont.*)

Name	RA$_{2000}$	DEC$_{2000}$	ID1	ID2	Spec. Type	M_V
EUVE J0531−068	05:31:35	−06:52.7	Parenago 742			12.7
2RE J0531−462	05:31:37	−46:24	UW Pic		CV AMHer	17
2RE J0532−030	05:32:06	−03:5.2	HBC 97		dMe	12.3
2RE J0532+094	05:32:17	+09:49.4	V998 Ori		M4e	11.5
EUVE J0532+510	05:32:34	+51:3.5	NO ID			
2RE J0533+015	05:33:45	+01:55.9	V371 Ori		M3Ve	11.7
2RE J0534−151	05:34:10	−15:16.8	HD 36869		F5	8
2RE J0534−021	05:34:16	−02:15.2	WD 0531−022		DA1.5	16.2
2RE J0536+111	05:36:35	+11:18.6	GJ 208		dMe	8.8
EUVE J0536−060	05:36:41	−06:05.2	HD 37209	TD1 4984	B1 V	5.72
2RE J0536−475	05:36:58	−47:57.1	HD 37572	SAO 217430	K0V	8.2
2RE J0537−393	05:37:03	−39:33	SAO 196024		K0	9.6
EUVE J0538−066	05:38:34	−06:38.0	2E0536.0−0642		X	
2RE J0539−193	05:39:23	−19:34.3	GSC5925.01547			11
2RE J0540−193	05:40:22	−19:39.6	SAO 150676		F5	8.7
2RE J0540−201	05:40:35	−20:18.1	TW Lep		F+G8III	7
2RE J0541+532	05:41:20	+53:29.5	DM+53° 934		K1V	6.2
EUVE J0543+103	05:43:01	+10:18.9	NO ID			
2RE J0544−222	05:44:26	−22:25.1	HD 38392	GJ 216B	K2V	6.1
2RE J0545−595	05:45:12	−59:55	SAO 234124	SRS 6622	G5	9.2
2RE J0550+000	05:50:39	+00:00	WD 0548+000	GD 257	DA1	14.77
2RE J0550−240	05:50:50	−24:08.7	WD 0550−241J		DA1	16.2
EUVE J0551−546	05:51:20	−54:40.9	NO ID			
2RE J0552−570	05:52:17	−57:8.2	HR 2072	HD 39937	F+G5III	5.9
2RE J0552+155	05:52:28	+15:53.4	WD 0549+158	GD 71	DA1.5	13.03
2RE J0553−815	05:53:03	−81:56.6	HD 42270		K0	9.1
2RE J0554+201	05:54:24	+20:16.7	χ^1 Ori	HD 39587	G0V	4.41
EUVE J0555−132	05:55:31	−13:17.9	NO ID			
2RE J0556−141	05:56:24	−14:10.6	HD 40136	GJ 225	F1V	3.7
EUVE J0557+120	05:57:21	+12:3.3	NO ID			
2RE J0557−380	05:57:49	−38:2.9	TY Col		G4V?	9.5
2RE J0558−373	05:58:12	−37:35	WD 0556−375		DA1	14.37
2RE J0600+024	06:00:04	+02:43	G 99−49	G 106−17	M4	11.33
2RE J0602−005	06:02:40	+00:00	BD−00° 1147		K0	8.6
2RE J0604−482	06:04:49	−48:26.9	HD 41824	TD1 6107	G6V	6.57
2RE J0604−343	06:04:50	−34:33.8			dMe	13.4
2RE J0605−482	06:05:02	−48:20.1	WD 0605−483J		DA1.5	15.9
EUVE J0607+472	06:07:36	+47:12.2	1RXSJ060732.5+471154		X	
EUVE J0608+590	06:08:39	+59:2.7	NO ID			
EUVE J0609−358	06:09:23	−35:49.6			dM4e	11.5
EUVE J0611+578	06:11:34	+57:49.8	NO ID			
EUVE J0611+482	06:11:44	+48:13.8	1RXSJ061145.4+481323		X	
2RE J0612−164	06:12:39	−16:48.7	BD−16° 1396		K0	9.1
EUVE J0612−376	06:12:39	−37:36.9	NO ID			
2RE J0613−274	06:13:10	−27:41.4	GSC6513.00291			13.3
2RE J0613−235	06:13:44	−23:51.7	HR 2225	HD 43162	G5	6.4
EUVE J0613−355	06:14:00	−35:36	NO ID			
2RE J0616−645	06:16:54	−64:57.4	GSC8901.00179			15.5
EUVE J0618+750	06:18:11	+75:5.7	GSC 04525−00194		dM3e	10.9
EUVE J0618−199	06:18:27	−19:59.7	NO ID			
2RE J0618−720	06:18:29	−72:2.6	CPD−71° 427		K3V	9.7
2RE J0619−032	06:19:10	−03:25.7	BD−03° 1386		G0	8.5
2RE J0620+132	06:20:50	+13:24.4	WD 0620+134J		DA1	14.5
2RE J0622−601	06:22:35	−60:13.3	HD 45270		G1V	6.6
2RE J0622−175	06:22:40	−17:57.8	β CMa	HD 44743	B1II/III	1.98
2RE J0623−374	06:23:13	−37:41.2	WD 0621−376	RE J0623−374	DA1	12.09
EUVE J0624−526	06:24:04	−52:40.7	α Car	TD1 7115	F0II	−0.72
EUVE J0624+229	06:24:10	+22:56.5	NO ID			
2RE J0625−600	06:25:56	−60:3.2	GSC8894.00426		dM5e	13.3
2RE J0626+184	06:26:09	+18:45.5	OU Gem		K0	6.73

Table A.1. (*cont.*)

Name	RA₂₀₀₀	DEC₂₀₀₀	ID1	ID2	Spec. Type	M_V
2RE J0629−024	06:29:24	−02:48.2	V577 Mon		M7	11.1
2RE J0631+500	06:31:02	+50:2.7	GJ 3395	StKM1−598	dMe	11.2
EUVE J0631−231	06:31:34	−23:11.2	2E0629.3−2308		X	
2RE J0632−050	06:32:58	−05:5.4	WD 0630−050		DA.5	15.54
EUVE J0632−232	06:32:21	−23:14.8	1WGAJ0632.2		X	
EUVE J0632−065	06:32:45	−06:32.7	HD 46361	BD−06° 1	G0	9.5
EUVE J0633−231	06:33:05	−23:11.3	2E0630.9−2307		X	
2RE J0633+200	06:33:41	+20:1.5	WD 0633+200J		DA1	17.5
2RE J0633+104	06:33:52	+10:41.9	WD 0631+107		DA2	13.82
EUVE J0634−698	06:34:37	−69:51.3	HD 47875	CD−69° 399	G3V	9.2
EUVE J0636−465	06:36:46	−46:35.8	NO ID			
2RE J0637−613	06:38:03	−61:31.9	HD 48189	TD1 7857	G1.5V	6.18
EUVE J0639−265	06:39:03	−26:35.4	HD 47787	SAO17209	K1 V	8.9
EUVE J0640−035	06:40:26	−03:32.5	HD 295290	BD −03° 1538	G0	9.1
EUVE J0642−046	06:42:01	−04:41.7	NO ID			
EUVE J0642−041	06:42:51	−04:08.6	HD 48332	TD1 7893	F5	6.65
EUVE J0642−174	06:42:51	−17:25.0	NO ID			
EUVE J0644−310	06:44:32	−31:03.3	HD 48917 B	TD1 8028	F	10.6
EUVE J0644−312	06:44:55	−31:13.7	NO ID			
2RE J0645−164	06:45:11	−16:42.4	α CMa B	Sirius B	DA2	8.44
EUVE J0647−506	06:47:58	−50:38.1	NO ID			
EUVE J0648−482	06:48:55	−48:17.4	NO ID			
2RE J0648−252	06:48:59	−25:23.4	WD 0648−253J		DA2	14.5
EUVE J0649−015	06:49:28	−01:35	HD 292477		F8	10
2RE J0650−003	06:50:50	−00:32.7	PMN J0650−0031	HD 49933	F2V	5.77
EUVE J0652−051	06:52:10	−05:10.0	HD 50281	2E 0649.8−0506	K3 V	6.57
EUVE J0652−596	06:52:47	−59:41.1	NO ID			
2RE J0653−430	06:53:48	−43:5.2	HD 51062		G3	8.9
EUVE J0653−564	06:53:58	−56:25.4	WD 0653−564		DA1.5	17.0
2RE J0654−020	06:54:15	−02:8.9	WD 0651−020	PG 0651−020	DA1.5	14.84
2RE J0658−285	06:58:40	−28:57.5	ε CMa	HD 52089	B2Iab	1.5
EUVE J0700−034	07:00:37	−03:20.0	NO ID			
2RE J0702+125	07:02:05	+12:58.4	RE J0702+125		Ke	9.9
EUVE J0702−064	07:02:06	−06:24.9	NO ID			
2RE J0702+255	07:02:23	+25:50.9	BD+26° 1435		G5	7.9
EUVE J0702−018	07:02:56	−01:47.6			dM2e	14.9
EUVE J0704+717	07:04:16	+71:44.2	NO ID			
EUVE J0710+184	07:10:18	+18:26.6	SAO 96579	AG+18° 703	K0	8.5
2RE J0710+451	07:10:59	+45:15.7	HD 54402		K0	7.7
2RE J0713−051	07:13:11	−05:11.6			dMe	12
2RE J0715−702	07:15:11	−70:24.8	WD 0715−704J		DA1	14.4
EUVE J0715+141	07:15:50	+14:10.3	NO ID			
EUVE J0717−500	07:17:11	−50:4.2	NO ID			
EUVE J0718−114	07:18:31	−11:27.5			M V:e	
2RE J0720−314	07:20:48	−31:46.9	WD 0718−316		DA+dM1e	14.82
EUVE J0723+500	07:23:09	+50:05	NO ID			
2RE J0723−274	07:23:21	−27:47.3	NO ID			
EUVE J0725−004	07:25:16	−00:26.2	2E0722.7−0020	BD−00° 1712	K5	8.9
EUVE J0726−022	07:26:01	−02:14.3	HD 58556	BD−01° 1707	G0	7.1
2RE J0728−301	07:28:51	−30:15.2	GJ 2060		M1V:e	10.4
2RE J0729−384	07:29:08	−38:48.8	y Pup		B5Vp+DA	5.41
2RE J0731+155	07:31:21	+15:56.3			K5e	9.7
2RE J0731+361	07:31:58	+36:14.1	VV Lyn		dM3e	10.6
EUVE J0734−234	07:34:20	−23:29.2	HR 2910	HD 60585	F6V	5.87
2RE J0734+315	07:34:39	+31:52.8	α Gem C		M1Ve	9.07
2RE J0734+144	07:34:56	+14:45.8			dMe	10.8
EUVE J0738+098	07:38:08	+09:49.5	NO ID			
2RE J0738+240	07:38:30	+23:59.9	1RXSJ073829.3+240014		dMe	12.7
2RE J0739+051	07:39:20	+05:13.8	α CMi A+B		F5IV−V	0.34
2RE J0743+285	07:43:20	+28:53.1	σ Gem		K1III	4.28

Table A.1. (*cont.*)

Name	RA$_{2000}$	DEC$_{2000}$	ID1	ID2	Spec. Type	M_V
2RE J0743+224	07:43:35	+22:42.5	SAO 79647		K0	10
2RE J0743−391	07:43:49	−39:11.1	WD 0743-391J		DA	16.3
2RE J0744+033	07:44:41	+03:33.5	YZ CMi		M4Ve	11.2
2RE J0749−764	07:49:16	−76:41.8			dM5e	11.8
2RE J0751+144	07:51:19	+14:44.7	PQ Gem		CV int.polar	14.5
EUVE J0803+164	08:03:50	+16:26	NO ID			
EUVE J0805−540	08:05:03	−54:4.3	NO ID			
2RE J0808+210	08:08:12	+21:6.1	BD+21° 1764A/B		K5e+dMe	8.1
2RE J0808+325	08:08:55	+32:50.2	GJ 1108A/B		M0.5Ve+M3Ve	10.3
2RE J0809−730	08:09:24	−72:59.5				
2RE J0813−073	08:13:48	−07:38.2	SAO 135659		K0	8.8
2RE J0815−190	08:15:06	−19:3.2	VV Pup		CV AMHer	14
2RE J0815−491	08:15:18	−49:13.9	IX Vel		CV nova−like	9.1
EUVE J0815−421	08:15:41	−42:6.1	NO ID			
EUVE J0817−827	08:18:02	−82:18:02			dM5e	
EUVE J0819−203	08:19:04	−20:21.6	SAO 175581	YZ 0 6554	A0	9.2
2RE J0823−252	08:23:35	−25:25.5	HD 70907		F3IV/V	8.8
2RE J0825−162	08:25:51	−16:24.8	GSC5998.01783		dM0e	12.9
2RE J0827+284	08:27:05	+28:44.2	WD 0834+288	PG 0824+289	DA1+dC	14.73
EUVE J0830−014	08:30:24	−01:27.4	NO ID			
2RE J0831−534	08:31:56	−53:40.3	WD 0830−535		DA1.5	14.48
EUVE J0832+057	08:32:00	+05:45.3	1RXSJ083147.3+054504		X	
2RE J0833−343	08:33:01	−34:38.3	HR 3385		K0III	6.4
EUVE J0834+040	08:34:44	+04:06	NO ID			
EUVE J0835+051	08:35:24	+05:08.2	AG+05° 1257	BD+05° 2004	F0	9.1
EUVE J0835−430	08:35:37	−43:03.5			M1e	18.4
EUVE J0836−449	08:36:27	−44:53.8			M1e	18.2
2RE J0838−430	08:38:56	−43:9.2	Vela SNR		SNR knot	
2RE J0839+650	08:39:16	+65:1.8	HD 72905	GJ 311	G1V	5.6
EUVE J0839−432	08:39:20	−43:16	NO ID			
EUVE J0840−724	08:40:17	−72:28.1	NO ID			
EUVE J0841+033	08:41:22	+03:20.9	WD 0841+033J	REJ084103+0320	DA1	15.0
2RE J0843−385	08:43:17	−38:53.6	HD 74576	GJ 320	K1V	6.6
EUVE J0843+033	08:43:28	+03:22.2	NO ID			
2RE J0845+485	08:45:49	+48:53	HD 74389 B		DA + A2V	15.5
2RE J0846+062	08:46:49	+06:24.9	ε Hya	HD 74874	G5III	3.38
EUVE J0852−192	08:52:17	−19:13.6	NO ID			
2RE J0853−074	08:53:12	−07:43.4	BD−07° 2647		G5	9.2
EUVE J0858−002	08:58:21	+00:00	NO ID			
2RE J0858+082	08:58:55	+08:27.9	G41−14		dMe	10.9
2RE J0859−274	08:59:43	−27:49	TY Pyx		G5V	6.9
2RE J0901−522	09:01:28	−52:29.5	GSC8177.00389			14.1
2RE J0902−040	09:02:18	−04:6.5	WD 0902−041J		DA2	13.19
EUVE J0905−492	09:05:45	−49:16.7	HD 78308	SAO 220840	F0IV	8.4
EUVE J0906−457	09:06:22	−45:47.5	MO 96−96		B	13.81
EUVE J0907−823	09:07:19	−82:18.1	HD 80258	SAO 258518	F7V	7.6
EUVE J0907+134	09:07:27	+13:27.2	NO ID			
2RE J0907+505	09:07:48	+50:57.9	WD 0904+511	PG 0904+512	DA1.5	16.13
EUVE J0908−258	09:08:35	−25:49.6	GJ 9287A	HD 78643	G0/G1V	6.77
EUVE J0908+326	09:08:18	+32:40.2	NO ID			
2RE J0908−370	09:08:19	−37:6.7	HD 78644		G3V	8.3
2RE J0909+423	09:09:54	+42:39.7	GSC2990.00341			14.9
EUVE J0911+394	09:11:27	+39:28.7	NO ID			
EUVE J0913−629	09:13:39	−62:59.8	NO ID			
EUVE J0914+022	09:14:06	+02:16.7	AG+02° 1243	BD+02° 2164	F2	8.3
2RE J0914+021	09:14:24	+02:18.9	θ Hya		DA+B9.5V	3.88
EUVE J0916+062	09:16:51	+06:14.6	NO ID			
2RE J0916−194	09:16:57	−19:46.7	WD 0916−197J		DA1+dM	17.3
EUVE J0920+506	09:20:51	+50:40.6	NO ID			
2RE J0922+401	09:22:29	+40:12.2	BF Lyn		K2V	7.63

Table A.1. (*cont.*)

Name	RA$_{2000}$	DEC$_{2000}$	ID1	ID2	Spec. Type	M_V
2RE J0924−234	09:24:51	−23:49.1	IL Hya		K2IV/Vp	7.9
EUVE J0925+278	09:25:30	+27:51.5	NO ID			
2RE J0925−531	09:25:36	−53:15.2	HD 81734		F7V	7.1
2RE J0930+103	09:30:37	+10:35.2	BD+11° 2052A/B		G5+G5	7.6
EUVE J0930−535	09:30:45	−53:32.2	HD 82533	SAO 237059	G5III	8.34
2RE J0932−111	09:32:25	−11:11	LQ Hya		K0Ve	7.8
2RE J0932+265	09:32:44	+26:59	HD 82443	SAO 80897	K0	7.01
EUVE J0933+154	09:33:55	+15:28.9	AG+15° 1077	BD +16° 1992	G0	8.9
2RE J0933+624	09:33:46	+62:49.8	HD 82286		G5	8.2
2RE J0934+695	09:34:27	+69:49.7	DK UMa		G4III−IV	4.6
EUVE J0939−283	09:39:18	−28:19.4	NO ID			
EUVE J0939+146	09:39:29	+14:37.8	NO ID			
2RE J0940+502	09:40:25	+50:20.5	WD 0937+505	PG 0937+506	DA1.5	16.06
EUVE J0946−328	09:46:06	−32:53.8	SAO 200715	HD 84726	A1IV	10.0
2RE J0951−144	09:51:24	−14:49.4	HR 3903	HD 85444	G7III	4.1
EUVE J0951+531	09:51:28	+53:8.7	SBSS 0948+533			14.5
EUVE J0954−150	09:54:33	−15:05	NO ID			
EUVE J0956+412	09:56:53	+41:14.9	PG 0953+415	QSO 0953+415	QSO	15.3
EUVE J0957+410	09:57:44	+41:02.6	HD 86146	TD1 14415	F6 Vs	5.1
2RE J0957+852	09:57:45	+85:29.9	WD 0957+854J	EXO 0947.3+8545	DA	16.4
2RE J0958−462	09:58:31	−46:25.4	LU Vel	LHS 2213	M5V	11.6
2RE J1000+243	10:00:03	+24:32.7	DH Leo		K0V	8.45
2RE J1014+210	10:14:18	+21:04	DK Leo		M0Ve	9.3
EUVE J1014+168	10:14:45	+16:48.9	HD 88776	BD +17° 2189	F	10.1
2RE J1016−052	10:16:30	−05:20.7	WD 1013−050		DA+dMe	14.3
EUVE J1017+443	10:17:41	+44:19.5	NO ID			
EUVE J1019+157	10:19:15	+15:44.2	NO ID			
2RE J1019+195	10:19:36	+19:52	AD Leo		M4.5Ve	9.43
2RE J1019−140	10:19:52	−14:7.8	WD 1017−138		DA1.5	14.93
EUVE J1023−002	10:23:54	+00:00	NO ID			
2RE J1024+262	10:24:01	+26:16.9	HD 90052	EUVE J1024+262	DA+G5	9.6
2RE J1024−302	10:24:44	−30:21.1	NO ID			
2RE J1027+322	10:27:14	+32:23.9	WD 1027+323J		DA1+G	13.2
2RE J1029+450	10:29:46	+45:6.7	WD 1026+453	PG 1026+454	DA	16.13
2RE J1031+823	10:31:02	+82:33.7	HD 90089	GJ 9330	F2V	5.3
2RE J1032+532	10:32:11	+53:29.2	WD 1032+534J		DA1.5	14.45
2RE J1033−114	10:33:44	−11:41.5	WD 1031−114	EG 70	DA2	13.0
EUVE J1034+073	10:34:12	+07:18.4	NO ID			
2RE J1034+393	10:34:40	+39:38.1	RX J1034.6+3938		AGN Seyf.	15.6
EUVE J1035−047	10:35:03	−04:43.8	NO ID			
2RE J1036+460	10:36:27	+46:8.2	WD 1033+464	GD 123	DA+dM	14.34
2RE J1043+490	10:43:12	+49:02	WD 1043+490J		DA1	16.23
2RE J1043+445	10:43:31	+44:52.6	WD 1040+451	PG 1040+451	DA1	16.96
2RE J1044+574	10:44:45	+57:44.6	WD 1041+580	PG 1041+580	DA1.5	14.60
2RE J1046−492	10:46:48	−49:25.3	μ Vel	HD 93497	G5III	2.69
EUVE J1047−442	10:47:08	−44:15.7	HD 93494	SAO 222327	A0V	7.7
EUVE J1049+124	10:49:18	+12:26.3	BD+13° 2309	AG+12° 1261	K0	10.3
EUVE J1050−207	10:50:21	−20:47.4	NO ID			
2RE J1051+561	10:51:06	+56:13.5	HD 93847		B9	7.4
2RE J1051+540	10:51:35	+54:04	EK UMa		CV	18
2RE J1054+560	10:54:05	+56:8.9	GSC3826.00232			11
2RE J1055+602	10:55:39	+60:28.1	DM UMa		K0IV	9.6
EUVE J1056+604	10:56:12	+60:25.7	1RXS J105624.6+601556		X	
2RE J1056+070	10:56:30	+07:0.8	CN Leo		M6Ve	13.5
EUVE J1056+602	10:56:47	+60:14.3	NO ID			
2RE J1058−384	10:58:22	−38:44.7	WD 1056−384		DA2	13.78
EUVE J1058+190	10:58:44	+19:1.1	NO ID			
EUVE J1058+602	10:58:52	+60:13.6	StKM 1−897		K6	12.39
2RE J1059+512	10:59:17	+51:24.3	LB 1919	WD 1059+514J	DA.5	16.78
2RE J1100+713	11:00:40	+71:38.4	WD 1057+719	PG 1057+719	DA1	14.95

Table A.1. (*cont.*)

Name	RA$_{2000}$	DEC$_{2000}$	ID1	ID2	Spec. Type	M_V
EUVE J1100+344	11:00:41	+34:24.5	NO ID			
EUVE J1102+226	11:02:04	+22:36.5	HD 95559	BD+23° 2297	G5	8.7
EUVE J1103+366	11:03:13	+36:38.6	LP 263−64		M3.5	13.67
EUVE J1103+359	11:03:21	+35:58.2	1RXS J110320.5+355900		X	
2RE J1104+450	11:04:25	+45:3.2	AN UMa	PG 1101+453	CV	15.5
2RE J1104+381	11:04:28	+38:12.3	Mrk 421	PKS 1101+384	BL Lac	13.5
2RE J1104−041	11:04:42	−04:13.3	BD−03° 3040A/B		G5e+M3Ve	7.6
EUVE J1105+588	11:05:37	+58:51.0	1RXS J110537.4+585128		X	
2RE J1111−224	11:11:42	−22:49.3	β Crt	HD 97277	DA1.5+A1V	4.48
EUVE J1112+358	11:12:34	+35:49.0	HD 97334	TD1 15449	G0 V	6.41
2RE J1112+240	11:12:39	+24:8.6	WD 1109+244	PG 1109+244	DA1.5	15.77
2RE J1116+445	11:16:46	+44:54.3	GSC3012.02468			10.9
EUVE J1118+314	11:18:13	+31:29.9	HD 98230	ξ UMa B	F8.5 V	4.87
EUVE J1118+404	11:18:39	+40:25.7	PG 1115+407	QSO 1115+4042	QSO	16.0
EUVE J1118−569	11:18:50	−56:58.8			dM3e	14.5
EUVE J1122+392	11:22:18	+39:12.2	AG+39° 1182	BD+39° 2437	K5	9.8
2RE J1122+434	11:22:54	+43:43.5	WD 1120+439	PG 1120+439	DA1	15.61
EUVE J1124+047	11:24:02	+04:46.9	NO ID			
2RE J1126−684	11:26:18	−68:40.7	N.Mus 1991		LMXB	13.2
2RE J1126+183	11:26:20	+18:39.1	WD 1123+189	PG 1123+189	DA1+dM	14.13
EUVE J1127+421	11:27:35	+42:06.2	NO ID			
2RE J1129+380	11:29:11	+38:9.2	WD 1126+384	GD 310	DA2	14.9
2RE J1128−025	11:28:14	−02:50.3	WD 1125−025	PG 1125−026	DA1.5	15.3
EUVE J1131−410	11:31:52	−41:2.8	V857 Cen	GJ 431	M3.5	12.9
EUVE J1131−346	11:31:53	−34:37.8			dM2e	11.8
2RE J1132+121	11:32:46	+12:11	BD+12° 2343		G5	8.9
EUVE J1134−615	11:34:48	−61:32	NO ID			
2RE J1135−630	11:35:50	−63:1.4	λ Cen	HD 100841	B9III+DA	3.1
EUVE J1137+510	11:37:47	+51:0.4	NO ID			
EUVE J1138−002	11:39:00	+00:00	IC 716	UGC 6612	Sb	14.9
2RE J1139−430	11:39:48	−43:8.2	GSC7752.00099			14.6
EUVE J1141+342	11:41:05	+34:12.0	HD 101501	TD1 15797	G8 V	5.1
EUVE J1142−011	11:42:32	−01:10.6	NO ID			
EUVE J1145−553	11:45:55	−55:20.6	NO ID			
EUVE J1146−461	11:46:31	−46:11.7	CPD−45° 5602	HD 102289	G	8.8
EUVE J1147+050	11:47:35	+05:01.7	REJ114+050		M V:e	12.1
2RE J1147+441	11:47:35	+44:19.1	PG 1145+446		MV:e	15.44
2RE J1147+125	11:47:47	+12:54.1	BPM 87617		K4	10.7
2RE J1147+201	11:47:57	+20:14	HD 102509	DQ Leo	F2IV+G5III	4.5
2RE J1148+183	11:48:01	+18:30.9	WD 1145+187	PG 1145+188	DA2	14.22
EUVE J1148−374	11:48:25	−37:29.2	CD−36° 7429	CCDMJ114−37A/B	M0	9.8
EUVE J1149−343	11:49:50	−34:23.7	NO ID			
EUVE J1149+287	11:49:59	+28:43.9	REJ1149+284	EU UMa	CV	17.0
2RE J1158+140	11:58:03	+14:2.2	BD+14° 2457		G5	8.4
EUVE J1201−365	12:01:01	−36:30.6	NO ID			
EUVE J1201+596	12:01:38	+59:40.4	SBSG 1158+599		blue	18.5 B
2RE J1201−034	12:01:45	−03:46	WD 1159−035	GW Vir	DO	14.87
EUVE J1203+445	12:03:08	+44:31.9	NGC 4051	1H 1205+44	Sy1	11.5
2RE J1212+773	12:12:26	+77:36.9	HR 4646	HD 106112	A5m	5.1
EUVE J1215−474	12:15:47	−47:25.1	NO ID			
2RE J1219+163	12:19:06	+16:33	BD+17° 2462		G2V	7.1
2RE J1222−390	12:22:36	−39:04.1	NO ID			
EUVE J1224−187	12:24:35	−18:47.6	NGC 4361	HD 107969	Plan Neb	12.9
2RE J1225+253	12:25:02	+25:33.8	BD+26° 2347		F8V+F8V	8.2
EUVE J1225−159	12:25:34	−15:59.5	LHS 2557	LP795−15	M	11.84
EUVE J1226−499	12:26:01	−49:54.4	NO ID			
EUVE J1227+069	12:27:34	+06:59.4	1WGA J1227+07		X	
EUVE J1229−134	12:29:58	−13:27.2	CCDMJ123−1324D			12.1
EUVE J1229+020	12:29:09	+02:03.2	3C 273	PG 1226+023	QSO	12.86
EUVE J1230+261	12:30:06	+26:11.7	NO ID			

Table A.1. (*cont.*)

Name	RA$_{2000}$	DEC$_{2000}$	ID1	ID2	Spec. Type	M_V
2RE J1235+233	12:35:20	+23:34.1	WD 1232+238	PG 1232+238	DA1	15.7
2RE J1236+475	12:36:46	+47:55.3	WD 1234+481	PG 1234+482	DA1	14.42
2RE J1237+264	12:37:36	+26:43.5	IC 3599		AGN Seyf.1	15.6
EUVE J1240−015	12:40:29	−01:31.7	NO ID			
2RE J1241−012	12:41:42	−01:26.6	γ Vir B	HD 110380	F0V	3.56
EUVE J1244−596	12:44:43	−59:41.1	NO ID			
2RE J1248+601	12:48:33	+60:18.5	Wo 9417	HD 111456	F6V	5.8
EUVE J1249+660	12:49:12	+66:5.4	DP Dra	GJ 487	M4	10.92
EUVE J1250+209	12:50:18	+20:54.2	NO ID			
EUVE J1251+225	12:51:46	+22:31.2	BD +23°2508	StKM 1−1038	M0	10.50
EUVE J1251+275	12:51:46	+27:30.6	HD 111812	31Com	G0 IIIp	4.94
2RE J1252−291	12:52:23	−29:14.7	EX Hya	1H 1251−291	CV	11.4
EUVE J1253+227	12:53:40	+22:44.4	NGP9F378−04218		Gal	19.7
2RE J1253+270	12:54:00	+27:5.3	NGC 4787		Gal, S0a	15.5
EUVE J1254−703	12:54:02	−70:20.5	NO ID			
EUVE J1254+224	12:54:06	+22:25.3	NO ID			
EUVE J1254−471	12:54:22	−47:05.6			dM3e	14.9
EUVE J1254−221	12:54:37	+22:08.1	HD 112196	TD1 16544	F8 V	6.97
EUVE J1255+258	12:55:37	+25:53.4	LoTr 5	HD 112313	Plan. Neb	8.7
EUVE J1255−678	12:55:44	−67:49.0	NO ID			
EUVE J1256+220	12:56:54	+22:00.5	WD 1254+223	GD153	DA1.5	13.40
EUVE J1257+087	12:57:06	+08:46.8	NO ID			
EUVE J1257−165	12:57:15	−16:30.1	NO ID			
EUVE J1257+351	12:57:48	+35:11.5	GJ 490A		M0.5	10.5
EUVE J1258−704	12:58:24	−70:29.1	NO ID			
EUVE J1259−701	12:59:34	−70:09.8	CPD−69° 1739	HD 112568	A3m...	9.0
2RE J1259+603	12:59:43	+60:39.7	LB 02513		bluestar	16.7
2RE J1300+122	13:00:45	12:23.2	GJ 494	BD+13° 2618	dM1.5	9.8
EUVE J1302+247	13:02:52	+24:46.3	NO ID			
EUVE J1303+224	13:03:04	+22:28.7	NO ID			
2RE J1303−405	13:03:21	−40:56.9	NO ID			
EUVE J1304+160	13:04:21	+16:1.7	NO ID			
EUVE J1305+024	13:05:15	+02:26.5	HD 113664	SAO 119759	A5	9.8
EUVE J1305−409	13:05:26	−40:54.3			dM1e	13.3
2RE J1307+535	13:07:56	+53:51.7	EV UMa		CV AMHer	17
EUVE J1308−687	13:08:41	−68:46.6	NO ID			
EUVE J1308−215	13:08:43	−21:33.5	NO ID			
2RE J1309+081	13:09:28	+08:15.1	GSC0884.00380			11.5
EUVE J1312+848	13:12:44	+84:52.4	NO ID			
EUVE J1315−679	13:15:12	−67:56.4	NO ID			
2RE J1316+290	13:16:23	+29:5.8	WD 1314+293	HZ 43	DA1	12.94
2RE J1316+092	13:16:48	+09:25.6	59 Vir		G0Vs	5.22
EUVE J1320+070	13:20:50	+07:1.3			K+K	9
2RE J1325−111	13:25:09	−11:11.3	Spica	α Vir	B1 IV	0.97
2RE J1332+223	13:32:44	+22:30	BD+23° 2581			9.67
EUVE J1334−083	13:34:44	−08:20.4	HD 118100	EQ Vir	K5 Ve	9.31
2RE J1334+371	13:34:49	+37:10.6	BH CVn		F2 IV	4.98
2RE J1336+694	13:36:14	+69:49.5	WD 1335+700	PG 1335+701	DA1.5	15.3
EUVE J1339−055	13:39:11	−05:30.3	NO ID			
EUVE J1343+076	13:43:30	+07:39.8	BD+08° 2755	HD 119478	K0	9.2
2RE J1344−612	13:44:00	−61:22.3	CPD−60° 4913	HD 119285	K2IV	7.9
2RE J1345−330	13:45:43	−33:2.8	HR 5168		F2V	4.2
EUVE J1346+051	13:46:08	+05:06.7	HT Vir	HD 119931	G0,var	7.16
EUVE J1348+079	13:48:55	+07:57.6	CR Boo	PG 1346+082	CV,var	13.0 V3
EUVE J1350−414	13:50:56	−41:26.6	PPM 318915	CD−40° 8199		10.9
EUVE J1352+062	13:52:18	+06:16.2	NO ID			
EUVE J1352+692	13:52:57	+69:17.7	Mrk 279	1H 1350+696	Seyfert	14.5
EUVE J1353+318	13:53:16	+31:48.9	NO ID			
EUVE J1355−249	13:55:55	−24:54.9	NO ID			
EUVE J1359+679	13:59:35	+67:56.9	NO ID			

Table A.1. (*cont.*)

Name	RA$_{2000}$	DEC$_{2000}$	ID1	ID2	Spec. Type	M_V
2RE J1405+100	14:05:03	+10:0.7	HD 123034	SAO 100840	G5	8.5
EUVE J1409−452	14:09:04	−45:17.9	V834 Cen	1H 1404−450	CV	14.2
EUVE J1412+292	14:12:04	+29:14	NO ID			
EUVE J1413+463	14:13:56	+46:21.5	BD+46° 1944	HD 124694	G0	7.19
2RE J1415−055	14:15:59	−05:59.1	ι Vir	HD 124850	F6III	4.1
2RE J1417−680	14:17:50	−68:2.3	HD 124672		F6V	7.3
EUVE J1418−216	14:18:57	−21:41.6	NO ID			
EUVE J1422+039	14:22:57	+03:54.1	NO ID			
2RE J1425+515	14:25:13	+51:50.8	θ Boo A	HD 126660	F7V	4.1
2RE J1426+500	14:26:44	+50:6.1	WD 1424+503		DA1.5	14
2RE J1426+025	14:26:47	+02:55.2	GSC0321.00127			13.7
2RE J1428−021	14:28:11	−02:13.4	HD 126868	GJ 550.2A/B	G2IV+G4V	4.8
2RE J1428+424	14:28:36	+42:41.3	1H 1430+423		AGN BL Lac	16.5
EUVE J1429−380	14:29:24	−38:04.1			CV	12.0
2RE J1429−624	14:29:48	−62:40.6	α Cen C	PM 14263−6228	M5Ve	11.05
2RE J1431+370	14:31:56	+37:6.4	WD 1429+373	GD 336	DA1	15.33
EUVE J1433−694	14:33:37	−69:26.9	NO ID			
2RE J1434−602	14:34:15	−60:24.7	CPD−59° 5631	HD 127535	K1IV	8.6
EUVE J1434−363	14:34:21	−36:22.0	NO ID			
EUVE J1434+357	14:34:27	+35:43.5	NO ID			
EUVE J1436−378	14:36:20	−37:49.3	SAO 205735	CD−37° 9540	F0	10.3
EUVE J1436−382	14:36:36	−38:16.9	NO ID			
EUVE J1438−432	14:38:46	−43:14.4	WT 486/487		M2.5/3	13.5
2RE J1438+641	14:38:57	+64:17.4	EK Dra		F8	7.54
EUVE J1439−240	14:39:32	−24:3.8	NO ID			
2RE J1439−605	14:39:43	−60:50.2	α Cen	HD 128621	G2V	−0.01
2RE J1440+750	14:39:58	+75:5.3	WD 1440+750J		DA1	15.22
EUVE J1441−373	14:41:06	−37:22.3	NO ID			
2RE J1442+352	14:42:06	+35:26	Mkn 478		AGN Seyf.1	14.6
2RE J1446+632	14:46:03	+63:29.1	WD 1446+634J		DA1	16
2RE J1450−321	14:50:21	−32:10.4	GSC7301.01288			14.6
2RE J1451+190	14:51:22	+19:6.1	χ Boo A		G8V	4.7
EUVE J1456−627	14:56:37	−62:43.6	NO ID			
EUVE J1500+749	15:00:04	+74:58.4	SAO 8142	PPM 8798	K2	9.2
2RE J1501−434	15:01:11	−43:40.9			dMe	11.9
2RE J1502+661	15:02:07	+66:12	WD 1501+664		DZ	
2RE J1503+473	15:03:46	+47:38.8	44 Boo	HD 133640	G2	5.9
EUVE J1506−050	15:06:17	−05:4.2	NO ID			
2RE J1507+761	15:07:52	+76:12.2	HD 135363	SAO 8175	G5	9.2
2RE J1511+615	15:11:51	+61:51.8	BV Dra		F7V	7.88
EUVE J1513+176	15:13:32	+17:37.7	NO ID			
EUVE J1515−623	15:15:12	−62:19.6	CPD−61° 4865	HD 134846	A0V	10.1
EUVE J1517+238	15:17:24	+23:49.4	NO ID			
2RE J1521+522	15:21:46	+52:21.6	WD 1520+525	PG 1520+525	DO	16.6
EUVE J1521−842	15:21:49	−84:13	NO ID			
EUVE J1521−168	15:21:52	−16:48.9	NO ID			
EUVE J1526+568	15:26:18	+56:49.2	NO ID			
EUVE J1529−471	15:29:12	−47:10.2	NO ID			
2RE J1529+802	15:29:20	+80:26.8	HD 139777	TD1 18446	G0IV−V	6.58
2RE J1529+483	15:29:44	+48:36.7	WD 1529+486J		DA1	15.08
EUVE J1533+337	15:33:58	+33:43.1	NO ID			
2RE J1535+600	15:35:23	+60:05.2	HD 139477	SAO 16775	K5	8
EUVE J1535−774	15:35:33	−77:25.5	WD 1535−774J		DA1	17.7
EUVE J1538−153	15:38:18	−15:20.4	NO ID			
2RE J1538−574	15:38:57	−57:42.2	V343 Nor		K0V	8.1
EUVE J1545−465	15:45:30	−46:35.5	HD 327428	CD−46° 10332	K0	9.8
2RE J1545−302	15:45:48	−30:20.5	SAO 206946		K2Ve	9.4
EUVE J1546+676	15:46:42	+67:38.5	NO ID			
2RE J1546−364	15:46:59	−36:46.7	WD 1543−366		DA1	15.81
EUVE J1547−450	15:47:57	−45:01.1	NO ID			

Table A.1. (*cont.*)

Name	RA$_{2000}$	DEC$_{2000}$	ID1	ID2	Spec. Type	M_V
EUVE J1549+260	15:49:42	+26:04.0	HD 141714	δ CrB	G3.5III	4.63
2RE J1553−421	15:53:27	−42:14.9	SAO 226339		G1V	7.5
EUVE J1556−043	15:56:53	−04:20.9	NO ID			
EUVE J1558+255	15:58:45	+25:34.3	HD 143313	MS Ser	K2 V	8.33
EUVE J1600−165	16:00:32	−16:35.4	NO ID			
2RE J1601+664	16:01:34	+66:48.1	AG Dra		K3IIIe	9.8
2RE J1601+512	16:01:45	+51:20.9	HD 144110		G5	8.5
2RE J1601+583	16:01:54	+58:33.5	θ Dra		F8IV	4
EUVE J1601−164	16:01:57	−16:24.8	NO ID			
EUVE J1603+052	16:03:16	+05:17.9	NO ID			
2RE J1603−574	16:03:34	−57:45.5	SAO 243278		G	8
2RE J1604−215	16:04:01	−21:55.5	HD 143937	SAO 184077	K0V	8.65
EUVE J1605+518	16:05:37	+51:48.8	NO ID			
2RE J1605+104	16:05:54	+10:41.5	AG+10° 1883		G8V	8.3
EUVE J1608−774	16:08:11	−77:28	NO ID			
2RE J1614+335	16:14:40	+33:51.5	σ CrB	TD1 19016	G0	5.64
2RE J1617+551	16:17:03	+55:16.5	CR Dra		M1Ve	9.96
EUVE J1618−179	16:18:51	−17:54.5	NO ID			
EUVE J1619−495	16:19:23	−49:32.2	NO ID			
2RE J1619−153	16:19:55	−15:38.6	Sco X−1	1H 1617−155	LMXB	12.2
2RE J1619+394	16:19:56	+39:42.2	HR 6091	HD 147365	F3V	5.5
EUVE J1621+081	16:21:13	+08:09.3	NO ID			
2RE J1621+185	16:21:18	+18:57.3	GSC1513.00024			14.8
2RE J1623−391	16:23:33	−39:13.7	WD 1620−391	CD−38° 10980	DA2	11.00
2RE J1625−490	16:25:14	−49:8.5	CD−48° 10809		G2IV	7.1
EUVE J1627+101	16:27:49	+10:11	NO ID			
EUVE J1629+780	16:29:16	+78:03.6	WD 1631+781	REJ1629+780	DA+dM	13.0
2RE J1636+525	16:36:14	+52:53.8	ADS 10129C		B9.5V	5.5
EUVE J1636−285	16:36:34	−28:32	NO ID			
2RE J1638+350	16:38:27	+34:59.8	WD 1636+351	PG 1636+351	DA1.5	14.98
EUVE J1639−576	16:39:51	−57:39.3	NO ID			
EUVE J1644+454	16:44:06	+45:28.7	NO ID			
EUVE J1649−439	16:49:20	−43:55.8	1RXSJ1649−4355		X	
2RE J1650+403	16:50:21	+40:37.6	WD 1650+406J		DA1	15.83
EUVE J1651−443	16:51:28	−44:21.6	1RXSJ1651−4422		X	
EUVE J1653+397	16:53:51	+39:45.1	Mrk 501	PKS 1652+398	BL Lac	13.8
2RE J1655−081	16:55:28	−08:19.8	V1054 Oph	1E 1652.7−0815	M3Ve	9.04
2RE J1656+650	16:56:04	+65:7.9	19 Dra	HD 153597	F6V,var	4.89
2RE J1656−390	16:56:45	−39:7.3	BPM 61550		M3Ve	11.4
2RE J1657+352	16:57:49	+35:20.6	Her X−1	1H 1656+354	LMXB	13
EUVE J1658+343	16:58:47	+34:19.7	WD 1657+343	PG 1657+343	DA1	16.12
2RE J1659+440	16:59:47	+44:1.3	WD 1658+440	PG 1658+440	DAP1.5	15.02
EUVE J1703−018	17:03:54	−01:52.9	NO ID			
EUVE J1705−018	17:05:13	−01:50.6	2E 1702.6−0142	1RXS J170507.9−014708	X	
EUVE J1706−450	17:06:43	−45:1.7	NO ID			
EUVE J1708+034	17:08:18	+03:28.7	NO ID			
EUVE J1708−111	17:08:29	−11:11.4	NO ID			
2RE J1711+664	17:11:29	+66:45.5	WD1711+667J		DA+dM	17.1
EUVE J1712−248	17:12:41	−24:48.2	NO ID			
EUVE J1715−266	17:15:22	−26:36.4	HD 155886	2EUVE 1715−26.5	K0 V	5.2
EUVE J1716−265	17:16:16	−26:32.8	HD 156026	2EUVE 1716−26.5	K5 V	6.3
2RE J1717−665	17:17:22	−66:56.9	V824 Ara		K1Vp	6.67
EUVE J1717+104	17:17:28	+10:28.3	SAO 102717	BD+10° 3176	F2	9.7
2RE J1717+102	17:17:36	+10:25.1	HD 156498		G5	8.4
EUVE J1721+278	17:21:58	+27:51.2	NO ID			
EUVE J1723+153	17:23:50	+15:23.4	NO ID			
2RE J1726+583	17:26:43	+58:37.6	WD 1725+586	PG 1725+587	DA1	15.5
2RE J1727−360	17:27:37	−36:0.1	WD 1724−359	RE J1727−355	DA1.5	15.46
2RE J1732+741	17:32:43	+74:13.9	DR Dra		K0III	6.55
2RE J1734+615	17:34:54	+61:52.6	GJ 684B	BD+61° 1678B	K3V	8.1

Table A.1. (*cont.*)

Name	RA$_{2000}$	DEC$_{2000}$	ID1	ID2	Spec. Type	M_V
2RE J1736+684	17:36:52	+68:45.4	ω Dra	HD 160922	F5V	4.8
2RE J1737+665	17:37:59	+66:53.8	WD 1738+669J		DA.5	14.61
2RE J1738+611	17:38:39	+61:14.6	HD 160934		K8V	10.28
EUVE J1745−883	17:45:03	−88:19	NO ID			
2RE J1746−703	17:46:13	−70:39.2	WD 1740−706	RE J1746−703	DA1	16.51
EUVE J1748+484	17:48:25	+48:30	NO ID			
2RE J1755+361	17:55:24	+36:11.7	V835 Her		G5	7.9
2RE J1757+291	17:57:46	+29:14.8	χ Her	TD1 21338	G8III	3.7
EUVE J1759+428	17:59:34	+42:50.4	NO ID			
2RE J1800+683	18:00:08	+68:36.1	WD 1800−685	KUV 18004+6836	DA1	14.74
2RE J1802+641	18:02:15	+64:15.8	G 227−22		dMe	12.9
EUVE J1805+204	18:05:25	+20:30.4	NO ID			
EUVE J1805+024	18:05:27	+02:29.7	70 Oph	TD1 21598	K0V	4.03
2RE J1805+212	18:05:51	+21:27	V772 Her		G0	7.07
2RE J1808+294	18:08:16	+29:41.6	V815 Her		G5	7.66
EUVE J1809+300	18:09:08	+30:01.9	HD 166435	TD1 21816	G0	6.8
2RE J1813+642	18:13:54	+64:24.1	36 Dra		F5V	5
EUVE J1816+498	18:16:14	+49:51.7	AM Her	1H 1814+498	CV	12.4
2RE J1816+541	18:16:18	+54:10.5			dKe	12
EUVE J1818−371	18:18:40	−37:10.7	HD 167820		G8 III	7.6
2RE J1820+580	18:20:28	+58:4.7			DA	13.95
2RE J1821+642	18:21:50	+64:21.7	WD 1821+643	DS Dra	Plan Neb	15.04
EUVE J1822+017	18:22:20	+01:42.2	HIC 90035	BD+01° 3657	K7V	
EUVE J1823−342	18:23:47	−34:14.4	V2539 Sgr		V	14.7
2RE J1829+345	18:29:03	+34:59	NO ID			
EUVE J1833−005	18:33:13	−00:32.8	1RXSJ1833−0034		X	
2RE J1833+514	18:33:53	+51:43.1	BY Dra		K6Ve	8.07
2RE J1834+184	18:34:20	+18:41.4	HD 171488	TD1 22739	G0V	7.3
EUVE J1834−006	18:34:40	−00:41.2	IRAS 18320−0044			
EUVE J1841−654	18:41:43	−65:29.1	NO ID			
EUVE J1844+255	18:44:42	+25:31.5	NO ID			
2RE J1844−741	18:44:54	−74:18.5	NO ID		CV	17.6
2RE J1845+682	18:45:08	+68:22.6	WD1845+683	KUV 18453+6819	DA1.5	15
EUVE J1845−648	18:45:40	−64:51.5	HD 172555	TD1 22956	A5IV−V	4.79
2RE J1846+191	18:46:10	+19:13	BPM 93474		K4	11.4
2RE J1847+015	18:47:38	+01:57.4	WD 1845+019	LAN 18	DA1.5	12.95
2RE J1847−221	18:47:57	−22:20.2	WD1847−223J	WD 1847−223	DA1.5	13.72
EUVE J1849−238	18:49:49	−23:50	V1216 Sgr	GJ 729	M4.5V:e	10.95
EUVE J1854−324	18:54:49	−32:25.2	NO ID			
2RE J1855+082	18:55:28	+08:24.4	GJ 735		M2Ve	10.1
EUVE J1855+084	18:55:39	+08:28.6	AG+08° 2416	BD +08° 3905	F8	10.2
2RE J1855+233	18:55:53	+23:33.7	V775 Her		K0	8.09
EUVE J1857−524	18:57:25	−52:28.7	NO ID			
EUVE J1858−299	18:58:11	−29:56.1	1RXSJ1858−2953		X	
EUVE J1906+169	19:06:08	+16:56.7	NO ID			
2RE J1906+274	19:06:24	+27:43.1	BD+27° 3245	HD 337518	K0	8.6
2RE J1907+301	19:07:31	+30:15.3	V478 Lyr		G8V	
EUVE J1908+524	19:08:19	+52:25.9	V1762 Cyg	HD 179094	K1IV	5.81
EUVE J1911−286	19:11:02	−28:40.0	CPD−28° 6801	CD−28° 15466	G5	9.1
EUVE J1911−606	19:11:54	−60:40.7	NO ID			
EUVE J1911+299	19:11:56	+29:57.3	NO ID			
2RE J1918+595	19:18:22	+59:59.9	LB 342		DA	14.3
EUVE J1925+031	19:25:30	+03:06.8	δ Aql	HD 182640	F0IV	3.40
2RE J1925−563	19:25:57	−56:33.8	NO ID			
EUVE J1928−500	19:28:27	−50:1.9	NO ID			
2RE J1929−230	19:29:41	−23:2.4	NO ID			
EUVE J1934−646	19:34:59	−64:39.1	NO ID			
EUVE J1935−192	19:35:29	−19:17.8	NO ID			
EUVE J1936+502	19:36:18	+50:14.1	θ Cyg	HD 185395	F4V	4.48
EUVE J1936+265	19:36:39	+26:34.5	NO ID			

Table A.1. (*cont.*)

Name	RA$_{2000}$	DEC$_{2000}$	ID1	ID2	Spec. Type	M_V
2RE J1938−461	19:38:34	−46:13.3	RX J19386−4612		CV	15.2
2RE J1943+500	19:43:40	+50:05	WD 1943+500J		DA1.5	14.62
EUVE J1947−110	19:47:12	−11:0.9	NO ID			
EUVE J1947+262	19:47:13	+26:17.6	NO ID			
2RE J1950+260	19:50:03	+26:8.5	GSC2144.00087			13.1
EUVE J1951−585	19:51:49	−58:32.6	NO ID			
EUVE J1953+444	19:53:56	+44:24.6	G208−45	GJ1245B	M5.5 V	13.99
EUVE J1954−239	19:54:11	−23:54.7	HD 188088	GJ 770	K3/K4 V	6.18
EUVE J1956−030	19:56:34	−03:4.3	NO ID			
EUVE J1957+298	19:57:25	+29:50.8	HD 333118		A7	10.0
EUVE J1958+550	19:58:43	+55:0.7	NO ID			
EUVE J1959+227	19:59:32	+22:43	NGC 6853		Plan Neb	7.5
2RE J2000−334	20:00:21	−33:42.2	TD1 25797		F7V	5.66
EUVE J2003+170	20:03:58	+17:2.7	HD 354613	SAO 105627	K0	9.2
2RE J2004−560	20:04:17	−56:3.2	WD 2000−562		DA1	15.05
EUVE J2005+544	20:05:04	+54:25.9	G230−26	GJ 781	M0	11.97
EUVE J2005+226	20:05:44	+22:40.5	QQ Vul	H2005+22	CV	14.86
2RE J2009−602	20:09:04	−60:25.9	WD 2004−605		DA1	13.4
EUVE J2010+426	20:10:12	+42:38.9	NO ID			
2RE J2013+400	20:13:09	+40:2.6	WD 2013+400J		DAO+dM4	14.02
EUVE J2013+467	20:13:34	+46:46.1	HD 192577		K	
2RE J2018−572	20:18:51	−57:22.1	WD 2014−575	L 0210-114	DA2	13.61
EUVE J2023+224	20:23:19	+22:26.8	NO ID			
2RE J2024−422	20:23:59	−42:24.3	κ^2 Sgr	TD1 26582	AV5+DA	5.64
2RE J2024+200	20:24:14	+20:01	WD 2024+200J		DA+dM	16.4
EUVE J2024+304	20:24:28	+30:24.3	NO ID			
EUVE J2028−114	20:28:46	−11:29.4	1RXSJ2028−112		X	
EUVE J2029+096	20:29:50	+09:41.1	EUVE J2029+0	HU Del	M4.5	13.04
2RE J2029+391	20:29:53	+39:13.9	WD 2028+390	GD 391	DA2	13.37
EUVE J2030+526	20:30:28	+52:37.4	V1974 Cyg	NOVA Cyg 1992	Nova	
EUVE J2030+177	20:30:52	+17:45.6	NO ID			
2RE J2031+051	20:31:14	+05:13.3	HD 195434		K0	11
2RE J2037+753	20:37:19	+75:36.2	VW Cep		K0V,var	7.38
EUVE J2041−368	20:41:15	−36:52.8	NO ID			
2RE J2041−322	20:41:53	−32:25.6	AT Mic		M4Ve	10.25
2RE J2045−312	20:45:09	−31:20.4	AU Mic		M0Ve	8.61
EUVE J2045+310	20:45:48	+31:0.9	NO ID			
EUVE J2046+153	20:46:10	+15:22.6	SAO 106469	YZ 0 1798	K0	8.5
2RE J2047−363	20:47:46	−36:35.7	SAO 212437		K0V	9.3
EUVE J2049−759	20:49:46	−75:55.0	CPD−76° 1447	HD 197388	G2/G3V	9.3
2RE J2051−134	20:51:19	−13:41.5	NO ID			
EUVE J2053+298	20:53:56	+29:52.6	NO ID			
EUVE J2054−750	20:54:18	−75:04.6	CPD−75° 1663		F8	10.4
EUVE J2055+164	20:55:28	+16:27.1	NO ID			
2RE J2055−170	20:55:47	−17:6.4	BD−17° 6127	HD 199143	F8V	7.1
EUVE J2057+310	20:57:17	+31:02.7	[SPB96] 2075		UV	14.2
2RE J2057+311	20:57:30	+31:11.5	Cyg.Loop SNR	SNR knot	SNR knot	
EUVE J2059+400	20:59:44	+40:05.6	V1396 Cyg	GJ815A	M1.5	10.1
EUVE J2100−426	21:00:07	−42:38.4	1RXSJ2100−4238		X	
2RE J2102+274	21:02:25	+27:48.5	ER Vul		G0V	7.27
EUVE J2104−770	21:04:31	−77:2.2	α Oct	TD1 27496	A7III	5.15
2RE J2106+164	21:06:51	+16:44.5	GSC1649.00256			12.9
2RE J2106+384	21:06:51	+38:44	HD 201091	GJ 820	K5V	5.2
2RE J2107−051	21:07:56	−05:16.5	HU Aqr		CV AMHer	15.3
EUVE J2110+454	21:10:39	+45:25.3	NO ID			
2RE J2112+500	21:12:46	+50:6.1	WD 2111+498	GD 394	DA.5,var	13.09
EUVE J2114+503	21:14:38	+50:18.4			CV	
EUVE J2115−586	21:15:46	−58:39.8	RBS 1735		CV	16
2RE J2116+735	21:16:51	+73:50.9	WD 2116+736	KUV 21168+7338	DA	14.87
EUVE J2117+342	21:17:10	+34:12.0	WD 2117+341	RXJ2117.1+3412	WD	13.2

Table A.1. (*cont.*)

Name	RA$_{2000}$	DEC$_{2000}$	ID1	ID2	Spec. Type	M_V
EUVE J2118+115	21:18:36	+11:34.0	1RXSJ2118+113		X	
EUVE J2119−385	21:19:57	−38:35.5	NO ID			
2RE J2121+402	21:21:00	+40:20.7	HD 203454	TD1 27993	F8V	6.4
EUVE J2121+104	21:21:23	+10:24.2	HD 203345	BD+09° 4786	F5	6.75
2RE J2122−164	21:22:14	−16:49.9	ι Cap	HD 203387	G7III	4.3
2RE J2125+282	21:24:59	+28:26	NO ID			
2RE J2126+192	21:26:25	+19:22.5	IK Peg	HR 8210 B	A8m+DA	6.07
EUVE J2126−694	21:26:42	−69:26.7	NO ID			
2RE J2127−221	21:27:42	−22:11.8	WD 2127−221J		DA1	14.6
EUVE J2131+233	21:31:01	+23:20.5	LO Peg	GJ 4199	K8	9.25
EUVE J2132+002	21:32:12	+00:14.1	BD−00° 4234	LDS 749A	K3Ve+K7Ve	9.89V?
EUVE J2132+101	21:32:31	+10:07.9	MKN 1513	ZW II 136	Sy1	14.64
2RE J2133+453	21:34:01	+45:35.8	ρ Cyg	TD1 28246	G8III	3.99
EUVE J2135+008	21:35:07	+00:51.9	NO ID			
2RE J2137−800	21:37:34	−80:5.9	NO ID			
EUVE J2137+016	21:37:42	+01:36.8	RXJ 2137.6+013		M	12.35
EUVE J2137+205	21:37:52	+20:35.2	NO ID			
EUVE J2139−273	21:39:14	−27:19.2	HD 205905		G4 IV−V	6.7
EUVE J2141−140	21:41:33	−14:3.6	42 Cap	TD1 28359	G2V	5.18
EUVE J2142−549	21:42:35	−54:54.3	NO ID			
2RE J2142+433	21:42:44	+43:35.4	SS Cyg		CV dw.nova	11.4
EUVE J2144+192	21:44:12	+19:17.3	HD 206808	SAO 107361	K2	8.6
2RE J2147−160	21:47:02	−16:07	δ Cap	HD 207098	A7IIIm	2.87
EUVE J2149−720	21:49:04	−72:05.9	AY Ind	CPD−72° 2640	M0.5	9.8
EUVE J2152−620	21:52:12	−62:02.5	HD 207575	TD1 28496	F6 V	7.3
2RE J2154−302	21:54:52	−30:29.6	WD 2151−307		DA2	14.4
EUVE J2155−618	21:55:12	−61:52.7	IDS2147−62A/B	CPD−62° 6277A/B	F0 IV	6.7
EUVE J2155+043	21:55:47	+04:22.6	NO ID			
2RE J2156−543	21:56:21	−54:38.5	WD 2156−546J		DA1	14.4
2RE J2156−414	21:56:34	−41:42.4	WD 2153−419		DA1	15.89
EUVE J2157−406	21:57:25	−40:36.2	NO ID			
2RE J2157−505	21:57:39	−50:59.8	GJ 841A		dM4e	10.4
EUVE J2157−400	21:57:53	−40:01.0	BPS CS22881−0015			
EUVE J2158−404	21:58:20	−40:24.5	1RXSJ21581−402		X	
2RE J2158+825	21:58:29	+82:52.9	V376 Cep		F5	7.49
2RE J2158−301	21:58:51	−30:13.8	PKS 2155−304	1H 2156−304	BL Lac	14
EUVE J2158−590	21:58:29	−59:01.2	HD 208496	TD1 28604	F3 V	6.2
EUVE J2200+548	22:00:52	+54:48.9	NO ID			
2RE J2201+281	22:01:12	+28:18.8	G 188−38		M?	12
EUVE J2202−412	22:02:29	−41:13.4	BPS CS 228−0031		sdB	15.81
EUVE J2203+482	22:03:18	+48:16.0	NO ID			
EUVE J2204+472	22:04:44	+47:16.4	HK Lac	HD 209813	K0 III	6.91
2RE J2206+634	22:06:34	+63:45.8	GSC 04271−01011			
2RE J2207+252	22:07:44	+25:20.2	WD 2207+253J		DA2	14.4
2RE J2208+454	22:08:40	+45:44.5	AR Lac	1H 2207+455	G2IV	6.11
EUVE J2210−300	22:10:28	−30:03.6	WD 2207−303	REJ221028−3005	DA2	14.8
2RE J2214−491	22:14:11	−49:19.5	WD 2211−445		DA1	11.71
2RE J2220+493	22:20:07	+49:30.3	SAO 51891		K0	8.1
2RE J2223+322	22:23:29	+32:28	GJ 856A/B		M0Ve	11.4
EUVE J2225+419	22:25:13	+41:59.6	HD 212669	SAO 51976	K5	7.7
EUVE J2225+167	22:25:20	+16:46.4	SAO 107903	BD+16° 4733	M0	9.6
EUVE J2226−167	22:26:33	−16:45.4	53 Aqr	HD 212698	G3V	6.35
EUVE J2228−000	22:28:50	−00:01.7	ζ Aqr	BD−00° 4365	F3III−IV	3.65
EUVE J2228+576	22:28:02	+57:41.6	DO Cep	1E 2226.1+5726	M4.5V:e	10.3
EUVE J2228+116	22:28:06	+11:41.3	NO ID			
EUVE J2228+635	22:28:06	+63:32.8	NO ID			
EUVE J2229+000	22:29:00	+00:01.5	PHL 1928			18.5
EUVE J2231+017	22:31:03	+01:47.8	NO ID			
2RE J2234+081	22:34:28	+08:12.8	GSC1151.00552			14.4
EUVE J2236+001	22:36:06	+00:07.4	PHL 329	RXJ2236.0+0007	DA	13.97

Table A.1. (*cont.*)

Name	RA$_{2000}$	DEC$_{2000}$	ID1	ID2	Spec. Type	M_V
2RE J2238−652	22:38:23	−65:22.7	GJ 865		dMe	11.5
EUVE J2238+022	22:38:30	+02:17.0	HD 214494	AG+02° 2867	G0	8.0
EUVE J2238−323	22:38:39	−32:20.1	NO ID			
2RE J2238−203	22:38:45	−20:37.1	FL Aqr	HD 214479	M2V:e	11.43
EUVE J2240+097	22:40:22	+09:43.6	NO ID			
EUVE J2242−475	22:42:55	−47:35	NO ID			
2RE J2244−321	22:44:43	−32:19.8	WD 2241−325	TON 572	DA1.5	15.6
2RE J2246+442	22:46:49	+44:20.1	EV Lac		M4.5V:e	10.09
2RE J2248−510	22:48:40	−51:9.5	IRAS F22456−5125		Seyf.1	15
EUVE J2249−209	22:49:06	−20:59.1	HD 215964	BD−21° 6309	G3/G5V	9.5
2RE J2249+583	22:49:55	+58:34.6	WD 2249+585J	LAN 23	DA1	14
EUVE J2252+406	22:52:13	+40:39.2	NO ID			
2RE J2300−070	23:00:35	−07:4.7	HD 217411 B	BD−07° 5906B	DA+G5	9.8
EUVE J2301−392	23:01:14	−39:16.8	NO ID			
EUVE J2302−560	23:02:35	−56:1.5	NO ID			
2RE J2308+572	23:08:31	+57:22.7	GSC4006.01709			12.5
2RE J2309+475	23:09:54	+47:57.4	KZ And		G5	7.91
EUVE J2310+206	23:10:39	+20:36.5	NO ID			
2RE J2312+104	23:12:23	+10:46.9	WD 2309+105	GD 246	DA1	13.11
2RE J2313+024	23:13:24	+02:40.2	SZ Psc		F8IV+K1IV	7.2
EUVE J2315−642	23:15:05	−64:15.1	NO ID			
2RE J2319+790	23:19:25	+79:0.5	V368 Cep		G9V	7.54
EUVE J2320+203	23:20:59	+20:20.6	NO ID			
EUVE J2321−269	23:21:16	−26:59.7	HD 220096	TD1 29801	G4V	5.64
EUVE J2322−469	23:22:24	−46:54.4	NO ID			
EUVE J2323−247	23:23:03	−24:46.6	NO ID			
EUVE J2323−268	23:23:19	−26:52.2	NO ID			
EUVE J2324+458	23:24:28	+45:51.1	NO ID			
2RE J2324−544	23:24:30	−54:41.7	WD 2321−549		DA1	15.20
EUVE J2327+048	23:27:50	+04:50.0	VZ Psc	BD+04° 5072	K5	10.3
2RE J2329+412	23:29:24	+41:28.3	G 190−28		dM3	11.89
EUVE J2329−477	23:29:35	−47:47.4	NO ID			
2RE J2331+195	23:31:51	+19:56.4	EQ Peg		M4V	10.32
EUVE J2334−472	23:34:02	−47:14.2	WD 2331−475	MCT 2331−4731	DA1	13.44
EUVE J2336−213	23:36:42	−21:18.7	NO ID			
EUVE J2336+480	23:36:50	+48:2.1	NO ID			
2RE J2337+462	23:37:33	+46:27.7	λ And	HD 222107	G8III	3.82
2RE J2339−691	23:39:31	−69:11.5	CPD−69° 3329A/B	•	G5IV+?	8
EUVE J2350−206	23:50:06	−20:41.8	NO ID			
2RE J2352+753	23:52:24	+75:32.3	GJ 909A/B		K3V+M2	6.4
2RE J2353−243	23:53:03	−24:31.9	WD 2350−248		DA1.5	15.44
2RE J2353−702	23:53:07	−70:23.5	HD 223816 B	CD−71 1808 B	DA+F5IV	8.8
2RE J2355+283	23:55:03	+28:38	II Peg		K0V	7.37
2RE J2355+250	23:55:42	+25:8.2	HD 224168		G5	8.7
EUVE J2356−327	23:56:22	−32:44.7	NO ID			
EUVE J2359−306	23:59:08	−30:37.7	1H 2354−315	H 2354−31.5	BL LAC	16.84

Table A.2. *All-sky survey source list with count rates (in counts per 1000 seconds) in the WFC S1, WFC S2, EUVE 100 Å, EUVE 200 Å, EUVE 400 Å and EUVE 600 Å filters.*

	WFC		EUVE			
Name	S1	S2	100 Å	200 Å	400 Å	600 Å
EUVE J0001+276				<16		
EUVE J0001+515			30±8			
EUVE J0002−495			21±7			110±46
EUVE J0003+242				60±15		
2RE J0003+433	46±7	17±7	43±9			
EUVE J0005−500			13±4	<8		
EUVE J0006+201			<11			
EUVE J0006+290			23±6	<10		
2RE J0007+331	208±15	58±9	219±18			
EUVE J0008+208			29±9	5±3		
2RE J0012+143	29±6	22±7				
2RE J0014+691	8±0	36±10				
EUVE J0016+198			15±6	<11		
2RE J0018+305	16±6	23±7				
EUVE J0019+064			5±2			
2RE J0024−741	14±5	32±9				
2RE J0029−632	660±36	1020±39	798±25	52±15		
2RE J0030−481	38±0	41±11	58±11		30±15	
2RE J0035−033	16±5	30±7				
2RE J0039+103	20±6	7±0				
EUVE J0040+307			23±7			81±34
2RE J0041+342	23±6	25±0	9.6			
2RE J0042+353	28±6	25±7	38±9			79±35
2RE J0043−175	37±10	70±11	71±13			192±82
2RE J0044+093	16±5	13±5	38±11	33±14	42±21	
EUVE J0045−287			40±13	55±21		
EUVE J0046−170			37±10	47±18		189±78
2RE J0047−115	71±8	28±8	80±14			
2RE J0047+241	30±6	41±7	41±10	28±11	44±21	
EUVE J0051−410			36±11			
2RE J0053−743	28±6	30±8	62±10			
2RE J0053−325	722±28	2193±45	1787±52	908±46		
2RE J0053+360	24±5	16±6	35±9			
2RE J0054−142	14±0	34±7				
EUVE J0057−223			64±12			
EUVE J0102−027			33±10			
EUVE J0103+623			22±5			
2RE J0103−725	5±0	28±6				
EUVE J0106−637			15±5			
2RE J0106−225	78±8	39±8	65±12	35±15		
2RE J0108−353	45±9	133±13	78±14	42±17	44±21	
EUVE J0113+624			25±7			65±30
EUVE J0113−738			13±6	45±10	26±11	
EUVE J0115+269			41±10			
2RE J0116−022	47±7	46±8	68±12			
EUVE J0122−567			12±9	<16		
EUVE J0122−603			23±6			
2RE J0122+004	20±5	28±6				
2RE J0122+072	83±9	92±12	145±15	28±10		
EUVE J0123+178			<5			
EUVE J0129−185					77±24	
EUVE J0134+536			31±9	30±10		
2RE J0134−160	270±14	693±24	683±28	137±19		
2RE J0135−295	44±5	55±7	79±13			
EUVE J0136+414			<10	<10		
2RE J0137+183	21±5	16±0				
2RE J0138+252	60±9	60±9	98±14			103±43

Table A.2. (*cont.*)

	WFC		EUVE			
Name	S1	S2	100 Å	200 Å	400 Å	600 Å
2RE J0138+675	5±2	12±3				
2RE J0138−175	25±5	18±5	93±13	28±14		
2RE J0140−675	257±19	211±17	52±7			75±37
EUVE J0142−241			30±9			155±60
2RE J0148−253	15±5	25±6				
2RE J0151+673	23±4	46±5	58±9			85±31
EUVE J0153+295			36±9			78±36
2RE J0153+603	6±2	14±4				
EUVE J0154+208			6±3	6±3		
2RE J0155−513	26±10	32±9	64±11			134±64
EUVE J0202+105			16±4	10±3		
2RE J0205+771	47±10	63±9	51±10			95±46
EUVE J0205+093				65±15		
EUVE J0208−100			3±2			
EUVE J0210−472			25±7			
EUVE J0211+043			6.4			
EUVE J0213+057			8.8			
EUVE J0213+368			12±4	<5		
2RE J0214+513	4±0	25±6				
EUVE J0215−095			13±4			
2RE J0218+143	20±0	48±12	50±11			
2RE J0222+503	14±3	17±5				
EUVE J0223+650				29±8		
EUVE J0223+371			31±9			
EUVE J0223−264			25±7			
2RE J0228−611	675±35	1104±36	1182±29	93±12		
2RE J0230−475	28±4	137±9	86±10	46±11		
2RE J0232−025	15±0	45±11				
2RE J0234−434	57±7	50±8	132±12	22±10		
2RE J0235+034	20±0	1006±38	1081±35	5359±75	750±48	426±54
2RE J0237−122	78±13	155±15	160±17	64±14		
2RE J0238−525	18±5	12±4				
2RE J0238−581	26±9	20±9	53±9	20±9		
EUVE J0239−196						
EUVE J0242−196			2.7			
2RE J0239+500	27±4	41±6	40±9		41±15	77±38
EUVE J0241−006			15±6	<10		
2RE J0241−525	25±5	14±4	38±7	22±9		
2RE J0243−375	27±5	44±7	61±9	16±7		68±28
EUVE J0245−185			13.3	3.6		
2RE J0248+310	119±11	132±16	116±15	177±9	18±5	
EUVE J0249+312			72±9	<10		
EUVE J0249+099			28±9			
EUVE J0250+312				9±4		
EUVE J0251+312			19±5			
EUVE J0251−373			21±6			
EUVE J0252−127			46±10	21±10	30±14	66±33
EUVE J0254−511			23±6	21±9		
2RE J0254−051	65±13	127±21	137±14	44±12		
2RE J0255+474	10±3	15±4				
EUVE J0256+080			19±8		35±16	265±57
EUVE J0304+029			46±11			
2RE J0308+405	144±9	212±11	277±20	42±13		111±50
EUVE J0309+496			<5	4±2		
EUVE J0309−009			31±10	27±13		100±45
EUVE J0310−443			10±5		47±13	
EUVE J0311−229			6±3			
EUVE J0311−318			61±11			90±33
2RE J0312−285	36±11	48±13	106±12	16±8		81±30

Table A.2. (*cont.*)

Name	WFC		EUVE			
	S1	S2	100 Å	200 Å	400 Å	600 Å
2RE J0312−442	25±6	27±6	36±7		22±10	
2RE J0314−223	195±20	266±28	419±21			
2RE J0317−853	45±11	168±16	105±8	21±8		60±27
EUVE J0318−363			23±6			
2RE J0319+032	41±9	47±18				
EUVE J0320−122				57±14		
2RE J0320−764	13±5	38±11				
2RE J0322−534	77±7	213±12	168±10	66±10		
2RE J0324+234	16±4	20±5				
EUVE J0325+012				56±15		
2RE J0326+284	93±8	120±9	631±28	62±15		
2RE J0327+094	18±4	22±5				
EUVE J0328+041			<5			
2RE J0328+040	18±4	11±4				
EUVE J0328−296			24±7			
2RE J0332−092	84±13	83±25	133±15	59±16		
2RE J0333+461	27±5	22±5	30±9		37±15	75±32
EUVE J0333−258			117±13	29±11		
EUVE J0335−257			1156±32	38±11	28±12	
2RE J0335+311	8±3	13±4				
EUVE J0335−348			33±7			
2RE J0336+003	286±14	349±15	580±26	72±12		87±41
EUVE J0336+004			3.1	3.2		
2RE J0337+255	90±8	110±9	137±9	19±4		
EUVE J0337−419			26±8			
EUVE J0346−575					36±10	
2RE J0348−005	256±14	509±20	430±23	280±24	218±28	540±65
EUVE J0349−477			24±7			
2RE J0350+171	309±16	1163±30	740±32	504±30		
EUVE J0352−492			2 3			
EUVE J0356−366			177±16	36±12		125±54
2RE J0357+283	46±9	55±10	54±11			
EUVE J0357+286			51±6			
2RE J0357−010	18±4	13±5				
EUVE J0359−398			34±9			
2RE J0402−001	34±6	41±7	50±12			88±42
EUVE J0404+310			38±11		34±16	
EUVE J0405+220			12.3			
EUVE J0407−525			26±6	18±7		
2RE J0407+380	34±6	19±6				
2RE J0409−711	17±4	25±6	16±3	13±5	19±7	
2RE J0409−075	79±8	111±10	107±14	24±12		
2RE J0410+592	16±5	23±5				
EUVE J0411+236			8±5	<6		
EUVE J0412+236			22.3	10.3		
EUVE J0414−624			15±3	18±6		
EUVE J0414+224			2.4			
2RE J0415−073	23±5	19±6	47±11	26±12		73±33
2RE J0415−402	51±9	29±10				
EUVE J0415+658			36±11			
2RE J0418+231	17±6	23±0	<5	5±3		
2RE J0419+153	29±8	23±0	10±5	<5		
EUVE J0419+217						
EUVE J0422+150			5±3	<5		
EUVE J0423+149			6±3	<8		
EUVE J0424+147			13±4	5±3		
EUVE J0425−068				43±12		
2RE J0425−571	21±5	4±0				
EUVE J0426−379			31±8	26±11		

Table A.2. (*cont.*)

Name	WFC		EUVE			
	S1	S2	100 Å	200 Å	400 Å	600 Å
2RE J0426+153	18±6	19±0	12±6	<6		
2RE J0427+740	90±8	42±6	151±14		30±13	
2RE J0428+165	15±5	27±0				
EUVE J0429+194				70±18		146±53
EUVE J0432+054			39±11			
2RE J0436+270	50±6	80±10				
2RE J0437−022	18±5	26±8				
EUVE J0439−342			27±7	19±9		
EUVE J0438+231			5±3	<5		
EUVE J0439−716			5±2		24±6	
EUVE J0440−418			31±7	24±9		55±26
2RE J0441+205	46±6	65±9	91±9	23±7		156±61
EUVE J0441+483			10.2	3.0		
2RE J0443−034	116±10	25±0	139±19	32±14		132±46
EUVE J0444−278			32±9	18±8		
EUVE J0444−262			28±8			
2RE J0446+763	16±5	13±4				
EUVE J0446+090			7±5	<10		
2RE J0447−275	11±4	32±7				
EUVE J0447−169			36±10	36±12		83±38
2RE J0449+065	37±5	57±10	56±7	30±6		
EUVE J0449+057			8±5	157±15		
EUVE J0449+236			11.2			
EUVE J0450−086			31±10			
EUVE J0451−391				30±8		
2RE J0453−421	53±9	18±7	44±8			
2RE J0453−555	41±8	30±0				
EUVE J0456+326			46±15	34±17		
2RE J0457−280	113±9	2289±34	1936±45	7216±85	1401±59	1933±90
EUVE J0457−063			<5			
2RE J0458+002	28±5	63±10	52±15			130±50
EUVE J0459−030			3±2			
2RE J0459+375	15±3	12±4				
2RE J0459−101	65±7	216±14	149±20	51±16		
2RE J0500−362	20±5	47±8				
2RE J0500−345	14±4	21±6	41±9			
EUVE J0500−057			6±4	<10		
EUVE J0501−064			<5			
2RE J0501+095	18±4	17±7				
2RE J0503−285	2±0	107±9	77±13	938±36	2454±83	4094±126
EUVE J0504−039			13±5			
2RE J0505−572	33±6	40±8				
2RE J0505+524	47±6	3208±44	2586±51	13032±113	2169±78	1951±96
EUVE J0505−181					76±24	
2RE J0506−213	33±5	35±7	47.1	20.4		
EUVE J0510−578			10±5	16±8	46±14	53±24
EUVE J0511+225			40±13			129±52
2RE J0512−004	135±10	479±21	309±32	73±22		
2RE J0512−115	19±4	20±7				
2RE J0512−414	147±9	231±12	196±15	22±8		71±28
2RE J0514−151	14±3	10±0				
2RE J0515+324	884±23	2753±42	2082±49	648±32		
2RE J0516+455	395±16	475±18	393±23	135±17		
2RE J0517−352	17±4	24±5				
EUVE J0520+363				52±14		112±42
2RE J0520−394	10±3	15±0				
2RE J0521−102	42±5	211±14	140±20	70±17		
EUVE J0521+395			28±9			81±39
2RE J0524+172	19±4	36±7	32±9			

Table A.2. (*cont.*)

Name	WFC		EUVE			
	S1	S2	100 Å	200 Å	400 Å	600 Å
2RE J0527−115	53±6	78±9	112±16	41±14		
EUVE J0527+654				41±11		
2RE J0528−652	125±4	138±5	177±4	16±3		
2RE J0530−191	8±3	16±5				
2RE J0531−462	20±4	7±0				
EUVE J0531−036			7.9			
EUVE J0531−068				149±12		
2RE J0532−030	15±4	15±5	10.4			
2RE J0532+094	23±4	17±5				
EUVE J0532+510					48±15	79±39
2RE J0533+015	17±4	17±0				
2RE J0534−151	8±3	19±5				
2RE J0534−021	13±3	34±7				
EUVE J0536−060			370±16			
2RE J0536+111	10±3	20±6				
2RE J0536−475	28±4	40±5	57±9			46±23
2RE J0537−393	10±3	15±0				
EUVE J0538−066			13±6			
2RE J0539−193	9±3	23±0				
2RE J0540−193	11±3	13±5				
2RE J0540−201	15±3	3±0				
2RE J0541+532	14±4	14±5				
EUVE J0543+103					56±17	
2RE J0544−222	7±3	22±5				
2RE J0545−595	13±3	10±3	20±3		8±4	
2RE J0550+000	92±7	86±9	133±16			
2RE J0550−240	21±4	16±5	41±9			64±32
EUVE J0551−546			13±4			
2RE J0552−570	6±2	9±3				
2RE J0552+155	598±19	2216±36	1547±43	1083±40	41±19	150±56
2RE J0553−815	23±7	25±7				
2RE J0554+201	47±6	36±7	97±14	70±16		
EUVE J0555−132					60±17	85±38
2RE J0556−141	8±3	15±5				
EUVE J0557+120					74±22	
2RE J0557−380	7±2	14±4				
2RE J0558−373	4±0	37±6				
2RE J0600+024	9±3	28±7	47±11		32±15	
2RE J0602−005	17±4	23±0				
2RE J0604−482	34±4	34±6	47±6			
2RE J0604−343	9±2	31±6				
2RE J0605−482	26±3	48±7	33±6			
EUVE J0607+472			3±2			
EUVE J0608+590			32±9			82±36
EUVE J0609−358			20±8			
EUVE J0611+578				16±8	63±17	
EUVE J0611+482			<4			
2RE J0612−164	15±3	29±0				
EUVE J0612−376			27±7		29±11	
2RE J0613−274	7±2	25±8				
2RE J0613−235	18±3	35±9				
EUVE J0613−355				54±12		82±36
2RE J0616−645	12±0	12±2				
EUVE J0618+750			33±8			
EUVE J0618−199				9±5		
2RE J0618−720	21±4	31±4	15±3			33±12
2RE J0619−032	10±3	26±10				
2RE J0620+132	14±3	33±0				
2RE J0622−601	7±2	10±3				

Table A.2. (*cont.*)

Name	WFC		EUVE			
	S1	S2	100 Å	200 Å	400 Å	600 Å
2RE J0622−175	15±3	22±0	1111±35		683±55	6378±145
2RE J0623−374	5±0	452±16	401±17	1936±37	102±20	
EUVE J0624−526			44±6			
EUVE J0624+229			3.5			
2RE J0625−600	10±2	11±3				
2RE J0626+184	26±4	34±11	42±11	37±16		123±55
2RE J0629−024	20±4	24±8	54±11	24±12		
2RE J0631+500	16±4	10±0				
EUVE J0631−231			15±5	23±4		
2RE J0632−050	142±9	255±15	275±19			
EUVE J0632−232			121±9	24±5		
EUVE J0632−065			6±4			
EUVE J0633−231			16±4	42±7		
2RE J0633+200	18±4	7±0				
2RE J0633+104	59±6	169±13	107±14	37±14		96±47
EUVE J0634−698			13±3		8±4	
EUVE J0636−465			18±5	22±8		
2RE J0637−613	34±3	32±4	47±4	9±4		
EUVE J0639−265			5±3			
EUVE J0640−035	9±3	32±8	24±6	10±6		
EUVE J0642−046			28±8	24±12		
EUVE J0642−041			6±3	<10		
EUVE J0642−174			31.5			
EUVE J0644−310			17±4	19±7		
EUVE J0644−312			6±3	21±4		
2RE J0645−164	3299±43	9370±75	7297±78	1854±45	34±13	
EUVE J0647−506					35±9	
EUVE J0648−482			26±6			
2RE J0648−252	82±7	276±14	190±15	122±17	35±13	
EUVE J0649−015			29±9	39±14		
2RE J0650−003	13±3	26±0	5±3			
EUVE J0652−051			12±4	10±5		
EUVE J0652−596			100±3			
2RE J0653−430	11±3	12±4				
EUVE J0653−564			12±3			
2RE J0654−020	100±8	299±15	231±15	100±13		
2RE J0658−285	21±4	28±6	1097±33	60±14	4473±100	78018±496
EUVE J0700−034			15.7			
EUVE J0702−018			30±9			
2RE J0702+125	74±7	112±11	124±15	28±13		
EUVE J0702−064			2.6			
2RE J0702+255	21±4	17±7				
EUVE J0704+717			10.2			
EUVE J0710+184			48±11	38±14		
2RE J0710+451	8±3	26±7				
2RE J0713−051	17±4	24±6				
2RE J0715−702	571±17	904±25	1221±18	48±6		30±15
EUVE J0715+141			23±10		31±15	106±30
EUVE J0717−500					44±13	
EUVE J0718−114			<5			
2RE J0720−314	109±9	429±21	285±19	117±18		
EUVE J0723+500			34±9			88±44
2RE J0723−274	329±16	1024±32	787±30	470±30		164±60
EUVE J0725−004	15±4	18±6	35±6	16±8		
EUVE J0726−022			<6			
2RE J0728−301	36±7	37±10	48±10			
2RE J0729−384	61±14	108±16	94±12			64±32
2RE J0731+155	10±3	24±6				
2RE J0731+361	26±5	20±6				

Table A.2. (*cont.*)

Name	WFC		EUVE			
	S1	S2	100 Å	200 Å	400 Å	600 Å
EUVE J0734−234			6.5			
2RE J0734+315	65±7	90±10	151±19	80±21		
2RE J0734+144	8±3	19±6				
EUVE J0738+098					115±32	
2RE J0738+240	6±3	38±7	4±2	<3		
2RE J0739+051	115±9	191±14	276±25	273±31		
2RE J0743+285	135±9	210±13	382±28	46±20		
2RE J0743+224	12±3	4±0				
2RE J0743−391	44±9	74±12	63±9	20±9		
2RE J0744+033	36±6	44±9				
2RE J0749−764	18±5	24±5	40±6	15±6	14±6	
2RE J0751+144	19±4	6±0				
EUVE J0803+164			21±10			67±21
EUVE J0805−540			14±5	27±9		98±37
2RE J0808+210	14±4	27±0				
2RE J0808+325	19±4	29±6				
2RE J0809−730	17±6	63±13	76±9			
2RE J0813−073	14±6	55±16				
2RE J0815−190	248±15	282±16				
2RE J0815−491	11±3	33±5				
EUVE J0815−421			19±6	20±10		113±47
EUVE J0817−827			20±6			
EUVE J0819−203			37±9			
2RE J0823−252	52±7	83±9	58±10	28±11		96±48
2RE J0825−162	12±5	21±6				
2RE J0827+284	48±8	55±9	96±14	29±14		
EUVE J0830−014			33±10			
2RE J0831−534	26±5	58±6	46±8		23±11	
EUVE J0832+057			7±4			
2RE J0833−343	16±4	26±0				
EUVE J0834+040			41±11			
EUVE J0835+051			16±3	6±3		
EUVE J0835−430			26±8			
EUVE J0836−449			23±7			
2RE J0838−430	23±5	7±0				
2RE J0839+650	16±5	18±5				
EUVE J0839−432			80±12		24±12	
EUVE J0840−724			22±6			
2RE J0841+032	228±15	447±21	340±23	47±15		
2RE J0843−385	13±3	13±0				
EUVE J0843+033			10±4			
2RE J0845+485	226±12	663±23	487±26	210±24	38±16	
2RE J0846+062	18±5	32±0	49±11	50±15		
EUVE J0852−192			3.8			
2RE J0853−074	23±5	16±5				
EUVE J0858−002			38±10			
2RE J0858+082	32±6	43±8				
2RE J0859−274	27±5	38±8	54±10	32±12		
2RE J0901−522	10±4	16±5				
2RE J0902−040	25±5	38±7				
EUVE J0905−492			29±8		27±13	
EUVE J0906−457					77±22	
EUVE J0907−823			34±7	12±6		
EUVE J0907+134			31±10			
2RE J0907+505	38±0	55±8	42±10	39±14	45±19	
EUVE J0908−258			3			
EUVE J0908+326			28±9			155±62
2RE J0908−370	17±4	23±5	35±9	33±14	40±17	
2RE J0909+423	43±11	45±0				

Table A.2. (*cont.*)

Name	WFC		EUVE			
	S1	S2	100 Å	200 Å	400 Å	600 Å
EUVE J0911+394			31±10			
EUVE J0913−629			13±4			
EUVE J0914+022			46±13	13±11		
2RE J0914+021	52±7	148±12	46±13	13±11		
EUVE J0916+062					73±23	
2RE J0916−194	58±7	23±6	87±12	26±12	41±19	96±43
EUVE J0920+506					117±37	
2RE J0922+401	48±8	95±13	97±14	41±15		
2RE J0924−234	18±4	13±4	31±9			
EUVE J0925+278			29±9			
2RE J0925−531	11±5	23±6				
2RE J0930+103	15±4	39±7				
EUVE J0930−535			25±6		31±14	
2RE J0932−111	42±6	45±7	77±13			
2RE J0932+265	8±3	30±7	44±11			139±59
EUVE J0933+154			<7	<5		
2RE J0933+624	24±5	32±6	61±11			112±47
2RE J0934+695	49±7	52±7	55±10			
EUVE J0939−283				52±18	100±28	
EUVE J0939+146			<6			
2RE J0940+502	52±10	107±17	48±11	36±14		
EUVE J0946−328			1.4			
2RE J0951−144	10±3	12±4				
EUVE J0951+531			33±9			
EUVE J0954−150			37±11		62±26	
EUVE J0956+412			9±5			
EUVE J0957+410			<5			
2RE J0957+852	66±5	50±6	86±9		27±12	
2RE J0958−462	30±6	36±7				
2RE J1000+243	44±6	49±8	61±12	38±13		
2RE J1014+210	17±4	14±0				
EUVE J1014+168			48±10	15±4		
2RE J1016−052	486±18	590±22	816±37	63±22		
EUVE J1017+443			38±10		40±18	97±47
EUVE J1019+157			29±9			
2RE J1019+195	46±6	66±9	97±14			
2RE J1019−140	36±6	78±8				
EUVE J1023−002			40±12	53±23		
2RE J1024+262	33±5	19±7	41±9			
2RE J1024−302	60±8	142±12	110±15	40±12		
2RE J1027+322	15±4	23±7				
2RE J1029+450	34±5	91±9	68±12	36±11		
2RE J1031+823	8±3	12±4				
2RE J1032+532	1073±24	2810±38	2322±46	806±31		
2RE J1033−114	38±7	85±9	102±18		85±35	
EUVE J1034+073					127±38	
2RE J1034+393	23±5	16±0				
EUVE J1035−047			48±13			428±136
2RE J1036+460	58±6	215±13	143±15	66±12		
2RE J1043+490	228±11	464±18	387±20	91±13		
2RE J1043+445	23±5	15±5				
2RE J1044+574	50±6	126±9	67±11			
2RE J1046−492	38±7	57±10	75±10	18±9		
EUVE J1047−442					55±16	89±42
EUVE J1049+124			43±13			354±138
EUVE J1050−207					92±27	79±37
2RE J1051+561	9±4	18±5				
2RE J1051+540	33±6	13±0	63±11			
2RE J1054+560	19±5	4±0				

Table A.2. (*cont.*)

Name	WFC		EUVE			
	S1	S2	100 Å	200 Å	400 Å	600 Å
2RE J1055+602	12±4	15±5				
EUVE J1056+604			18±8			
2RE J1056+070	12±0	34±7				
EUVE J1056+602			<8			
2RE J1058−384	48±8	131±12	101±13	29±12		
EUVE J1058+190			57±15			
EUVE J1058+602			10±6			
2RE J1059+512	593±22	1052±27	1017±30	101±13		
2RE J1100+713	150±20	362±19	277±17	43±12		83±36
EUVE J1100+344			41±11			
EUVE J1102+226	11±5	38±9	3±2			
EUVE J1103+366			<5			
EUVE J1103+359			7±5	<6		
2RE J1104+450	180±11	262±16	211±16			
2RE J1104+381	85±9	16±0	154±10	17±5		
2RE J1104−041	28±5	27±7				
EUVE J1105+588			11±7			
2RE J1111−224	151±13	459±22	316±21	65±17	40±20	
EUVE J1112+358			6±4	<10		
2RE J1112+240	88±12	92±12	151±21			
2RE J1116+445	14±5	30±7				
EUVE J1118+314	78±10	92±14	20±4	10±3	68±33	
EUVE J1118+404			<10			
EUVE J1118−569			17±5			
EUVE J1122+392			<10			
2RE J1122+434	13±5	41±8				
EUVE J1124+047					65±18	
2RE J1126−684	19±8	52±15				
2RE J1126+183	184±12	627±21	445±33	278±36		
EUVE J1127+421			14.1			
2RE J1129+380	6±0	29±8	15±5	8±5		
2RE J1128−025	22±5	49±8				
EUVE J1131−410			53±9	27±11		
EUVE J1131−346			24±8			
2RE J1132+121	17±4	14±0				
EUVE J1134−615			32±9	20±9		
2RE J1135−630	21±5	12±0				
EUVE J1137+510			26±7			85±38
EUVE J1138−002			37±12			
2RE J1139−430	20±9	32±11				
EUVE J1141+342			13±4	11±4		
EUVE J1142−011				40±17	109±34	
EUVE J1145−553			18±5			
EUVE J1146−461			1			
EUVE J1147+050			24±7	<6		
2RE J1147+441	8±0	36±8	32±7	22±8		171±67
2RE J1147+125	13±4	19±6				
2RE J1147+201	30±5	24±6				
2RE J1148+183	37±6	99±10	68±14	48±15		
EUVE J1148−374			7±3			
EUVE J1149−343			15±7		55±17	98±44
EUVE J1149+287	149±11	112±10	125±9	<6	82±41	
2RE J1158+140	22±5	25±6				
EUVE J1201−365			43±9			
EUVE J1201+596			5.2			
2RE J1201−034	20±5	20±6	43±11	27±13		
EUVE J1203+445			48±12			
2RE J1212+773	13±0	18±5				
EUVE J1215−474			23±6			72±33

Table A.2. (*cont.*)

Name	WFC		EUVE			
	S1	S2	100 Å	200 Å	400 Å	600 Å
2RE J1219+163	14±4	16±6				
2RE J1222−390	15±6	23±9				
EUVE J1224−187			42±10			
2RE J1225+253	19±4	17±0				
EUVE J1225−159			<5			
EUVE J1226−499			23±8			
EUVE J1227+069			<6			
EUVE J1229−134			8±4			
EUVE J1229+020			13±4	<6		
EUVE J1230+261			38±10			
2RE J1235+233	15±4	17±5				
2RE J1236+475	541±20	1448±31	951±33	396±27		
2RE J1237+264	17±4	5±0				
EUVE J1240−015					85±23	
2RE J1241−012	37±9	47±14	82±13	27±12		
EUVE J1244−596						199±43
2RE J1248+601	13±4	14±5				
EUVE J1249+660			24±6			
EUVE J1250+209			37±10	32±13		
EUVE J1251+225			13±5	<10		
EUVE J1251+275	16±4	42±7	8±3	6±3		
2RE J1252−291	58±9	51±10	82±12			
EUVE J1253+227			<6	<6		
2RE J1253+270	5±0	10±4				
EUVE J1254−703			2.4			
EUVE J1254+224			<8	7±5		
EUVE J1254−471			31±10			
EUVE J1254+221			<4			
EUVE J1255+258	7±3	54±7	9±4			
EUVE J1255−678			16±5			
EUVE J1256+220	1448±32	5019±55	4229±45	3261±37	235±35	532±77
EUVE J1257+087			29±9			
EUVE J1257−165			31±9			
EUVE J1257+351	18±4	22±5	8±4	<5		
EUVE J1258−704			4.7			
EUVE J1259−701				3.7		
2RE J1259+603	13±4	16±5				
2RE J1300+122	28±8	22±7				
EUVE J1302+247			4.7	4.3		
EUVE J1303+224			7.8			
2RE J1303−405	11±0	30±7				
EUVE J1304+160			31±9	26±12		
EUVE J1305+024			18±8		66±19	112±44
EUVE J1305−409			28±11			
2RE J1307+535	54±7	28±5				
EUVE J1308−687			<6			
EUVE J1308−215				48±13	26±13	
2RE J1309+081	4±0	29±7				
EUVE J1312+848				21±5		
EUVE J1315−679			<8			
2RE J1316+290	14559±100	37036±142	33876±178	21367±144	1667±73	2825±135
2RE J1316+092	19±6	22±6	53±11	40±12		
EUVE J1320+070			33±9			
2RE J1325−111	25±6	24±0				
2RE J1332+223	24±5	25±6	44±10			98±44
EUVE J1334−083	24±6	15±7	27±6	<8		
2RE J1334+371	33±5	48±6	158±14	31±12		
2RE J1336+694	11±3	25±5				
EUVE J1339−055			94±9	<10		

Table A.2. (*cont.*)

Name	WFC		EUVE			
	S1	S2	100 Å	200 Å	400 Å	600 Å
EUVE J1343+076			1.9			
2RE J1344−612	26±7	8±0				
2RE J1345−330	16±4	13±5				
EUVE J1346+051			12.3			
EUVE J1348+079			1.0			
EUVE J1350−414			130±13			
EUVE J1352+062					7.5	
EUVE J1352+692			33±6			
EUVE J1353+318			28±8			
EUVE J1355−249			40±12	43±15		
EUVE J1359+679					35±10	101±34
2RE J1405+100	11±4	19±6	30±8			
EUVE J1409−452			282±20			
EUVE J1412+292			19±7		56±15	
EUVE J1413+463			6.9			
2RE J1415−055	14±5	13±5				
2RE J1417−680	9±3	14±5				
EUVE J1418−216			43±12			
EUVE J1422+039			41±13			154±63
2RE J1425+515	38±5	33±4	60±9			91±40
2RE J1426+500	33±4	49±5	38±8	18±9		
2RE J1426+025	13±4	13±5				
2RE J1428−021	10±3	12±5				
2RE J1428+424	18±4	12±9				
EUVE J1429−380			4.2			
2RE J1429−624	33±5	38±7	39±12		47±20	162±81
2RE J1431+370	28±5	30±6				
EUVE J1433−694			26±9			
2RE J1434−602	22±5	17±0				
EUVE J1434−363			1			
EUVE J1436−382			1			
EUVE J1434+357			27±7			
EUVE J1436−378			34±10			78±38
EUVE J1438−432			8±3	5±3		
2RE J1438+641	19±3	28±4	31±7			
EUVE J1439−240			36±11			
2RE J1439−605	125±10	287±15	459±34	575±48		
2RE J1440+750	98±10	107±9	116±9			58±28
EUVE J1441−373			<6			
2RE J1442+352	66±7	18±0				
2RE J1446+632	31±5	36±5	34±8			
2RE J1450−321	25±8	15±0				
2RE J1451+190	45±7	70±9	129±18	64±18		
EUVE J1456−627			4.6			
EUVE J1500+749			23±5	15±7		
2RE J1501−434	21±5	23±6				
2RE J1502+661	4782±37	7267±43	6762±52	34±10		255±89
2RE J1503+473	106±7	112±8	178±11	48±8		72±36
EUVE J1506−050			39±11		24±12	
2RE J1507+761	77±0	48±8	27±5	24±8		
2RE J1511+615	13±3	17±3	25±6	17±7	20±10	
EUVE J1513+176					47±15	
EUVE J1515−623			6.3			
EUVE J1517+238			32±10			
2RE J1521+522	45±5	41±6	75±8			
EUVE J1521−842			26±8			
EUVE J1521−168			5.9			
EUVE J1526+568			10±4	26±6	25±10	
EUVE J1529−471			38±10		27±13	

Table A.2. (*cont.*)

Name	WFC		EUVE			
	S1	S2	100 Å	200 Å	400 Å	600 Å
2RE J1529+802	15±3	21±3	6±3	5±3	17±8	
2RE J1529+483	20±4	47±7	32±6	16±6		
EUVE J1533+337						131±34
2RE J1535+600	8±0	9±2	18±5		20±9	
EUVE J1535−774			70±15			
EUVE J1538−153			36±10			
2RE J1538−574	26±5	39±6	47±11			
EUVE J1545−465			38±9			
2RE J1545−302	27±5	26±6				
EUVE J1546+676			15±4		28±12	
2RE J1546−364	73±8	61±9				
EUVE J1547−450			4±3			
EUVE J1549+260			10±4	<6		
2RE J1553−421	19±5	20±5				
EUVE J1556−043			32±9			
EUVE J1558+255			<5			
EUVE J1600−165			<4			
2RE J1601+664	10±2	4±0				
2RE J1601+512	20±4	25±5	32±7			
2RE J1601+583	21±3	11±3				
EUVE J1601−164					54±15	
EUVE J1603+052			<6			
2RE J1603−574	48±7	32±7				
2RE J1604−215	11±4	24±6	33±9	25±12		
EUVE J1605+518			26±6			
2RE J1605+104	23±5	18±0				
EUVE J1608−774			45±13			
2RE J1614+335	208±10	331±13	429±19	76±13		
2RE J1617+551	15±3	16±4	30±6			
EUVE J1618−179			33±10			
EUVE J1619−495			59±15	32±13		
2RE J1619−153	36±6	81±8	314±22	248±23		59±28
2RE J1619+394	12±4	14±5				
EUVE J1621+081			20±4	<8		
2RE J1621+185	17±5	15±5				
2RE J1623−391	232±15	577±22	349±28	82±17		
2RE J1625−490	23±6	18±0				
EUVE J1627+101			34±9			
EUVE J1629+780	1008±15	1472±18	1617±21	82±8	13±6	60±25
2RE J1636+525	12±4	46±11				
EUVE J1636−285						410±79
2RE J1638+350	57±7	54±7	62±9		20±9	72±28
EUVE J1639−576			55±16			
EUVE J1644+454			20±5	16±7	14±7	
EUVE J1649−439			<4			
2RE J1650+403	19±5	34±6				
EUVE J1651−443			3±2			
EUVE J1653+397			31±7			
2RE J1655−081	74±8	116±12	131±19	35±15		
2RE J1656+650	7±2	9±2	14±3	14±4		37±14
2RE J1656−390	41±8	32±9				
2RE J1657+352	93±8	14±0	180±12	16±7		52±22
EUVE J1658+343			28±7			64±24
2RE J1659+440	64±7	235±11	164±11	97±10		
EUVE J1703−018					80±24	
EUVE J1705−018			9±4			
EUVE J1706−450			49±16			
EUVE J1708+034					67±22	
EUVE J1708−111					107±30	

Table A.2. (*cont.*)

Name	WFC		EUVE			
	S1	S2	100 Å	200 Å	400 Å	600 Å
2RE J1711+664	10±1	6±0	16±2	10±3		
EUVE J1712−248					93±28	
EUVE J1715−266			15±6			
EUVE J1716−265			6±4			
2RE J1717−665	60±14	77±14	94±16	20±10		
EUVE J1717+104			56±16	46±19		
2RE J1717+102	27±7	23±8				
EUVE J1721+278			39±11		41±17	
EUVE J1723+153			54±15	41±19		
2RE J1726+583	32±4	10±0	53±4		11±5	
2RE J1727−360	48±8	89±11	60±17	34±16		
2RE J1732+741	18±2	9±2	26±3			
2RE J1734+615	6±0	12±3				
2RE J1736+684	11±1	14±1	16±2	8±3		28±11
2RE J1737+665	3±1	5±1				
2RE J1738+611	15±3	20±3	17±3	12±4	8±4	
EUVE J1745−883			22±6	19±9		
2RE J1746−703	141±20	274±26	281±18	38±12		
EUVE J1748+484				15±7		133±32
2RE J1755+361	30±4	46±6	34±7		21±9	
2RE J1757+291	21±5	28±7	41±8			
EUVE J1759+428			23±6			44±22
2RE J1800+683	53±2	21±1	67±2	6±2		35±7
2RE J1802+641	3±1	5±1				
EUVE J1805+204			7.8			
EUVE J1805+024			77±12	62±15		
2RE J1805+212	43±8	42±13	59±10			
2RE J1808+294	55±6	86±14	61±9	20±9	35±14	54±25
EUVE J1809+300	26±4	51±13	25±7			
2RE J1813+642	6±1	5±1				
EUVE J1816+498			4233±36	94±9		
2RE J1816+541	6±1	5±2				
EUVE J1818−371			13±4			
2RE J1820+580	103±4	116±4	133±6		8±4	
2RE J1821+642	11±1	33±0	13±2			
EUVE J1822+017			31±9			
EUVE J1823−342			13±6	<6		
2RE J1829+345	9±2	3±0				
EUVE J1833−005			<4			
2RE J1833+514	52±3	60±4	61±6	14±6	17±7	
2RE J1834+184	39±5	33±6	43±10			
EUVE J1834−006			5±3			
EUVE J1841−654			8			
EUVE J1844+255			30±8			
2RE J1844−741	117±22	74±23	150±11			93±34
2RE J1845+682	30±2	29±2	45±3			27±9
EUVE J1845−648			42±8			
2RE J1846+191	11±3	19±0				
2RE J1847+015	243±13	816±26	634±22	234±16		
2RE J1847−221	31±8	66±11	89±15	42±21	36±15	
EUVE J1848−223			32±9			
EUVE J1849−238			35±8	14±6		
EUVE J1854−324			18±9		86±24	
2RE J1855+082	19±4	14±5	13±6			
2RE J1855+233	36±4	58±7	64±11	32±13		
EUVE J1857−524			31±9		43±18	
EUVE J1858−299			4±3			
EUVE J1906+169			29±8	32±13		
2RE J1906+274	28±5	24±6				
2RE J1907+301	43±6	46±7	81±11			

Table A.2. (*cont.*)

Name	WFC		EUVE			
	S1	S2	100 Å	200 Å	400 Å	600 Å
EUVE J1908+524			19±5		14±7	
EUVE J1911−286			4±3			
EUVE J1911−606			1.8			
EUVE J1911+299			26±8			68±33
2RE J1918+595	11±2	17±3	20±4	20±6		
EUVE J1925+031			3	3		
2RE J1925−563	710±38	697±40	890±26			
EUVE J1928−500			31±9			
2RE J1929−230	16±6	28±9				
EUVE J1934−646			24±7			102±48
EUVE J1935−192			39±10	27±13		
EUVE J1936+502			10.5			
EUVE J1936+265			21	4		
2RE J1938−461	424±21	302±18	142±15			
2RE J1943+500	22±4	29±4	25±5			55±22
EUVE J1947−110			34±10	47±18		
EUVE J1947+262			1.1			
2RE J1950+260	25±7	49±19				
EUVE J1951−585			26±8	24±12		
EUVE J1953+444			6±3	6±4		
EUVE J1954−239			17±4	10±4		
EUVE J1956−030			40±12			
EUVE J1957+298			<6			
EUVE J1958+550				42±9		72±30
EUVE J1959+227			60±17		51±25	
2RE J2000−334	38±8	33±9	58±13			
EUVE J2003+170			54±16			
2RE J2004−560	17±0	45±10	34±9			73±35
EUVE J2005+544			6±2	<13		
2RE J2005+224	454±19	147±13	466±21	8±4		
2RE J2009−602	1320±46	2478±66	2255±42	325±21		
EUVE J2010+426			5.3			
2RE J2013+400	63±6	34±5	111±11			59±28
EUVE J2013+467			6±3			
2RE J2018−572	26±9	43±10	34±9			
EUVE J2023+224				84±21		
2RE J2024−422	62±10	191±19	134±17	32±14		94±44
2RE J2024+200	24±5	18±5				
EUVE J2024+304			31±8			
EUVE J2028−114			4±3			
EUVE J2029+096			3±2			
2RE J2029+391	24±5	30±6	41±8	18±9	35±13	
EUVE J2030+526			129±11	92±14	30±13	
EUVE J2030+177			37±11			
2RE J2031+051	22±5	22±6				
2RE J2037+753	26±4	34±4	57±6	16±5	10±5	
EUVE J2041−368					84±25	
2RE J2041−322	46±9	90±14	99±20	41±19		131±57
2RE J2045−312	161±14	119±17	181±26	74±24		
EUVE J2045+310			65±12			
EUVE J2046+153			39±10			
2RE J2047−363	25±6	118±15				
EUVE J2049−759			6			
2RE J2051−134	4±0	29±7				
EUVE J2053+298			77±16			
EUVE J2054−750			10			
EUVE J2055+164			30±9	29±12		
2RE J2055−170	44±7	55±8				
2RE J2057+311	25±6	24±0	262±18	94±20		
EUVE J2059+400			9±3			

Table A.2. (*cont.*)

Name	WFC		EUVE			
	S1	S2	100 Å	200 Å	400 Å	600 Å
EUVE J2100−426			<5			
2RE J2100+400	15±3	20±5	39±9			110±45
2RE J2102+274	53±7	56±9	64±12			
EUVE J2104−770			30±6			
2RE J2106+164	31±7	19±0				
2RE J2106+384	12±3	21±0				
2RE J2107−051	24±5	6±0				
EUVE J2110+454			23±7			
2RE J2112+500	402±13	1292±21	996±35	529±26		
EUVE J2114+503			60±3			
EUVE J2115−586			59±12	31±12		
2RE J2116+735	83±7	82±10	110±8			
EUVE J2117+342				126±18		
EUVE J2119−385				84±23		
EUVE J2118+115			4±2			
2RE J2121+402	18±4	11±4	30±8	3		
EUVE J2121+104			5±2			
2RE J2122−164	22±5	16±5				
2RE J2125+282	16±4	18±6	32±9	44±16		
2RE J2126+192	263±16	757±25	626±28	392±30		
EUVE J2126−694			27±8			
2RE J2127−221	57±8	95±10	78±14	33±16		
EUVE J2131+233			42±10			147±54
EUVE J2132+002			5.8			
EUVE J2132+101			2			
2RE J2133+453	9±3	9±3	27±8			
EUVE J2135+008			8.0			
2RE J2137−800	28±11	38±11				
EUVE J2137+016			4±2	<10		
EUVE J2137+205				79±19		116±54
EUVE J2139−273			3±2			
EUVE J2141−140			5±4	<8		
EUVE J2142−549			39±13	37±16		
2RE J2142+433	375±13	61±6				
EUVE J2144+192				75±19	35±16	
2RE J2147−160	45±9	63±10	69±14	39±19		
EUVE J2149−720			32±14			
EUVE J2152−620			4±2			
2RE J2154−302	30±6	84±10	91±14			
EUVE J2155−618			<5			
EUVE J2155+043			47±12			
2RE J2156−543	1401±34	3693±59	2181±57	725±39		
2RE J2156−414	157±12	193±13	260±15	<20		
EUVE J2157−406			<5	<8		
2RE J2157−505	30±6	39±7				
EUVE J2157−400			<6	12±7		
EUVE J2158−404			<6	<8		
2RE J2158+825	52±8	77±9	111±11	23±9	18±9	
2RE J2158−301	180±12	17±0	333±21			
EUVE J2158−590			12±4			
EUVE J2200+548					47±14	
2RE J2201+281	7±3	11±3				
EUVE J2202−412			<5			
EUVE J2203+482			11±7	<8		
EUVE J2204+472			11±7			
2RE J2206+634	12±4	17±5	16±5		19±9	
2RE J2207+252	23±4	55±6	34±10			175±67
2RE J2208+454	119±8	193±11	106±11			
2RE J2210−300	67±10	140±14	91±11	39±9	37±18	
2RE J2214−491	12±4	1089±29	1097±34	5021±73	107±21	

Table A.2. (*cont.*)

Name	WFC		EUVE			
	S1	S2	100 Å	200 Å	400 Å	600 Å
2RE J2220+493	27±5	33±5	51±7			
2RE J2223+322	24±5	29±6				
EUVE J2225+419			28±9			
EUVE J2225+167			48±12			
EUVE J2226−167			50±12	42±19		
EUVE J2228−000			20±9			
EUVE J2228+576			25±7	20±7		
EUVE J2228+116			44±11	43±18	46±21	
EUVE J2228+635			17±5			
EUVE J2229+000			15±9			
EUVE J2231+017				85±22		
2RE J2234+081	22±5	9±0				
EUVE J2236+001			12±6	<8		
2RE J2238−652	16±5	44±9				
EUVE J2238+022			<5			
EUVE J2238−323					63±20	
2RE J2238−203	83±16	102±16	160±17	50±18		
EUVE J2240+097			39±12		69±28	
EUVE J2242−475			33±9	28±14		
2RE J2244−321	26±0	76±13	13±4			154±74
2RE J2246+442	117±9	148±10	107±13	37±14		
2RE J2248−510	18±5	10±0				
EUVE J2249−209			34±10	<5		
2RE J2249+583	71±8	17±6	120±11			
EUVE J2252+406			35±9			
2RE J2300−070	42±9	43±9	69±14	32±16		
EUVE J2301−392				34±17	81±21	
EUVE J2302−560			14±7	57±15		
2RE J2308+572	3±0	26±6				
2RE J2309+475	49±8	25±6	59±10	28±9	33±15	
EUVE J2310+206			59±16	55±27		
2RE J2312+104	1672±36	4653±56	3601±75	1328±55	33±16	
2RE J2313+024	28±7	22±0				
EUVE J2315−642			30±8		24±11	
2RE J2319+790	62±7	81±6	96±8	16±6		49±21
EUVE J2320+203			44±13			310±149
EUVE J2321−269			40±11	39±17		
EUVE J2322−469			<11			
EUVE J2323−247			39±11			
EUVE J2323−268			16±8		71±21	127±57
EUVE J2324+458				41±10		
2RE J2324−544	308±22	667±30	601±27	169±22		
EUVE J2327+048			4±2			
2RE J2329+412	18±6	16±6	37±9			85±39
EUVE J2329−477			32±9	41±14		162±74
2RE J2331+195	33±9	81±11	105±16			
2RE J2334−471	41±0	396±31	287±17	687±25		
EUVE J2336−213					109±36	243±98
EUVE J2336+480			23±7			
2RE J2337+462	150±13	174±13	288±18	31±10		71±32
2RE J2339−691	20±5	18±0				
EUVE J2350−206			49±13			
2RE J2352+753	8±2	18±3				
2RE J2353−243	45±10	71±10				
2RE J2353−702	80±8	711±23	428±21	929±32		
2RE J2355+283	131±12	203±17	144±15	33±12		110±52
2RE J2355+250	13±5	21±7				
EUVE J2356−327			12.9			
EUVE J2359−306			14.7			

Table A.3. *Composite EUVE deep survey source list;* c *and* d *are the Deep Survey 100 Å and 200 Å bands respectively.*

Name	RA$_{2000}$	DEC$_{2000}$	ID1	ID2	Sp	M_V	c	d
EUVE J0008+594	00:08:18	+59:27.1	NO ID				14±8	
EUVE J0028−631	00:28:22	−63:06.9	NO ID				<8	
EUVE J0030−633	00:30:59	−63:21.9	NO ID				10±3	
EUVE J0031−629	00:31:40	−62:58.2	HD 2885	HR 127	A2 V+	4.54	13±5	
EUVE J0032−630	00:32:51	−63:02.4	HD 3003	TD1 292	A0 V	5.08	<6	
EUVE J0033−633	00:33:06	−63:18.1	1WGA J0033−6317		X		4±2	
EUVE J0042+041	00:42:22	+04:10.4	TD1 389	HD 3972	F8V	7.58	12.6±1.7	
EUVE J0048+052	00:48:23	+05:17.3	TD1 447	HD 4628	K2V	5.75	4.8±0.7	
EUVE J0113+075	01:13:46	+07:34.6	ζ Psc B	HD 7345	F7V	6.32	7.7±0.9	
EUVE J0218+146	02:18:55	+14:34.3	WD 0216+143	PG 0216+143	DA	14.53	96.9±6.4	
EUVE J0232+150	02:32:56	+15:02.3	29 Ari	TD1 1493	F8V	6.00	24.8±1.2	
EUVE J0318+184	03:18:24	+18:24.6			dK8	11.8	2.2±0.5	
EUVE J0319+033	03:19:29	+03:18.4	AG+03 377	BD+02° 521	G0	8.5	<5	
EUVE J0320+030	03:20:53	+03:03.8	NO ID				<5	
EUVE J0330+188	03:30:57	+18:48.1	SAO 93468	BD+18 492	F8	7.8	3.7±0.8	
EUVE J0418+215	04:18:23	+21:34.8	51 Tau	TD1 2995	F0V	5.60	4.1±0.7	
EUVE J0419+217	04:19:41	+21:45.9			dM2e	11.7	12.5±1.4	
EUVE J0424+154	04:24:58	+15:28.1						
EUVE J0424+158	04:24:51	+15:52.2	Cl* Melotte 25 VA33	BPM 85733	K7	11.66		
EUVE J0424+217	04:24:17	+21:44.3	TD1 3121	HD 27808	F8V	7.14	9.7±0.8	
EUVE J0425+152	04:25:53	+15:17.4						
EUVE J0425+155	04:25:51	+15:30.9	CL Melotte 25 1366	LP 415−1458	M0	12.38		
EUVE J0425+159	04:25:38	+15:55.8	Cl Melotte 25 57	HD 27991	F7 V	6.46		<30
EUVE J0425−572	04:25:41	−57:15.1	EUVEJ0424.6−5714	RX J0425.6−5714	CV	19	71±8	
EUVE J0426+155	04:26:08	+15:31.0	Cl Melotte 25 59	HD 28034	G0	7.49		
EUVE J0426−572	04:26:03	−57:12.7	1H 0419−577	1ES 0425−57.3	Sy	15.1	79±8	
EUVE J0427+154	04:27:05	+15:24.0	Cl* Melotte 25 VA41			16.13		
EUVE J0427+155	04:27:36	+15:34.9	CL Melotte 25 65	HD 28205	G0	7.42		
EUVE J0428+158	04:28:41	+15:52.3	Cl Melotte 25 72	HD 28319	A7 III	3.4	10±4	
EUVE J0428+159	04:28:38	+15:57.4	θ¹ Tau	HD 28307	K0IIIb	3.80		
EUVE J0428+162	04:28:51	+16:17.2	Cl Melotte 25 190	HD 285806	K7 V	10.70	8±4	
EUVE J0429+161	04:29:00	+16:09.4	Cl Melotte 25 75	HD 28363	F8	6.59	7±4	
EUVE J0429+163	04:29:01	+16:20.5	Cl Melotte 25 VA 512	2E 0426.1+1614	M1.5	14.28	<6	
EUVE J0429+219	04:29:02	+21:54.8	HD 28343	SAO 76626	K7V	8.27	4.8±0.7	
EUVE J0430+156	04:30:07	+15:38.1	Cl Melotte 25 80	HD 28485	F0V	5.58	±6	
EUVE J0430+157	04:30:35	+15:43.5	Cl Melotte 25 182	HD 28545	K0	8.99	<10	
EUVE J0430+161	04:30:46	+16:08.0	Cl Melotte 25 85	HD 28568	F2	6.51		22±9
EUVE J0436−472	04:36:23	−47:15.5	2EUVE J0436−47.2	1RXSJ043617−471	X		<8	
EUVE J0436−472	04:36:57	−47:15.9	NO ID				7±4	
EUVE J0437−475	04:37:13	−47:32.8	1RXS J043711.9−473142		X		<10	
EUVE J0437−472	04:37:30	−47:12.6	1ES 0435−472	RX J04374−4711	Sy1	15.3	35±9	
EUVE J0438−474	04:38:47	−47:29.2	1RXS J043847.0−472751		X		<8	
EUVE J0453+222	04:53:04	+22:14.2	HD 30973	SAO 76806	K5V	8.77	7.6±1.4	
EUVE J0455−285	04:55:41	−28:33.8	GJ 2037	HD 31560	K3V	8.12		
EUVE J0531−036	05:31:24	−03:41.4	HD 36395	GJ 205	M1.5	7.92		63±23
EUVE J0532−031	05:32:05	−03:06.5	HBC 97	2RE J0532−030	K7e	11.52		64±29
EUVE J0536+234	05:36:53	+23:25.9	HD 245358	SAO 77318	G0	8.9	14.0±1.2	
EUVE J0634+179	06:34:08	+18:00.0	1E 0631.2+1802	EXO 063111+1801.9	dMe	?		
EUVE J0539+231	05:39:32	+23:06.6	NO ID				6.7±1.0	
EUVE J0551+242	05:51:10	+24:14.8	NO ID				41.9±7.7	
EUVE J0604+232	06:04:08	+23:15.5	1 Gem	TD1 5900	G5III	4.15	5.4±0.8	
EUVE J0651+228	06:51:26	+22:48.3	NO ID				8.4±0.8	
EUVE J0659−293	06:59:32	−29:18.9	1WGA J0659.5−2918		X		<6	
EUVE J0659−292	06:59:55	−29:17.7	NO ID				22±5	
EUVE J0700−288	07:00:04	−28:49.3	WT 1539			12.5	<6	
EUVE J0706+226	07:06:18	+22:40.4	HD 53532	SAO 79056	G0	8.23	2.6±0.5	
EUVE J0720+219	07:20:11	+21:58.7	δ Gem A	TD1 9732	F0IV	3.53	19.2±1.2	
EUVE J0722+217	07:22:13	+21:44.0	NO ID				14.1±1.9	
EUVE J0727+214	07:27:40	+21:26.8	63 Gem	TD1 10122	F5V	5.24		44.2+8.6
EUVE J0729−381	07:29:30	−38:08.2	HD 59704		F7 V	7.8	8±4	
EUVE J0750+148	07:50:34	+14:50.4	HD 63648	TD1 11121	F2	7.2	<5	
EUVE J0815−492	08:15:18	−49:13.1	V* IX Vel	SAO 219684	CV	9.44		

Table A.3. (*cont.*)

Name	RA$_{2000}$	DEC$_{2000}$	ID1	ID2	Sp	M_V	c	d
EUVE J0942+700	09:42:28	+70:03.1	LHS2176/2178		M2/3	10.6/11.2	<13	
EUVE J0951+131	09:51:38	+13:09.0	AG+13 989	PPM 126860	F2	9.6	5.9±1.6	
EUVE J1019+194	10:19:42	+19:28.7	HD 89449	TD1 14746	F6 IV	4.80	11±6	
EUVE J1034+396	10:34:38	+39:38.4	Z 1031.6+3953	KUG 1031+398	Sy1	15.6	35±8	
EUVE J1102+717	11:02:11	+71:44.8						
EUVE J1104+382	11:04:44	+38:14.9	HD 95976	BD+39° 2418	F2	7.4	5±3	
EUVE J1105+382	11:05:50	+38:13.1	WD 1103+384	PG 1103+384	DA	17.2	5±3	
EUVE J1221+172	12:21:21	+17:13.3						
EUVE J1332+372	13:32:37	+37:12.4	NGP9 F270−0321988		Gal		<15	
EUVE J1334+372	13:34:17	+37:15.5	2E 1332.0+3730	QSO 1332+375	QSO	18.2	6±2	
EUVE J1335+371	13:35:38	+37:06.7	NO ID				<6	
EUVE J1348+263	13:48:36	+26:22.1	MS 1346.2+2637	EXO 1346.2+2637	QSO	18.5	<6	
EUVE J1348+265	13:48:36	+26:31.1	PGC 94626	2E 1346.2+2646	Sy2	16.5	<6	
EUVE J1348+265	13:48:55	+26:35.3	EXSS 1346.5+2650	1WGAJ1348.8+263	X		18±6	
EUVE J1349+269	13:49:07	+26:58.0	HD 120476	BD+27° 2296	K2	7.04	15±6	
EUVE J1418+245	14:18:54	+24:31.2						
EUVE J1453+191	14:53:23	+19:09.3	HD 131511	BD+19° 2881	K2 V	6.01	28±8	
EUVE J1531−194	15:31:25	−19:28.6	NO ID				7.3±1.7	
EUVE J1623−392	16:23:57	−39:12.0	HR 6094	HD 147513	G5V	5.40V?		
EUVE J1712−231	17:12:12	−23:09.3	SAO 185139	CD−23 13200	G5	9.57	3.3±0.7	
EUVE J1716−231	17:16:21	−23:10.1	NO ID				2.5±0.6	
EUVE J1717−231	17:17:45	−23:11.6	NO ID				9.1±0.9	
EUVE J1931−215	19:31:27	−21:34.3			dMe		8.7±0.9	
EUVE J2027+525	20:27:45	+52:30.2						
EUVE J2053−175	20:53:31	−17:33.9			K	10.4	6.8±0.8	
EUVE J2055−171	20:55:50	−17:07.3	RE J2055−170	HD 199143	F8V	7.1	85.9±2.5	
EUVE J2056−171	20:56:02	−17:10.8			dK7e	10.6	100.5±2.6	
EUVE J2141−140	21:41:33	−14:02.5	42 Cap	TD1 28359	G2V	5.18	38.1±1.9	
EUVE J2154−304	21:54:53	−30:29.3	RE J2154−302		DA:	14.4 :		
EUVE J2213−111	22:13:12	−11:10.4	HD 210803	NC 56	G5	9.31	14.2±1.1	
EUVE J2233−096	22:33:18	−09:36.9	StKM 1−2018		M3	12.41	18.7±3.3	
EUVE J2237−085	22:37:25	−08:31.7	HD 214318	SAO 146207	G5	9.2	3.1±0.7	
EUVE J2300−070	23:00:30	−07:03.0	RE J2300−070	HD 217411	G5	9.8	107.4±7.0	
EUVE J2330−473	23:30:59	−47:23.0	CPD−48 10901	HD 221260	K1/K2V	9.5		
EUVE J2338+462	23:38:00	+46:12.2	HD 222143	BD+45° 4288	G5	6.4	15±4	
EUVE J2343−145	23:43:01	−14:34.2	BD−15 6477	HD 222685	F5V	8.9		
EUVE J2348−281	23:48:56	−28:06.6	HD 223352	TD1 30097	A0 V	4.57	25±6	
EUVE J2348−276	23:48:40	−27:39.8	LHS 4016		M2.5	12.4	11±5	
EUVE J2349−278	23:49:22	−27:51.5	HD 223408	CPD−28°7723	F6/F7 V	7.1	13±5	

References

Abbott, M.J., *et al.* (1996). *Astrophys. J. Suppl.*, **107**, 451.

Adams, G.P., Rochester, G.K., Sumner, T.J. and Williams, O.R. (1987). *J. Phys. E: Sci. Instrum.*, **20**, 1261.

Aller, L.H. (1959). *Publ. Ast. Soc. Pac.*, **71**, 324.

Anders, E. and Grevesse, N. (1989). Abundance of the elements: meteoritic and solar. *Geochim. Cosmochim. Acta*, **53**, 197.

Antiochos, S.K., Haisch, B.M. and Stern, R.A. (1986). *Astrophys. J. Lett.*, **307**, L55.

Antunes A., Nagase, F. and White, N.E. (1994). *Astrophys. J. Lett.*, **436**, L83.

Arabadjis, J.S. and Bregman, J.N. (1999). *Astrophys. J.*, **514**, 607.

Arnaud, K.A. (1996). Astronomical Data Analysis Software and Systems V. *ASP Conference Series*, volume 101, eds. G. Jacoby and J. Barnes, p. 17.

Arnaud, M. and Raymond, J.C. (1992). *Astrophys. J.*, **398**, 394.

Arnaud, M. and Rothenflug, R. (1985). *Astron. Astrophys. Suppl.*, **60**, 425.

Aschenbach, B. (1988). *Applied Optics*, **27**, 1404.

Ayres. T.R. (1996). In Astrophysics in the Extreme Ultraviolet, eds. S. Bowyer and R.F. Malina, Kluwer, 113.

Ayres, T.R., Osten, R.A. and Brown, A. (1999). *Astrophys. J.*, **526**, 445.

Barber, C.R., Warwick, R.S., McGale, P.A., Pye, J.P. and Bertram, D. (1995). *Mon. Not. Roy. Astron. Soc.*, **273**, 93.

Barbera, M., Micela, G., Sciortino, S., Harnden, R.F., Jr and Rosner, R. (1993). *Astrophys. J.*, **414**, 846.

Bannister, N.P. (2001). PhD Thesis, University of Leicester.

Bannister, N.P., *et al.* (1999). 11th European Workshop on White Dwarfs. *ASP Conference Series*, volume 169, eds. J.-E. Solheim and E. Meistas, p. 188.

Bannister, N.P., Barstow, M.A., Holberg, J.B. and Bruhweiler, F.C. (2001). 12th European Workshop on White Dwarfs. *ASP Conference Series*, volume 226, eds. J.L. Provencal, H. Shipman, J. MacDonald and S. Goodchild, p. 105.

Barstow, M.A. (1989). In *White Dwarfs*, ed. G. Wegner, Springer-Verlag, Berlin, p. 156.

Barstow, M.A. and Hubeny, I. (1998). *Mon. Not. Roy. Astron. Soc.*, **299**, 379.

Barstow, M.A. and Pounds, K.A. (1988). In *Hot Thin Plasmas in Astrophysics*, ed. R. Pallavicini, Kluwer, Dordrecht, p. 359.

Barstow, M.A. and Sansom A.E. (1990). *Proc. SPIE*, **1344**, 244.

Barstow, M.A. and Sion, E.M. (1994). *Mon. Not. Roy. Astron. Soc.*, **271**, L52.

Barstow, M.A., Fraser, G.W. and Milward, S.R. (1985). *Proc. SPIE*, **597**, 352.

Barstow, M.A., Holberg, J.B., Grauer, A.D. and Winget, D.E. (1986). *Astrophys. J. Lett.*, **306**, L25.

Barstow, M.A., Kent, B.J., Whiteley, M.J. and Spurret, P.H. (1987). *J. Mod. Opt.*, **34**, 1491.

Barstow, M.A., Bromage, G.E., Pankiewicz, G.S., Gonzalez-Riestra, R., Denby, M. and Pye, J.P. (1991). *Nature*, **353**, 635.

Barstow, M.A., Pye, J.P., Bromage, G.E., Fleming, T.A., Guinan, E. and Dorren, J.D. (1992a). In *Stellar Chromospheres, Coronae and Winds*, eds. C.S. Jeffery and R.E.M. Griffin, Cambridge University Press, Cambridge.

Barstow, M.A., Schmitt, J.H.M.M., Clemens, J.C., Pye, J.P., Denby, M., Harris, A.W. and Pankiewicz, G.S. (1992b). *Mon. Not. Roy. Astron. Soc.*, **255**, 369.

Barstow, M.A., *et al.* (1993). *Mon. Not. Roy. Astron. Soc.*, **264**, 16.

Barstow, M.A., Holberg, J.B., Fleming, T.A., Marsh, M.C., Koester, D. and Wonnacott D. (1994a). *Mon. Not. Roy. Astron. Soc.*, **270**, 499.

Barstow, M.A., Holberg, J.B. and Koester, D. (1994b). *Mon. Not. Roy. Astron. Soc.*, **268**, L35.

Barstow, M.A., Holberg, J.B. and Koester, D. (1994c). *Mon. Not. Roy. Astron. Soc.*, **270**, 516.

Barstow, M.A., Holberg, J.B., Werner, K., Buckley, D.A.H. and Stobie, R.S. (1994d). *Mon. Not. Roy. Astron. Soc.*, **267**, 653.

Barstow, M.A., *et al.* (1994e). *Mon. Not. Roy. Astron. Soc.*, **267**, 647.

Barstow, M.A., Holberg, J.B., Koester, D., Nousek, J.A. and Werner, K. (1995a). In *White Dwarfs*, eds. D. Koester and K. Werner, Springer-Verlag, Berlin, p. 302.

Barstow, M.A., Holberg, J.B. and Koester, D. (1995b). *Mon. Not. Roy. Astron. Soc.*, **274**, L31.

Barstow, M.A., O'Donoghue, D., Kilkenny, D., Burleigh, M.R. and Fleming, T.A. (1995c). *Mon. Not. Roy. Astron. Soc.*, **273**, 711.

Barstow, M.A., *et al.* (1995d). *Mon. Not. Roy. Astron. Soc.*, **272**, 531.

Barstow, M.A., Holberg, J.B., Hubeny, I., Lanz, T., Bruhweiler, F.C. and Tweedy, R.W. (1996a). *Mon. Not. Roy. Astron. Soc.*, **279**, 1120.

Barstow, M.A., Hubeny, I., Lanz, T., Holberg, J.B. and Sion, E.M. (1996b). In *Astrophysics in the Extreme Ultraviolet*, eds. S. Bowyer and R.F. Malina, Kluwer, Dordrecht, p. 203.

Barstow, M.A., Dobbie, P.D., Holberg, J.B., Hubeny, I. and Lanz, T. (1997). *Mon. Not. Roy. Astron. Soc.*, **286**, 58.

Barstow, M.A., Hubeny, I. and Holberg, J.B. (1998). *Mon. Not. Roy. Astron. Soc.*, **299**, 520.

Barstow, M.A., Hubeny, I. and Holberg, J.B. (1999). *Mon. Not. Roy. Astron. Soc.*, **307**, 884.

Barstow, M.A., *et al.* (2000). *Mon. Not. Roy. Astron. Soc.*, in press.

Becker, R. H. (1981). *Astrophys. J.*, **251**, 626.

Becker, W. and Trumper, J. (1993). *Nature*, **365**, 528.

Beeckmans, F. and Burger, M. (1977). *Astron. Astrophys.*, **61**, 815.

Beichmann, C.A., Helou, G. and Walker, D.W. (1988). NASA RP (Reference Publication), Washington: NASA, ed. C.A. Beichmann *et al.* (volume 1); G. Helou (volume 7); D.W. Walker (volume 7).

Bell, K.L. and Kingston, A.E. (1967). *Mon. Not. Roy. Astron. Soc.*, **136**, 241.

Bell, K.L. and Kingston, A.E. (1988). *Proc. Phys. Soc.*, **90**, 31.

Bergeron, P., Kidder, K.M., Holberg, J.B., Liebert, J., Wesemael, F. and Saffer, R.A. (1991). *Astrophys. J.*, **372**, 267.

Bergeron, P., Saffer, R.A. and Liebert, J. (1992). *Astrophys. J.*, **394**, 228.

Bergeron, P., Liebert, J. and Fulbright, M.S. (1995). *Astrophys. J.*, **444**, 810.

Berghöfer, T.W., Vennes, S. and Dupuis, J. (2000). *Astrophys. J.* **538**, 854.

Beuermann, K. and Thomas, H.-C. (1993). *Adv. Space Sci.*, **12**, No 12, 115.

Beuermann, K., *et al.* (1987). *Astron. Astrophys.*, **175**, L9.

Bleeker, J.A.M., Heise, J., Tanaka, Y., Hayakawa, S. and Yamashita, K. (1978). *Astron. Astrophys.*, **69**, 145.

Bloch, J.J. (1996). In *Astrophysics in the Extreme Ultraviolet*, eds. S. Bowyer and R.F. Malina, Kluwer, Dordrecht, p. 7.

Bloch, J.J., *et al.* (1986). *Astrophys. J. Lett.*, **308**, L59.

Bloch, J.J., Edwards, B.C., Priedhorsky, W.C., Roussel-Dupré, D.C., Smith, B.W., Siegmund, O.H., Carone, T., Cully, S., Rodriguez-Bell, T., Warren, J.K. and Vallerga, J.V. (1994). *Proc. SPIE*, **2280**, 297.

Bohlin, R.C. (1973). *Astrophys. J.*, **182**, 139.

Bohlin, R.C., Savage, B.D. and Drake, J.F. (1978). *Astrophys. J.*, **224**, 132.

Böhm-Vitense, E. (1980). *Astrophys. J. Lett.*, **239**, L79.

Böhm-Vitense, E. (1993). *Astron. J.*, **106**, 1113.

Bowyer, S. (1996). In *Astrophysics in the Extreme Ultraviolet, Proceedings of Colloquium No 152 of the International Astronomical Union*, ed. S. Bowyer and R. F. Malina, Berkeley, California, March 27–30 1995, Kluwer, Dordrecht, p. 225.

Bowyer, S. and Malina, R.F. (1991a). In *Extreme Ultraviolet Astronomy*, eds. R.F. Malina and S. Bowyer, Pergamon, New York, p. 397.

Bowyer, S. and Malina, R.F. (1991b). *Adv. Space Res.*, **11**, No. 11, 205.

Bowyer, S., Freeman, J., Lampton, M. and Paresce, F. (1977a). Apollo-Soyuz Test Project Summary Science Report (Vol 1), NASA SP-412, p. 71.

Bowyer, S., Margon, B., Lampton, M., Paresce, F. and Stern, R. (1977b). Apollo-Soyuz Test Project Summary Science Report (Vol 1), NASA SP-412, p. 49.

Bowyer, S., Lieu, R., Lampton, M., Lewis, J., Wu, X., Drake, J.J. and Malina, R.F. (1994). *Astrophys. J. Suppl.*, **93**, 569.

Bowyer, S., Lieu, R., Sidher, S.D., Lampton, M. and Knude, J. (1995). *Nature*, **375**, 212.

Bowyer, S., Lampton, M., Lewis, J., Wu, X., Jelinksy, P. and Malina, R.F. (1996). *Astrophys. J. Suppl.*, **102**, 129.

Bowyer, S., Berghöfer, T.W. and Korpela, E.J. (1999). *Astrophys. J.*, **526**, 592.

Brandt, W.N., Pounds, K.A. and Fink, H. (1995). *Mon. Not. Roy. Astron. Soc.*, **273**, L47.

Brickhouse, N., Edgar, R., Kaastra, K., Kallman, T., Liedahl, D., Masai, K., Monsignori-Fossi, B., Petre, R., Sanders, W., Savin, D.W. and Stern, R. (1995). Report on the Plasma Codes. *Legacy*, the journal of the High Energy Astrophysics Science Archive Research Center, No. 6, p. 4.

Brickhouse, N.S. (1996). In *Astrophysics in the Extreme Ultraviolet*, eds. S. Bowyer and R.F. Malina, Kluwer, Drodrecht, p. 105.

Brickhouse, N.S. and Dupree, A.K. (1998). *Astrophys. J.*, **502**, 918.

Brickhouse, N.S., Raymond J.C. and Smith, B.W. (1995). *Astrophys. J. Suppl.*, **97**, 551.

Broadfoot, A.L., *et al.* (1977). *Space Sci. Rev.*, **21**, 183.

Brown, A. (1994). In Cool Stars, Stellar Systems and the Sun; Eighth Cambridge Workshop. *ASP Conference Series*, volume 64, ed. J.-P. Caillault, p. 23.

Bruhweiler, F.C. (1996). In *Astrophysics in the Extreme Ultraviolet*, eds. S. Bowyer and R.F. Malina, Kluwer, Dordrecht, p. 261.

Bruhweiler, F.C. and Kondo, Y. (1981). *Astrophys. J.*, **269**, 657.

Bruhweiler, F.C. and Kondo, Y. (1982). *Astrophys. J. Lett.*, **260**, L91.

Bruhweiler, F.C. and Kondo, Y. (1983). *Astrophys. J.*, **269**, 657.

Burleigh, M.R. (1999). 11th European Workshop on White Dwarfs. eds. J.-E. Solheim and E.G. Meistas, *ASP Conference Series*, volume 169, p. 249.

Burleigh, M.R. and Barstow, M.A. (1998). *Mon. Not. Roy. Astron. Soc.*, **295**, L15.

Burleigh, M.R. and Barstow, M.A. (1999). *Astron. Astrophys.*, **341**, 795.

Burleigh, M.R., Barstow, M.A. and Dobbie, P.D. (1996). *Astron. Astrophys.*, **317**, L27.

Burleigh, M.R., Barstow, M.A. and Fleming, T.A. (1997). *Mon. Not. Roy. Astron. Soc.*, **287**, 381.

Burleigh, M.R., Barstow, M.A. and Holberg, J.B. (1998). *Mon. Not. Roy. Astron. Soc.*, **300**, 511.

Caillault, J.-P. and Helfand, D.J. (1985). *Astrophys. J.*, **289**, 279.

Cash, W., Bowyer, S. and Lampton, M. (1978). *Astrophys. J. Lett.* **221**, L87.

Cash, W., Bowyer, S. and Lampton, M. (1979). *Astron. Astrophys.*, **80**, 67.

Cassinelli, J.P., Cohen, D.H., MacFarlane, J.J., Sanders, W.T. and Welsh, B.Y. (1994). *Astrophys. J.*, **421**, 705.

Cassinelli, J.P., Cohen, D.H., MacFarlane, J.J., Drew, J.E., Lynas-Gray, A.E., Hoare, M.G., Vallerga, J.V., Welsh, B.Y., Vedder, P.W. and Hubeny, I. (1995). *Astrophys. J.*, **438**, 932.

Cassinelli, J.P., Cohen, D.H., MacFarlane, J.J., Drew, J.E., Lynas-Gray, A.E., Hubeny, I., Vallerga, J.V., Welsh, B.Y. and Hoare, M.G. (1996). *Astrophys. J.*, **460**, 949.

Catalan, M.S., Sarna, M.J., Jomaron, C.M. and Smith R.C. (1995). *Mon. Not. Roy. Astron. Soc.*, **275**, 153.

Catura, R.C., Acton, L.W. and Johnson, H.M. (1975). *Astrophys. J. Lett.* **196**, L47.

Chakrabarti, S., Bowyer, S., Paresce, F., Franke, J.B. and Christensen, A.B. (1982). *Appl. Opt.*, **21**, 3417.

Chakrabarti, S., Kimble R. and Bowyer S. (1984). *J. Geophys. Res.*, **89**, 5660.

Chan, W., Cooper, G. and Brion, C. (1991). *Phys. Rev. (A)*, **44**, 186.

Chayer, P., LeBlanc, F., Fontaine, G., Wesemael, F., Michaud, G. and Vennes, S. (1994). *Astrophys. J. Lett.*, **436**, 161.

Chayer, P., Fontaine, G. and Wesemael, F. (1995a). *Astrophys. J. Suppl.*, **99**, 189.

Chayer, P., Vennes, S., Pradhan, A.K., Thejll, P., Beauchamp, A., Fontaine, G. and Wesemael, F. (1995b). *Astrophys. J.*, **454**, 429.

Chayer, P., Kruk, J.W., Ake, T.B., Dupree, A.K., Malina, R.F., Siegmund, O.H.W., Sonneborn, G. and Ohl, R.G. (2000). *Astrophys. J. Lett.*, **538**, L91.

Cheng, K.-P. and Bruhweiler, F.C. (1990). *Astrophys. J.*, **364**, 573.

Chiang, J., Reynolds, C.S.,.Blaes, O.M., Nowak, M.A., Murray, N., Madejski, G., Marshall, H.L. and Magdziarz, P. (2000). *Astrophys. J.*, **528**, 292.

Christian, D., Drake, J.J., Patterer, R.J., Vedder, P.W. and Bowyer, S. (1996). *Astron. J.*, **112**, 751.

Christian, D.J. (2000). *Astrophys. J.*, **119**, 19.

Christian, D.J., Vennes, S., Thorstensen, J.R. and Mathioudakis, M. (1996). *Astron. J.*, **112**, 258.

Clemens, J.C., *et al.* (1992). *Astrophys. J.*, **391**, 773.

Code, A.D., Davis, J., Bless, R.C. and Hanbury Brown, R. (1976). **203**, 417.

Cohen, D.H., MacFarlane, J.J., Owocki, S.P., Cassinelli, J.P. and Wang, P. (1996). *Astrophys. J.*, **460**, 506.

Cohen, D.H., Hurwitz, M., Cassinelli, J.P. and Bowyer S. (1998). *Astrophys. J. Lett.*, **500**, L511.

Cooke, B.A., *et al.* (1992). *Nature*, **355**, 61.

Cooper, R.G. (1994). PhD Thesis, University of Delaware.

Cordova, F.A. and Mason, K.O. (1984). *Mon. Not. Roy. Astron. Soc.*, **206**, 879.

Cordova, F.A., Chester, T.J., Mason, K.O., Kahn, S.M. and Garmire, G.P. (1984). *Astrophys. J.*, **278**, 739.

Cox, D.P. (1996). In *Astrophysics in the Extreme Ultraviolet*, eds. S. Bowyer and R.F. Malina, Kluwer, Dordrecht, p. 247.

Cox, D.P. and Snowden, S.L. (1986). *Adv. Space Res.*, **6**, 97.

Craig, N., *et al.* (1996). *Astron. J.*, **110**, 1304.

Cruddace, R.G., Paresce, F., Bowyer, S. and Lampton, M. (1974). *Astrophys. J.*, **187**, 497.

Culhane, J.L., White, N.E., Parmar, A.N. and Shafer, R.A. (1990). *Mon. Not. Roy. Astron. Soc.*, **243**, 424.

Cully, S.L., Fisher, G., Abbott, M.J. and Siegmund, O.H.W. (1994). *Astrophys. J.*, **435**, 449.

Czerny, B. and Elvis, M. (1987). *Astrophys. J.*, **321**, 305.

Dahlem, M., Kreysing, H.-C., White, S., Engels, D., Condon, J.J. and Voges, W. (1995). *Astron. Astrophys.*, **295**, L13.

Diamond, C.J., Jewell, S.J. and Ponman, T.J. (1995). *Mon. Not. Roy. Astron. Soc.*, **274**, 589.

Dickey, J.M. and Lockman, F.J. (1990). *Ann. Rev. Astr. Astrophys.*, **28**, 215.

Dobbie, P.D., Barstow M.A., Burleigh, M.R. and Holberg, J.B. (1999). *Astron. Astrophys.*, **346**, 163.

Donati, J.-F., Brown, S.F., Semel, M., Rees, D.E., Dempsey, R.C., Matthews, J.M., Henry, G.W. and Hall, D.S. (1992). *Astron. Astrophys.*, **265**, 669.

Doschek, G.A. (1991). In *Extreme Ultraviolet Astronomy*, eds. R.F. Malina and S. Bowyer, Pergamon, New York, p. 94.

Drake, J.J. (1996). In Cool Stars, Stellar Systems and the Sun. *ASP Conference Series*, volume 109, ed. R. Pallavicini and A.K. Dupree, p. 203.

Drake, J.J. (1999). *Astrophys. J. Suppl.*, **122**, 269.

Drake, J.J. and Smith, G. (1993). *Astrophys. J.*, **412**, 797.

Drake, J.J., Brown, A., Bowyer, S., Jelsinsky, P. and Malina, R.F. (1994a). In Cool Stars, Stellar Systems and the Sun; Eighth Cambridge Workshop. *ASP Conference Series*, volume 64, ed. J.-P. Caillault, p. 35.

Drake, J.J., Brown, A., Patterer, R.J., Vedder, P., Bowyer, S. and Guinan, E.F. (1994b). *Astrophys. J. Lett.*, **421**, L43.

Drake, S.A., Singh, K.P., White, N.E. and Simon, T. (1994). *Astrophys. J. Lett.*, **436**, L87.

Drake, J.J., Laming, J.M., Widing, K.G., Schmitt, J.H.M.M., Haisch, B. and Bowyer, S. (1995a). *Science*, **267**, 1470.

Drake, J.J., Laming, J.M. and Widing, K.G. (1995b). *Astrophys. J.*, **443**, 393.

Drake, J.J., Stern, R.A., Stringfellow G., Mathioudakis, M., Laming, J.M. and Lambert, D.L. (1996). *Astrophys. J.*, **469**, 828.

Dreizler, S. and Werner, K. (1993). *Astron. Astrophys.*, **278**, 199.

Dreizler, S. and Werner, K. (1996). *Astron. Astrophys.*, **314**, 217.

Dreizler, S. and Wolff, B. (1999). *Astron. Astrophys.*, **348**, 189.

Drew, J.E., Denby, M. and Hoare, M.G. (1994). *Mon. Not. Roy. Astron. Soc.*, **266**, 917.

Dupree, A.K. (1996). In Cool Stars, Stellar Systems and the Sun. *ASP Conference Series*, volume 109, ed. R. Pallavicini and A.K. Dupree, p. 237.

Dupree, A.K., Brickhouse, N.S., Doschek, G.A., Green, J.C. and Raymond, J.C. (1993). *Astrophys. J.*, **418**, L41.

Dupuis, J., Vennes, S., Bowyer, S., Pradhan, A.K. and Thejll, P. (1995). *Astrophys. J.*, **455**, 574.

Dupuis, J., Vennes, S. and Bowyer, S. (1997). In *White Dwarfs*, eds. J. Isern, M. Hernanz and E. Garcia-Berro, Dordrecht, Kluwer, p. 277.

Edelson, R.A., *et al.* (1995). *Astrophys. J.*, **438**, 120.

Edelstein, J., Forster, R.S. and Bowyer, S. (1995). *Astrophys. J.*, **454**, 442.

Edvardsson, B., Andersen, J., Gustafsson, B., Lambert, D.L., Nissen, P.E. and Tomkin, J. (1993). *Astron. Astrophys.*, **275**, 101.

Edwards, B.C., Priedhorsky, W.C. and Smith, B.W. (1991). *Geophys. Res. Lett.*, **18**, 2161.

Edwards, B.C., Bloch, J.J., Roussel-Dupré, D., Pfafman, T.E. and Ryan, S. (1996). In *Astrophysics in the Extreme Ultraviolet*, eds. S. Bowyer and R.F. Malina, Kluwer, Dordrecht, p. 465.

Eggen, O.J. (1983). *Mon. Not. Roy. Astron. Soc.*, **204**, 391.

Eggen, O.J. (1984). *Astron. J.*, **89**, 1350.

ESA (1997). *Hipparcos and Tycho Catalogues*, ESA SP-1200.

EUVE Guest Observer Center User Guide. (1993). *EUVE* Guest Observer Center Version 0.1.

Fabian, A.C., Arnaud, K.A., Bautz, M.W. and Tawara, Y. (1994). *Astrophys. J.*, **436**, 63.

Fano, U. (1961). *Phys. Rev.*, **124**, 1866.

Favata, F., Micela, G. and Sciortino, S. (1993a). In *Physics of Solar and Stellar Coronae: G.S. Vaiana Memorial Symposium*, eds. J. Linsky and S. Serio, Kluwer, Dordrecht, p. 287.

Favata, F., Barbera, M., Micela, G. and Sciortino, S. (1993b). In *Physics of Solar and Stellar Coronae: G.S. Vaiana Memorial Symposium*, eds. J. Linsky and S. Serio, Kluwer, Dordrecht, p. 303.

Fawcett, B.C. and Cowan, R.D. (1975). *Mon. Not. Roy. Astron. Soc.*, **171**, 1.

Fawcett, B.C., Cowan, R.D., Kononov, E.Y. and Hayes, R.W. (1972). *J. Phys. B: At. Mol. Opt. Phys.*, **5**, 1255.

Feldman, U., Mandelbaum, P., Seely, J.F., Doschek, G.A. and Gursky, H. (1992). *Astrophys. J. Suppl.*, **81**, 387.

Fernley, J., Taylor, K. and Seaton, M. (1987). *J. Phys. B: At. Mol. Opt. Phys.*, **20**, 6457.

Finley, D.S. (1996). In *Astrophysics in the Extreme Ultraviolet*, eds. S. Bowyer and R.F. Malina, Kluwer, Dordrecht, p. 223.

Finley, D.S., Green, J., Bowyer, S. and Malina, R.F. (1986a). *Proc. SPIE*, **640**, 91.

Finley, D.S., Jelinsky, P., Bowyer, S. and Malina, R.F. (1986b). *Proc. SPIE*, **628**, 176.

Finley, D.S., Koester, D. and Basri, G. (1997). *Astrophys. J.*, **488**, 375.

Fleming, T.A., Liebert, J.L. and Green, R.F. (1986). *Astrophys. J.*, **308**, 176.

Fleming, T.A., Schmitt, J.H.M.M., Barstow, M.A. and Mittaz, J.P.D. (1991). *Astron. Astrophys.*, **246**, L47.

Fleming, T.A., Barstow, M.A., Sansom, A.E., Holberg, J.B., Liebert, J. and Tweedy, R.W. (1993a). In *White Dwarfs: Advances in Observation and Theory*, ed. M.A. Barstow, Kluwer, Dordrecht, p. 155.

Fleming, T.A., Giampapa, M.S., Schmitt, J.H.M.M. and Bookbinder, J.A. (1993b). *Astrophys. J.*, **410**, 387.

Foster, R.S., Edelstein, J. and Bowyer, S. (1996). In *Astrophysics in the Extreme Ultraviolet*, eds. S. Bowyer and R.F. Malina, Kluwer, Dordrecht, p. 437.

Fraser, G.W. (1983). *Nucl. Instrum. Methods*, **206**, 445.

Fraser, G.W. (1989). *X-ray Detectors in Astronomy*, Cambridge University Press, Cambridge.

Fraser, G.W. and Mathieson, E. (1981a). *Nucl. Instrum. Methods*, **179**, 591.

Fraser, G.W. and Mathieson, E. (1981b). *Nucl. Instrum. Methods*, **184**, 537.

Fraser, G.W. and Pearson, J.F. (1984). *Nucl. Instrum. Methods A*, **219**, 199.

Fraser, G.W., Whiteley, M.J. and Pearson, J.F. (1985). *Proc. SPIE*, **597**, 343.

Fraser, G.W., Barstow, M.A. and Pearson, J.F. (1988). *Nucl. Instrum. Methods Phys. Res. (A)*, **273**, 667.

Frisch, P.C. (1994). Science, **265**, 1423.

Frisch, P. and York, D. (1983). *Astrophys. J. Lett.*, **271**, L59.

Fruscione, A. (1996a). *Astrophys. J.*, **459**, 509.

Fruscione, A. (1996b). In *Astrophysics in the Extreme Ultraviolet*, eds. S. Bowyer and R.F. Malina, Kluwer, Dordrecht, p. 381.

Fruscione, A., Bruhweiler, F., Cheng, K.P., Hall, C.R., Kafatos, M., Ramos, E. and Kondo, Y. (1996) In *UV and X-Ray Spectroscopy of Astrophysical and Laboratory Plasmas*, eds. K. Yamashita and T. Watanabe, Universal Academy Press, Tokyo, p. 351.

Gabriel, A.H., Fawcett, B.C. and Jordan, C. (1966). *Proc. Phys. Soc.*, **87**, 825.

Gear, C.W. (1969) *Proc. Skytop Conv. USAEC Conf. No. 670301*, 552.

Genova, R., Bowyer, S., Vennes, S., Lieu, R., Henry, J.P. and Gioia, I. (1995). *Astron. J.*, **110**, 788.

Giacconi, R., *et al.* (1979). *Astrophys. J.*, **230**, 540.

Giommi, P., *et al.* (1987). *IAU Circ.*, 4486.

Gladstone, G.R., McDonald, J.S., Boyd, W.T. and Bowyer, S. (1994). *Geophys. Res. Lett.*, **21**, 461.

Gliese, W. (1969). Veröf. Astron. Rechen-Inst., Heidelb., No. 22.

Gliese, W. and Jahreiss, H. (1991). *Third Catalogue of Nearby Stars*, Veröf. Astron. Rechen-Inst., Heidelb. (supplied on NSSDC CD-ROM).

Golub, L., Harnden, F.R., Jr, Pallavicini, R., Rosner, R. and Vaiana, G.S. (1982). *Astrophys. J.*, **253**, 242.

Gondhalekar, P.M., Pounds, K.A., Sembay, S., Sokoloski, J., Urry, C.M., Matthews, L. and Quenby, J. (1992). In *Physics of Active Galactic Nuclei*, eds. W.J. Duschl and S.J. Wagner, Springer-Verlag, Heidelberg, p. 52.

Gorenstein, P., Gursky, H., Harnden, F.R., DeCaprio, A. and Bjorkholm, P. (1975). *IEEE Trans. Nucl. Sci.*, **NS-22**, 616.

Green, J., Finley, D.S., Bowyer, S. and Malina, R.F. (1986). *Proc. SPIE*, **628**, 172.

Green, J., Jelinsky P., Bowyer S. (1990). *Astrophys. J.*, **359**, 499.

Green, R.F., Schmidt, M. and Liebert, J.L. (1986). *Astrophys. J. Suppl.*, **61**, 305.

Gregorio, A., Stalio, R., Broadfoot, L., Castelli, F., Hack, M. and Holberg, J. (2001). *Astron. Astrophys.*, in press.

Grewing, M. (1975). *Astron. Astrophys.*, **38**, 399.

Gry, C., York, D.G. and Vidal-Madjar A. (1985). *Astrophys. J.*, **296**, 593.

Gry, C., Lemonon, L., Vidal-Madjar A., Lemoine, M. and Ferlet, R. (1995). *Astron. Astrophys.*, **302**, 497.

Gunderson, K., Green, J.C. and Wilkinson, E. (1999). 11th European Workshop on White Dwarfs. *ASP Conference Series*, volume 169, eds. J.-E. Solheim and E. Meistas, p. 559.

Gunderson, K., Wilkinson, E., Green, J.C. and Barstow, M.A. (2001a). 12th European Workshop on White Dwarfs. *ASP Conference Series*, volume 226, eds. J.L. Provencal, H. Shipman, J. MacDonald and S. Goodchild, p. 117.

Gunderson, K., Wilkinson, E., Green, J.C. and Barstow, M.A. (2001b). *Astrophys. J.*, in press.

Güdel, M., Linsky, J.L., Brown, A. and Nagase, F. (1999). *Astrophys. J.*, **511**, 405.

Haisch, B.M., Linsky, J.L., Lampton, M., Paresce, F., Margon, B. and Stern, R. (1977). *Astrophys. J. Lett.*, **213**, L119.

Haisch, B., Bowyer, S. and Malina, R. (1994) In Cool Stars, Stellar Systems and the Sun; Eighth Cambridge Workshop. *ASP Conference Series*, volume 64, ed. J.-P. Caillault, p.3.

Halpern, J.P. and Marshall, H.M. (1996). *Astrophys. J.*, **464**, 760.

Hameury, J-M., King, A.R. and Lasota, J.-P. (1988). *Astron. Astrophys.*, **195**, L12.

Hameury, J-M., King, A.R. and Lasota, J.-P. (1991). *Astron. Astrophys.*, **248**, 525.

Hammann, F., Korista, K. and Morris, S.L. (1993). *Astrophys. J.*, **415**, 541.

Hanbury Brown R., Davis, J. and Allen, L.R. (1974). *Mon. Not. Roy. Astron. Soc.*, **167**, 121.

Harwit, M. (1981). *Cosmic Discovery*, Basic Books, New York.

Heise, J., *et al.* (1985). *Astron. Astrophys.*, **148**, L14.

Heise, J., Paerels, F., Bleeker, J.A.M. and Brinkman, A.C. (1988). *Astrophys. J.*, **334**, 958.

Henry, P., Cruddace, R., Paresce, F., Bowyer, S. and Lampton, M. (1975). *Astrophys. J.*, **195**, 107.

Henry, P., Bowyer, S., Lampton, M., Paresce, F. and Cruddace, R. (1976a). *Astrophys. J.*, **205**, 426.

Henry, P., Bowyer, S., Rapley, C.G. and Culhane, J.L. (1976b). *Astrophys. J. Lett.*, **209**, L29.

Hettrick, M.C., Bowyer, S., Malina, R.F., Martin, C. and Mrowka, S. (1985). *Appl. Opt.*, **24**, 1737.

Hoare, M.G., Drew, J.E. and Denby, M. (1993). *Mon. Not. Roy. Astron. Soc.*, **262**, L19.

Hoare, M.G., Drake, J.J., Werner, K. and Dreizler, S. (1996). *Mon. Not. Roy. Astron. Soc.*,

Hodgkin, S.T. and Pye, J.P. (1994). *Mon. Not. Roy. Astron. Soc.*, **267**, 840.

Hodgkin, S.T., Barstow, M.A., Fleming, T.A., Monier, R. and Pye, J.P. (1993). *Mon. Not. Roy. Astron. Soc.*, **263**, 229.

Hoffleit, D. and Jaschek, C. (1982). *Bright Star Catalogue*, University Press, New Haven.

Holberg, J.B. (1984). In NASA Report Goddard Space Flight Center Local Interstellar Medium, No. 81, pp. 91–95.

Holberg, J.B. (1986). *Astrophys. J.*, **311**, 969.

Holberg, J.B. (1990a). In *IAU Colloquium No. 123, Observatories in Earth Orbit and Beyond*, Kluwer, Dordrecht, p. 47.

Holberg, J.B. (1990b). In *Extreme Ultraviolet Astronomy*, R. Malina and S. Bowyer, eds., Pergamon Press, New York, p. 8.

Holberg, J.B., Sandel, B.R., Forrester, W.T., Broadfoot, A.L., Shipman, H.L. and Barry, D.C. (1980). *Astrophys. J. Lett.*, **242**, L119.

Holberg, J.B., Forrester, W.T., Shemansky, D.E. and Barry, D.C. (1982). *Astrophys. J.*, **257**, 656.

Holberg, J.B., Wesemael, F. and Basile, J. (1986). *Astrophys. J.*, **306**, 629.

Holberg, J.B., Ali, B., Carone, T. E. and Polidan, R. (1991). *Astrophys. J.*, **375**, 716.

Holberg, J.B., *et al.* (1993). *Astrophys. J.*, **416**, 803.

Holberg, J.B., Hubeny, I., Barstow, M.A., Lanz, T., Sion, E.M. and Tweedy, R.W. (1994). *Astrophys. J. Lett.*, **425**, L105.

Holberg, J.B., Barstow, M.A., Bruhweiler, F.C. and Sion, E.M. (1995). Astrophys. J., **453**, 313.

Holberg, J.B., Barstow, M.A. and Green, E.M. (1997a). *Astrophys. J. Lett.*, **474**, L127.

Holberg, J.B., Barstow, M.A., Lanz, T. and Hubeny, I. (1997b). *Astrophys. J.*, **484**, 871.

Holberg, J.B., Barstow, M.A. and Sion, E.M. (1998). *Astrophys. J. Suppl.*, **119**, 207.

Holberg, J.B., Barstow, M.A., Bruhweiler, F.C., Hubeny, I. and Green, E.M. (1999). *Astrophys. J.*, **517**, 850.

Hubeny, I. (1988). *Comp. Phys. Comm.*, **52**, 103.

Hubeny, I. and Lanz, T. (1992). *Astron. Astrophys.*, **262**, 501.

Hubeny, I. and Lanz, T. (1995). *Astrophys. J.*, **439**, 875.

Hubeny, I. and Lanz, T. (1996). In *Astrophysics in the Extreme Ultraviolet*, eds. S. Bowyer and R.F. Malina, Kluwer, Dordrecht, p. 381.

Hummer, D.G. and Mihalas, D. (1988). *Astrophys. J.*, **331**, 794.

Hummer, D.G., Berrington, K.A., Eissner, W., Pradhan, A.K., Saraph, H.E. and Tully, J.A. (1993). *Astron. Astrophys.*, **279**, 298.

Humphery, A., Cabral, R., Brisette, R., Carroll, R., Morris, J. and Harvey, P. (1978). *IEEE Trans. Nucl. Sci.*, **NS-25**, 445.

Hunter, W.R. (2000). In *Vacuum Ultraviolet Spectroscopy*, eds. J.A.R. Samson and D.L. Ederer, Academic Press, San Diego, p. 183.

Hurwitz, M. and Bowyer, S. (1996). In *Astrophysics in the Extreme Ultraviolet*, eds. S. Bowyer and R.F. Malina, Kluwer, Dordrecht, p. 601.

Hurwitz, M., Sirk, M., Bowyer, S. and Ko, Y-K. (1997). *Astrophys. J.*, **477**, 390.

Hutchinson, I., Warwick, R.S. and Willingale, R. (2001). *Mon. Not. Roy. Astron. Soc.*, in press.

Hwang, C.-Y. and Bowyer, S. (1997). *Astrophys. J.*, **475**, 552.

Iben, I. Jr and MacDonald, J. (1985). *Astrophys. J.*, **296**, 540.

Iben, I. Jr and Tutukov, A.V. (1984). *Astrophys. J.*, **282**, 615.

Jeffries, R.D. and Jewell, S.J. (1993). *Mon. Not. Roy. Astron. Soc.*, **264**, 106.

Jeffries, R.D. and Stevens, I.R. (1996). *Mon. Not. Roy. Astron. Soc.*, **279**, 180.

Jeffries, R.D., Bertram, D. and Spurgeon, B.R. (1995). *Mon. Not. Roy. Astron. Soc.*, **276**, 397.

Jelinsky, P., Martin, C., Kimble, R., Bowyer, S. and Steele, G. (1983). *Appl. Opt.*, **22**, 1227.

Jelinksy, P., Vallerga, J.V. and Edelstein, J. (1995). *Astrophys. J.*, **442**, 653.

Jelinky, S.R., Siegmund, O.H. and Mir, J.A. (1996). *Proc. SPIE*, **2808**, 617.

Jenkins, E.B. (1978). *Astrophys. J.*, **210**, 845.

Jensen, K.A., Swank, J.H., Petre, R., Guinan, E.F., Sion, E.M. and Shipman, H.E. (1986). *Astrophys. J. Lett.*, **309**, L27.

Johnson, H.M. (1981). *Astrophys. J.*, **243**, 234.

Johnson, H.R. and Ake, T.B. (1986). ESA SP-263, 395.

Johnston, S., *et al.* (1993). *Nature*, **361**, 613.

Jones, M.H. and Watson, M.G. (1992). *Mon. Not. Roy. Astron. Soc.*, **257**, 633.

Jordan, C. (1996). In *Astrophysics in the Extreme Ultraviolet*, eds. S. Bowyer and R.F. Malina, Kluwer, Dordrecht, p. 81.

Jordan, C., Ayres, T.R., Brown, A., Linsky, J.L. and Simon, T. (1987). *Mon. Not. Roy. Astron. Soc.*, **225**, 903.

Jordan, S. and Koester, D. (1986). *Astron. Astrophys. Suppl.*, **65**, 367.

Jordan, S., Koester, D., Wulf-Mathies, C. and Brunner, H. (1987). *Astron. Astrophys.*, **185**, 253.

Jordan, S., Wolff, B., Koester, D. and Napiwotzki, R. (1994). *Astron. Astrophys.*, **290**, 834.

Jordan, S., Koester, D. and Finley, D. (1996). In *Astrophysics in the Extreme Ultraviolet*, eds. S. Bowyer and R.F. Malina, Kluwer, Dordrecht, p. 235.

Jordan, S., Napiwotzki, R., Koester D. and Rauch, T. (1997). *Astron. Astrophys.* **318**, 461.

Kaastra, J.S. (1992). An X-ray spectral code for optically thin plasmas, Internal SRON-Leiden Report, updated version 2.0.

Kaastra, J.S., Roos, N. and Mewe, R. (1995). *Astron. Astrophys.*, **300**, 25.

Kahn, S.M., Wesemael, F., Liebert, J., Raymond, J.C., Steiner, J.E. and Shipman, H.L. (1984). *Astrophys. J.*, **278**, 255.

Kartje, J.F., Königl, A., Hwang, C.-Y. and Bowyer S. (1997). *Astrophys. J.*, **474**, 630.

Katsova, M.M., Drake, J.J. and Livshits, M.A. (1996). In *Astrophysics in the Extreme Ultraviolet*, eds. S. Bowyer and R.F. Malina, Kluwer, Dordrecht, p. 175.

Keenan, F.P. (1996). In *Astrophysics in the Extreme Ultraviolet*, eds. S. Bowyer and R.F. Malina, Kluwer, Dordrecht, p. 595.

Kellogg, E., Henry, P., Murray, S., Van Speybroeck, L. and Bjorkholm, P. (1976). *Rev. Sci. Instrum.*, **47**, 282.

Kent, B.J., Reading, D.H., Swinyard, B.M., Graper, E.B. and Spurret, P.H. (1990). *Proc. SPIE*, **1344**, 255.

Keski-Kuha, R.A.M., Fleetwood, C.M. and Robichaud, J. (1997). *Appl. Opt.*, **36**, 4409.

Kidder, K.M. (1991). PhD Thesis, University of Arizona.

Kimble, R.A., *et al.* (1993a). *Astrophys. J.*, **404**, 663.

Kimble, R.A., Davidsen, A.F., Long, K.S. and Feldman, P.D. (1993b). *Astrophys. J. Lett.*, **408**, L41.

Kimble, R.A., *et al.* (1999). *Proc SPIE*, **3764**, 209.

Kirkpatrick, P. and Baez, A.M. (1948). *J. Opt. Soc. Am.*, **38**, 766.

Koester, D. (1989). *Astrophys. J.*, **342**, 999.

Koester, D. (1996). In *Astrophysics in the Extreme Ultraviolet*, eds. S. Bowyer and R.F. Malina, Kluwer, Dordrecht, p. 185.

Koester, D. and Schoenberner, D. (1986). *Astron. Astrophys.*, **154**, 125.

de Kool, M. and Ritter, H. (1993). *Astron. Astrophys.*, **267**, 397.

de Korte, P.A.J., Bleeker, J.A.M., Deerenberg, A.J.M., Hayakawa, S., Yamishita, K. and Tanaka, Y. (1976). *Astron. Astrophys.*, **48**, 235.

de Korte, P.A.J., *et al.* (1981a). *Space Sci. Rev.*, **30**, 495.

de Korte, P.A.J., Giralt, R., Coste, J.W., Frindel, S., Flamand, J. and Contet, J.J. (1981b). *Appl. Opt.*, **20**, 1080.

Königl, A. (1996). In *Astrophysics in the Extreme Ultraviolet*, eds. S. Bowyer and R.F. Malina, Kluwer, Dordrecht, p. 27.

Königl, A., Kartje, J.F., Bowyer, S., Kahn, S.M. and Hwang, C.-Y. (1995). *Astrophys. J.*, **446**, 598.

Kowalski, M.P., *et al.* (1999). *Appl. Opt.*, in press.

Krautter, J., Gelman, H. and Starrfield, S. (1992). *IAUC*, 5550.

Kudritzki, R.F., Puls, J., Gabler, R. and Schmitt, J.H.M.H. (1991). In *Extreme Ultraviolet Astronomy*, eds. R.F. Malina and S. Bowyer, Pergamon, New York, p. 130.

Kurucz, R.L. (1979). *Astrophys. J. Suppl.*, **40**, 1.

Kurucz, R. L. (1988). In *IAU Transactions*, volume XXB, ed. M. McNally, Kluwer, Dordrecht, p. 168.

Kurucz, R.L. (1992). In *Model Atmopheres for Population Synthesis*, eds. B. Barbuy and A. Renzini, Kluwer, Dordrecht, p. 441.

Labov, S. (1988). PhD Thesis, University of California, Berkeley.

Laming, J.M. and Drake, J.J. (1999). *Astrophys. J.*, **516**, 324.

Laming, J.M., Drake, J.J. and Widing, K.G. (1995). *Astrophys. J.*, **443**, 416.

Laming, J.M., Drake, J.J. and Widing, K.G. (1996). *Astrophys. J.*, **462**, 948.

Lampton, M., Margon, B., Paresce, F., Stern, R. and Bowyer, S. (1976a). *Astrophys. J. Lett.*, **203**, L71.

Lampton, M., Margon, B. and Bowyer, S. (1976b). *Astrophys. J. Lett.*, **208**, L177.

Landini, M. and Monsignori-Fossi, B.C. (1990). *Astron. Astrophys. Suppl.*, **82**, 229.

Landini, M., Monsignori-Fossi, B.C., Pallavicini, R. and Piro, L. (1986). *Astron. Astrophys.*, **157**, 217.

Landsman, W., Simon, T. and Bergeron, P. (1993). *Publ. Ast. Soc. Pac.*, **105**, 841.

Lanz, T. and Hubeny, I. (1995). *Astrophys. J.*, **439**, 905.

Lanz, T., Barstow, M.A., Hubeny, I. and Holberg, J.B. (1996). *Astrophys. J.*, **473**, 1089.

Lapington, J.S., Sanderson, B., Worth, L.B.C. and Tandy, J.A. (2000). *Nucl. Instrum. Methods Phys. Res. (A)*, in press.

Leger, L.J., Visentine, J.T. and Kuminez, J.F. (1984). *AIAA*, 84-0458.

Lemen, J.R., Mewe, R., Schrijver, C.J. and Fludra, A. (1989). *Astrophys. J.*, **341**, 474.

Levine, A., Petre, R., Rappaport, S., Smith, G.C., Evans, K.D. and Rolfe, D. (1979). *Astrophys. J. Lett.*, **228**, L99.

Liedahl, D.A., Paerels, F., Hur, M.Y., Kahn, S.M., Fruscione, A. and Bowyer, S. (1996). In *Astrophysics in the Extreme Ultraviolet*, eds. S. Bowyer and R.F. Malina, Kluwer, Dordrecht, p. 57.

Lieu, R., Bowyer, S., Lampton, M., Jelinsky, P. and Edelstein, J. (1993). *Astrophys. J. Lett.*, **417**, L41.

Lieu, R., Mittaz, J.P.D., Bowyer, S., Lockman, F.J., Hwang, C.-Y. and Schmitt, J.H.M.M. (1996). *Astrophys. J. Lett.*, **458**, L5.

Lieu, R., Bonamente, M., Mittaz, J.P.D., Durret, F., Dos Santos, S. and Kaastra, J.S. (1999). *Astrophys. J. Lett.*, **527**, L77.

Linick, S.H. and Holberg, J.B. (1991). *J. Brit. Interplanetary Soc.*, **44**, 513.

Linsky, J.L. and Haisch, B.M. (1979). *Astrophys. J. Lett.*, **229**, L27.

Linsky, J.L., Bornmann, P.L., Carpenter, K. G., Hege, E.K., Wing, R.F., Giampapa, M.S. and Worden, S.P. (1982). *Astrophys. J.*, **260**, 670.

Livshits, M.A. and Katsova, M.M. (1996). In *Astrophysics in the Extreme Ultraviolet*, eds. S. Bowyer and R.F. Malina, Kluwer, Dordrecht, p. 171.

Lloyd, H.M., *et al.* (1992). *Nature*, **356**, 222.

Long, K.S. (1996). In *Astrophysics in the Extreme Ultraviolet*, eds. S. Bowyer and R.F. Malina, Kluwer, Dordrecht, p. 301.

Lyu, C.-H. and Bruhweiler, F.C. (1996). *Astrophys. J.*, **459**, 216.

MacDonald, J. (1996). In *Astrophysics in the Extreme Ultraviolet*, eds. S. Bowyer and R.F. Malina, Kluwer, Dordrecht, p. 395.

MacDonald, J. and Vennes, S. (1991). *Astrophys. J.*, **373**, L51.

Maeder, A. and Meynet, G. (1988). *Astron. Astrophys. Suppl.*, **76**, 411.

Malina, R.F., Bowyer, S. and Basri, G. (1982). *Astrophys. J.*, **262**, 717.

Malina, R.F., *et al.* (1994). *Astron. J.*, **107**, 751.

Malinovksi, M. and Heroux, L. (1973). *Astrophys. J.*, **181**, 1009.

Manning, R.A., Jeffries, R.D. and Willmore, A.P. (1996). *Mon. Not. Roy. Astron. Soc.*, **278**, 577.

Margon, B., Szkody, P., Bowyer, S., Lampton, M. and Paresce, F. (1978). *Astrophys. J.*, **224**, 167.

Marr, G. and West, J. (1976). *Atom. Nucl. Data Tables*, **18**, 497.

Marsh, M.C., *et al.* (1997a). *Mon. Not. Roy. Astron. Soc.*, **286**, 369.

Marsh, M.C., *et al.* (1997b). *Mon. Not. Roy. Astron. Soc.*, **287**, 705.

Marshall, F. (1982). PhD Thesis, Massachusetts Institute of Technology.

Marshall, H.L., Carone, T.E., Shull, J.M., Malkan, M.A. and Elvis, M. (1996). *Astrophys. J.*, **457**, 169.

Marshall, H.L., *et al.* (1997). *Astrophys. J.*, **479**, 222.

Marshall, H.L. (1996). In *Astrophysics in the Extreme Ultraviolet*, eds. S. Bowyer and R.F. Malina, Kluwer, Dordrecht, p. 63.

Martens, P.C.H. (1988). *Astrophys. J.*, **330**, L131.

Martin, C., Jelinsky, P., Lampton, M., Malina, R.F. and Anger, H.O. (1981). *Rev. Sci. Instrum.*, **52**, 1067.

Mason, H.E. (1996). In *Astrophysics in the Extreme Ultraviolet*, eds. S. Bowyer and R.F. Malina, Kluwer, Dordrecht, p. 561.

Mason, K.O., Cordova, F.A., Watson, M.G. and King, A.R. (1988). *Mon. Not. Roy. Astron. Soc.*, **232**, 779.

Mason, K.O., *et al.* (1995). *Mon. Not. Roy. Astron. Soc.*, **274**, 1194.

Mathioudakis, M., *et al.* (1995). *Astron. Astrophys.*, **302**, 422.

Mauche, C.W. (1997). *Astrophys. J. Lett.*, **476**, L85.

Mauche, C.W. (1999). In Annapolis Workshop on Magnetic Cataclysmic Variables. *ASP Conference Series*, eds. C. Hellier and K. Mukai, volume 157, p. 157.

Mauche, C.W. and Raymond, J.C. (2000). *Astrophys. J.*, **541**, 924.

Mauche, C.W., Raymond, J.C. and Mattei, J.A. (1996). *Astrophys. J.*, **446**, 842.

McCammon, D., Burrows, D.N., Sanders, W.T. and Kraushaar, W.L. (1983). *Astrophys. J.*, **269**, 107.

McCook, G.P. and Sion, E.M. (1987). *Astrophys. J. Suppl.*, **65**, 603.

McCook, G.P. and Sion, E.M. (1999). *Astrophys. J. Suppl.*, **121**, 1.

McDonald, K., Craig, N., Sirk, M.M., Drake, J.J., Fruscione, A., Vallerga, J.V. and Malina, R.F. (1994). *Astron. J.*, **108**, 1843.

McKee, C. and Ostriker, J. (1977). *Astrophys. J.*, **218**, 148.

Meier, R.R. and Mange, P. (1970). *Planet. Space. Sci.*, **18**, 803.

Meier, R.R. and Prinz, D.K. (1971). *J. Geophys. Res.*, **76**, 4608.

Meier, R.R. and Weller, C.S. (1974). *J. Geophys. Res.*, **79**, 1575.

Mewe, R., Heise, J., Gronenschild, E.H.B.M., Brinkman, A.C., Schrijver, J. and den Boggende, A.J.F. (1975). *Astrophys. J. Lett.*, **202**, L67.

Mewe, R., Gronenschild, E.H.B.M. and van den Oord, G.H.J. (1985). *Astron. Astrophys. Suppl.*, **62**, 197.

Mewe, R., Lemen, J.R. and van den Oord, G.H.J. (1986). *Astron. Astrophys. Suppl.*, **65**, 511.

Mewe, R., Kaastra, J.S., Schrijver, C.J., van den Oord, G.H.J. and Alkemade, F.J.M. (1995). *Astron. Astrophys.*, **296**, 477.

Mewe, R., Kaastra, J.S., White, S.M. and Pallavicini, R. (1996). *Astron. Astrophys.*, **315**, 170.

Meyer, J.-P. (1985). *Astrophys. J. Suppl.*, **57**, 173.

Meyer, W.F. (1943). *Publ. Ast. Soc. Pac.*, **46**, 202.

Mitrou, C.K., Mathioudakis, M., Doyle, J.G. and Antonopoulou, E. (1997). *Astron. Astrophys.*, **317**, 147.

Monsignori-Fossi, B.C., Landini, M., Fruscione, A. and Dupuis, J. (1995). *Astrophys. J.*, **449**, 376.

Monsignori-Fossi, B.C., Landini, M., Del Zanna, G. and Bowyer, S. (1996). *Astrophys. J.*, **466**, 427.

Morvan, E., Vauclair, G. and Vauclair, S. (1986). *Astron. Astrophys.*, **163**, 145.

Najarro F., Kudritzki R.P., Cassinelli J.P., Stahl O. and Hillier D.J. (1996). *Astron. Astrophys.*, **306**, 892.

Nandra, K., Pounds, K.A. and Stewart, G.C. (1991). *Mon. Not. Roy. Astron. Soc.*, **248**, 760.

Nandra, K., *et al.* (1993). *Mon. Not. Roy. Astron. Soc.*, **260**, 504.

Napiwotzki, R. (1995). In *White Dwarfs*, eds. D. Koester and K. Werner, Springer-Verlag, Berlin, p. 132.

Napiwotzki, R., Barstow, M.A., Fleming, T., Holweger, H., Jordan, S. and Werner, K. (1993). *Astron. Astrophys.*, **278**, 478.

Napiwotzki, R., Jordan, S., Bowyer, S., Hurwitz, M., Koester, D., Weidemann, V., Lampton, M. and Edelstein, J. (1995). *Astron. Astrophys.*, **300**, L5.

Nelson, N. and Young, A. (1970). *Publ. Ast. Soc. Pac.*, **82**, 699.

Ögelman, H., Beuermann, K. and Krautter, J. (1984). *Astrophys. J. Lett.*, **287**, L31.

Ögelman, H., Krautter, J. and Beuermann, K. (1987). *Astron. Astrophys.*, **177**, 110.

van den Oord, G.H.J., Schrijver, C.J., Mewe, R. and Kaastra, J.S. (1996). In Cool Stars, Stellar Systems and the Sun. *ASP Conference Series*, volume 109, eds. R. Pallavicini and A.K. Dupree, p. 231.

Osborne, J. P. (1987). *Astrophys. Space Sci.*, **130**, 207.

Osborne, J. P. (1988). *Mem. S. A. It.*, **59**, 1.

Osborne, J. P., *et al.* (1987). *Astrophys. J.*, **315**, L123.

Osten, R.A. and Brown, A. (1999). *Astrophys. J.*, **515**, 746.

Ottmann, R., Schmitt, J.H.M.M. and Kürster, M. (1993). *Astrophys. J.*, **413**, 710.

Owens, A., Page, C.G., Sembay, S. and Schaefer, B.E. (1993). *Mon. Not. Roy. Astron. Soc.*, **260**, L25.

Owocki, S.P., Castor, J.I. and Rybicki, G.B. (1988). *Astrophys. J.*, **335**, 914.

Oza, D. (1986). *Phys. Rev. (A)*, **33**, 824.

Paerels, F., Hur, M.Y. and Mauche, C.W. (1996a). In *Astrophysics in the Extreme Ultraviolet*, eds. S. Bowyer and R.F. Malina, Kluwer, Dordrecht, p. 309.

Paerels, F., Hur, M.Y., Mauche, C.W. and Heise, J. (1996b). *Astrophys. J.*, **464**, 884.

Paerels, F.B.S. and Heise, J. (1989). *Astrophys. J.*, **339**, 1000.

Paerels, F.B.S., Bleeker, J.A.M., Brinkman, A.C., Gronenschild, E.H.B.M. and Heise, J. (1986a). *Astrophys. J.*, **308**, 190.

Paerels, F.B.S., Bleeker, J.A.M., Brinkman, A.C. and Heise, J. (1986b). *Astrophys. J. Lett.*, **309**, L33.

Paerels, F.B.S., Bleeker, J.A.M., Brinkman, A.C., Gronenschild, E.H.B.M. and Heise, J. (1988). *Astrophys. J.*, **329**, 849.

Paerels, F.B.S., Bleeker, J.A.M., Brinkman, A.C., Heise, J. and Dijkstra, J. (1990). *Astron. Astrophys. Suppl.*, **85**, 1021.

Pallavicini, R. (1989). *Astron. Astrophys. Rev.*, **1**, 177.

Pallavicini, R., Golub, L., Rosner, R., Vaiana, G.S., Ayres, T. and Linsky, J.L. (1981). *Astrophys. J.*, **248**, 279.

Pallavicini, R., Tagliaferri, G. and Stellar, L. (1990). *Astron. Astrophys.*, **228**, 403.

Paresce, F. (1984). *Astron. J.*, **89**, 1022.

Paresce, F., Kumar, S. and Bowyer, S. (1972). *Planet. Space. Sci.*, **20**, 297.

Parker, E.N. (1988). *Astrophys. J.*, **330**, 474.

Pauldrach, A.W.A. and Puls, J. (1990). *Astron. Astrophys.*, **237**, 409.

Petre, R., Shipman, H.L. and Canizares, C.R. (1986). *Astrophys. J.*, **304**, 356.

Ponman, T.J., Belloni, T., Duck, S.R., Verbunt, F., Watson, M.G. and Wheatley, P.J. (1995). *Mon. Not. Roy. Astron. Soc.*, **276**, 495.

Pottasch, S.R. (1963). *Astrophys. J.*, **137**, 945.

Pounds, K.A., *et al.* (1991). *Mon. Not. Roy. Astron. Soc.*, **253**, 364.

Pounds, K.A., *et al.* (1993). *Mon. Not. Roy. Astron. Soc.*, **260**, 77.

Pradhan, A. K. (1996). In *Astrophysics in the Extreme Ultraviolet*, eds. S. Bowyer and R.F. Malina, Kluwer, Dordrecht, p. 569.

Press, W.H., Teulosky S.A., Vetterling W.T. and Flannery B.P. (1992). Numerical Recipes, 2nd edition, Cambridge University Press, Cambridge.

Priedhorsky, W.C., *et al.* (1988). *Proc. SPIE*, **982**, 188.

Pringle, J.E. (1997). *Mon. Not. Roy. Astron. Soc.*, **178**, 195.

Pringle, J.E., Bateson, F.M., Hassall, B.J.M., Heise, J. and van der Woerd, H. (1987). *Mon. Not. Roy. Astron. Soc.*, **225**, 73.

Pye, J.P. and McHardy, I.M. (1988). In *Activity in Cool Star Envelopes*, Astrophysics and Space Science Library No. 143, eds. O. Havnes, B.R. Petterson, J.H.M.M. Schmitt and J.E. Solheim, Kluwer, Dordrecht, p. 231.

Pye, J.P., *et al.* (1995). *Mon. Not. Roy. Astron. Soc.*, **274**, 1165.

Randich, S., Gratton, R. and Pallavicini, R. (1993). *Astron. Astrophys.*, **273**, 194.

Rappaport, S., Petre, R., Kayat, M.A., Evans, K.D., Smith, G.C. and Levine, A. (1979). *Astrophys. J.*, **227**, 285.

Rappaport, S., Joss, P.C. and Webbink, R.F. (1982). *Astrophys. J.*, **254**, 616.

Raymond, J.C. (1988). In *Hot Thin Plasmas in Astrophysics*, ed. R. Pallavicini, Kluwer, Dordrecht, p. 3.

Raymond, J.C. and Doyle, J.G. (1981). *Astrophys. J.*, **245**, 1141.

Raymond, J.C. and Smith, B. (1977). *Astrophys. J. Suppl.*, **35**, 419.

van Regemorter, H. (1962). *Astrophys. J.*, **136**, 906.

Reid, N. and Wegner, G. (1988). *Astrophys. J.*, **335**, 953.

Robinson, E.L., Clemens, J.C. and Hine, B.P. (1988). *Astrophys. J. Lett.*, **331**, L29.

Rogerson, J.B., York, D.G., Drake, J.F., Jenkins, E.B., Morton, D.C. and Spitzer, L. (1973). *Astrophys. J. Lett.*, **181**, L110.

Rosen, S.R., *et al.* (1996). *Mon. Not. Roy. Astron. Soc.*, **280**, 1121.

Roussel-Dupré, D., Bloch, J., Ryan, S., Edwards, B., Pfafman, T., Ramsey, K. and Stem, S. (1996). In *Astrophysics in the Extreme Ultraviolet*, eds. S. Bowyer and R.F. Malina, Kluwer, Dordrecht, p. 485.

Rucinski, S.M. (1998). *Astron. J.*, **115**, 303.

Rucinski, S.M., Mewe, R., Kaastra, J.S., Vilhu, O. and White, S.M. (1995). *Astrophys. J.*, **449**, 900.

Rumph, T., Bowyer, S. and Vennes, S. (1994). *Astron. J.*, **197**, 2108.

Samson, J.A.R. (1967). *Techniques in Vacuum Ultraviolet Spectroscopy*, Wiley, New York.

Sandel, B.R., *et al.* (2000). *Space. Sci. Rev.*, **91**, 197.

Sanders, W.T., Kraushaar, W.L., Nousek, J.A. and Fried, P.M. (1977). *Astrophys. J. Lett.*, **217**, L87.

Saraph, H.E. and Storey, P.J. (1996). *Astron. Astrophys. Suppl.*, **115**, 151.

Savage, B.D. and Jenkins, E.B. (1972). *Astrophys. J.*, **172**, 491.

Schindler, M., Stencel, R.E., Linsky, J.L., Basri, G.S. and Helfand., D. (1982). *Astrophys. J.*, **263**, 269.

Schmitt, J.H.M.M. (1988). In *Hot Thin Plasmas in Astrophysics*, ed. R. Pallavicini, Kluwer, Dordrecht, p. 109.

Schmitt, J.H.M.M., Snowden, S.L., Aschenbach, B., Hasinger, G., Pfeffermann, E., Predehl, P. and Trumper, J. (1991). *Nature*, **349**, 583.

Schmitt, J.H.M.M., Drake, J.J., Haisch, B.M. and Stern, R.A. (1996a). *Astrophys. J.*, **467**, 841.

Schmitt, J.H.M.M., Drake, J.J., Stern, R.A. and Haisch, B.M. (1996b). *Astrophys. J.*, **457**, 882.

Schmitt, J.H.M.M., Drake, J.J. and Stern, R.A. (1996c). *Astrophys. J. Lett.*, **465**, L51.

Schmitt, J.H.M.M., Stern, R.A., Drake, J.J. and Kürster, M. (1996d). *Astrophys. J.*, **464**, 898.

Schuh, S. (2000). Diploma Thesis, Universität Tübingen.

Schuh, S., Dreizler, S. and Wolff, B. (2001). 12th European Workshop on White Dwarfs. *ASP Conference Series*, eds. J.L. Provencal, H. Shipman, J. MacDonald and S. Goodchild, p. 79.

Sciama, D.W. (1990). *Astrophys. J.*, **364**, 549.

Sciama, D.W. (1993). *Modern Cosmology and the Dark Matter Problem*, Cambridge University Press, Cambridge.

Schrijver, C.J., Mewe, R., van den Oord, G.H.J. and Kaastra, J.S. (1995). *Astron. Astrophys.*, **302**, 438.

Schwarzschild, K. (1905). *Abh. Wiss. Göttingen*, BD IV, No. 2.

Seaton, M.J. (1962). In *Atomic and Molecular Processes*, ed. D. Bates, Academic Press, New York, p. 374.

Sfeir, D.M., Lallement, R., Crifo, F. and Welch, B.Y. (1999). *Astron. Astrophys. Suppl.*, **346**, 785S.

Shafer, R.A., Haberl, F., Arnaud, K.A. and Tennant, A.F. (1991). ESA TM-09

Shemi, A. (1995). *Mon. Not. Roy. Astron. Soc.*, **275**, 7.

Shipman, H.L. (1976). *Astrophys. J. Lett.*, **206**, L67.

Shipman, H.L. and Geczi, J. (1989). In *White Dwarfs*, ed. G. Wegner, Springer-Verlag, Berlin, p. 134.

Shobbrook, R.R. (1973). *Mon. Not. Roy. Astron. Soc.*, **101**, 257.

Shull, J.M. and van Steenberg, M. (1982a). *Astrophys. J. Suppl.*, **48**, 95.

Shull, J.M. and van Steenberg, M. (1982b). *Astrophys. J. Suppl.*, **49**, 352.

Siegmund, O.H.W., Clothier, S., Thorton, J., Lemen, J., Harper, R., Mason, I. and Culhane, J.L. (1983a). *IEEE Trans. Nucl. Sci.*, **NS-30**, 503.

Siegmund, O.H.W., Lampton, M., Bixler, Y., Bowyer, S. and Malina, R.F. (1983b). *IEEE Trans. Nucl. Sci.*, **NS-33**, 724.

Siegmund, O.H.W., Coburn, K. and Malina, R.F. (1985). *IEEE Trans. Nucl. Sci.*, **NS-32**, 443.

Siegmund, O.H.W., Lampton, M., Chakrabarti, S., Bowyer S. and Malina, R.F. (1986a). *Proc. SPIE*, **627**,

Siegmund, O.H.W., Vallerga, J.V., Everman, E., Labov, S. and Bixler, J. (1986b). *Proc. SPIE*, **687**, 117.

Siegmund, O.H.W., Everman, E., Vallerga, J.V., Sokolowski, J. and Lampton, M. (1987). *Appl. Opt.*, **26**, 3607.

Siegmund, O.H.W., Everman, E., Vallerga, J.V. and Lampton, M. (1988). *Proc. SPIE*, **868**, 18.

Siegmund, O.H.W., *et al.* (1997). *Proc. SPIE*, **3114**, 283.

Simon, T. and Drake, S.A. (1989). *Astrophys. J.*, **346**, 303.

Sims, M.R., *et al.* (1990). *Opt. Eng.*, **26**, 649.

Singh, K.P., White, N.E. and Drake, S.A. (1996). *Astrophys. J.*, **456**, 766.

Sion, E.M. (1986). *Publ. Ast. Soc. Pac.*, **98**, 821.

Sion, E.M., Guinan, E.F. and Wesemael, F. (1982). *Astrophys. J.*, **255**, 232.

Sion, E.M., Holberg, J.B., Barstow, M.A. and Kidder, K.M. (1995). *Publ. Ast. Soc. Pac.*, **107**, 232.

Sion, E.M., Schaefer, K.G., Bond, H.E., Saffer, R.A. and Cheng, F.H. (1998). *Astrophys. J. Lett.*, **496**, L26.

Slavin, J.D. (1989). *Astrophys. J.*, **346**, 718.

Slemp, W.S., Santos-Mason, B., Sykes, G.F. and Witte, W.G. (1985). *AIAA*, 85-0421.

Smith, B.W., Bloch, J.J. and Roussel-Dupré, D. (1989). *Proc. SPIE*, **1160**, 171.

Smith, B.W., Bloch, J.J. and Roussel-Dupré, D. (1990). *Opt. Eng.*, **29**, 592.

Smith, B.W., Pfafman, T.E., Bloch, J.J. and Edwards, B.C. (1996). In *Astrophysics in the Extreme Ultraviolet*, eds. S. Bowyer and R.F. Malina, Kluwer, Dordrecht, p. 283.

Snowden, S.L., *et al.* (1995). *Astrophys. J.*, **454**, 643.

Soderblom, D. and Mayor, M. (1993). *Astrophys. J. Lett.*, **405**, L5.

Spitzer, L., Drake, J.F., Jenkins, E.B., Morton, D.C., Rogerson, J.B. and York, D.G. (1973). *Astrophys. J. Lett.*, **181**, L116.

Stalio, R., Gregorio, A. and Trampus, P. (1999). *J. Ital. Astron. Soc.*, **70/2**, 349.

Stencel, R.E., Neff, J.A. and McClure, R.D. (1984). In NASA CP-2349, p. 400.

Stern, R. and Bowyer, S. (1979). *Astrophys. J.*, **230**, 755.

Stern, R. and Bowyer, S. (1980). *Astron. Astrophys.*, **83**, L1.

Stern, R., Wang, E. and Bowyer, S. (1978). *Astrophys. J. Suppl.*, **37**, 195.

Stern, R.A. (1992). In *Frontiers of X-Ray Astronomy, Proceedings of the 28th Yamada Conference* (Nagoya, Japan), eds. K. Koyama and Y. Tanaka, Universal Academy Press, Tokyo, p. 259.

Stern, R.A., Zolcinski, M-C., Antiochos, S.K. and Underwood, J.H. (1981). *Astrophys. J.*, **249**, 647.

Stern, R.A., Uchida, Y., Tsuneta, S. and Nagase, F. (1992). *Astrophys. J.*, **400**, 321.

Stern, R.A., Lemen, J.R., Schmitt, J.H.M.M. and Pye, J.P. (1995a). *Astrophys. J. Lett.* **444**, L45.

Stern, R.A., Schmitt, J.H.M.M. and Kahabka, P.T. (1995b). *Astrophys. J.*, **448**, 683.

Storey, P.J., Mason, H.E. and Saraph, H.E. (1996). *Astron. Astrophys.*, **309**, 677.

Stringfellow, G.S. and Bowyer, S. (1993). *IAUC*, 5803.

Stringfellow, G.S. and Bowyer, S. (1996). In *Astrophysics in the Extreme Ultraviolet*, eds. S. Bowyer and R.F. Malina, Kluwer, Dordrecht, p. 401.

Sumner, T.J., Quenby, J.J., Lieu, R., Daniels, J., Willingale R. and Moussas, X. (1989). *Mon. Not. Roy. Astron. Soc.*, **238**, 1047.

Swank, J.H. and White, N.E. (1980). In *Cool Stars, Stellar Systems and the Sun*, ed. A. Dupree, SAO Special Report 389, p. 47.

Swank, J.H., White, N.E., Holt, S.S. and Becker, R.H. (1981). *Astrophys. J.*, **246**, 208.

Szkody, P., Vennes, S., Sion, E.M., Long, K.S. and Howell, S.B. (1997). *Astrophys. J.*, **487**, 916.

Tagliaferi, G., Doyle, J.G. and Giommi, P. (1990). *Astron. Astrophys.*, **231**, 131.

Taylor, R.C., Hettrick, M.C. and Malina, R.F. (1983). *Rev. Sci. Instrum.*, **54**, 171.

van Teeseling, A. and Verbunt, F. (1994). *Astron. Astrophys.*, **292**, 519.

van Teeseling, A., Heise, J. and Paerels, F. (1994). *Astron. Astrophys.*, **281**, 119.

van Teeseling, A., Kaastra, J.S. and Heise, J. (1996). *Astron. Astrophys.*, **312**, 186.

Terman, J.L. and Taam, R.E. (1996). *Astrophys. J.*, **458**, 692.

Thomas, G.E. and Anderson, D.E. (1976). *Planet. Space. Sci.*, **20**, 297.

Thomas, R.J. and Neupert, W.M. (1994). *Astrophys. J. Suppl.*, **91**, 461.

Thorstensen, J.R., Vennes, S. and Shambrook A. (1994). *Astron. J.*, **108**, 1924.

Timothy, J.G. (1986). *Proc. SPIE*, **691**, 35.

Timothy, J.G. and Bybee, R.L. (1986). *Proc. SPIE*, **687**, 109.

Tomkin, J., Lambert, D.L. and Lemke, M. (1993). *Mon. Not. Roy. Astron. Soc.*, **265**, 581.

Turner, T.J. and Pounds, K.A. (1989). *Mon. Not. Roy. Astron. Soc.*, **240**, 833.

Tweedy, R.W., Holberg, J.B., Barstow, M.A., Bergeron, P., Grauer, A.D., Liebert, J. and Fleming, T.A. (1993). *Astron. J.*, **105**, 1938.

Uesugi, A. and Fukuda, I. (1982). *Revised Catalogue of Stellar Rotation Velocities*, Kyoto University Press, Kyoto.

Urry, C.M., *et al.* (1997). *Astrophys. J.*, **486**, 799.

Uttley, P., McHardy, I.M., Papdakis, I.E., Cagnoni, I. and Fruscione, A. (2000). *Mon. Not. Roy. Astron. Soc.* **312**, 880.

Vaiana, G.S., *et al.* (1981). *Astrophys. J.*, **245**, 163.

Vallerga, J.V., Siegmund, O.H.W., Everman, E. and Jelinsky, P. (1986). *Proc. SPIE*, **689**, 138.

Vallerga, J.V., Vedder, P.W and Welsh, B.Y. (1993). *Astrophys. J. Lett.*, **414**, L65.

Vallerga, J.V., Vedder, P.W. and Siegmund, O.H.W. (1992). *Proc. SPIE*, **1742**, 392.

Vauclair, G. (1989). In *White Dwarfs*, ed. G. Wegner, Springer-Verlag, Berlin, p. 176.

Vedder, P.W., Vallerga, J.V., Sigemund, O.H.W., Gibson, J.L. and Hull, J. (1989). *Proc. SPIE*, **1159**, 392.

Vedder, P.W., Vallerga, J.V., Jelinsky, P., Marshall, H.L. and Bowyer, S. (1996). In *Astrophysics in the Extreme Ultraviolet*, eds. S. Bowyer and R.F. Malina, Kluwer, Dordrecht, p. 120.

Vennes, S. (1996). In *Astrophysics in the Extreme Ultraviolet*, eds. S. Bowyer and R.F. Malina, Kluwer, Dordrecht, p. 193.

Vennes, S. (2000). *Astron. Astrophys.*, **354**, 995.

Vennes, S. and Fontaine, G. (1992). *Astrophys. J.*, **401**, 288.

Vennes, S. and Thorstensen, J.R. (1994). *Astrophys. J. Lett.*, **433**, L29.

Vennes, S. and Thorstensen, J.R. (1995). In *White Dwarfs*, eds. D. Koester and K. Werner, Springer-Verlag, Berlin, p. 313.

Vennes, S., Pelletier, C., Fontaine, G. and Wesemael, F. (1988). *Astrophys. J.*, **331**, 876.

Vennes, S., Chayer, P., Fontaine, G. and Wesemael, F. (1989). *Astrophys. J.*, **390**, 590.

Vennes, S., Thejll, P. and Shipman, H.L. (1991). In *White Dwarfs*, eds. G. Vauclair and E.M. Sion, Kluwer, Dordrecht, p. 235.

Vennes, S., Chayer, P., Thorstensen, J.R., Bowyer, S. and Shipman, H.L. (1992). *Astrophys. J. Lett.*, **392**, L27.

Vennes, S., Dupuis, J., Rumph, T., Drake, J., Bowyer, S., Chayer, P. and Fontaine, G. (1993). *Astrophys. J. Lett.*, **410**, L119.

Vennes, S., Dupuis, J., Bowyer, S., Fontaine, G., Wiercigroch, A., Jelinsky, P., Wesemael, F. and Malina, R.F. (1994). *Astrophys. J. Lett.*, **421**, L119.

Vennes, S., Mathioudakis, M., Doyle, J.G., Thorstensen, J.R. and Byrne, P.B. (1995a). *Astron. Astrophys.*, **299**, L29.

Vennes, S., Szkody, P., Sion, E.M. and Long, K.S. (1995b). *Astrophys. J.*, **445**, 921.

Vennes, S., Bowyer, S. and Dupuis, J. (1996a). *Astrophys. J. Lett.*, **461**, L103.

Vennes, S., Chayer, P., Hurwitz M. and Bowyer S. (1996b). *Astrophys. J.*, **468**, 989.

Vennes, S., Berghöfer, T.W. and Christian, D.J. (1997a). *Astrophys. J.*, **491**, L85.

Vennes, S., Dupuis, J., Bowyer, S. and Pradhan, A.K. (1997b). *Astrophys. J. Lett.*, **482**, L73.

Vennes, S., Christian, D.J. and Thorstensen, J.R. (1998). *Astrophys. J.*, **502**, 763.

Vidal-Madjar, A., Allard, N.F., Koester, D., Lemoine, M., Ferlet, R. Bertin, P., Lallement, R. and Vauclair, G. (1994). *Astron. Astrophys.*, **287**, 175.

Vilhu, O. (1984). *Astron. Astrophys.*, **133**, 117.

Visentine, J.T., Leger, L.J., Kuminez, J.F. and Spiker, I.K. (1985). *AIAA*, 85-0415.

Walter, F.M. and Bowyer, C.S. (1981). *Astrophys. J.*, **245**, 671.

Walter, F.M., Cash, W., Charles, P.A. and Bowyer, C.S. (1980). *Astrophys. J.*, **236**, 212.

Walter, F.M., Gibson, D.M. and Basri, G.S. (1983). *Astrophys. J.*, **267**, 665.

Warner, B. (1995). In *Cataclysmic Variable Stars*, Cambridge University Press, Cambridge.

Warren, J.K. (1998). PhD Thesis, Physics Department, University of California, Berkeley.

Warren, J.K., Sirk, M.M. and Vallerga, J.V. (1995). *Astrophys. J.*, **445**, 909.

Warwick, R.S. (1994). In *Frontiers of Space and Ground-based Astronomy: The Astrophysics of the 21st Century*, eds. W. Wamsteker, M.S. Longair and Y. Kondo, Astrophysics and Space Science Library Volume 187, Kluwer, Boston, p. 57.

Warwick, R.S., Barber, C.R., Hodgkin, S.T. and Pye, J.P. (1993) *Mon. Not. Roy. Astron. Soc.*, **262**, 289.

Watson, M.G. (1993). *Adv. Space Res.*, **13**, No 12, 125.

Watson, M.G., Marsh, T.R., Fender, R.P., Barstow, M.A., Still, M., Page, M., Dhillon, V.S. and Beardmore, A.P. (1996). *Mon. Not. Roy. Astron. Soc.*, **281**, 1016.

Webbink, R.F., Guinan, E.F., Koch, R.H. Kondo, Y., Etzell, P.B. and Thrash, T.A. (1992). *Bull AAS*, 181.0703.

Wegner, G. (1974). *Mon. Not. Roy. Astron. Soc.*, **166**, 271.

Weidemann, V. (1990). *Ann. Rev. Astron. Astrophys.*, **28**, 103.

Welsh, B.Y. (1991). *Astrophys. J.*, **373**, 556.

Welsh, B.Y., Vallerga, J.V., Jelinsky, P.N., Vedder, P.W., Bowyer, C.S. and Malina, R.F. (1989). *Proc. SPIE*, **1160**, 554.

Welsh, B.Y., Sfeir, D.M., Sirk, M.M and Lallement, R. (1999). *Astron. Astrophys.*, **352**, 308.

Weller, C.S. (1981). *Proc. SPIE*, **279**, 216.

Weller, C.S. and Meier, R.R. (1974). *J. Geophys. Res.*, **79**, 1572.

Werner, K. (1986). *Astron. Astrophys.*, **161**, 177.

Werner, K. (1992). In *The Atmospheres of Early Type Stars*, eds. U. Heber and C.S. Jeffery, Springer-Verlag, Berlin, p. 273.

Werner, K. and Dreizler, S. (1994). *Astron. Astrophys.*, **286**, L31.

Werner, K. and Heber, U. (1993). In *White Dwarfs: Advances in Observation and Theory*, ed. M.A. Barstow, Kluwer, Dordrecht, p. 303.

Werner, K. and Wolff, B. (1999). *Astron. Astrophys.*, **347**, L9.

Werner, K., Heber, U. and Hunger K. (1991). *Astron. Astrophys.*, **244**, 437.

Werner, K., Dreizler, S., Heber, U. and Rauch, T. (1996). In *Astrophysics in the Extreme Ultraviolet*, eds. S. Bowyer and R.F. Malina, Kluwer, Dordrecht, p. 229.

Wesemael, F., van Horn, H.M., Savedoff, M.P. and Auer, L.H. (1980). *Astrophys. J. Suppl.*, **43**, 159.

West, R.G., Sims, M.R. and Willingale R. (1994). *Planet. Space Sci.*, **42**, 71.

West, R.G., Willingale, R., Pye, J.P. and Sumner, T.J. (1996). In *Astrophysics in the Extreme Ultraviolet*, eds. S. Bowyer and R.F. Malina, Kluwer, Dordrecht, p. 289.

Wheatley, P.J. (1996). In *Astrophysics in the Extreme Ultraviolet*, eds. S. Bowyer and R.F. Malina, Kluwer, Dordrecht, p. 337.

White, N.E. (1996). In Cool Stars, Stellar Systems and the Sun. *ASP Conference Series*, volume 109, eds. R. Pallavicini and A.K. Dupree, p. 193.

White, N.E. and Marshall, F.E. (1983). *Astrophys. J. Lett.*, **268**, L117.

Wheatley, P.J., Verbunt, F., Belloni, T., Watson, M.G., Naylor, T., Ishida, M., Duck, S.R and Pfeffermann, E. (1996). *Astron. Astrophys.*, **307**, 137.

White, N.E., Culhane, J.L., Parmar, A.N., Kellett, B.J., Kahn, S., van den Oord, G.H.J. and Kuipers, J. (1986). *Astrophys. J.*, **301**, 262.

White, N.E., Shafer, R.A., Horne, K., Parmar, A.N. and Culhane, J.L. (1990). *Astrophys. J.*, **350**, 776.

White, N.E., Arnaud, K., Day, C.S.R., Ebisawa, K., Gotthelf, E.V., Mukai, K., Soong, Y., Yaqoob, T. and Atunes, A. (1994). *Pub. Ast. Soc. Japan*, **46**, L97.

Whiteley, M.J., Pearson, J.F., Fraser, G.W. and Barstow, M.A. (1984). *Nucl. Instrum. Methods*, **224**, 287.

Willingale, R. (1988). *Appl. Opt.*, **27**, 1423.

Wilkinson, E., Green, J.C., and Cash, W. (1992). *Astrophys. J.*, **397**, L51.

Wilson, G.A. (1990). PhD Thesis, Australian National University, Mount Stromlo and Siding Spring Observatories.

Windt, D. L., *et al.* (1988a). *Appl. Opt.*, **27**, 246.

Windt, D. L., *et al.* (1988b). *Appl. Opt.*, **27**, 279.

van der Woerd, H.J. (1987). PhD Thesis, Universitate van Amsterdam.

van der Woerd, H. and Heise, J. (1987). *Mon. Not. Roy. Astron. Soc.*, **225**, 131.

van der Woerd, H., Heise, J. and Bateson, F. (1986). *Astron. Astrophys.*, **186**, 252.

Wolff, B., Jordan, S. and Koester, D. (1995). *Astron. Astrophys.*, **307**, 149.

Wolff, B., Koester, D., Dreizler S. and Haas, S. (1998). *Astron. Astrophys.*, **329**, 1045.

Wolter, H. (1952a). *Ann. Physik*, **10**, 94.

Wolter, H. (1952b). *Ann. Physik*, **10**, 286.

Wonnacott, D., Kellett, B.J. and Stickland, D. (1993). *Mon. Not. Roy. Astron. Soc.*, **262**, 277.

Wood, B.E. and Linsky, J.L. (1997). *Astrophys. J.*, **474**, L39.

Wood, B.E. Brown, A., Linsky, J.L. Kellet, B.J., Bromage, G.E., Hodgkin, S.T. and Pye. J.P. (1994). *Astrophys. J. Suppl.*, **93**, 287.

Wood, M.A. (1992). *Astrophys. J.*, **386**, 539.

Wood, M.A. (1995). In White Dwarfs, Proceedings of the 9th European Workshop on White Dwarfs Held at Kiel, Germany, 29 August–1 September 1994. *Springer Lecture Notes in Physics*, volume 443, eds. Detlev Koester and Klaus Werner, Springer-Verlag, Berlin, p.41.

Wolley, Sir R., Epps, E.A., Penston, M.J. and Pocock, S.B. (1970). *R. Obs. Ann.*, No. 5.

Wu, H.H. and Broadfoot, A.L. (1977). *J. Geophys. Res.*, **82**, 759.

Young, J.M., Weller, C.S., Johnson, C.Y. and Holmes, J.C. (1971). *J. Geophys. Res.*, **76**, 3710.

Zombeck M.W. (1990). *Handbook of Space Astronomy and Astrophysics*, Cambridge University Press, Cambridge.

Index

387